MAINTAINABILITY

NEW DIMENSIONS IN ENGINEERING

Editor
RODNEY D. STEWART

**MAINTAINABILITY: A KEY TO EFFECTIVE
SERVICEABILITY AND MAINTENANCE
MANAGEMENT**
Benjamin S. Blanchard
Dinesh Verma
Elmer L. Peterson

SYSTEM ENGINEERING MANAGEMENT
Benjamin S. Blanchard

LOGISTICS ENGINEERING
Linda L. Green

DESIGN TO COST
Jack V. Michaels
William P. Wood

COST ESTIMATING, SECOND EDITION
Rodney D. Stewart

MAINTAINABILITY:
A Key to Effective Serviceability and Maintenance Management

BENJAMIN S. BLANCHARD
Virginia Polytechnic Institute and State University
DINESH VERMA
Virginia Polytechnic Institute and State University
ELMER L. PETERSON
INTEC *Associates, Inc.*

A Wiley-Interscience Publication
JOHN WILEY & SONS, INC.
New York • Chichester • Brisbane • Toronto • Singapore

Copyright © 1995 by John Wiley & Sons, Inc.

Library of Congress Cataloging-in-Publication Data:

Blanchard, Benjamin S.
    Maintainability : a key to effective serviceability and
maintenance management / Benjamin S. Blanchard, Dinesh Verma, Elmer
L. Peterson.
      p. cm. — (New dimensions in engineering)
    Includes bibliographical references and index.
    ISBN 0-471-59132-7
    1. Maintainability (Engineering) I. Verma, Dinesh.
II. Peterson, Elmer L. III. Title. IV. Series.
TS174.B52   1994
620'.0045—dc20   94-13474

Dedicated to Dot, Padma, and Marie . . . for Their Patience and Support Through the Course of This Endeavor!

PREFACE

Current trends indicate that, in general, the complexity of systems is increasing with the introduction of new technologies, the length of time that it takes to develop and acquire a new system is getting longer, and many systems (or products) in use today are not meeting the needs of the consumer in terms of performance, quality, and overall cost effectiveness. At the same time, there is a greater degree of international cooperation and exchange, and competition is increasing worldwide.

When evaluating past experiences associated with the development of systems, a majority of the problems noted have been the direct result of not applying the "systems approach" in meeting the desired objectives. The overall requirements for the system in question were not well defined from the beginning; the perspective in terms of fulfilling a consumer need has been relatively "short term" in nature; and in many instances the approach has been to "design it now and fix it later!" In essence, the system design and development process has suffered from the lack of both good early planning and the subsequent definition of requirements in a complete and methodical manner with a comprehensive, well-integrated, and disciplined design process to follow. This overall philosophy has turned out to be rather costly in the long term.

In analyzing long-term costs (i.e., life-cycle costs), experience has indicated that a large percentage of the projected cost for a given system can be attributed to the maintenance and support activities associated with keeping that system available and operating throughout its planned utilization period. For many large-scale systems, the cost of system maintenance and support often ranges from 60 to 75% of the total overall life-cycle cost for that system. In evaluating "cause-and-effect" relationships, much of this cost can be traced back to many of the design and management decisions made during the early stages of the conceptual design phase. Some examples include: (1) selecting new technologies to meet performance requirements without addressing system maintenance and support considerations; (2) packaging equipment and/or software in such a way that it cannot be maintained easily; (3) selecting nonstandard unreliable components; (4) not defining the maintenance concept and whether we are to design for two or three levels of maintenance; and (5) specifying the placement and layout of control panels without considering human engineering principles. These and comparable factors have not been adequately considered in the early system design process; thus, many of the systems in use today are not easily "maintainable," and the costs of system maintenance and support are high!

This text deals with the subject of "maintainability," one of many considerations that must be addressed throughout the system design and development process. "Maintainability" is a characteristic of design, whereas "maintenance" is the result of design. Maintainability characteristics such as accessibility, component standardization, modularization, testability, mobility, interchangeability, serviceability, and safety, appropriately incorporated into the system design configuration, can result in fewer maintenance and logistic support requirements for the system later on, as well as a reduction in the life-cycle cost for that system. Ideally, maintainability (along with the consideration of performance factors, reliability, human factors, supportability, safety, producibility, disposability, and other parameters) should be addressed in the design and development of new systems from the beginning, as part of the system engineering process. It is during the early stages of the conceptual and preliminary design phases that the greatest impact on system life-cycle cost can be realized. Hence, proper consideration of maintainability at this time can be significant. On the other hand, for those systems already developed and being utilized by the consumer, the potential for improvement (in terms of increased effectiveness and reduced life-cycle cost) through the incorporation of additional maintainability features in the existing design is great! By reducing the maintenance and support costs for a given system, the user may gain significantly relative to increased profits and improving his/her "competitive" position internationally. Thus, the application of maintainability principles and practices, as defined throughout this text, pertains not only to the development and acquisition of new systems, but to the evaluation and subsequent improvement of existing systems in use.

The material in this text is presented in four sections. Chapter 1 provides an introduction to the subject of maintainability—What is maintainability? Why is maintainability important? How can the principles of maintainability be applied in the system life cycle? Additionally, this chapter includes some common terms and definitions. Chapters 2 and 3 describe the planning requirements for implementing a maintainability program, the various approaches that can be followed in developing a maintainability organization, and the many interfaces that exist with other organizational elements in a typical program. Included within are the identification of maintainability tasks, the development of a work breakdown structure, program scheduling and cost estimating requirements, and supplier activities. Chapter 4 discusses the "measures" of maintainability, that is, those quantitative factors that are usually applied when measuring the degree to which maintainability characteristics have been initially specified and later incorporated into the design of a system/ product. An understanding of these measures is based on an understanding of reliability factors, human factors, serviceability and logistic support factors, availability and dependability factors, economic factors, and so on. Chapters 5–12 describe the application of maintainability principles and practices throughout the system life cycle. This commences with the definition of system operational requirements and the maintenance concept during conceptual design and continues with the maintainability analysis (functional analysis and the allocation of maintainability requirements, the conductance of trade-off studies, and the establishment of specific design criteria), the participation of maintainability personnel in the day-to-day design process, maintainability prediction, the maintenance task analysis (MTA), formal design review, maintainability test and demonstration, and the application of maintainability considerations for system/product improvement. Finally, five appendices are included to support many of the activities described throughout these chapters.

Included within the material describing maintainability activities in the system life cycle (Chapters 5–12) are many practical problem exercises, ''real-world'' illustrations, and examples demonstrating the application of analysis methods, techniques, and analytical models in the completion of maintainability tasks. This includes the accomplishment of allocations and predictions, level of repair analyses, maintenance task analyses, life-cycle cost analyses, and the use of computerized tools for evaluation purposes.

In summary, the intent of the text is to describe maintainability in terms of its objectives and applications, both in the system development process and in the evaluation and improvement of existing systems. We feel that this text can be utilized effectively in the formal academic classroom at the undergraduate and graduate levels, and in support of continuing education seminars and workshops. Questions and problems appear at the end of each chapter, a few computer exercises are included along with the identification of software sources in Appendix C, and a recommended bibliography is given in Appendix E. Additionally, this text was developed with the hope that it will serve as a useful reference for the practicing engineer and manager in industry and government today.

BENJAMIN S. BLANCHARD
DINESH VERMA
ELMER L. PETERSON

September 1994

CONTENTS

MAINTAINABILITY

1
INTRODUCTION TO MAINTAINABILITY

Maintainability (*M*) is an inherent characteristic of system or product design. It pertains to the ease, accuracy, safety, and economy in the performance of maintenance actions. A system (or product) should be designed such that it can be maintained without large investments of time, at the least cost, with a minimum impact on the environment, and with a minimum expenditure of resources (e.g., personnel, materials, facilities, and test equipment). One goal is to maintain a system effectively and efficiently in its intended environment, without adversely affecting the mission of that system. Maintainability is the "ability" of an item to be maintained, whereas maintenance constitutes a series of actions necessary to restore or retain an item in an effective operational state. Maintainability is a design parameter. Maintenance is required as a consequence of design.

1.1 THE SCOPE OF MAINTAINABILITY

Maintainability, as a characteristic of design, can be expressed in terms of maintenance frequency factors, maintenance times and labor-hour factors, and maintenance cost. More specifically, maintainability may be defined as:

1. A characteristic of design and installation which is expressed as the probability that an item will be retained in or restored to a specified condition within a given period of time, when maintenance is performed in accordance with prescribed procedures and resources.[1]
2. A characteristic of design and installation which is expressed as the probability that maintenance will not be required more than *x* times in a given period,

---

[1] This definition is often used in the defense sector. Refer to MIL-STD-721C, "Definition of Effectiveness Terms for Reliability, Maintainability, Human Factors, and Safety," Department of Defense, Washington, DC, 1980.

1

when the system is operated in accordance with prescribed procedures by personnel with the proper skills. This may be analogous to reliability when the latter deals with the overall frequency of maintenance.

3. A characteristic of design and installation which is expressed as the probability that the maintenance cost for a system or product will not exceed y dollars per a designated period of time, when the system is operated and maintained in accordance with prescribed procedures. Cost must address such factors as resource consumption and environmental impacts, as well as their dollar equivalents.

Maintainability requires the consideration of many different factors, involving all aspects of a system, and the measures of maintainability often include a combination of the following:

1. Mean time between maintenance (MTBM), which includes both preventive (scheduled) and corrective (unscheduled) maintenance requirements. It includes consideration of reliability mean time between failure (MTBF).

2. Mean time between replacement (MTBR) of an item due to a maintenance action (usually generates a spare part requirement).

3. Maintenance downtime (MDT), or the total time during which the system (or product) is not in condition to perform its intended function. MDT includes the mean active maintenance time ($\overline{M}$), logistics delay time (LDT), and administrative delay time (ADT). ($\overline{M}$) is a function of the mean corrective maintenance time ($\overline{M}ct$), which is equivalent to mean time to repair (MTTR), and the mean preventive maintenance time ($\overline{M}pt$).

4. Turnaround time (TAT), or that element of maintenance time needed to service, repair, and/or check out an item for recommitment. This constitutes the time that it takes an item to go through the complete cycle from its installation in the operational system, through the maintenance shop, and into the spares inventory ready for use.

5. Maintenance labor hours per system/product operating hour (MLH/OH). This is equivalent to the term maintenance man-hours/operating hour (MMH/OH), often referred to in the literature.

6. Maintenance cost per system/product operating hour (Cost/OH). Maintenance cost should be considered in the context of life-cycle cost (LCC).

These measures represent only a sampling of the factors often included within the overall scope of maintainability, and are derived (along with other selected key parameters) in Chapter 4.

Maintainability in design is accomplished through the functional packaging and interchangeability of system components, the utilization of commercially available and reliable standard parts, the incorporation of condition monitoring, and good diagnostic provisions for unscheduled maintenance, including the necessary accessibility for the rapid removal and replacement of components, providing good labeling for rapid part identification, incorporating provisions for ease of handling and transportability, use of quick-release fasteners to allow for the rapid removal of access panels, and so on. Maintainability is actually "common sense in design," with the maintenance technician in mind. These and other characteristics are discussed throughout this text.

The principles of maintainability are applicable to any type of system or product, for example, large systems, small systems, commercial and defense products, manufacturing systems, communication systems, transportation systems, and software-intensive systems. An easily maintainable system/product is less costly in terms of both its initial production and its follow-on sustaining maintenance and support. If an equipment item is functionally packaged with interchangeable components, the proper accessibility provisions, effective built-in test and condition monitoring capabilities, and so on, this should lead toward the simplification of assembly and test requirements in the manufacture of this item. During the consumer use phase, a highly maintainable system can be repaired rapidly, with a minimum expenditure of supporting resources, without causing detrimental effects on the environment, and without inducing additional faults in the process. Again, a reduction in total cost is realized.

In any event, incorporating maintainability characteristics into a design leads to a reduction in overall life-cycle cost. This in turn has a positive impact on the marketplace, the level of customer satisfaction increases, future contracts and sales evolve, and the profitability for a given industrial firm engaged in such practices is enhanced.

1.2 MAINTAINABILITY IN THE SYSTEM LIFE CYCLE

The life cycle refers to the entire spectrum of activity for a given system (or product), commencing with the identification of a consumer need and extending through system design and development, production and/or construction, operational use, sustaining maintenance and support, and system retirement and phaseout. Since the activities in each phase interact significantly with activities in the other phases, it is essential that one consider the overall life cycle when addressing maintainability or any other system characteristic.

The specific phases in the life cycle (and the duration of each) may vary somewhat depending on the nature, complexity, and purpose of the system. Consumer needs may change, obsolescence may occur, and the levels of effort may be different, depending on the type of system and where it fits into the overall hierarchial structure of activities and events. However, the life cycle for most systems follows the illustration in Figure 1.1.[2]

Referring to the figure, an airplane, a ground transportation vehicle, an electronic product, a hydraulic pump, or an equivalent item may progress through con-

---

[2]Blanchard, B. S., *System Engineering Management,* John Wiley & Sons, New York, 1991.

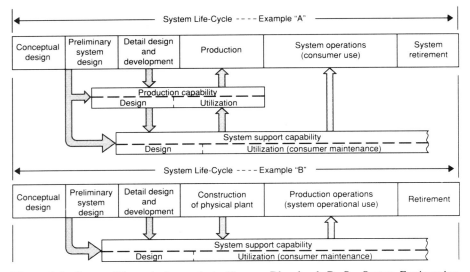

Figure 1.1. System life cycle (examples). (*Source:* Blanchard, B. S., *System Engineering Management,* John Wiley & Sons, N.Y., 1991, p. 4).

ceptual design, preliminary design, detail design, production, and so on, as reflected by the series of activities for Example "A." When evaluating this example further, the top row of activities is applicable to those elements of the system that relate directly to the fulfillment of some specific consumer need (e.g., a radar set developed to perform a specific function). At the same time, two other closely related life cycles of activity must also be considered and on a concurrent basis. The design, construction, and operation of the production capability, which has a significant impact on the operation of the prime elements of the system (i.e., the radar set), should also be addressed along with the system maintenance and support capability. Further, these activities should be addressed initially during the conceptual and preliminary design of those prime elements of the system represented by the top row. While all of these activities may be presented through an illustrated single flow (which is the general approach used in subsequent chapters of this text), the breakout here is intended to emphasize the importance of addressing *all* aspects of the total system process, and the various interactions that may occur.

Example "B" in Figure 1.1 covers the major program phases associated with a manufacturing capability, a processing plant, a satellite ground tracking facility, or the equivalent, where the construction of a "one-of-a-kind" system configuration is required. Again, the maintenance and support capability is identified separately to indicate the importance of considering the two life cycles concurrently, along with their interaction effects.

Although the approaches conveyed in the figure are not new, past practices associated with the procurement and acquisition of systems have favored a more sequential approach; that is, design the system considering major performance characteristics only (input–output, signal flow, range, accuracy) and build a prototype, then determine how the system/product is to be produced, and finally determine how the system is to be supported. In such instances, in evolving from the design phase to the production stage, many changes are often required to modify the proposed design configuration to a "producible" entity. Further, the "fixed" design approach may force the introduction of a manufacturing process that is neither desirable from a quality and reliability standpoint, nor economical. As one progresses into the next phase (i.e., system operation by the consumer), experience has indicated that many systems in use today are not meeting the basic requirements initially intended, nor are they cost effective in terms of their operation and support.

This "design it now and fix it later" approach has turned out to be rather costly, as the incorporation of changes "downstream" in the life cycle is an expensive process. While the objective is to acquire the appropriate relationship between some measure of system effectiveness and cost, trends in recent years have resulted in an imbalance between the two, as illustrated in Figure 1.2. The complexities of many systems have been increasing, primarily owing to the advent of new technologies in design. At the same time, with the ever-increasing emphasis on "performance" at the sacrifice of key design parameters such as reliability, maintainability, and quality, the overall effectiveness of these systems has been decreasing and the costs have been going up. This is occurring at a time when competition is increasing, there is a greater degree of international cooperation and exchange, and the requirements for producing a well-integrated, cost-effective, high-quality system are even greater than in the past.

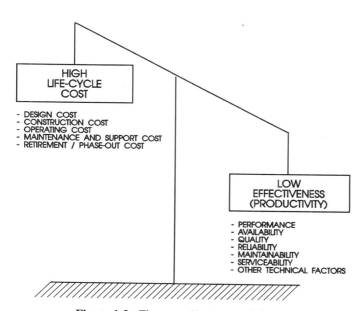

Figure 1.2. The cost-effectiveness balance.

When dealing with the aspect of cost, one often addresses only the short-term costs, or those expenses associated with the initial procurement of a system or product. Design, development, and manufacturing costs are usually fairly well known, as there is some historical basis for the prediction of such! However, the long-term costs associated with system operation and support are often hidden; yet experience has indicated that these costs often constitute a large percentage of the total life-cycle cost for a given system (e.g., up to 75% of the total cost). In essence, the lack of total cost visibility, as projected in the "iceberg effect" illustrated in Figure 1.3, prevails.

Additionally, when looking at "cause-and-effect relationships," one finds that a major portion of the projected life-cycle cost for a system stems from the consequences of decisions made during the early phases of advance planning and conceptual design. Those decisions pertaining to the utilization of new technologies, the selection of components and materials, the identification of equipment packaging schemes and diagnostic routines, the selection of a manufacturing process and maintenance support policies, and so on have a great impact on life-cycle cost. Referring to the general projections in Figure 1.4, there is a large "commitment" to life-cycle cost in the early phases of system/product development. In the figure, there are three different projections, presented in a generic manner, which may vary

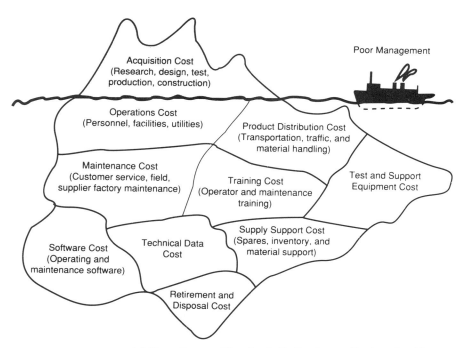

Figure 1.3. Total cost visibility. (*Source:* Blanchard, B. S., *System Engineering Management*, John Wiley & Sons, N.Y., 1991, p. 7).

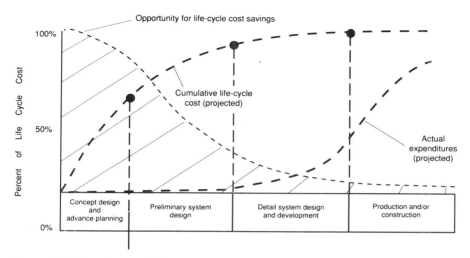

Figure 1.4. Commitment of life-cycle cost. (*Source:* Blanchard, B. S., *System Engineering Management,* John Wiley & Sons, N.Y., 1991, p. 8).

with the system in question. Although the actual expenditures on a given project will accumulate slowly at first, building up during the latter phases of design and into production, the commitment to life-cycle cost will be larger during the early stages of system development. For some systems, from 60 to 70% of the projected life-cycle cost is "locked in" by the end of preliminary design. In other words, the maintenance and support costs for a system, which often constitute a large percentage of the total, can be highly impacted by early design decisions.

With regard to past practices, a short-term, rather limited view of system requirements has been predominant, as compared to a total integrated life-cycle approach to the system design, development, and evaluation process. While the technical characteristics of system performance have received emphasis in system design and construction, very little attention has been directed toward such design characteristics as reliability, maintainability, serviceability, supportability, human factors, environmental factors, and the like. In particular, when reliability and maintainability are not considered during design, high maintenance and support costs result downstream. Additionally, these rather extensive maintenance and support requirements have had a definite degrading impact on overall system effectiveness (or productivity). The objective is to attain the proper balance among the elements identified in Figure 1.2.

Hence, the design and development process (i.e., the selection of technologies, equipment packaging schemes, components, procurement, and construction) for future systems must (1) consider *all* elements of the system on a totally integrated and concurrent basis, and (2) view the system from a long-term, live-cycle perspective. Those systems already operating should be evaluated continuously on the basis of life-cycle cost; possible high-cost contributors should be identified along with their causes; and modifications to the existing design should be incorporated to improve effectiveness and overall productivity. In all instances, the aspects of system/product maintenance and support must be inherent considerations throughout!

From the standpoint of implementation, maintainability must be a "built-in" requirement in each phase of the system life cycle, and must be properly integrated with other design parameters such as reliability, supportability, producibility, quality, and the like. Referring to Figure 1.5, maintainability can be initially specified, in terms of both qualitative and quantitative requirements, during the conceptual design phase (refer to block 1). These requirements must, of course, relate directly to the mission or the functions that the system is expected to perform. Given these basic requirements at the top level, maintainability factors can be allocated, or apportioned, down to the subsystem level or below, depending on the degree to which design control is desired. Given a requirement at the *system* level, what requirements should be specified at the *subsystem* level to ensure that all subsystems, when combined, will meet the basic overall system-level requirement? This top-down allocation of requirements is an integral part of the traditional systems engineering process described in a number of textbooks.[3]

Having established the basic input criteria for design (i.e., the specification of qualitative and quantitative requirements), the design activity continues with the iterative process of analysis, synthesis, optimization, and the ultimate development of prototype models of the system (and/or its various components) for the purposes of testing and evaluation (Figure 1.5, blocks 2 and 3). Maintainability characteristics must be considered when conducting design trade-off studies; design data/documentation must be reviewed to ensure that maintainability is adequately reflected in the proposed design configuration; prototype models are tested and evaluated to ensure that the resultant product does indeed meet the initially specified customer requirements; and recommendations are initiated as necessary in the event that corrective action is required, or for the purposes of product improvement. In the fulfillment of these objectives, various analytical methods, checklists, and tools are utilized to aid the accomplishment of maintainability analyses (trade-offs), modeling and prediction, detailed task analysis, design review, and the completion of a maintainability demonstration test.

During the production/construction and utilization phases (Figure 1.5, blocks 4 and 5), an ongoing test and evaluation effort is conducted to ensure that the system configuration meets all initially specified requirements. Factory and field data are collected and analyzed, and recommendations for corrective action and/or product improvement are initiated as appropriate.

The steps described thus far reflect the implementation of maintainability requirements applicable in the acquisition (or bringing into being) of any new system or product. This applies to an airplane, an electronic system, a manufacturing capa-

---

[3] At this point, the reader may wish to review the basic steps in the "system engineering process," as well as the objectives of "concurrent engineering." The implementation of maintainability objectives is a requirement for each! Refer to Appendix E for selected references in each area.

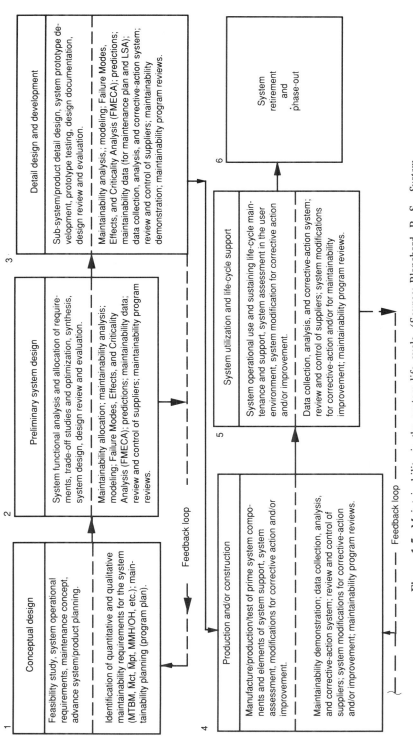

Figure 1.5. Maintainability in the system life cycle. (*Source*: Blanchard, B. S., *System Engineering Management*, John Wiley & Sons, N.Y., 1991, p. 99).

bility, or the equivalent. Any time that a new consumer need is identified, one should proceed through the basic "process" illustrated in Figure 1.5. Maintainability must be addressed throughout all phases on a life-cycle basis. However, it is important to ensure that the level of effort associated with the accomplishment of maintainability-related tasks is appropriately "tailored"—not too much or too little effort! A proper balance is necessary in the integration of maintainability requirements along with those of basic functional design (electrical, mechanical, structural), reliability, human factors, producibility, and quality.

For those systems already in being and in the consumer utilization phase, an ongoing iterative "evaluation" effort can be initiated to determine the "high-cost" contributors using a life-cycle cost analysis. In many instances, upon determining "cause-and-effect" relationships, one finds that a high-cost "driver" is the result of some unreliable component or an unmaintainable equipment item. In these instances, incorporating design modifications for reliability and/or maintainability improvement could lead to increased system effectiveness and a reduction in life-cycle cost. Again, utilizing selected maintainability tools, such as analysis and prediction models, can help in the accomplishment of this objective. From experience, we believe that the implementation of maintainability requirements, particularly with regard to many of our commercial factory operations today, can provide significant benefits.

1.3   THE LANGUAGE OF MAINTAINABILITY

With the objective of further defining maintainability, its relationships with other parameters, and how it fits into the overall system life cycle, it seems appropriate to direct some attention to a few additional terms. These terms were selected to provide the reader with the rudiments of the "systems" language.

1.3.1   System Engineering

Broadly defined, *system engineering* is "the effective application of scientific and engineering efforts to transform an operational need into a defined system configuration through the top-down iterative process of requirements definition, functional analysis and allocation, synthesis, optimization, design, test, and evaluation." The system engineering process, in its evolving of functional detail and design requirements, has as its goal achieving the proper balance between operational (i.e., performance), economic, and logistics factors.

The Department of Defense (DOD) defines system engineering as "the application of scientific and engineering efforts to (1) transform an operational need into a description of system performance parameters and a system configuration through the use of an iterative process of definition, synthesis, analysis, design, test, and evaluation; (2) integrate related technical parameters to ensure compatibility of all physical, functional, and program interfaces in a manner that optimizes the total system definition and design; and (3) integrate reliability, maintainability, safety, survivability, human, and other such factors into the total engineering effort to meet cost schedule, and technical performance objectives."[4]

Maintainability, as a characteristic of design, must be integrated with the many other parameters of design, and maintainability program requirements must be included within the overall context of system engineering program requirements.

1.3.2   Concurrent Engineering

"A systematic approach to the integrated, concurrent design of products and their related processes, including manufacture and support. This approach is intended to cause the developers, from the outset, to consider all elements of the product life cycle from conception through disposal, including quality, cost, schedule, and user requirements."[5] Concurrent engineering, like system engineering, is reflected in Figure 1.1. To include the manufacturing and support requirements, the various life cycles must be viewed on a *concurrent basis,* and maintainability principles are applicable in all areas.

1.3.3   Integrated Logistic Support (ILS)

Integrated logistic support is a management function that provides the initial planning, funding, and controls which help to assure that the ultimate consumer (or user) receives a system that will not only meet performance requirements, but can be supported expeditiously and economically through its programmed life cycle. A major ILS objective is to assure the integration of the various elements of support (i.e., manpower and personnel, training and training support, spare and repair parts and related inventories, test and support equipment, maintenance facilities, transportation and handling, computer resources, and technical data).

Of a more specific nature, the DOD defines ILS as "a disciplined, unified, and iterative approach to the management and technical activities necessary to (1) integrate support considerations into system and equipment design; (2) develop support requirements that are related consistently to readiness objectives, to design, and to each other; (3) acquire the required support; and (4) provide the required support during the operational phase at minimum cost."[6]

Included within the concept of ILS is the element of "design for supportability," and the requirements in this area include considering maintainability characteristics in the design. Thus, from an organizational perspective, one often finds a very close relationship between the maintainability function and logistic support.

---

[4] MIL-STD-499, Military Standard, "System Engineering Management," Department of Defense, Washington, DC; and *System Engineering Management Guide,* Defense Systems Management College (DSMC), Fort Belvoir, VA, January 1990.
[5] Winner, R. I., J. P. Pennell, H. E. Bertrand, and M. M. G. Slusarczuk, *The Role of Concurrent Engineering in Weapons System Acquisition,* Institute of Defense Analysis, Report R-338, December 1988. Refer to Appendix E for some additional references.
[6] Department of Defense Instruction 5000.2, "Defense Acquisition Management Policies and Procedures," Part 7, February 23, 1991.

1.3.4 System/Product Support Elements

The elements of support must be developed on an integrated basis with all other segments of the system. These elements are described below.

1. *Maintenance planning*. This includes all planning and analysis associated with the establishment of requirements for the overall support of a system throughout its programmed life cycle. Maintenance planning constitutes a sustaining level of activity commencing with the development of the maintenance concept in the conceptual design phase and continuing through the accomplishment of maintenance and logistic support analyses during design and development, the procurement and acquisition of support items, and the consumer-use phase, where an ongoing system/product support capability is required to sustain operations. Maintenance planning ensures the integration of the other elements of support.

2. *Manpower and personnel*. This category includes all personnel necessary for the initial distribution, installation, checkout, and sustaining maintenance of the system/product throughout its period of operational use by the consumer. Maintenance personnel at each level of maintenance (e.g., organization, intermediate, and depot or supplier) are specified in terms of quantity and skill-level requirements.

3. *Training and training support*. Training requirements may include short-term continuing education (seminars, workshops, etc.), long-term formal education (degree-granting programs), on-the-job training, or a combination thereof. Additionally, there is a requirement for "initial" training to familiarize personnel with a specific system/product, and "replenishment" training to cover attrition and replacement personnel. Training, in this instance, is directed toward the upgrading of the appropriate quantity of personnel to the skill levels defined for the sustaining maintenance and support of the system throughout its life cycle. Training support includes all equipment (simulators, mockups, fixtures, special devices, workstations), data (training manuals, drawings, databases), and special computer programs to support maintenance/training activities.

4. *Spares and repair parts*. This category includes all spares (repairable units, assemblies, modules), repair parts (nonrepairable components and piece parts), consumables (lubricants, fuels, gases), special supplies, and related inventories needed to support the prime operating equipment, test and support equipment, transportation and handling equipment, training equipment, facilities, and software. Within this category are the activities associated with provisioning documentation, procurement functions, material flow and warehousing, distribution, and the acquisition of spare/repair parts for all levels of maintenance.

5. *Test and support equipment*. This category includes all tools, special condition-monitoring equipment, diagnostic and checkout equipment, metrology and calibration equipment, servicing and handling equipment, and maintenance stands required to support scheduled and unscheduled maintenance actions associated with the system/product. Test and support equipment requirements at each level of maintenance must be addressed as well as the overall requirements for test "traceability" to a transfer, secondary, and/or primary standard. Test and support equipment may

be classified as "peculiar" or "nonstandard" (newly designed items), or as "common and standard" (existing items already in use and in the inventory).

6. *Maintenance facilities.* This includes all special facilities required for the performance of scheduled and unscheduled maintenance functions at each level. Physical plant, real estate, portable buildings, housing, intermediate maintenance shops, calibration laboratories, special "clean rooms," and supplier and special overhaul facilities must be considered. Capital equipment and utilities (heat, power, energy requirements, environmental controls, communications and built-in data/video networks, etc.) are generally included as part of facilities.

7. *Packaging, handling, storage, and transportation.* This category includes all special materials, containers (reusable and disposable), and supplies necessary to support the packaging, preservation, storage, handling, and/or transportation of prime mission-oriented equipment, test and support equipment, spares and repair parts, personnel, technical data, and mobile facilities. In essence, this category covers the initial distribution of products and the transportation of personnel and materials for maintenance purposes.

8. *Computer resources.* This facet of support refers to all computer equipment and accessories, software, program tapes/disks, databases, and so on necessary to perform system maintenance functions at each level. This includes both condition monitoring programs and maintenance diagnostic aids.

9. *Technical data.* This includes system installation and checkout procedures, inspection and calibration procedures, overhaul procedures, modification instructions, facilities information, drawings, and specifications necessary to perform system operation and maintenance functions. Such data cover not only the prime mission-oriented equipment, but also test and support equipment, transportation and handling equipment, training equipment, and facilities.

For most systems, the support requirements throughout the life cycle are significant! The prime operating elements of the system (i.e., that part of the system designed to perform the designated consumer-oriented function) must be designed with maintenance and support in mind, and the various elements of support must be compatible with the prime equipment and with each other. A major objective of maintainability is to design a system such that it can be maintained without large investments of time, at the least cost, with minimum or no environmental impact, and with a minimum expenditure of supporting resources. Thus, the principles of maintainability must be applied to both the prime operating elements of the system and the various elements of support in an integrated manner.

1.3.5   Reliability[7]

Reliability can be defined as the probability that a system or product will perform in a satisfactory manner for a given period of time when used under specified operating

[7] Many of these terms are also defined in Blanchard, B. S., *Logistics Engineering and Management,* 4th ed., Prentice Hall, Inc., Englewood Cliffs, NJ, 1992.

conditions in a given environment. This definition stresses the elements of "probability," "satisfactory performance," "time," and "operating conditions." These four elements are extremely important, since each plays a significant role in determining system/product reliability.

Probability, the first element in the reliability definition, is usually stated as a quantitative expression representing a fraction or a percent signifying the number of times that an event occurs (successes), divided by the total number of trials. For instance, a statement that the probability of survival (P_s) of an item for 80 hours is 0.75 (or 75%) indicates that the item will function properly for at least 80 hours, 75 times out of 100 trials. When there are a number of supposedly identical items operating under similar conditions, it can be expected that failures will occur at different points in time; thus, failures are described in probabilistic terms. The fundamental definition of reliability is heavily dependent on the concepts derived from probability theory.

Satisfactory performance, the second element is the reliability definition, indicates that specific criteria must be established to describe what is considered as satisfactory operation. A combination of qualitative and quantitative factors defining the functions that the system is to accomplish, usually presented in the context of a system specification, is required.

The third element, time, is one of the most important since it represents a measure against which the degree of performance can be related. One must know the "time" parameter in order to assess the probability of completing a mission or a given function as scheduled. Of particular interest is being able to predict the probability of an item surviving, without failure, for a designated period of time (sometimes designated as *R* or *P*). Also, reliability is frequently defined in terms of mean time between failure (MTBF), mean time to failure (MTTF), or mean time between maintenance (MTBM); thus, the aspect of time is critical in reliability measurement.[8]

The *specified operating conditions* under which we expect a system or product to function constitute the fourth significant element of the basic reliability definition. These conditions include environmental factors such as geographical location where the system is expected to operate, the operational profile, the transportation profile, temperature cycles, humidity, vibration, shock, and so on. Such factors must not only address the conditions for the period when the system is operating, but the conditions for the periods when the system (or elements thereof) is in a storage mode or being transported from one location to another. Experience has indicated that the transportation, handling, and storage modes are sometimes more critical from a reliability standpoint than the conditions experienced during actual system operational use.

These four elements are critical to determining the reliability of a system or product. System reliability (or unreliability) is a key factor in determining the frequency of maintenance and maintenance priorities which, in turn, have a significant

---

[8] While "time" is being emphasized here, it should be noted that reliability may also be expressed in terms of cycles of operation or some other mission-related parameter. It is important that the measure of reliability be directly related to the function to be performed.

impact on the requirements for system support and ultimate life-cycle cost. Reliability modeling, predictions, and analyses constitute a major input in the design of a system for maintainability.

1.3.6   Maintenance

Maintenance includes all actions necessary for retaining a system or product in, or restoring it to, a desired operational state. Maintenance may be categorized as follows:

1. *Corrective maintenance.* This includes all unscheduled maintenance actions performed, as a result of system/product failure, to restore the system to a specified condition. The corrective maintenance cycle includes failure identification and verification (based on some symptom), localization and fault isolation, disassembly to gain access to the faulty item, item removal and replacement with a spare or repair in place, reassembly, checkout, and condition verification. Also, unscheduled maintenance may occur as a result of a suspected failure, even if further investigation indicates that no actual failure occurred (i.e., false alarms). Unscheduled maintenance may be measured in terms of frequency ($MTBM_u$), elapsed time ($\overline{M}ct$ or MTTR), and labor hours per operating hour (MLH/OH). The term, "Emergency Maintenance," is sometimes used to cover activities in this area.

2. *Preventive maintenance.* This includes all scheduled maintenance actions performed to retain a system or product in a specified operational condition. Scheduled maintenance covers periodic inspections, condition monitoring, critical-item replacements (prior to failure), periodic calibration, and the like. In addition, servicing requirements (e.g., fueling and lubrication) may be included under scheduled maintenance. Some maintenance actions will result in system downtime, whereas others can be accomplished while the system is operating or in standby status. Scheduled maintenance may be measured in terms of frequency ($MTBM_s$), downtime when applicable ($\overline{M}pt$), and labor hours (MLH/OH).

3. *Predictive maintenance.* This often refers to a condition-monitoring preventive-maintenance program where direct monitoring methods are used to determine the exact status of equipment, for predicting possible degradation, and for the purposes of highlighting areas where maintenance is desired. To establish requirements in this area it is necessary to know how various system components fail (i.e., physics of failure), and to have available the use of such test methods as vibration signature analysis, thermography, and tribology. The objective is to predict when failures will occur and to take preventive measures accordingly.[9]

4. *Maintenance prevention.* This term, primarily used in the context of the concept of "Total Productive Maintenance (TPM)," refers to an effort leading toward "maintenance-free design." Basically, this constitutes the design and development of equipment for reliability and maintainability with the objective of minimizing maintenance downtime and the requirement for support resources, improving pro-

---

[9] Mobley, R. K., *An Introduction to Predictive Maintenance,* Van Nostrand Reinhold, New York, 1990.

ductivity (primarily in the commercial factory environment), and reducing life-cycle cost.[10]

5. *Adaptive maintenance.* This term primarily pertains to computer software and the changes in processing or the data environment.

6. *Perfective maintenance.* This refers basically to the changes in computer software for enhancing performance, packaging, or maintainability.

1.3.7 Maintenance Level

Correction and preventive maintenance may be accomplished on the system itself (prime mission-oriented elements) and/or any component of the system at the site where the system is operated by the consumer, in an intermediate shop relatively near the operational site, and/or at the manufacturer's facility or a remote depot. *Maintenance level* pertains to the division of functions and tasks for each area where maintenance is performed. Task complexity, personnel skill-level requirements, frequency of occurrence, special facility needs, economic criteria, and so on dictate to a great extent the specific functions to be accomplished at each level. In support of further discussions throughout this text, maintenance levels are classified as "organizational," "intermediate," and "factory" or "depot" in most instances.

1.3.8 Maintenance Concept

The *maintenance concept* (as defined in this text) begins with a series of statements and/or illustrations defining the input criteria to which the system should be designed. It develops into a description of the planned levels of maintenance, major functions to be accomplished at each level and the organizational responsibilities, "on-equipment" versus "off-equipment" maintenance, basic support policies, design criteria associated with the elements of support (e.g., built-in test versus external test, personnel skills to which to design), effectiveness factors (e.g., MTBM, MDT, $\overline{M}ct$, $\overline{M}pt$, MLH/OH, or MMH/OH, cost/MA), and anticipated maintenance environmental requirements. A preliminary maintenance concept is developed during the conceptual design phase, is continually updated, and is a prerequisite to system/product design and development. The maintenance concept also serves as the basis for the initial definition of maintainability design criteria and provides input in the development of the maintenance plan. Refer to Chapter 5 for additional coverage.

1.3.9 Maintenance Plan

The *maintenance plan* (as compared to the maintenance concept) is a detailed plan specifying the methods and procedures to be followed for the sustaining support of

---

[10]Nakajima, S., *An Introduction to Total Productive Maintenance (TPM),* Productivity Press, Inc., Cambridge, MA, translated into English in 1988.

the system/product throughout its programmed life cycle. It includes the identification, acquisition procedures, and utilization of the various elements of support (e.g., test equipment, spare/repair parts, personnel) necessary to maintain the system in full operational status throughout the consumer-use period. The maintenance plan is developed from the maintenance analysis (or the logistic support analysis), and is usually formulated during the detail design and development phase. In the defense sector, this plan is generally incorporated into the integrated logistic support plan (ILSP).

1.3.10 Total Productive Maintenance (TPM)

Total productive maintenance (TPM), a concept originally developed by the Japanese, constitutes an integrated, top-down, system-oriented, life-cycle approach to maintenance, with the objective of maximizing productivity. Total productive maintenance is directed primarily to the commercial manufacturing environment and:

1. Promotes the overall effectiveness and efficiency of equipment in the factory. It includes *maintenance prevention* (MP) and *maintainability improvement* (MI), which consider the appropriate incorporation of reliability and maintainability characteristics in design.
2. Establishes a complete preventive maintenance program for factory equipment based on life-cycle criteria (similar to the reliability-centered maintenance approach used in establishing preventive maintenance requirements).
3. Is implemented on a "team" basis involving various departments to include engineering, production operations, and maintenance.
4. Involves every employee in the company, from the top management to the workers on the shop floor. Even equipment operators are responsible for the care and maintenance of the equipment they operate.
5. Is based on the promotion of preventive maintenance through "motivational management" (the establishment of autonomous small-group activities for the maintenance and support of equipment).

Total productive maintenance, often defined as "productive maintenance" implemented by all employees, is based on the principle that equipment improvement must involve everyone in the organization, from line operators to top management. The objective is to eliminate equipment breakdowns, speed losses, minor stoppages, and so on. It promotes defect-free production, just-in-time (JIT) production, and automation. The concept of TPM promotes *continuous improvement in maintenance*.[11]

---

[11] Nakajima, S., Ed., *Total Productive Maintenance (TPM) Development Program,* Productivity Press, Inc., Cambridge, MA, translated into English in 1989.

1.3.11   Maintenance Management

Maintenance management refers to the application of the appropriate planning, organization and staffing, program implementation, and control methods to a maintenance activity. This may pertain to the management of a maintenance organization in the commercial factory, or the management of a sustaining maintenance and support activity responsible to ensure that the system being utilized by the consumer is effectively and efficiently maintained throughout its programmed life cycle (in accordance with the requirements of the detailed maintenance plan described in section 1.3.9). Specific activities include the description of tasks to be accomplished, the identification of organizational responsibilities, the development of a work breakdown structure, the scheduling and development of cost projections, program review and reporting requirements, and so on. Maintenance management functions are discussed further throughout this text.

1.3.12   Human Factors

Human factors pertain to the human element of the system and the interfaces between the human being, equipment, facilities, and associated software. The objective is to assure complete compatibility between the system physical and functional design features and the human element in the operation, maintenance, and support of the system. Considerations in design must be given to anthropometric factors (e.g., the physical dimensions of the human being, both in a static position and in a dynamic situation), human sensory factors (e.g., vision, touch, and hearing capabilities), physiological factors (impacts from environmental forces), psychological factors (e.g., human needs, wants, expectations, attitude, and motivation), and their interrelationships. Human factors (like reliability and maintainability) must be considered early in system development through functional analysis, detailed operator and maintenance task analysis, operational sequence diagramming, error analysis, safety analysis, and related design support activities. Operator and maintenance personnel requirements (i.e., personnel quantities and skill levels) and training needs evolve from the task analysis effort. Human factors design criteria must be included in the determination of maintainability requirements, particularly with regard to those maintenance functions involving the human being.[12]

1.3.13   Producibility

Producibility is a measure of the relative ease and economy of producing a system or a product. The characteristics of design must be such that an item can be produced easily and economically, using conventional and flexible manufacturing methods and processes without sacrificing function, performance, effectiveness, or quality. Some major objectives in designing for producibility are noted below:

---

[12] This area of activity may also be included under "human engineering," "ergonomics," "engineering psychology," and/or "system psychology."

1. The quantity and variety of components utilized in system design should be held to a minimum. Common and standard items should be selected where possible, and there should be a number of different supplier sources available throughout the planned life cycle of the system/product.

2. The materials selected for constructing the system should be standard, available in the quantities desired and at the appropriate times, and should possess characteristics for easy fabrication and processing. The design should preclude the specification of peculiar shapes requiring extensive machining and/or the application of special manufacturing methods.

3. The design configuration should allow for the easy assembly (and disassembly as required) of system elements, that is, equipment, units, assemblies, and modules. Assembly methods should be simple, repeatable, and economical and should not require the utilization of special tools and devices or high personnel skill levels.

4. The design configuration should be simplistic to the extent that the system (or product) can be produced by more than one supplier, using a given data package and conventional manufacturing methods and processes. The design should be compatible with the application of computer-aided design (CAD) and computer-aided manufacturing (CAM) technology where appropriate.

The basic underlying objectives are "simplicity" and "flexibility" in design. More specifically, the goal is to *minimize* the use of critical materials and critical processes, the use of proprietary items, the use of special production tooling, the application of unrealistic tolerances in fabrication and assembly, the use of special test systems, the use of high personnel skills in manufacturing, and the length of production/procurement lead times. These objectives are consistent with maintainability goals in design, particularly with regard to assembly and disassembly tasks, the use of standard parts, condition monitoring and diagnostic provisions, and the use of standard test equipment. If a product is designed for producibility, it is likely to be maintainable, and vice versa.

1.3.14   System Effectiveness (SE)

System effectiveness can be expressed as one or more figures of merit representing the extent to which the system is able to perform its intended function. The figures of merit used may vary considerably depending on the type of system and its mission requirements, and should consider the following:

1. *System performance parameters,* such as the capacity of a power plant, range or weight of an airplane, destructive capability of a weapon, quantity of letters processed through a postal system, amount of cargo delivered by a transportation system, quantity of products manufactured by a production capability, and accuracy of a radar system.

2. *Availability,* or the measure of the degree to which a system is in an operable and committable state at the start of a mission when the mission is called for at an unknown random point of time. This is often called *operational readiness.* Availability is a function of operating time (reliability) and downtime (maintainability).

3. *Dependability,* or the measure of the system operating condition at one or more points during the mission, given the system condition at the start of the mission (i.e., availability). Dependability is a function of operating time (reliability) and downtime (maintainability).

A combination of these (and perhaps other) measures represents the technical characteristics of a system, as opposed to cost and the economic aspects.

1.3.15 Life-Cycle Cost (LCC)

Life-cycle cost involves all costs associated with the system life cycle:

1. *Research and development (R&D) cost:* the cost of feasibility studies; system analyses; detail design and development; fabrication, assembly, and test of engineering models; initial system test and evaluation; and associated documentation.

2. *Production and construction cost:* the cost of fabrication, assembly, and test of operating systems (production models); operation and the sustaining maintenance and support of the manufacturing capability; facility construction; and the acquisition of an *initial* system support capability (e.g., test and support equipment, spare/repair parts, and technical documentation).

3. *Operation and maintenance cost:* the cost of system operation, and the sustaining maintenance and support of the system through its planned life cycle (e.g., manpower and personnel, spare/repair parts and related inventories, test and support equipment, transportation and handling, facilities, software, modifications, and technical data).

4. *System retirement and phaseout cost:* the cost of phasing the system and its components out of the inventory because of obsolescence or wearout; recycling of items for further use; condemnation and the disposal of materials.

Life-cycle costs may be categorized many different ways, depending on the type of system and the sensitivities desired in cost-effectiveness measurement. The objective is to provide *total cost visibility* (refer to Figure 1.3).[13]

---

[13]Two references that address life-cycle cost in detail are Fabrycky, W. J. and B. S. Blanchard, *Life-Cycle Cost And Economic Analysis,* Prentice-Hall, Englewood Cliffs, NJ, 1991; and Michaels, J. V. and W. P. Wood, *Design To Cost,* John Wiley & Sons, New York, 1989.

1.3.16   Cost Effectiveness (CE)

The development of a system or product that is cost effective, within the constraints specified by operational and maintenance requirements, is a prime objective. Cost effectiveness relates to the measure of a system in terms of mission fulfillment (system effectiveness) and total life-cycle cost, that is, a proper balance of the factors identified in Figure 1.6. Cost effectiveness, which is similar to the standard cost–benefit analysis factor used for decision-making purposes in many industrial and business applications, can be expressed in different terms (one or more figures of merit), depending on the specific mission or system parameters that one wishes to measure. One may see such ratios as availability to life-cycle cost, reliability to life-cycle cost, system capacity to life-cycle cost, overall equipment effectiveness

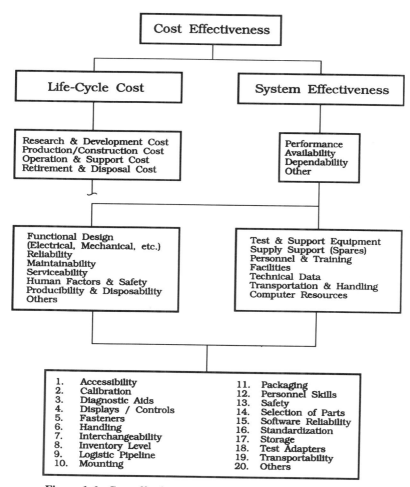

Figure 1.6. Cost effectiveness and supporting evaluation factors.

to life-cycle cost, and life-cycle cost to facility space as measures of cost effectiveness. Chapter 4 includes additional discussion of this subject.

1.3.17   Total Quality Management (TQM)

Total quality management (TQM) can be described as a totally integrated management approach that addresses system/product quality during all phases of the life cycle and at each level in the overall system hierarchy. It provides a before-the-fact orientation to quality, and it focuses on system design and development activities as well as production, manufacturing, assembly, construction, maintenance and support, and related functions. Total quality management is a unification mechanism linking human capabilities to engineering, production, and support processes. Some specific characteristics of TQM are:

1. Total customer satisfaction is the primary objective, as compared to the practice of accomplishing as little as possible in conforming to the minimum requirements.
2. Emphasis is placed on the iterative practice of "continuous improvement" as applied to engineering, production, and maintenance and support processes. The objective is to seek improvement on a day-to-day basis, as compared to the often-imposed last-minute single thrust initiated to force compliance with some standard or program requirement (the Japanese version of this approach, known as "Kaizen," is increasing in popularity).
3. In support of item 2, an individual understanding of processes, the effects of variation, the application of process control methods, and so on, is required. If individual employees are to be contributors relative to continuous improvement, they must be knowledgeable about the various processes and their inherent characteristics.
4. Total quality management emphasizes a total organizational approach, involving every group in the organization, not just the quality control department. Individual employees are motivated from within and are recognized as being key contributors to meeting TQM objectives.

As part of the initial system design and development effort, consideration must be given to (1) the design of the processes that will be used to produce the system (i.e., manufacture the components), and (2) the design of the support capability that will be used to provide the necessary ongoing maintenance and support of that system. As illustrated in Figure 1.1, these facets of program activity interact, and the results (in terms of ultimate customer satisfaction) will depend heavily on the level of quality attained.[14]

---

[14] In recent years, the Department of Defense (DOD) has been advocating TQM across the board. Two good references are DOD 5100.51G, *Total Quality Management: A Guide for Implementation*, Department of Defense, Washington, DC; and RAC SOAR-7, *A Guide for Implementing Total Quality Management, Rome* Air Development Center, New York, 1990. There are some additional references in Appendix E covering commercial applications.

1.4   MAINTAINABILITY AND RELATED INTERFACES

Maintainability is one of many characteristics in design, and the interfaces between maintainability and other requirements for a given program or project are numerous. By way of summary, the following points are emphasized:

1. Maintainability requirements must be (a) initially planned for and included within the overall planning documentation for a given program or project; (b) specified in the top-level specification for the applicable system/product; (c) designed in through the iterative process of functional analysis, requirements allocation, trade-off and optimization, synthesis, and component selection; and (d) measured in terms of adequacy through system test and evaluation. This basic process is illustrated in Figure 1.7.

2. Maintainability is an inherent part of, and should be implemented within the context of the overall system engineering and concurrent engineering processes. Maintainability requirements must be closely integrated with other design characteristics. Figure 1.8 conveys some of the major design interfaces. Of particular significance are the relationships with reliability, human factors, and supportability requirements. The completion of maintainability program tasks is highly dependent on these factors (and vice versa).

3. The incorporation of maintainability characteristics in design (or lack of) has a significant impact on both sides of the balance illustrated in Figure 1.2, that is, system effectiveness and life-cycle cost. The ultimate effectiveness and efficiency in the implementation of a maintenance program for any system (whether it be a transportation system, a communications system, or a manufacturing capability), and the management thereof, is highly dependent on the incorporation of maintainability characteristics in design. An overall objective is to provide the proper balance between technical characteristics and economic factors, as shown in Figure 1.9.

4. The successful implementation of a maintainability program is highly dependent on the application of TQM methods and practices associated with both the

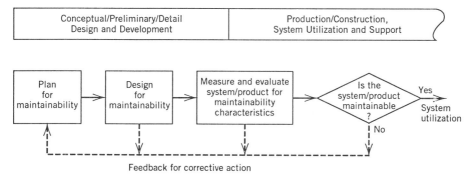

Figure 1.7. Maintainability requirements.

technical aspects of design and the conductance of maintainability tasks as part of a project organization. The interfaces are numerous, and there is a great need for the establishment of good communications throughout!

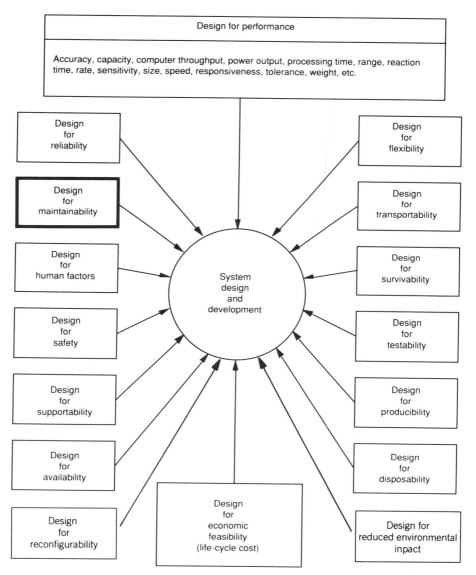

Figure 1.8. System design integration. (*Source:* Blanchard, B. S., *System Engineering Management,* John Wiley & Sons, N.Y., 1991, p. 7).

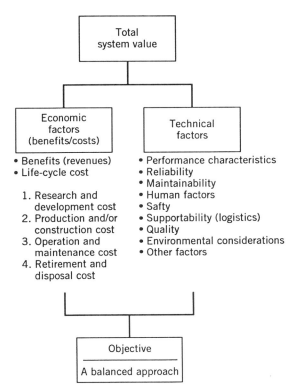

Figure 1.9. The balance of effectiveness characteristics.

1.5 SUMMARY

The purpose of this chapter is to introduce the subject of maintainability by defining a few key terms, identifying its measures, and describing some of the interrelationships that exist between maintainability and other characteristics of a system. Given a familiarization with the nomenclature, the reader should now be prepared to discuss the requirements for planning, organizing, and implementing a maintainability program in response to a typical project need.

QUESTIONS AND PROBLEMS

1. Describe *maintainability*. What are its objectives? How is it measured? Why is maintainability important in system design and development?

2. Describe the differences between *maintainability* and *maintenance*.

3. What are the differences between MTBF and MTBM? MTBM and MTBR? $\overline{M}ct$ and MDT? MDT and TAT?

4. Describe *system* engineering and *concurrent* engineering. How does maintainability relate to each?

5. Describe integrated logistic support (ILS) and total productive maintenance (TPM). How does maintainability relate to each?

6. Identify the major elements of system/product support. How do these elements interrelate with each other? Include some specific examples.

7. Identify and briefly describe some of the characteristics in design that enhance the "maintainability" of a system/product.

8. Refer to Figure 1.2. How does maintainability impact each side of the balance? Include some specific examples.

9. Describe the relationships between *maintainability* and *reliability; maintainability* and *human factors; maintainability* and *supportability*. Include a specific example in each case.

10. Describe the difference between the *maintenance concept* and the *maintenance plan*.

11. Describe the relationships between *maintainability* and *producibility*.

12. Define *cost effectiveness*.

13. What is meant by *life-cycle cost?* What is included? How may life-cycle costing be utilized in the decision-making process throughout system design and development?

14. Describe *total quality management (TQM)*. How does it impact maintainability?

15. Select a system of your choice and develop a flow diagram showing the evolutionary steps of requirements definition, design and development, test and evaluation, production/construction, operational utilization, sustaining support, and retirement. Identify and briefly describe the appropriate maintainability functions in each phase of activity.

2
PLANNING FOR MAINTAINABILITY

The key to the successful implementation of any program is early planning. Planning for the maintainability activities, as discussed in Figure 1.5, commences at program inception. As the need for a system is identified and feasibility studies are accomplished in selecting a technical design approach, requirements are being established to define a program structure that can be implemented to bring the system into being. The definition of maintainability requirements is an inherent part of this process.

Planning commences with the definition of program requirements. Maintainability functions and tasks are identified, a work breakdown structure (WBS) is developed, program schedules are generated, cost estimates are prepared, an organizational structure is defined, key policies and procedures are described, and a detailed maintainability program plan is prepared and implemented. The initial plan, usually developed during the conceptual design phase when the maintainability requirements for system design are established, covers all maintainability activities throughout the system life cycle, including those pertaining to changes and system modifications accomplished later on. The maintainability program plan must directly support any higher-level plan, such as the system engineering management plan (SEMP), and should complement other related program plans, for example, the reliability program plan, the human factors program plan, and the integrated logistic support plan (ILSP).

This chapter covers maintainability planning, the first step in the implementation of program requirements. The material presented leads directly into the organization for maintainability, presented in Chapter 3.

2.1   THE SYSTEM/PRODUCT LIFE CYCLE

The major phases of the system/product life cycle are identified in Figure 1.5. Although the nomenclature and characteristics may vary somewhat depending on the nature of the system, it is necessary that a baseline be developed for the purposes of further discussion and the establishment of maintainability program requirements. As described in Chapter 1, maintainability activities are accomplished in each phase; however, these activities must be "tailored" (or adapted) to the specific system being developed.[1]

Figure 2.1 presents the basic program phases from a different perspective, with the addition of some major milestones that are considered as being critical. In the development of a given system, various configuration baselines are established as one progresses from the initial identification of a need to the development of a fully operational system being utilized by the consumer in the field. Through this process, top-level system/product requirements are defined, functional analyses and allocations are performed, and system definition progresses down to the lowest level. These various degrees of design definition are formally evaluated through the iterative process of design review, and are documented through a series of specifications (i.e., the system Type "A" specification leading to the development Type "B" specification, and so on).

Inherent within the systems engineering process are the considerations for maintainability (and related design requirements). Maintainability qualitative and quantitative requirements at the *system* level must be included in the system specification (Type "A"); design requirements at the *subsystem* level must be included in the development specification (Type "B"); requirements for the procurement of a *standard item* already in the inventory must be included in the product specication (Type "C"); and so on. At the same time, maintainability program activities and management requirements must be reflected in the overall program management plan (or equivalent), and must be integrated into the design process described in the SEMP.[2]

Basically, the requirements for maintainability must evolve from the "top!" They must be properly integrated into the definition of system-level requirements, and allocated down to the depth necessary to provide "design-to" criteria as an input to the development and/or procurement of various system components. These technical *design* requirements must be appropriately covered through the applicable specifications and standards governing the design of the system/product. At the same time, those *program* activities necessary to ensure that maintainability requirements are, indeed, considered in the development of new systems/products must be addressed in the top-level management and systems engineering plans and documentation.

---

[1] Figure 1.5 is presented to illustrate a "process!" A lengthy or extensive cycle of activities is not intended, as one progresses through the same basic series of steps that apply in the development of *any* system. While the steps are appropriate across the board, the depth of coverage will vary from one situation to the next. Too much effort, or not enough, could be quite costly. Thus, the aspect of ''tailoring'' to the need is critical.

[2] A rather in-depth discussion of program specifications (i.e., specification categories, hierarchy, proposed content) and the system engineering management plan (SEMP) is contained in Blanchard, B. S., *System Engineering Management,* John Wiley & Sons, New York, 1991.

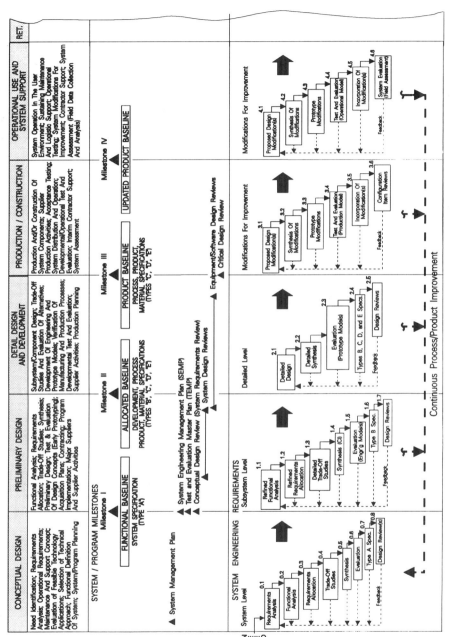

Figure 2.1. The systems acquisition process and major milestones.

29

2.2   MAINTAINABILITY PROGRAM REQUIREMENTS

Maintainability, as an inherent characteristic of system/product design, must be addressed at program inception when system-level requirements are established. Appropriate qualitative and quantitative maintainability factors must be properly integrated with performance, reliability, human, supportability, environmental, and other factors. Consumer requirements are analyzed and defined, maintainability requirements are balanced with other factors, feasibility analysis and trade-off studies are conducted, and the results are described through the system specification (Type "A").

Given the need to "design the system/product for maintainability," a formal program may be established to facilitate this objective. This program should address the major activities illustrated in Figure 1.7; that is, the planning for maintainability, the design for maintainability, the measurement and evaluation of the system/product for the incorporation of maintainability characteristics in design, and the process of feedback for corrective action and/or improvement as feasible. More specifically, a program can be developed to include the following tasks:

1. *Maintainability program plan.* To develop a maintainability program plan which identifies, integrates, and assists in the implementation of all management tasks applicable to fulfilling maintainability program requirements. This plan includes a description of the basic requirements for maintainability in system/product design, the specific tasks to be accomplished (statement of work), the maintainability organization, organizational responsibilities and interfaces, supplier requirements, task schedules and milestones, applicable policies and procedures, and projected resource requirements. The plan should directly support the SEMP, and should complement other program plans to include the reliability program plan, the human factors program plan, and the integrated logistic support plan (ILSP). The maintainability program plan is initially prepared during the conceptual design phase (refer to Figure 2.1).

2. *Supplier requirements, review, and control.* To establish the initial maintainability program requirements for suppliers or subcontractors, and to accomplish the necessary program review, evaluation, feedback, and control of supplier/subcontractor program activities. Supplier program plans are developed in response to the requirements in the overall maintainability program plan for the system.

3. *Maintainability design participation.* To participate, as a member of the "design team," in the day-to-day design process by providing direct assistance relative to establishing and interpreting maintainability design criteria, assisting in the selection of components, reviewing and evaluating design data for the incorporation of maintainability characteristics, and so on. This task involves a sustaining level of effort throughout the design process, and is dependent of the other design-related tasks described in this section (refer to Section 7.4)

4. *Maintainability modeling.* To develop a maintainability mathematical model for making initial numerical allocations and, for subsequent estimates (predictions), to evaluate maintainability characteristics in design.

5. *Maintainability allocation.* To allocate, or apportion, top system-level requirements to the lower indenture levels in the system hierarchial structure. This is accomplished down to the depth necessary to provide specific criteria as an input to design; for example, given a system-level requirement, what design requirement should be included in the development specification for Unit "B" of that system? (refer to Section 6.2.2).

6. *Maintainability analysis.* To accomplish various design-related studies pertaining to equipment packaging schemes, fault-isolation and diagnostic provisions, built-in versus external test equipment, levels of repair, component standardization, manual provisions versus automation, producibility considerations, environmental considerations, and so on. Maintainability mathematical models, level of repair analysis models, reliability-centered maintenance models, and life-cycle cost models are utilized as required (refer to Chapter 6).

7. *Failure mode, effects, and criticality analysis (FMECA).* To identify potential design weaknesses through a systematic analysis, considering all possible ways in which a component can fail (the modes of failure), the possible causes for each failure, the likely frequency of occurrence, the criticality of failure, the effects of each failure on system operation (and on various components), and any corrective action that should be initiated to prevent (or reduce the probability of) the potential problem from occurring in the future. The objective is to determine maintainability design requirements as a result of anticipated corrective and/or preventive maintenance needs (refer to Section 7.2.2).

8. *Maintainability prediction.* To estimate the maintainability characteristics in design, measured quantitatively ($\overline{Mct}$, $\overline{Mpt}$, MMH/OH, or whatever is appropriate when considering the initially specified requirements), as a result of a review of design drawings, parts lists, mockups, and so on. This is accomplished periodically throughout the system design and development process to determine whether the specified system requirements are likely to be met, given the proposed design configuration at that time (refer to Chapter 8).

9. *Maintenance engineering task analysis (MTA).* To accomplish a detailed task analysis for the purposes of (1) evaluating the system/product design for maintainability characteristics, and (2) determining the system support resource requirements in terms of spare/repair parts, test and support equipment, personnel quantities and skill levels, facilities, data, transportation and handling equipment, and computer resources. The objective is to not only evaluate the prime elements of the system for maintainability, but to evaluate and assist in the definition of the support capability as well. The results are used in the development of the maintenance plan and in the preparation of logistic support analysis (LSA) data (refer to Chapter 9).

10. *Maintainability demonstration.* To plan and implement a program where testing is accomplished to verify that the system/product maintainability requirements have been met. Testing may be accomplished at the component level, using analytical methods (i.e., computer-aided design), employing a preproduction prototype model (either sequential testing of a "fixed" sample size), and/or on the system when it is in operational use at a customer site. Maintainability characteristics (i.e., $\overline{Mct}$, $\overline{Mpt}$, MMH/OH) are measured and compared with the initially specified re-

quirements. Areas of noncompliance are noted, and recommendation for corrective action are initiated as appropriate (refer to Chapter 11).

11. *Data collection, analysis, and corrective-action system.* To establish a closed-loop system for data collection, feedback, analysis, and corrective action. Recommendations for maintainability improvement are prepared, reviewed, and system modifications are incorporated where appropriate.

12. *Maintainability program reviews.* To conduct periodic program reviews and formal design reviews at designated milestones, for example, conceptual design review, system design reviews, equipment/software design reviews, and critical design review. Participate in supplier/subcontractor reviews as appropriate. The objective is to ensure that maintainability requirements in design will be achieved (refer to Chapter 10).

While the task titles, specific content of each, and so on will vary from program to program, these 12 basic tasks are representative of a typical system developmental effort. It is important to ensure that the appropriate level of effort is applied—not too much or too little!

2.3 DEVELOPING THE MAINTAINABILITY PROGRAM PLAN

Referring to Figure 2.2, maintainability planning commences during the conceptual design phase when the overall requirements for the system are first established based on an identified consumer need. These requirements are expanded, through advance system planning, and included in a top-level program management plan (PMP). This, in turn, leads to the development of a SEMP or the equivalent. The maintainability program plan is then developed with the objective of describing the management approach and the tasks to be implemented in the conduct of a formal maintainability program effort.

Development of the maintainability program plan should, of course, be responsive to meeting the design requirements that are included in the system Type "A" specification. Additionally, the plan must support the integrated, team-oriented, life-cycle design approach described in the SEMP. Maintainability represents one of the many design disciplines that is included within the system engineering process. Finally, the plan must complement (be mutually supportive of) other closely related plans such as the reliability program plan, the human factors program plan, the ILSP, and so on.

Preparation of the maintainability program plan is the responsibility of the "Maintainability Program Manager" (i.e., the individual assigned the responsibility for implementing program requirements), and may be accomplished by the consumer/user (i.e., the "customer") or by a major contractor, depending on the program. The relationships between the consumer, prime contractors or major producers, subcontractors, suppliers, and so on, particularly for large-scale systems, may take the form illustrated in Figure 2.3. In such instances, the consumer/user is

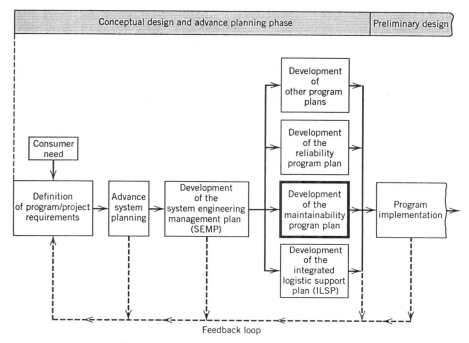

Figure 2.2. The early system planning process.

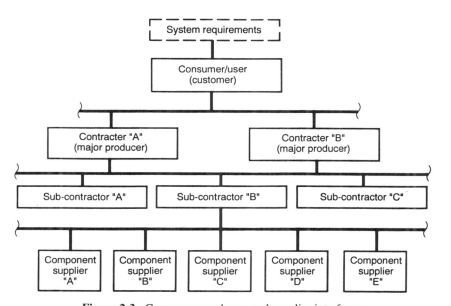

Figure 2.3. Consumer, producer, and supplier interfaces.

responsible for the plan, but may delegate the overall program management and implementation responsibility to a prime contractor.

In the event that the consumer prepares the maintainability program plan for the overall system, then Contractor "A" and Contractor "B" must each prepare a maintainability program plan covering their respective maintainability program activities, each in response to the higher-level plan. On the other hand, if the program management and implementation responsibility is delegated to Contractor "A" (for example), responsibility for the preparation of the maintainability program plan, and for implementing the activities defined therein, will be at this level.

This discussion may initially appear to be rather trivial! However, if the maintainability program plan is to be meaningful and to accomplish its objectives, it must be developed directly from the top-level program management plan and the SEMP. Further, the responsibility for the maintainability program plan, and for the accomplishment of the functions and tasks defined within, must be clearly defined and be supported by the program manager (or program director). When the maintainability program responsibility is delegated to Contractor "A" in Figure 2.3, then Contractor "A" must be given both the *responsibility* and the *authority* to perform all system-level maintainability functions on behalf of Contractor "B," as well as for all subcontractors and applicable suppliers.

If maintainability requirements for the system are defined by the consumer, through the preparation of a "Request for Proposal (RPF)" or an "Invitation for Bid (IFB)," and the contractor is assigned the responsibility (and authority) for program implementation, then the contractor will prepare the maintainability program plan and include it as part of the proposal to the consumer. The information content should be tailored to the system requirements, the program size and complexity, and the procurement or acquisition phase.

To indicate the nature of the information included in the maintainability program plan, a few selected areas have been identified for consideration:

1. *Purpose and scope.* Identify the plan objective(s) and describe the overall requirements for the maintainability program. Include a brief description of the system/product being developed.

2. *Maintainability program tasks (statement of work).* Describe in detail each program task, task input requirements, expected task output results (e.g., accomplishments, report, document, and completed product), task schedule and major milestones, and projected cost and/or man-loading.

3. *Organization.* Present an organizational chart showing the overall structure of the applicable industrial firm or government agency. Include a detailed organizational breakdown of the maintainability group, show a work breakdown structure (WBS), identify the group personnel and related backgrounds and experience, and indicate the personnel assigned by task.

4. *Organizational interfaces.* Describe other organizations where major interfaces exist [the customer, suppliers, systems engineering, basic design functions,

reliability engineering, human factors, configuration management, data management, logistic support (logistic support analysis, spares/repair parts provisioning, technical publications), test and evaluation, etc.]. Show relationships and identify the lines of communication.

5. *Subcontractor/supplier activity.* Describe the maintainability program activity for subcontractors and suppliers, identify the applicable organizational relationships, and define the procedures for review and control.

6. *Policies and procedures.* Reference or incorporate management policy directives which implement maintainability activity on a company or agency-wide basis. Reference detailed maintainability operating procedures. Describe specific techniques covering maintainability allocation, prediction, task analysis, and demonstration methods. Discuss design liaison functions to cover participation of trade-off studies, design review and evaluation, utilization of design aids, and data collection, analysis, and corrective action. The objective is to provide the customer with the *assurance* that the organization responsible for maintainability program implementation will be effective in fulfilling its goals.

7. *Program review, evaluation, and control.* Describe the methods to be used for program reviews (for cost and schedule status) and technical design reviews, and the approach for feedback and control. How are proposed changes to be evaluated and incorporated (when approved), and the appropriate corrective action initiated? Include coverage of (or reference to) the risk management plan.

8. *Maintenance concept.* Describe and illustrate the basic system-level maintenance concept. Elaborate in sufficient detail to cover operational and support concepts, qualitative and quantitative maintainability/maintenance goals, levels of maintenance and organizational responsibilities, personnel factors, spares/repair part factors, test and support equipment criteria, facility data, cost constraints, and so on. This area of coverage is not only an important aspect of the maintainability program plan, but serves as the basis for the preparation of the ILSP.

9. *Maintainability design criteria.* Describe or reference specific design features (i.e., criteria) that are directly applicable to the system being developed. This may relate to quantitative and qualitative factors associated with equipment packaging, accessibility, diagnostic provisions, mounting, interchangeability, selection of components, and so on. These criteria constitute an *input* to the design process.

10. *Technical communications.* Include a brief description of each deliverable data item, and the required date of delivery.

11. *References.* Include a list of applicable standards, specifications, plans, and other references pertinent to the fulfillment of maintainability program requirements.

These topic areas, along with references to applicable reports and documentation, represent the basic ingredients of a maintainability program plan. Once again, the objective is *not* to convey the requirement for an excessive amount of work, but to *tailor* these requirements to the particular needs of the program.

2.3.1 Development of a Work Breakdown Structure

Given a specific statement of work, the next step in the planning process is the development of a WBS[3]—a product-oriented family tree identifying the activities, functions, tasks, subtasks, work packages, and so on that must be performed to complete a program. It displays and defines the system (or product) to be developed, and portrays all of the elements of work to be accomplished. The WBS is *not* an organizational chart in terms of project personnel assignments and responsibilities, but does represent an organization of work packages prepared for the purposes of program planning, budgeting, contracting, and reporting.

Figure 2.4 illustrates an approach to the development of the WBS. During the early stages of system planning, a summary work breakdown structure (SWBS) is usually prepared by the customer and included in a RFP or an IFB. This structure, developed from the top down primarily for budgetary and reporting purposes, covers all program functions for the acquisition phase and generally includes three levels of activity:

1. *Level 1*—identifies the total program scope of work for the system to be developed, produced, and delivered to the customer. Level 1 is the basis for the authorization and "go-ahead" (or release) for all program effort.
2. *Level 2*—identifies the various projects, or categories of activity, that must be completed in response to program requirements. It also may include major elements of the system and/or significant project activities (e.g., subsystems, equipment, software, elements of support, program management, and test and evaluation). Program budgets are generally prepared at this level.
3. *Level 3*—identifies the activities, functions, major tasks, and/or components of the system that are directly subordinate to the Level 2 items. Program schedules are generally prepared at this level.

As program planning progresses and individual contract negotiations are consumated, the SWBS is developed further and adapted to a particular contract or procurement action, resulting in a contract work breakdown structure (CWBS). Referring to Figure 2.3, for example, the customer may develop the SWBS with the objective of initiating program work activity. This structure will usually reflect the integrated efforts of all organizational entities assigned to the project, and should not be related to any single department, group, or section. The SWBS, included in the customer's RFP, is the basis for the definition of all internal and contracted work to be performed on a given program. Through the subsequent preparation of proposals, contract negotiations, and related processes, Contractor "A" may be se-

---

[3] The subject of WBS and work packaging are covered in most texts dealing with project management. A good reference is Kerzner, H., *Project Management: A Systems Approach to Planning, Scheduling, and Controlling,* 3rd ed., Van Nostrand Reinhold, New York, 1989. In the defense sector, an excellent source is MIL-STD-881A, Military Standard, "Work Breakdown Structure for Defense Materiel Items," Department of Defense, Washington, DC.

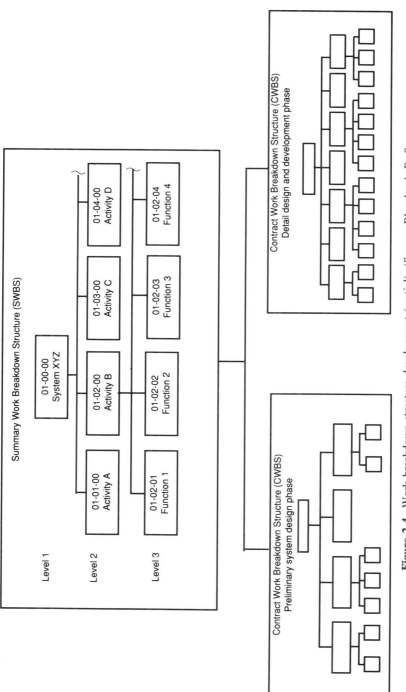

Figure 2.4. Work breakdown structure development (partial). (*Source:* Blanchard, B. S., *System Engineering Management*, John Wiley & Sons, N.Y., 1991, p. 209).

lected to accomplish all work associated with the preliminary system design phase, while Contractor "B" may be selected to complete all work associated with the detail design and development phase. From the definition of individual statements of work, a CWBS is developed to identify the elements of work for each program phase. The CWBS is tailored to a specific contract (or procurement action) and may be applicable to prime contractors, subcontractors, and suppliers as shown in Figure 2.3.

The WBS constitutes a top-down hierarchial breakout of project activities that can be further divided into functions, tasks, subtasks, levels of effort, and so on. Conversely, detailed tasks (with starting and ending dates) can be combined into work packages, and work packages can be integrated into functions and activities, with the accumulation of all work being reflected at the top program or system level.

In developing a WBS, care must be exercised to ensure that (1) a continuous flow of work-related information is provided from the top down, (2) all applicable work is represented, (3) enough levels are provided to cause the identification of well-defined work packages for cost/schedule control purposes, and (4) the duplication of work effort is eliminated. If the WBS does not contain enough levels, management visibility and the integration of work packages may prove to be difficult. However, if too many levels exist, too much time may be wasted in performing program review and control actions.

Figure 2.5 presents an example of a SWBS covering the development of an aircraft system. The same approach can be used in developing a SWBS for an electronic system, a hydraulic system, a production capability, and the like. As program requirements become defined through a contractual (or procurement) arrangement, the SWBS can be readily converted into a CWBS to reflect the actual work under that contract. The CWBS, as it appears in a contractual document, may also be presented at three levels to provide a good baseline for planning purposes, while allowing some flexibility within the contractor's organization. An expansion of the CWBS can be accomplished as necessary to provide for internal cost/schedule control.

Figure 2.6 shows a partial expansion of the SWBS to the fourth level; that is, those work packages under 3A1300 in Figure 2.5. The purpose is to recognize the major maintainability program tasks described in Section 2.2, and to provide a breakout of these tasks in a CWBS format to the extent necessary for proper cost/schedule visibility. Because of the fact that the nature and depth of the tasks accomplished may vary from one phase to the next, the CWBS may change in structure as one progresses through a system developmental effort.

The elements of the WBS may represent an identifiable item of equipment or software, a deliverable data package, an element of logistic support, a human service, or a combination thereof. The WBS elements should be selected to permit the initial structuring of budgets and the subsequent tracking of technical performance measures (i.e., quantitative technical characteristics of the system) against cost. Thus, in expanding the WBS to progressively lower levels, the requirements for day-to-day task management must be balanced against the overall reporting requirements for the program. In essence, program activities are broken down to the lowest

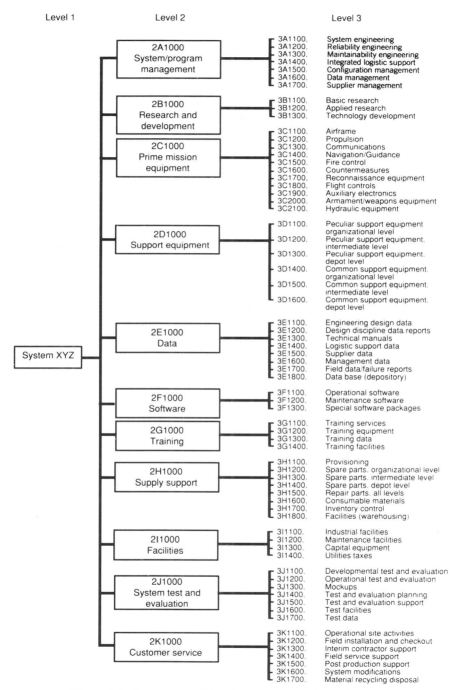

Figure 2.5. Summary work breakdown structure (aircraft system).

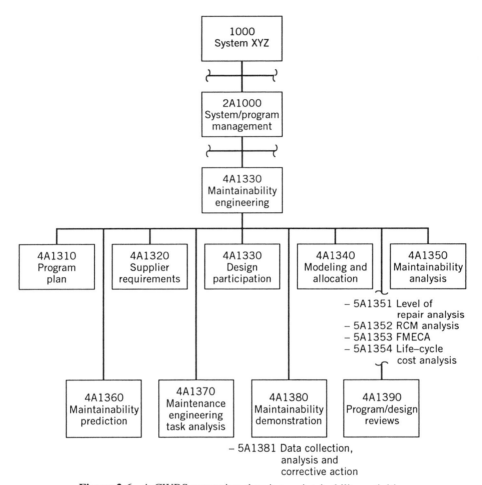

Figure 2.6. A CWBS expansion showing maintainability activities.

level that can be associated with both an organization and a cost account, as illustrated in Figure 2.7. From this schedules are developed, cost estimates are generated, accounts are established, and program activities are monitored for the purposes of schedule and cost control.

In developing the WBS, it is essential that a good comprehensive "WBS Dictionary" be prepared. This constitutes a document containing the terminology and definition of each element of the WBS. Traceability must be maintained from the top down, and all applicable work must be included. This is facilitated by assigning a number to each work package in the WBS. Referring to Figure 2.4, the total program is represented by 01-00-00, and the numbers are broken down for activities, functions, tasks, subtasks, and so on. In Figure 2.5, a slightly different numbering

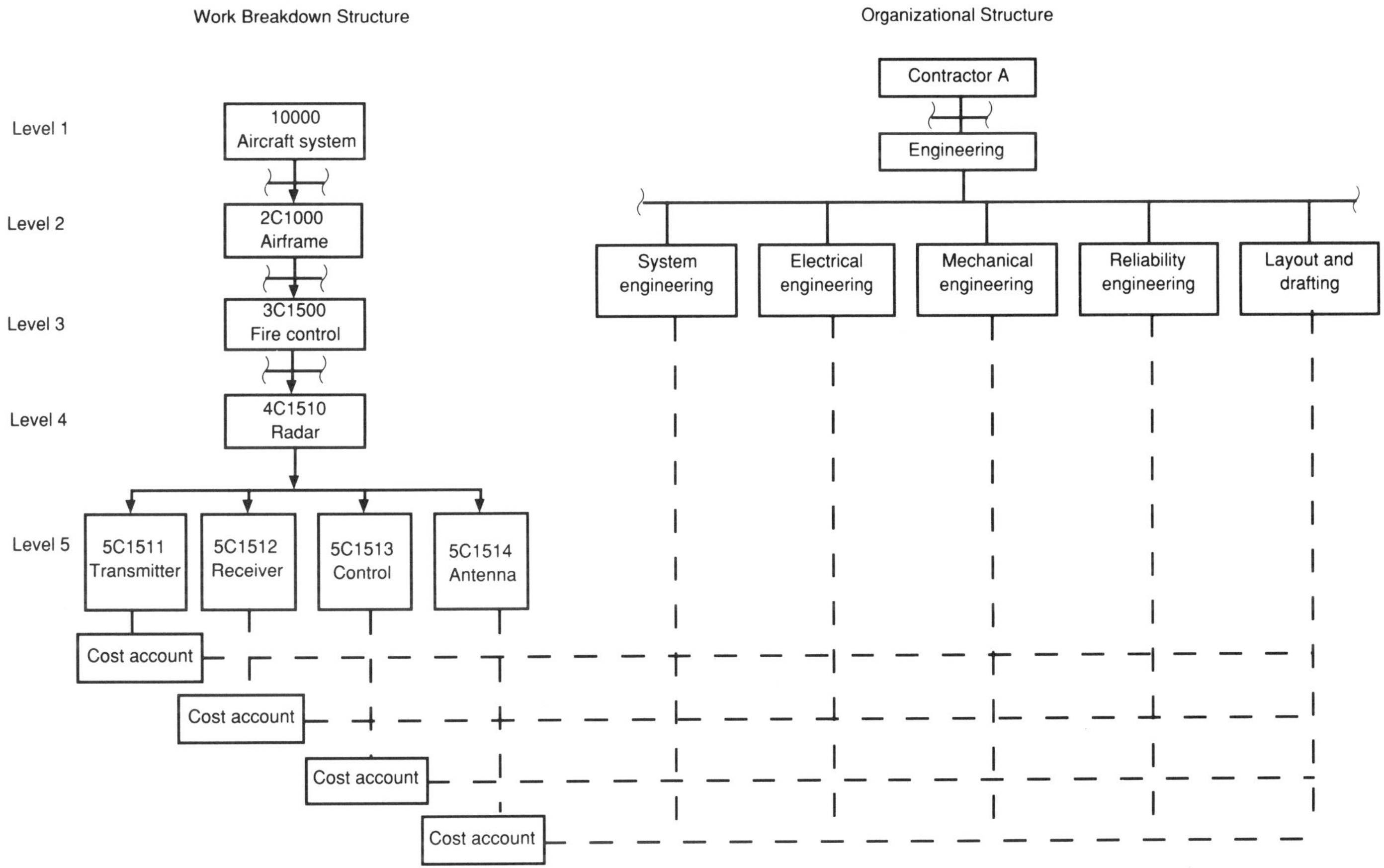

Organizational Structure

Work Breakdown Structure

Figure 2.7. Organizational integration with CWBS. (*Source:* Blanchard, B. S., *System Engineering Management*, John Wiley & Sons, N.Y., 1991, p. 214).

system is used. Although the numbering systems will vary for different programs (and with different contractors), it is important to ensure that both activities and budgets/costs can be traced, both upward and downward. In the initial generation of a CWBS by a contractor during the preparation of a proposal, budgets may be allocated downward to specific tasks. After contract award, as tasks are being accomplished, costs are being incurred and charged to the appropriate cost account. These costs are then collected upward for reporting purposes. The WBS provides the vehicle for measuring work-package progress in terms of both schedule and cost.

In summary, the WBS provides many benefits, some of which are noted:

1. The total program, or system, can easily be described through the logical breakout of elements into nicely definable work packages.
2. The discipline associated with the development of the WBS provides a greater probability that *every* program activity will be included.
3. The WBS is an excellent vehicle for linking program objectives and activities with available resources.
4. The WBS faciliates the initial allocation of budgets and the subsequent collecting and reporting of costs.
5. The WBS provides an excellent matrix for the assignment of tasks and work packages to various organizational departments, groups, and/or sections. Responsibility assignments can be readily identified. Individual work packages are then managed and reported at the first-line supervisor level, with the results summarized to higher WBS levels.
6. The WBS is an excellent vehicle for the reporting of system technical performance measures (TPMs), or quantitative system characteristics, against schedule and cost.

Finally, the WBS is an excellent tool for the promotion of program communications at various levels. As such, it must be updated to reflect program and/or system changes, consistent with configuration management actions.

2.3.2   Scheduling of Maintainability Tasks

In line with the statement of work (SOW) and the WBS, individual program tasks are presented in terms of a time line (i.e., a beginning time and an ending time). Schedules are developed to reflect the work requirements throughout all phases of a program.

Schedule planning commences with the identification of major program milestones at the top level and proceeds downward through successively lower levels of detail. A "program master schedule" is initially prepared, laying out the major program activities on the basis of elapsed time. This serves as the frame of reference for a family of subordinate schedules, developed to cover subdivisions of work as represented by the WBS. Progress against a given schedule is measured at the work-package level, and task status information is related to the appropriate cost account identified by the WBS element and the responsible organization (refer to Figure 2.7).

Program task scheduling may be accomplished using one or a combination of techniques. Some of the more common methods include the simple bar chart, the milestone chart, the combined milestone/bar chart, program network/cost charts, the Gantt chart, the line of balance, and so on. Two of these are discussed herein.

One of the most popular scheduling methods is the combined *milestone/bar chart*. Figure 2.8 presents the primary maintainability program tasks described in Section 2.2. The bars, of course, reflect major activities and the milestones represent the start/completion dates for specific program outputs. This chart serves as the basis for the assignment of resources and the development of cost projections.

The second method described herein is the program network, which may include the *Program Evaluation and Review Technique (PERT)*, the *Critical Path Method (CPM)*, and/or various combinations of these. Both PERT and CPM are ideally suited for early planning where precise task time data are not readily available, and the aspects of probability are introduced to help define risk, leading to improved decision making. These techniques provide visibility and enable management to control one-of-a-kind projects as opposed to repetitive functions such as the manufacture of many like products. Further, the network approach is effective in showing the interrelationships of combined activities.[4]

Figure 2.9 represents an example of a network diagram consisting of 25 "events" and 51 major "activities." Events are usually designated by circles and are considered as checkpoints showing specific milestones, that is, dates for starting a task, completing a task, and delivering an item under contract. Activities are represented by the lines between the circles, indicating the work that needs to be accomplished to complete an event. Work can start on the next activity only after the preceding event has been completed. The letters on the activity lines relate to the activity descriptions in Figure 2.10. Figure 2.11 provides "time-related" information and an example of the calculations for a given program network. The numbers in Columns 3–5, applied to the activity lines in Figure 2.9, indicate the time required in days, weeks, or months. The first number reflects an optimistic time estimate, the second number indicates the expected time, and the third number constitutes a pessimistic time estimate.[5]

In applying PERT/CPM to a project, one must identify all interdependent events

---

[4] Two good references covering scheduling methods are Kerzner, H., *Project Management: A Systems Approach to Planning, Scheduling, and Controlling,* 3rd ed., Van Nostrand Reinhold, New York, 1989; and Ullmann, J. E. Ed., *Handbook of Engineering Management,* John Wiley & Sons, New York, 1986.

[5] The level of detail and depth of network development (i.e., the number of events and activities included) are based on the criticality of tasks and the extent to which program evaluation and control are desired. Milestones that are critical in meeting the objectives of the program should be included along with activities that require extensive interaction for successful completion. We had experience dealing with PERT/CPM networks including from 10 to 700 events. The number of events/activities will, of course, vary with the project. Figure 2.9 shows significant maintainability program activities, as they are tied in with a few system engineering events.

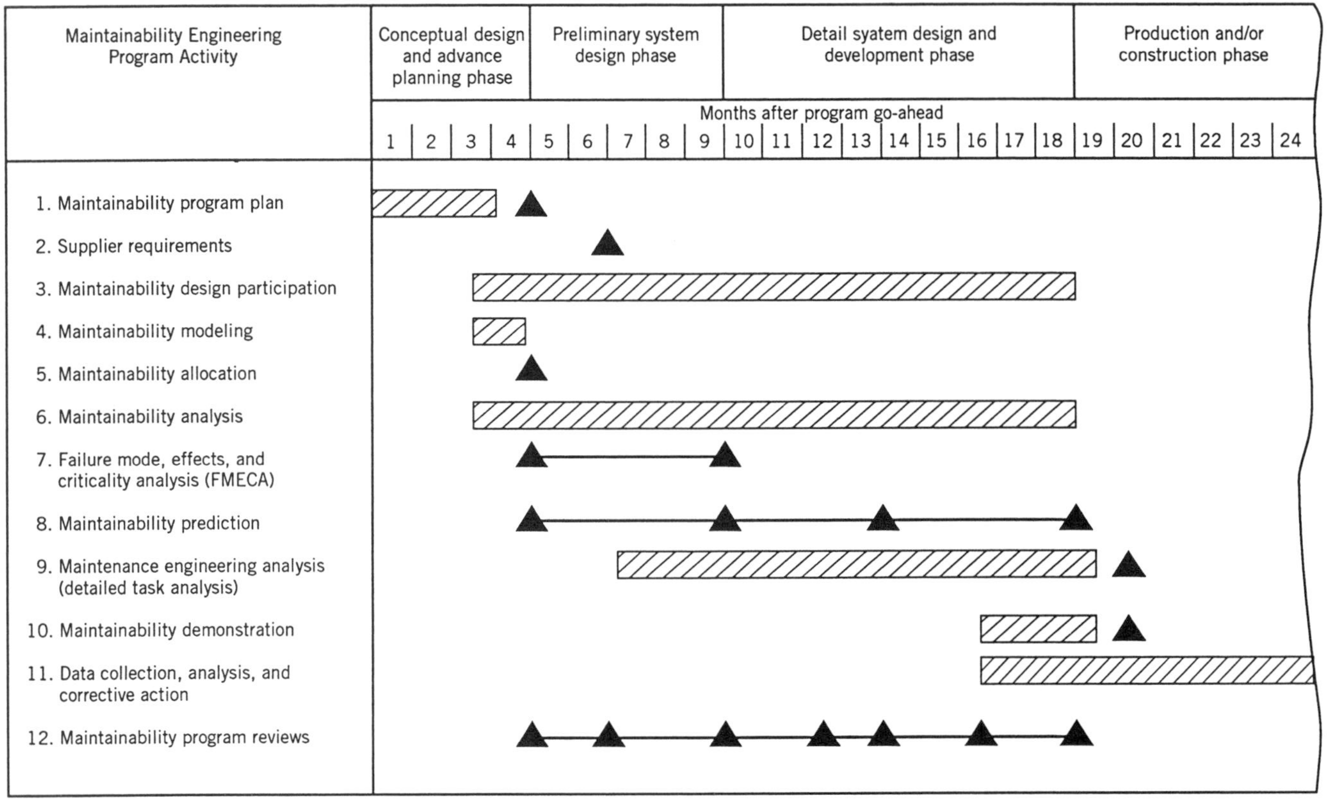

Figure 2.8. Major maintainability activities and milestones.

44

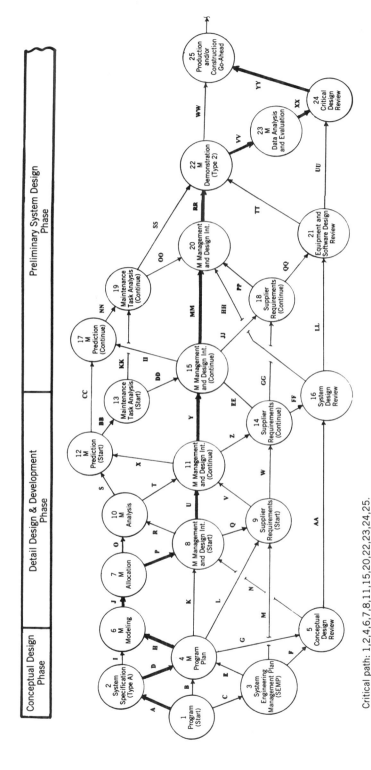

Figure 2.9. Maintainability program network.

Critical path: 1,2,4,6,7,8,11,15,20,22,23,24,25.

Activity	Description of Program Activity
A	Perform a needs analysis, conduct feasibility studies, define system operational requirements and the maintenance concept, define the system in functional terms. $\underline{M}$ requirements for system design are included.
B	Conduct advance planning, identify basic program functions to be performed, and prepare $\underline{M}$ program plan.
C	Develop the system engineering management plan (SEMP) based on system-level requirements. $\underline{M}$ program requirements for the system are included.
D	"Tailor" the system-level technical design requirements for incorporation within the $\underline{M}$ program plan. This includes definition of the maintenance concept, top-level maintenance functions, and $\underline{M}$ design criteria.
E	Incorporate system-level program requirements into the $\underline{M}$ program plan. The $\underline{M}$ program plan must "track" the SEMP.
F	Prepare the system engineering management plan (SEMP) for the Conceptual Design Review.
G	Prepare the $\underline{M}$ program plan for the Conceptual Design Review.
H	Develop a $\underline{M}$ model for making initial numerical allocations, and for subsequent analyses and predictions to evaluate system/product maintainability. As design progresses, maintainability top-down functional block diagrams, logic troubleshooting flow diagrams, and so on, are developed and used as a basis for accomplishing periodic predictions, preparing data for the logistic support analysis, maintenance task analysis, and testability analysis. This should evolve directly from the system-level maintenance concept, top-level maintenance functional flow diagrams, and appropriate reliability analysis data.
I	Incorporate the necessary maintenance concept and functional analysis requirements into the $\underline{M}$ model.
J	Accomplish a $\underline{M}$ allocation (or apportionment) of top-level requirements to the sub-system and below.
K	Develop the necessary organizational and related infrastructure in preparation for the accomplishment of the required program management and design integration tasks.
L	Define the specific supplier $\underline{M}$ program requirements, and prepare statements of work, standards, detailed specifications, and procedures.
M	Determine system-level requirements for overall supplier activity. Prepare "B", "C", "D", and "E" level specifications as applicable. Significant $\underline{M}$ requirements are included, along with the requirements pertaining to other design disciplines.
N	Translate the results from the Conceptual Design Review to the $\underline{M}$ design integration activity (e.g., approved design data, recommendations for improvement and/or corrective action).
O	Translate the results of $\underline{M}$ allocation into specific design criteria (i.e., bounds, and constraints) for the purpose of evaluation and analysis.
P	Translate the results of $\underline{M}$ allocation into specific criteria as an input to the prime design activity.
Q	Translate the results of $\underline{M}$ allocation, as it applies to various system components, into specific supplier "design-to" requirements.
R	Review and evaluate available design data. Accomplish $\underline{M}$ analysis. This may include initial design trade-off studies; failure modes, effect, and criticality analysis; level of repair analysis; life-cycle cost analysis; and related activities. While many of these tools are used throughout the program, the objective at this stage is to further develop $\underline{M}$ criteria as an input to system/product design.
S	Accomplish a $\underline{M}$ prediction based on the results of reliability predictions and the $\underline{M}$ analysis.
T	Prepare and submit recommendations for design improvement as a result of the $\underline{M}$ analysis.
U	Accomplish preliminary design and related design integration activities.
V	Provide supplier design data for integration into the overall system design activity.
W	Design and develop supplier components as specified.
X	Review and evaluate available design data. Accomplish a $\underline{M}$ prediction based on the configuration described.

Figure 2.10. Description of program network activities (sheet 1).

and activities for each phase of the project. Events are related to program milestone dates that are based on management objectives. Figure 2.10 describes the major activities reflected by the lines in Figure 2.9. Managers and programmers work with various engineering organizations to define these objectives and identify tasks and subtasks. When this is accomplished to the necessary level of detail, networks are developed, starting with a summary network and working down to detailed networks covering specific segments of a program. The network in Figure 2.9 is a

Activity	Description of Program Activity
Y	Continue to accomplish design and related design-integration activities.
Z	Translate design requirements and transfer the results of design integration activities to the appropriate component suppliers.
AA	Conduct the necessary planning and prepare for the System Design Review.
BB	Accomplish a detailed maintenance task analysis based on available design data and $\underline{M}$ prediction.
CC	Accomplish an updated $\underline{M}$ prediction based not only on the latest design configuration, but utilizing better field data and improved modeling capabilities.
DD	Prepare and submit recommendations for design improvement as a result of the detailed maintenance task analysis.
EE	Provide supplier design data for integration into the overall system design activity.
FF	Prepare supplier data, documentation, and/or prototype models for the System Design Review.
GG	Design and develop supplier components as specified.
HH	Translate the results from the System Design Review to the $\underline{M}$ design integration activity (e.g., approved design data, recommendations for corrective action or design improvement).
II	Accomplish a $\underline{M}$ prediction based on the latest design configuration and the results of design integration activities.
JJ	Translate the design requirements and transfer the results of design integration activities to the appropriate component suppliers.
KK	Update the maintenance task analysis based not only on the latest design configuration, but utilizing better field data, logistic support analyses, and improved modeling capabilities.
LL	Conduct the necessary planning and prepare for the Equipment/Software Design Review.
MM	Accomplish detail design and development and related design integration activities.
NN	Accomplish a detailed maintenance task analysis based on available design data and $\underline{M}$ prediction.
OO	Prepare and submit recommendations for design improvement as a result of the detailed maintenance task analysis.
PP	Provide supplier design data, documentation, and/or prototype models for integration into the overall system design activity.
QQ	Prepare supplier data, documentation, and/or prototype models for the Equipment/Software Design Review.
RR	Prepare for and conduct a $\underline{M}$ demonstration (as part of Type 2 testing).
SS	Determine $\underline{M}$ demonstration requirements from the detailed maintenance task analysis.
TT	Translate the results from the Equipment/Software Design Review to ensure that the model used in $\underline{M}$ demonstration is of the latest approved design configuration. Initiate modifications as necessary to upgrade the proposed demonstration model.
UU	Conduct the necessary planning and prepare for the Critical Design Review.
VV	Collect data and analyze the results of the $\underline{M}$ demonstration. Initiate recommendations for corrective action as required.
WW	Refurbish the $\underline{M}$ demonstration model for return to the active inventory.
XX	Prepare $\underline{M}$ demonstration report, and provide for evaluation at the Critical Design Review.
YY	Evaluate the results of the Critical Design Review and initiate system/product modifications for corrective action as necessary. Complete the $\underline{M}$ program requirements for system design and development.

Figure 2.10. Description of program network activities (sheet 2).

detailed network supporting the maintainability program requirement which, in turn, evolves from a higher-level summary network.

When actually constructing networks, one starts with an end objective (i.e., Event 25 in Figure 2.9) and works backward in developing the network until Event 1 is identified. Each event is labeled, coded, and checked in terms of program time frame. Activities are then identified and checked to ensure that they are properly sequenced. Some activities can be performed on a concurrent basis, and others must be accomplished in series. For each completed network, there is "one beginning event" and "one ending event," and all activities must lead to the ending event.

The next step in developing a network is to estimate activity times and to relate these times in terms of probability of occurrence. An example of the calculations that support a typical PERT/CPM network is presented in Figure 2.11 and described below.

1	2	3	4	5	6	7	8	9	10	11	12
Event #	Previous #	T_a Weeks	T_b Weeks	T_c Weeks	T_e Weeks	σ^2	TE Weeks	TL Weeks	TS Weeks	TC Weeks	Prob. (%)
25	24	6	8	10	8.000	0.444	136.000	136.000	Crit.	120	0.000
	22	3	4	5	4.000	0.111	119.833	136.000	16.170	125	0.038
24	23	3	4	6	4.170	0.250	128.000	128.000	Crit.	130	3.310
	21	2	3	4	3.000	0.111	91.500	128.000	36.500	135	37.970
23	22	7	8	9	8.000	0.111	123.833	123.833	Crit.	136	50.000
22	21	2	4	6	4.000	0.444	92.500	115.833	23.330	140	88.960
	20	12	16	18	15.670	1.000	115.833	115.833	Crit.	148	99.990
	19	2	3	4	3.000	0.111	92.667	115.833	23.170		
21	18	10	12	14	12.000	0.444	88.500	111.833	23.330		
	16	2	3	4	3.000	0.111	63.333	111.833	48.500		
20	19	2	4	6	4.000	0.444	93.670	100.167	17.670		
	18	4	6	8	6.000	0.444	82.500	96.167	14.830		
	15	26	30	36	30.330	2.778	100.170	100.170	Crit.		
	16	1	2	3	2.000	0.111	62.330	100.170	37.830		
19	17	10	12	13	11.830	0.250	89.670	96.170	6.500		
	13	14	16	20	16.330	1.000	81.330	96.170	14.830		
18	15	5	6	7	6.000	0.111	75.830	94.170	18.330		
	14	20	24	30	24.330	2.778	76.500	94.170	17.670		
17	15	6	8	10	8.000	0.444	77.830	84.330	6.500		
	12	6	8	10	8.000	0.444	61.170	84.330	23.170		
16	14	7	8	10	8.167	0.250	60.333	98.170	37.830		
	5	2	3	4	3.000	0.111	16.500	98.170	81.667		
15	14	2	3	4	3.000	0.111	55.170	69.830	14.667		
	13	2	4	6	4.000	0.444	69.000	69.830	0.830		
	11	18	22	24	21.667	1.000	69.830	69.830	Crit.		
14	11	3	4	5	4.000	0.111	52.170	66.830	14.670		
	9	12	16	20	16.000	1.778	46.170	66.830	20.670		
13	12	10	12	13	11.830	0.250	65.000	65.830	0.833		
12	11	3	5	7	5.000	0.444	53.170	54.000	0.833		
	10	2	3	4	3.000	0.111	43.500	54.000	10.500		
11	10	2	4	6	4.000	0.444	44.500	48.170	3.667		
	9	2	3	4	3.000	0.111	33.170	48.170	15.000		
	8	16	20	24	20.000	1.778	48.170	48.170	Crit.		
10	8	10	12	16	12.333	1.000	40.500	44.170	3.667		
	7	1	2	3	2.000	0.111	26.170	44.170	18.000		
9	8	1	2	3	2.000	0.111	30.170	45.170	15.000		
	4	7	8	10	8.167	0.250	19.670	45.170	25.500		
	3	4	6	8	6.000	0.444	12.000	45.170	33.170		
8	7	2	4	6	4.000	0.444	28.170	28.170	Crit.		
	5	1	2	3	2.000	0.111	15.500	28.170	12.670		
	4	3	4	5	4.000	0.111	15.500	28.170	12.670		
7	6	3	4	6	4.167	0.250	24.167	24.167	Crit.		
6	4	6	8	13	8.500	1.361	20.000	20.000	Crit.		
	2	2	4	7	4.167	0.694	12.500	20.000	7.500		
5	4	1	2	3	2.000	0.111	13.500	26.167	12.670		
	3	1	2	3	2.000	0.111	8.000	26.167	18.167		
4	3	2	3	4	3.000	0.111	9.000	11.500	2.500		
	2	2	3	5	3.167	0.250	11.500	11.500	Crit.		
	1	6	8	9	8.333	1.000	8.333	11.500	3.167		
3	1	4	6	8	6.000	0.444	6.000	8.500	2.500		
2	1	6	8	12	8.333	1.000	8.333	8.333	0.000		

Figure 2.11. Example of program network calculations.

1. *Column 1:* List each event, starting from the last event and working backward to the beginning (i.e., from Event 25 to Event 1 in Figure 2.9).

2. *Column 2:* List all previous events that lead into, or are shown as being prior to, the event listed in Column 1 (e.g., Events 22 and 24 lead into Event 25).

3. *Columns 3–5:* Determine the optimistic time (t_a), the most likely time (t_b), and the pessimistic time (t_c) in weeks or months for each activity. Optimistic time

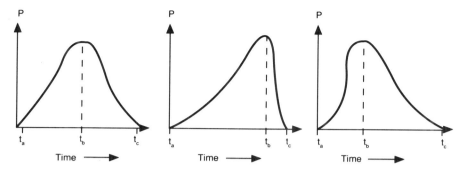

Figure 2.12. Sample distribution curves. (*Source:* Blanchard, B. S., *System Engineering Management,* John Wiley & Sons, N.Y., 1991, p. 225).

means that there is very little chance that the activity can be completed before this time, while pessimistic time means that there is little likelihood that the activity will take longer. The most likely time (t_b) is located at the highest probability point or the peak of the distribution curve. These times may be predicted by someone who is experienced in estimating. The time estimates may follow different distribution curves, where P represents the probability factor. The three-time estimates are illustrated in Figure 2.12.

4. *Column 6:* Calculate the expected or mean time, t_e, from

$$t_e = \frac{t_a + 4t_b + t_c}{6} \qquad (2.1)$$

5. *Column 7:* In any statistical distribution, one may wish to determine the various probability factors for different activity times. Thus, it is necessary to compute the variance (σ^2) associated with each mean value. The square root of the variance, or the standard deviation, is a measure of the dispersion of values within a distribution, and is useful in determining the percentage of the total population sample that falls within a specified band of values. The variance is calculated from Equation (2.2):

$$\sigma^2 = \left(\frac{t_c - t_a}{6}\right)^2 \qquad (2.2)$$

6. *Column 8:* The earliest expected time for the project *TE,* is the sum of all times, t_e, for each activity, along a given network path, or the cumulative total of the expected times through the preceding event remaining on the same path throughout the network. When several activities lead to an event, the highest time value (t_e) will be used. For example, in Figure 2.9, the highest value for *TE* (if one were to check all network paths) is 136.00 weeks for Path 1, 2, 4, 6, 7, 8, 11, 15, 20, 22, 23, 24, 25, and this is the value selected for Event 25. The *TE* values for Events

22 and 24, and so on, are calculated in a similar manner, working backward to Event 1.

7. *Column 9:* The latest allowable time for an event, *TL,* is the latest time for completion of the activities that immediately precede the event. *TL* is calculated by starting with the latest time for the last event (i.e., where *TE* equals 136,000 in Figure 2.11) and working backward subtracting the expected time (t_e) for each activity, remaining on the same path. The *TL* values for Events 22, 24, and so on are calculated in a similar manner.

8. *Column 10:* The slack time, *TS,* is the difference between the latest allowable time *(TL)* and the earliest expected time *(TE):*

$$TS = TL - TE \tag{2.3}$$

9. *Columns 11–12:* TC refers to the required scheduled time for the network based on the actual need. Assume that management specifies that the project reflected in Figure 2.9 must be completed in 130 weeks. It is now necessary to determine the likelihood, or probability *(P),* that this will occur. This probability factor is determined as follows:

$$Z = \frac{TC - TE}{\sqrt{\Sigma \text{Path Variances}}} \tag{2.4}$$

where *Z* is related to the area under the normal distribution curve, which equates to the probability factor. The "path variance" is the sum of the individual variances along the longest path, or the critical path, in Figure 2.9 (i.e., Path 1, 2, 4, 6, 7, 8, 11, 15, 20, 22, 23, 24, 25).

$$Z = \frac{130 - 136}{\sqrt{3.266}} = -1.837$$

Referring to normal distribution tables, the calculated value of -1.837 represents an area of approximately 0.0331; that is, the probability of meeting the scheduled time of 130 weeks is 3.31%. If the management requirement is 135 weeks, then the probability of success would be approximately 37.9%; or if 148 weeks were specified, the probability of success would be around 99.9%.

When evaluating the resultant probability value (Column 12 of Figure 2.11), management must decide on the range of factors allowable in terms of risk. If the probability factor is too low, additional resources may be applied to the project to reduce the activity schedule times and improve the probability of success. On the other hand, if the probability factor is too high (i.e., there is practically no risk involved), this may indicate that excess resources are being applied, some of which could be diverted elsewhere. Management must assess the situation and establish a goal.

Referring to Figure 2.9, the critical path, which is reflected by the dark line (i.e., Path 1, 2, 4, 6, 7, 8, 11, 15, 20, 22, 23, 24, 25), includes the series of activities requiring the greatest amount of time for completion. These are *critical* activities where slack times are zero, and a slippage of schedule in any one of these activities will cause a schedule delay in the overall program. Thus, these activities must be closely monitored and controlled throughout the program.

The network paths representing other program activities shown in Figure 2.9 include slack time *(TS)*, which constitutes a measure of program scheduling flexibility. The slack time is the interval of time where an activity could actually be delayed beyond its earliest scheduled start without necessarily delaying the overall program completion time. The availability of slack time will allow for a possible reallocation of resources. Program scheduling improvements may be possible by shifting resources from activities with slack time to activities along the critical path.

As an additional point relative to program schedules, a hierarchy of individual networks may be developed following a pattern similar to the WBS development approach illustrated in Figure 2.4. To provide the proper monitoring and control actions, scheduling may be accomplished at different levels.

The utilization of the PERT/CPM scheduling technique offers a number of advantages:

1. It is readily adaptable to advanced planning and essentially forces the detailed definition of tasks, task sequences, and task interrelationships. All levels of management and engineering are required to think through and evaluate the entire project carefully.

2. With the identification of task interrelationships, it tends to force the initial definition and subsequent management and control of the interfaces between customer and contractor, organizations within the contractor's structure, and between the contractor and various suppliers. Management and engineering gain a greater appreciation of the project in terms of total resource requirements.

3. It enables management and engineering to predict with some degree of certainty the probably time that it will take to achieve an objective. Areas of program risk/uncertainty can be readily identified.

4. It enables the rapid assessment of progress and allows for the early detection of possible delays and problems.

The implementation of PERT/CPM in a comprehensive and timely manner is possible since the technique is particularly adaptable to computer methods. In fact, a number of computer models and associated software are available for network scheduling.

Finally, as an extension, the PERT/CPM network may include cost by superimposing a cost structure on the time schedule. When implementing this technique, there is always the time–cost option, which enables management to evaluate alternatives relative to the allocation of resources for activity accomplishment. In many instances, time can be saved by applying more resources. Conversely, cost may be reduced by extending the time to complete an activity.

The time–cost option can be attained through the following general steps:

1. For each activity in the network, determine possible alternative time and cost estimates (and cost slope) and select the lowest cost alternative.
2. Calculate the critical path for the network. Select the lowest cost option for each network activity, and check to ensure that the total of the incremental activity times does not exceed the allowable overall program completion time. If the calculated value exceeds the program time permitted, review the activities along the critical path and select the alternative with the lowest cost slope. Reduce the time value to be compatible with the program requirement.
3. After the critical path has been established in terms of the lowest cost option, review all network paths with slack time, and shift activities to extend the times and reduce costs wherever possible. Activities with the steepest cost–time slopes should be addressed first.

PERT/CPM–COST has proven to be a very useful technique in the planning of program events and activities, and it allows for the necessary program cost–schedule status monitoring and control requirements accomplished throughout system development.

2.3.3   Cost Estimating and Control

Good cost control is important to all organizations regardless of the size. This is particularly true in our current environment where resources are limited and competition for these resources is high.

Cost control starts with the initial development of cost estimates for a given program and continues with the functions of cost monitoring, the collection and analysis of data, and the initiation of corrective action before it is too late. Cost control implies good overall cost management which includes cost estimating, cost accounting, cost monitoring, cost analysis, reporting, and the necessary feedback and control function. More specifically, the following activities are applicable.[6]

1. *Define elements of work.* Develop a SOW in response to customer and program requirements (refer to Section 2.3). Detailed project tasks are identified and described in the maintainability program plan, and task schedules are developed as discussed in Section 2.3.2.

---

[6] Two references covering the subject of cost estimating are Stewart, R.D. and R. M. Wyskida, *Cost Estimator's Reference Manual,* John Wiley & Sons, New York, 1987; and Ostwald, P. F., *Cost Estimating,* 2nd ed., Prentice-Hall, Englewood Cliffs, NJ, 1984. The emphasis here is primarily oriented to the costing of internal project tasks identified in the WBS versus the development of cost-estimating relationships for life-cycle cost analysis purposes. Cost control is covered in most texts on program/project management. One good reference is Kerzner, H., *Project Management: A Systems Approach to Planning, Scheduling, and Controlling,* 3rd ed., Van Nostrand Reinhold, New York, 1989.

2. *Integrate tasks into the work breakdown structure (WBS).* Combine project tasks into work packages, and integrate these elements of work by relating each to the specific block in the WBS. These work packages and WBS blocks are then related to organizational groups, branches, departments, suppliers, and so on. The WBS is structured and coded in such a manner that project costs can be initially allocated (or targeted) and then collected against each block. Costs may be accumulated both vertically and horizontally to provide summary figures for various categories of work. Work breakdown structure objectives are described in Section 2.3.1 (refer to Figures 2.5 and 2.6 for maintainability functions in the WBS).

3. *Develop cost estimates for each project task.* Prepare a cost projection for each project task, develop the appropriate cost accounts, and relate the results to elements of the WBS as shown in Figure 2.7. For maintainability, the areas of work in the CWBS, Figure 2.6, may be tied directly to the appropriate cost accounts.

4. *Develop a cost data collection and reporting capability.* Develop a method for cost accounting (i.e., the collection and presentation of costs), data analysis, and the reporting of cost data for management information purposes. Figure 2.13 presents an extract from a report covering both schedule and cost data. Estimated costs may be related directly to each of the tasks presented in the combined activity–milestone chart in Figure 2.8, or to the activity lines in the PERT/CPM network illustrated in Figure 2.9 and described in Figure 2.10. In this instance, cost data are related to a network schedule format. Major areas of concern are highlighted in terms of both schedule and cost, that is, potential cost overruns or schedule slippages.

5. *Develop a procedure for evaluation and corrective action.* Inherent within the overall requirement for cost control is the provision for feedback and corrective action. As deficiencies are noted, potential areas of risk are identified, potential changes are assessed, and (if approved) modifications for corrective action are initiated in an expeditious manner.

In estimating maintainability costs, one needs to "tailor" the tasks recommended in Section 2.2 to the appropriate program need. For some large system developmental efforts involving a significant amount of new equipment and/or software design, a full-scale level of activity may include a department head, several senior engineers, a number of junior engineers and analysts, secretarial assistance, and so on. For smaller programs involving the design and development of a single equipment item, maintainability requirements may be fulfilled by one or two individuals, or may even be accomplished by personnel assigned to perform other functions as well (e.g., reliability, human factors, logistic support). In the evaluation of an existing system such as a manufacturing capability, one individual may be assigned for a relatively short period of time to initiate the evaluation effort and to prepare recommendations for design improvement.

In any event, the goal is to participate in the system development and evaluation process (as a member of the design or management "team") to the extent necessary to fulfill the specified requirements in this area. The requirements for maintainabil-

Network/Cost Status Report

	Project: System XYZ					Contract Number: 6BSB-1002					Report Date: 6/1/95	
Item/Identification					Time Status				Cost Status			
WBS. No.	Cost Account	Beginning Event	Ending Event	Exp. Elap. Time (te) (weeks)	Earliest Completion Date	Latest Completion Date	Slack (Weeks)	Actual Date Completed	Cost Est. ($)	Actual Cost to Date ($)	Latest Revised Est. ($)	Overrun (Underrun) ($)
4A1350	3310	8	10	12.33	10.00	16.00	-3.67	4/4/95	2500	2250	2250	(250)
4A1360	3762	10	8	3.00	2.00	4.00	10.50		4500	4650	5000	500
4A1330	3521	7	12	4.00	2.00	6.00	0		6750	5150	6750	0

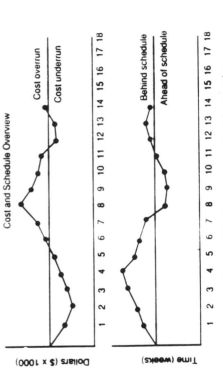

Figure 2.13. Program cost/schedule reporting.

ity must be properly integrated into the overall program cost/schedule control process, and the steps described above provide some guidance in this direction.[7]

2.3.4   Maintainability Documentation/Reports

The extent of the maintainability documentation effort, and the type and number of reports required, is a function of the overall magnitude of the program, the requirements for maintainability in design, the tasks to be accomplished, the relationships between the customer and the prime contractor, the number of suppliers and their relationships with the contractor, schedule, and cost. In some instances, maintainability coverage may be included in plans and reports covering a higher level of activity (e.g., system engineering management plan, integrated logistic support plan, design assurance plan). For large programs, there may be a significant number of plans and reports developed to cover maintainability requirements only. In any event, the program documentation and report requirements must support the tasks identified in the SOW and described in Section 2.2.3.

With the objective of providing a "shopping list" for planning purposes, a number of plans and reports have been selected for discussion. These requirements relate to the program tasks presented in the combined activity–milestone schedule in Figure 2.8.

1. *Maintainability program plan*—usually submitted to the customer in the conceptual design and advanced planning phase, or 30 days after program go-ahead.
2. *Maintainability model and allocation*—included as part of the maintainability program plan, or submitted 30 days after approval of the plan.
3. *Maintainability trade-off studies*—the documentation of the results of design trade-off studies, accomplished as part of the maintainability analysis, prepared on an "as required" basis to justify design recommendations and decisions.
4. *Failure mode, effects, and criticality analysis (FMECA)*—prepared and submitted during the preliminary system design phase, or included within reliability, safety, logistic support, and/or other related program documentation. Reliability, safety, and integrated logistic support (ILS) program requirements usually include the FMECA as a requirement.
5. *Maintainability predictions*—usually prepared throughout the system design and development process, as the design evolves from one configuration baseline to another, and submitted as an input for review during the formal design review process.
6. *Maintenance engineering task analysis*—prepared during the preliminary system design and detail design and development phases, based on available

---

[7] Some additional coverage of maintainability program activity is presented in Chapter 3.

design data, and submitted as necessary for the purposes of design evaluation. Maintenance task analysis data are also a requirement in a typical ILS program.

7. *Maintainability demonstration plan*—prepared and included in an overall system-level test and evaluation master plan during the conceptual design phase, or submitted 30–90 days before the start of a maintainability demonstration during the detail design and development phase.

8. *Maintainability demonstration report*—covers the results of the maintainability demonstration, and is usually submitted 30 days after the completion of the demonstration test.

9. *Failure reports and recommendations for design improvement*—individual reports, submitted as necessary, covering maintainability recommendations for corrective action and/or for system/product improvement. These may be combined with similar reporting requirements for other programs such as reliability engineering, human factors, safety, and so on.

10. *Maintainability program final report*—a final report, covering program results and accomplishments, submitted 30–60 days after the completion of the formal maintainability program.

As stated previously, the documentation and report requirements must be "tailored" to the specific needs of the maintainability program. In some instances, a full-scale system design and development effort may require all of the documentation/reports identified herein. On the other hand, the evaluation of an existing system/product already in the inventory may only require the accomplishment of a maintenance engineering task analysis.

2.4 MAJOR INTERFACES WITH OTHER FUNCTIONS

The completion of maintainability tasks for a typical project is highly dependent on the establishment of successful interface relationships with many other functions, including various aspects of program management, the basic design activities responsible for the development of the system/product, system engineering, reliability engineering, human factors engineering, logistics, test and evaluation, manufacturing or production, major contributing suppliers, and/or other comparable organizational entities that constitute part of the overall design team effort. Although not all-inclusive, Figure 2.14 identifies some of these major interfaces, and the interface relationships described below represent a sample of the dependency requirements that exist throughout the development of a system. The proper integration of maintainability requirements, along with those of reliability, human factors, logistic support, and so on, is essential if the output results are to be effective. Thus, a thorough understanding of the major interface relationships between maintainability and other functions is necessary.

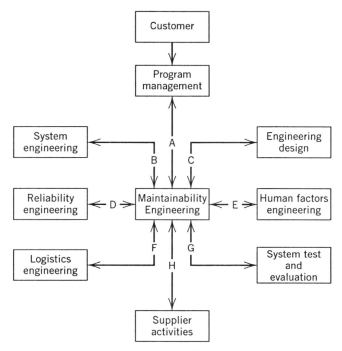

Figure 2.14. Major maintainability interfaces (example).

1. *Program management (Figure 2.14, "A").* The development of a maintainability program plan is highly dependent on, and must directly support, the requirements specified in the top-level system management plan for the system being developed (i.e., the SMP identified in Figure 2.1). Subsequently, there are the obvious ongoing program management interfaces that must exist throughout the design and development, production, and follow-on phases of activity (e.g., cost and schedule control). Since many program managers may not fully comprehend the benefits or be familiar with the requirements associated with the accomplishment of maintainability functions, it is essential that the appropriate level of expertise representing the maintainability organization be available and participate in the preparation of the SMP. The initial interpretation of customer requirements for maintainability is a critical input in the identification of the functions/tasks to be accomplished throughout the program.

2. *System engineering (Figure 2.14, "B").* The system engineering function is responsible for the definition and integration of top-level design requirements for the overall system and for the ultimate description of these requirements through the preparation of the system type "A" specification. Additionally, the program activities necessary to ensure that design objectives will be met are described in the

SEMP (refer to Figure 2.1 and events 2 and 3 in Figure 2.9). The maintainability program plan must include coverage of (a) the technical characteristics of design or the specific criteria supporting the qualitative and quantitative maintainability requirements described in the system specification, and (b) the maintainability program tasks to be accomplished in support of the applicable tasks described in the SEMP. To ensure that the proper requirements are reflected from the top down, the maintainability organization should be properly represented during the initial preparation of the "A" specification and the SEMP.

3. *Design engineering (Figure 2.14, "C").* Maintainability is one of many characteristics in design, and the design effort is the result of a "team" approach. Maintainability requirements in the form of qualitative and quantitative criteria must be provided as an input to the design process; maintainability is considered (along with many other characteristics) in the accomplishment of design trade-off studies; maintainability analyses and predictions are prepared for the purposes of assessment. Maintainability personnel should participate (or be adequately represented) in the day-to-day informal design review and evaluation process and in formal design reviews. Maintainability participates in test and evaluation activities, and maintainability is involved in the data collection, analysis, and corrective action process. In essence, the major interfaces constitute the continuing exchange of ideas, design data and documentation, reports, and so on. The maintainability organizational role is to actively participate and *contribute* to the system design on an "as-needed" basis (refer to activities "U," "Y," and "MM," Figure 2.9).

4. *Reliability engineering (Figure 2.14, "D").* Reliability and maintainability are very complementary in terms of design objectives—establishing the proper balance among reliability and maintainability factors to meet higher-level system design goals, the day-to-day participation in design, the accomplishment of analyses and predictions, participation in test and evaluation activities—and in the data collection and corrective action process. Reliability analysis and prediction reports are a required input to the completion of maintainability predictions and the detailed task analysis. The results of maintainability analyses are necessary to verify reliability requirements in design. Further, numerous tasks (the FMECA, design review and evaluation, etc.) should be accomplished jointly. Of all of the activity interfaces discussed throughout this text, the relationships with reliability are probably the closest.

5. *Human factors engineering (Figure 2.14, "E").* Referring to Section 1.3.12, human engineering requirements often include the identification and description of human functions and tasks for the system and its elements, including those designated for the purposes of maintenance. Criteria associated with the human (e.g., anthropometric factors and sensory factors) and the specification of personnel quantities and skill levels constitute an input to the completion of the detailed task analysis and to the determination of maintenance training requirements. The results of the detailed task analysis may be used as an input to the development of operational sequence diagrams covering maintenance activities, and in the final verification of

personnel and training requirements for the system as an entity. Again, the interfaces are numerous.

6. *Logistics engineering (Figure 2.14, "F")*. The "design for supportability" objective, described in Section 1.3.3, can be partially accomplished through the maintainability design-related activities identified in Section 2.2. The engineering functions within the overall spectrum of logistics are directly comparable in many instances. Further, the identification of specific system support requirements (i.e., manpower and personnel, spares and repair parts, test and support equipment, facilities, maintenance software, and data) is accomplished in the detailed task analysis which, in turn, is not only a maintainability requirement but a logistics program requirement as well. A number of program tasks can be accomplished jointly, and the results of maintainability analyses and predictions constitute an input in the development of logistic support analysis documentation.

7. *Test and evaluation (Figure 2.14, "G")*. The overall requirements for maintainability verification and demonstration must be integrated with the test and evaluation requirements for the system from the beginning. Further, as the design progresses, the specific tasks to be demonstrated (along with the necessary supporting resources) will be determined through the detailed task analysis. Thus, there is an ongoing interface relationship with the system test and evaluation organization.

8. *Supplier activities (Figure 2.14, "H")*. Referring to Section 2.3, the extent of the interface relationships with suppliers may vary significantly depending on the type of system being developed, the degree of internal versus external work to be accomplished, and so on. On one hand, for large programs all of the maintainability tasks specified in Section 2.3 may be applicable. On the other hand, for smaller efforts only one or two of the tasks may be applicable. The major interfaces in this area are discussed further in Section 2.5.

The objective here is to identify a few significant areas where maintainability interface relationships exist. These and other interfaces will become more apparent as one progresses through the chapters of this text.

2.5   SUPPLIER MANAGEMENT

The term "supplier," as used throughout this text, refers to a broad category of organizations who provide system components to the producer. Referring to Figure 2.3, Component Supplier "D" is a supplier for Subcontractor "B," who is a supplier for Contractor "A," who is a supplier for the ultimate consumer, and so on. For large systems, there may be a layering of suppliers, as reflected by the structure illustrated in Figure 2.15.

The components provided by various suppliers may range from a large element of the system (e.g., a facility, an intermediate maintenance shop full of test equipment, Unit "B" of System "XYZ") to a small nonrepairable item (e.g., a resistor, fastener, bracket, or cable). In some instances, the component may be newly devel-

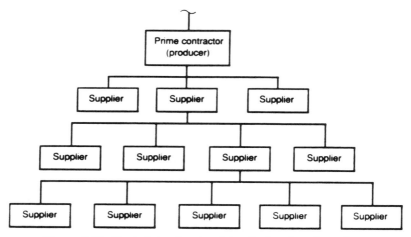

Figure 2.15. Typical structure involving the layering of suppliers.

oped and require design activity. The supplier will design and produce the component in the quantities desired. In other cases, the supplier will serve as a manufacturing source for the component. The supplier will produce the desired quantities from a given data set. In other words, the component design has been completed (whether by the same firm or some other), and the basic service being provided is "production." A third scenario may involve the supplier as an inventory source for one or more common and standard "off-the-shelf" components. There are no design or production activities, but just distribution and materials handling functions. As one can see, individual supplier roles can vary significantly.

The quantity of suppliers, their location, and the nature of their activities are a function of the type and complexity of the system being developed, along with the methods of operation implemented by the customer and contractor organizations. For items requiring new design, a development specification Type "B," along with a SOW, will be prepared by the producer and directed to the selected supplier for compliance. Qualitative and quantitative maintainability design parameters applicable to the item being developed should be included in the "B" specification (e.g., a specific $\overline{M}ct$ or MTBM value, guidelines for packaging and accessibility, a self-test thoroughness figure of merit), and the program tasks that must be accomplished by the supplier are covered in the SOW. The requirements specified by the customer for the system and imposed on the contractor, through the "A" specification, must be passed on to the various suppliers; however, these requirements must be appropriately *tailored* to the particular need.

For suppliers involved in the design and development of a new equipment item or an element of software, such as the components of System "XYZ" shown in Figure 2.16, the appropriate maintainability program tasks may include:

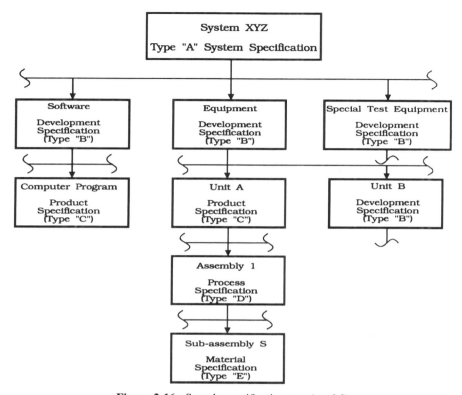

Figure 2.16. Sample specification tree (partial).

1. The preparation of an abbreviated *maintainability program plan* covering the activities related to the development of the equipment/software in question. This plan needs to be integrated with the *M* plan for System "XYZ."

2. *Maintainability analysis* covering significant design-related trade-off studies pertaining to the selection of component parts, equipment/packaging schemes, diagnostic capability, levels of repair, and so on.

3. *Maintainability prediction* providing periodic assessments of the basic equipment/software design configuration in terms of the $\overline{M}ct$, MTBM, MMM/OH, or equivalent factors that are included in the "B" specification.

4. *Maintenance detailed task analysis* describing the maintainability characteristics in design and identifying the resources necessary to support the equipment/software through the planned life cycle.

The tasks assigned to a given supplier must relate to the maintainability goals in the "B" specification, and must directly support the overall program requirements described in the maintainability program plan for System "XYZ." Special care must be exercised to ensure that (1) there is a top-down *traceability* of requirements through the applicable levels of specifications illustrated in Figure 2.16, and (2) that *all* program activities are accomplished effectively and efficiently without the introduction of overlapping requirements and duplication of effort.

For those components that have been designated as common and standard "off-the-shelf" items, maintainability parameters must be initially included in the product specification Type "C" which, in turn, is used in subsequent procurement actions. For example, if the component to be acquired is "repairable," a $\overline{M}ct$ or MMH/OH factor might be specified as a contingency factor for a procurement decision. The contractor will prepare the "C" specification, a RFP will be generated, and each supplier qualified and interested in responding will submit a proposal. Usually included within each supplier proposal is a maintainability prediction, a maintenance task analysis, and/or an assessment of the item's maintainability based on experience in the field. The contractor will then evaluate each proposal and make a selection. The maintainability organization must be represented not only during the preparation of the initial specification and the associated SOW, but in the supplier selection process.

With regard to supplier evaluation and control, the contractor must incorporate supplier activities as part of the initial planning process. For large design and development efforts, individual selected design reviews may be conducted periodically at the supplier's facility, with the results of these reviews being included in the higher-level reviews conducted at the contractor's plant. For smaller programs, the review process may not be as formal, with the results of the supplier's effort being integrated into the evaluation of a larger element of the system. When addressing projects involving the manufacture and production of components, the contractor's primary concern is that of incoming inspection and quality control. It is essential that the characteristics designed into the component, or "advertised" in an off-the-shelf item, be maintained throughout.

In essence, supplier evaluation and control, with the appropriate feedback and corrective-action process, are merely extensions of the program planning, review, and control activities initiated by the customer and imposed on the contractor for the system. The contractor must, in turn, impose certain requirements on the supplier. Large suppliers must impose the necessary controls on smaller suppliers in the event that a "layering of suppliers" exists as shown in Figure 2.16. The objective is to ensure that maintainability requirements are initially specified and allocated from the top, and subsequently fulfilled from the bottom up.

QUESTIONS AND PROBLEMS

1. Maintainability planning commences early at program inception with the definition of overall system requirements. Why is it essential that this planning activity start as soon as possible? What is likely to happen if maintainability planning is initiated at a later time in the life cycle?

2. Briefly describe the relationship between the maintainability program plan and the SEMP; the reliability program plan; and the ILSP?

3. Assume that you have just been assigned to implement a maintainability program for the design and development of a new system. Identify the program phases and describe the tasks that you would include in such a program.

4. Identify and describe some of the "commonalities" that exist between the tasks that are usually included in a maintainability program, a reliability program, a human factors program, and a logistic support program.

5. Briefly describe the differences between the "A," "B," "C," "D," and "E" specifications. System-level maintainability design requirements are usually included in which specification type? If you were to specify maintainability requirements associated with a manufacturing process, what category of specification would you prepare?

6. Refer to Question 3. Develop a detailed outline (i.e., table of contents) for an M program plan covering the acquisition of the new system.

7. One of the maintainability tasks described in Section 2.2 is "design participation." In your new function as the maintainability department manager, how would you ensure that this task is accomplished in an effective manner? Describe the steps that you would take.

8. What is meant by "design criteria?" Provide some examples of both "qualitative" and "quantitative" criteria.

9. What is meant by "tailoring?" How would you apply the concept in the implementation of a maintainability program? Provide some examples.

10. What is the purpose of a WBS? What is the difference between a WBS, a SWBS, and a CWBS? how do work packages relate to the WBS? Construct a CWBS for a program of your choice.

11. Describe the relationships between the WBS, program functions/tasks, cost estimates, cost control numbers, and organizational responsibilities.

12. Of the numerous scheduling methods available, the "combined milestone/bar chart" and the "network" approaches are discussed. Describe the advantages/disadvantages of each method. Which method would you prefer in managing a maintainability activity? Why?

13. Refer to Figure 2.9. What is meant by the "critical path?" slack time? *te?* What would happen if the supplier activity "W" slipped to 24 weeks?

14. Refer to Figure 2.9. Assume that the program manager specified a new project completion requirement of 145 days, what is the anticipated probability of success?

15. The following data are available:

Event	Previous Event	t_a	t_b	t_c
8	7	20	30	40
	6	15	20	35
	5	8	12	15
7	4	30	35	50
	3	3	7	12
6	3	40	45	65
	2	25	35	50
5	2	55	70	95
4	1	10	20	35
3	1	5	15	25
2	1	10	15	30

   a. Construct a PERT/CPM chart from the data above.
   b. Determine the values for standard deviation, *TE, TL, TS, TC,* and P (when *PC* = 100).
   c. What is the critical path? What does this value mean?

16. When employing PERT/COST, the time–cost option applies. What is meant by the time–cost option? How can it affect the critical path?

17. Assume that as a prime contractor you have a need to acquire a small communication system. In surveying the market, you locate several suppliers who can meet your need with some relatively minor design effort. What type of *M* requirements would you specify? How would you specify these requirements? Given that a particular supplier has been selected, how would you ensure that these requirements will be met?

18. The following data are available:

Event	Previous Event	t_a	t_b	t_c
8	7	10	20	30
	4	12	14	16
7	6	4	10	15
	5	18	26	34
6	3	11	23	35
5	2	23	30	45
4	2	20	25	30
3	1	18	24	32
2	1	3	9	15

   a. Construct a PERT/CPM chart from the data above.
   b. Determine the values for standard deviation, *TE, TL, TS, TC,* and *P* (when *PC* = 70).

19. Imagine that you have been charged with the development of a Maintainability Program Plan. Enumerate your input requirements in order to complete your task successfully.

20. What are the ingredients of a robust cost control system? Discuss how the WBS supports the cost control objective. Prepare a flow chart depicting all of the activities that constitute a good cost control system/procedure.

21. Discuss the interface of a maintainability engineer with regard to the other design, engineering, and management disciplines.

3
ORGANIZATION FOR MAINTAINABILITY

The management aspects of maintainability commenced with the definition of the maintainability planning requirements described in Chapter 2. Maintainability program functions and tasks were identified, a WBS was developed, schedule and cost requirements were formulated, and plans and specifications were addressed. It is now appropriate to present this material in the context of an organizational structure.

Organization constitutes the combining of resources in such a manner as to fulfill some specific need. Organizations involve a group of individuals of varying levels of expertise combined into a social structure of some type to accomplish one or more functions. Organizational structures vary with the functions to be performed, and the results depend on the established goals and objectives, the resources available, the communications and working relationships of the individual participants, personal motivation, and many other factors. The ultimate objective, of course, is to achieve the most effective utilization of human, material, and monetary resources through the establishment of decision-making and communications processes designed to accomplish specific objectives.

The fulfillment of maintainability objectives is highly dependent on the proper mix of resources and the development of good communications. The uniqueness of tasks and the many different interfaces that exist require not only good communication skills, but an understanding of the system as an entity and the many design disciplines that contribute to its development. Maintainability is only one of these design disciplines; however, the successful implementation of maintainability program functions requires a thorough understanding of not only system-level requirements, but also the many organizational interfaces that exist. This chapter deals with maintainability organization, its functions, organizational interfaces, and the personnel skills necessary to meet its objectives.[1]

---

[1] The level of discussion of organizational concepts in this chapter is rather cursory in nature, and is intended to provide the reader with an overview of some key points with respect to maintainability. A more in-depth coverage of organization and management theory is provided in several of the references listed in Appendix E.

3.1 ORGANIZATIONAL RESPONSIBILITIES

To properly address the subject of "organization for maintainability," one needs to understand the environment in which maintainability engineering functions are performed. Although this may vary somewhat depending on the size of the project and the stage of design and development, this discussion is primarily directed to a large project operation, characterized by the procurement and acquisition of many large-scale systems. By addressing large projects and some of the associated challenges, it is hoped that a better understanding of the role of maintainability engineering in a somewhat complex situation will be provided. The reader must, of course, tailor and adapt to his/her program requirements.

For a relatively large project, maintainability program requirements may appear at several organizational levels, as shown in Figure 2.3. The consumer (or the customer) may establish a maintainability organization to accomplish many of the tasks described in Chapter 2, or these tasks may be relegated to a prime contractor (the major producer) through some form of contractual structure. The question is— who is responsible for and has the authority to perform the major functions associated with the design of the overall system? Or, who is responsible to fulfill the basic requirements of systems engineering?

In some instances, the consumer may assume full responsibility for the overall design and integration of the system. Top system-level design, the development of system operational requirements and the maintenance concept, preparation of the system Type "A" specification, preparation of the SEMP, preparation of the ILSP, and the completion of major systems engineering tasks are accomplished by the consumer organization. In this situation, a maintainability organizational structure must be established for the accomplishment of such program tasks as the development of design requirements for inclusion in the system specification, the preparation of the maintainability program plan, participation in the early ongoing conceptual design activities, the accomplishment of top-level maintainability analysis tasks, and so on.

In other cases, while the consumer provides overall guidance in terms of initiating the procurement and acquisition process (i.e., issuing a general statement of work or an equivalent contractual document), the producer (or prime contractor) is held responsible for system design and integration. In this situation, completion of maintainability program requirements is accomplished by the producer, with supporting tasks being conducted by individual suppliers as required. In other words, while both the producer and the consumer have established maintainability organizations, the basic responsibility (and authority) for fulfilling the objectives described throughout this text lies with the producer's organization. This is the model that will serve as the basis for much of the discussion that follows.

3.1.1 Consumer/Customer Functions and Activities

The consumer/customer organization may range from one or a small group of individuals to an industrial firm, a commercial business, an academic institution, a government agency, or a military service such as the US Air Force (or the US Army or US Navy). The consumer may be the ultimate "user" of the system in question, or the procuring agency for a user. In any event, there may be a variety of approaches and organizational relationships involved in the design and development of new systems.

The objective is to identify the overall "Program Manager," and to pinpoint the responsibility and authority for "Maintainability Engineering." In the past, there have been many instances where the procuring agency has initiated a contract with an industrial firm (i.e., the producer) for the design and development of a large system, but has not delegated complete responsibility (or authority) for systems engineering and the establishment of maintainability requirements from the top down. The company has been held responsible for the design, development, and delivery of a system in response to certain specified requirements. However, the customer has not always provided the producer with the necessary data and/or controls to allow the development effort to proceed in accordance with good system engineering practices. At the same time, the customer has not accomplished the necessary functions of systems engineering management, including the proper considerations for maintainability as an integral part of the total engineering design effort. The net result has been the development of a system without the incorporation of the desired maintainability characteristics in design; that is, the system design configuration is not reliable, not maintainable, not supportable, not cost effective, and not responsive to the needs of the customer or, more specifically, the needs of the ultimate user.

The fulfillment of maintainability engineering objectives is highly dependent on a *commitment* from the top down! These objectives must be recognized from the beginning by the customer, and an organizational entity needs to be established to ensure that these objectives are met. The customer must create the appropriate environment and take the lead by initiating either one of the following courses of action:

1. Accomplish systems engineering functions within the customer's organizational structure and establish maintainability design requirements through the completion of early trade-off studies, the development of the maintenance concept, the establishment of maintainability design criteria, the incorporation of maintainability design requirements in the system specification, and the development of the maintainability program plan. The complete job of system-level design and integration is accomplished by the customer or procuring agency.

2. Accomplish systems engineering functions, and the establishment of maintainability design requirements, within an industrial firm. The conductance of early feasibility studies, the definition of system operational requirements and the development of the maintenance concept, the establishment of maintainability design criteria and the incorporation of such in the system specification, and the preparation of the maintainability program plan have been delegated to the producer.

As the split in authority, responsibilities, and subsequent duties may not be as clean cut as inferred, it is important that the responsibility for maintainability be established at the beginning. The customer must clarify system objectives and program functions, and the requirements and responsibilities for maintainability must be well defined.

If the maintainability engineering responsibility is delegated to the producer (i.e., the second option above), the consumer must support this decision by providing the necessary top-down guidance and managerial backing. Responsibilities must be properly delineated, system-level design data generated through early customer activities must be made available to the producer (e.g., the results of feasibility studies or the documentation covering system operational requirements), and the producer must be given the necessary leeway relative to making decisions at the system level. The challenge for the customer is to prepare a good comprehensive, well-written, and clear statement of work to be implemented by the producer. The successful implementation of maintainability program requirements in the design and development of a new system is highly dependent on this essential top-down guidance and support from the start!

3.1.2   Producer/Contractor Functions and Activities

For the majority of large-scale projects, the producer (or contractor) will undertake the bulk of the effort associated with the design and development of a new system (i.e., the second option identified in Section 3.1.1). The customer will specify system-level program requirements through the preparation of a RFP or an IFB, as described in Section 2.2, and various industrial firms will respond in terms of a formal proposal. Included within the producer's proposal is a description of the proposed organizational structure, organizational interfaces, staffing levels, and responsibilities.

Functional Organization Structure.[2] The primary building block for most organizational patterns is the *functional* structure reflected in Figure 3.1. This approach, sometimes referred to as the "classical" or "traditional" approach, involves the grouping of specialists or disciplines into separately identifiable entities. The intent is to perform similar activities within one organizational component. For example, all engineering work would be the responsibility of one executive, all production or manufacturing work would be the responsibility of another executive, and so on.

Figure 3.2 shows a further breakout of engineering activities, where the maintainability engineering organization is shown within the design assurance division. In reality, the maintainability organization may be included within the systems engineering organization, design engineering, integrated logistic support, or production operations in the factory environment when "maintainability improvement" is a

---

[2] See Blanchard, B. S., *Systems Engineering Management,* John Wiley & Sons, Inc., New York, 1991.

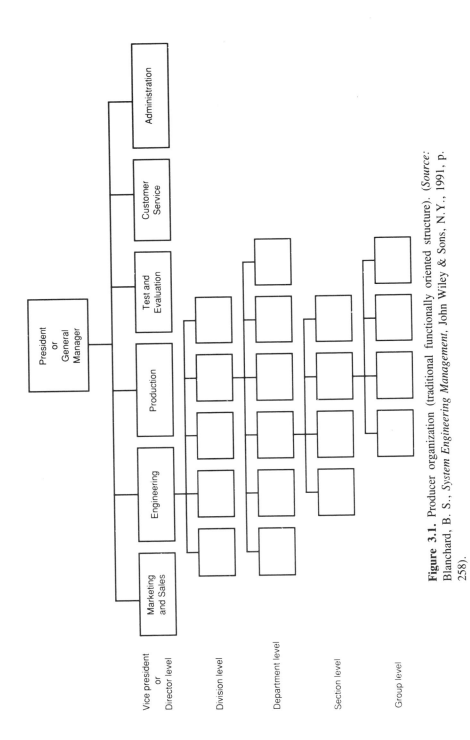

Figure 3.1. Producer organization (traditional functionally oriented structure). (*Source:* Blanchard, B. S., *System Engineering Management*, John Wiley & Sons, N.Y., 1991, p. 258).

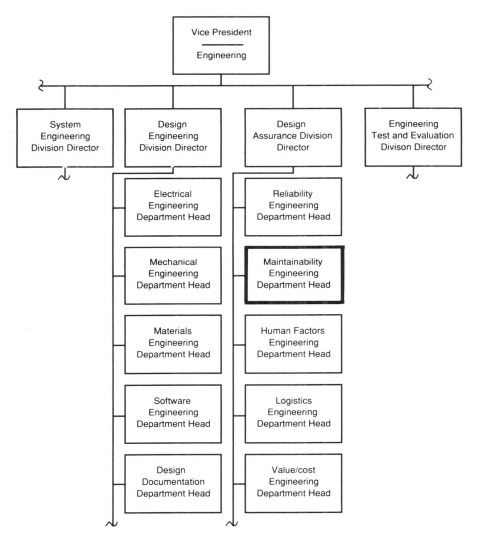

Figure 3.2. Breakout of engineering organizational activities. (*Source:* Blanchard, B. S., *System Engineering Management,* John Wiley & Sons, N.Y., 1991, p. 259).

major objective.[3] The location of the maintainability engineering function will vary from one company to the next. However, it is essential that this organizational component be positioned such that the requirements for maintainability can be ef-

---

[3] An example where the maintainability function might be included within the production operations organization is when the concept of TPM is being implemented and "maintainability improvement" is a major objective.

fectively and efficiently integrated into the design and development of systems in a timely manner.[4]

Figure 3.3 presents an example of a maintainability organization for a large-scale project. The activities identified represent an extension of the tasks described in Section 2.2, and include the supporting management functions of developing a WBS, scheduling tasks, cost estimation and control, and so on. The organization (and its associated tasks) must, of course, be tailored to the particular program requirement. While the figure reflects a full-scale effort, there may be many programs where only a few of these activities are necessary.

As for any organizational structure, there are advantages and disadvantages. Figure 3.4 identifies some of the pros and cons associated with the pure functional approach illustrated in Figure 3.1. As shown, the president (or general manager) has under his/her control all of the functional entities necessary to design and develop, produce, deliver, and support a system. Each department maintains a strong concentration of technical expertise and, as such, a project can benefit from the most advanced technology in the field. Additionally, levels of authority and responsibility are clearly defined, communication channels are well structured, and the necessary controls over budgets and costs can easily be established. In general, this organizational structure is well suited for a single project operation, large or small.

On the other hand, the pure functional organization may not be appropriate for large multiproduct firms or agencies. When there are a large number of different projects, each competing for special attention and the appropriate resources, there are some disadvantages. The main problem is that there is no strong central authority or individual responsibility for the total project (i.e., specific customer focus). As a result, the integration of activities that cross functional lines becomes difficult. Conflicts occur as each functional activity struggles for power and resources, and decisions are often made on the basis of what is best for the functional group rather than what is best for the project. Further, the decision-making processes are sometimes slow and tedious since all communications must be channeled through upper-level management. Basically, projects may fall behind and suffer in the classical functional organization structure.

Product-Line/Project Organization Structure. As industrial firms grow and there are more products being developed, it is often convenient to classify these products into common groups and to establish a product-line organization structure as shown at the top of Figure 3.5. A company may become involved in the development of communication systems, transportation systems, and electronic test and support equipment. Where there is functional commonality, it may be appropriate to organize the company into three divisions, one for each product line. In such instances, each division (which may be geographically separated) will be self-sufficient relative to system design and support. Further, each may serve as a functional entity with operations similar to those described above.

---

[4]Location of the maintainability engineering function within the producer's overall organizational structure must consider the many interfaces (and interface requirements) that are described in Section 2.4.

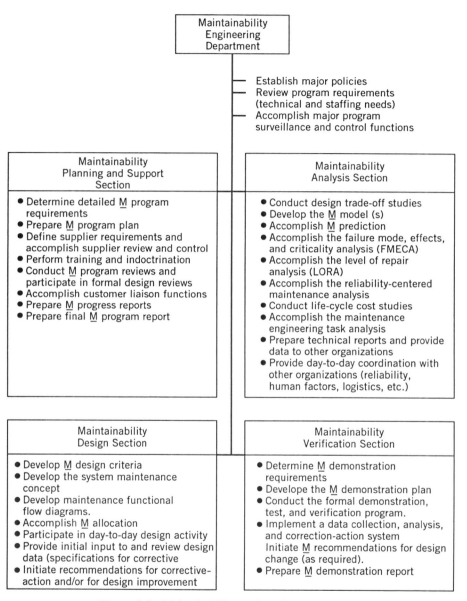

Figure 3.3. Maintainability engineering organization.

Advantages
1. Enables the development of a better technical capability for the organization. Specialists can be grouped to share knowledge. Experiences from one project can be transferred to other projects through personnel exchange. Cross-training is relatively easy.
2. The organization can respond quicker to a specific requirement through the careful assignment (or reassignment) of personnel. There are a larger number of personnel in the organization with the required skills in a given area. The manager has a greater degree of flexibility in the use of personnel and a broader manpower base with which to work. Greater technical control can be maintained.
3. Budgeting and cost control is easier due to the centralization of areas of expertise. Common tasks for different projects are integrated, and it is easier not only to estimate costs but to monitor and control costs.
4. The channels of communication are well established. The reporting structure is vertical, and there is no question as to who is the "boss."

Disadvantages
1. It is difficult to maintain an identity with a specific project. No single individual is responsible for the total project or the integration of its activities. It is hard to pinpoint specific project responsibilities.
2. Concepts and techniques tend to be functionally oriented with little regard toward project requirements. The "tailoring" of technical requirements to a particular project is discouraged.
3. There is little customer orientation or focal point. Response to specific customer needs is slow. Decisions are made on the basis of the strongest functional area of activity.
4. Because of the group orientation relative to specific areas of expertise, there is less personal motivation to excel and innovation concerning the generation of new ideas is lacking.

Figure 3.4. A functional organization—advantages and disadvantages. (*Source:* Blanchard, B. S., *System Engineering Management,* John Wiley & Sons, N.Y., 1991, p. 261).

In divisions where large systems are being developed, the product-line responsibilities may be subdivided into projects, as illustrated at the bottom of Figure 3.5. In such cases, the project will be the lowest independent entity, and maintainability program requirements will be accomplished in each project as required.

A project organization is one that is solely responsive to the planning, design and development, production, and support of a single system or a large product. It is time-limited and directly oriented to the life cycle of a particular system, and the commitment of personnel and material is purely for the purposes of accomplishing tasks peculiar to that system. Each project will contain its own management structure, its own engineering functions, its own production capability, its own support

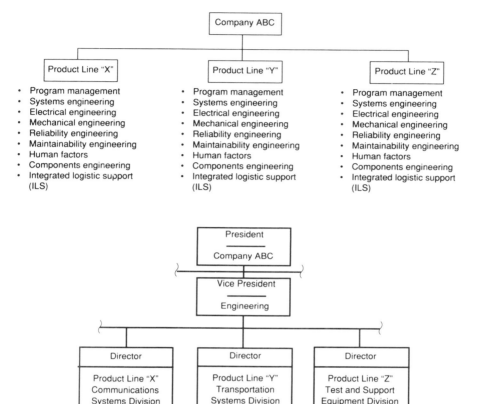

Figure 3.5. Typical product-line/project organization structures. (*Source:* Blanchard, B. S., *System Engineering Management,* John Wiley & Sons, N.Y., 1991, pp. 261–262).

function, and so on. The project manager has the authority and responsibility for all aspects of the project, whether it is a success or a failure.

In the case for both the product line and the project structures, the activities are organized as presented in Figure 3.5 (top). The lines of authority and responsibility for a given project are clearly defined, and there is no question as to priorities. On the other hand, there is potential for the duplication of activities within a given firm, and this can turn out to be quite costly. Emphasis is on individual projects as compared to the overall functional approach illustrated in Figure 3.1. Some of the ad-

Advantages

1. The lines of authority and responsibility for a given project are clearly defined. Project participants work directly for the project manager, communication channels within the project are strong, and there is no question as to priorities. A good project orientation is provided.

2. There is a strong customer orientation, a company focal point is readily identified, and the communication processes between the customer and the contractor are relatively easy to maintain. A rapid response to customer needs is realized.

3. Personnel assigned to the project generally exhibit a high degree of loyalty to the project, there is strong motivation, and personal morale is usually better with product identification and affiliation.

4. The required personnel expertise can be assigned and retained exclusively on the project without the time sharing that is often required under the functional approach.

5. There is greater visibility relative to all project activities. Cost, schedule, and performance progress can be easily monitored, and potential problem areas (with the appropriate follow-on corrective action) can be identified earlier.

Disadvantages

1. The application of new technologies tends to suffer without strong functional groups and the opportunities for technical interchange between projects. As projects go on and on, those technologies that are applicable at project inception continue to be applied on a repetitive basis. There is no perpetuation of technology, and the introduction of new methods and procedures is discouraged.

2. In contractor organizations where there are many different projects, there is usually a duplication of effort, personnel, and the use of facilities and equipment. The overall operation is inefficient and the results can be quite costly. There are times when a completely decentralized approach is not as efficient as centralization.

3. From a managerial perspective, it is difficult to effectively utilize personnel in the transfer from one project to another. Good qualified workers assigned to projects are retained by project managers for as long as possible (whether they are being effectively utilized or not), and the reassignment of such personnel usually requires approval from a higher level of authority which can be quite time-consuming. The shifting of personnel in response to short-term needs is essentially impossible.

4. The continuity of an individual's career, his or her growth potential, and the opportunities for promotion are often not as good when assigned to a project for an extended period of time. Project personnel are limited in terms of opportunities to be innovative relative to the application of new technologies. The repetitiousness of tasks sometimes results in stagnation.

Figure 3.6. A product-line/project organization—advantages and disadvantages. (*Source:* Blanchard, B. S., *System Engineering Management,* John Wiley & Sons, N.Y., 1991, p. 263).

vantages and disadvantages of product-line/project structures are presented in Figure 3.6.

Matrix Organization Structure. The matrix organization structure is an attempt to combine the advantages of the pure functional organization and the pure project organization. In the functional organization, technology is emphasized while project-oriented tasks, schedules, and time constraints are often sacrificed. For the pure project, technology tends to suffer since there is no single group for the planning and development of such! Matrix management is an attempt to acquire the greatest amount of technology, consistent with project schedules, time and cost constraints, and related customer requirements. Figure 3.7 presents a typical matrix organization structure.

Each project manager reports to a vice president, and has the overall responsibility and accountability for project success. At the same time, the functional departments are responsible for maintaining technical excellence and for ensuring that all available technical information is exchanged between projects. The functional managers, who also report to a vice president, are responsible to ensure that their personnel are knowledgeable about the latest accomplishments in their respective fields. In this case, maintainability personnel may be permanently based in a functional "home" department (from a disciplinary perspective), but temporarily assigned to a given project for the accomplishment of the required maintainability tasks associated with that project.

The matrix organization, in its simplest form, can be considered a two-dimensional entity with projects representing potential profit centers and the functional departments identified as cost centers. For small industrial firms, the two-dimensional structure may reflect the preferred organizational approach because of the flexibility allowed. The sharing of personnel and the shifting back and forth are often inherent characteristics. On the other hand, for large corporations with many product divisions, the matrix becomes a multidimensional structure.

As the number of projects and functional departments increase, the matrix structure can become quite complex. To ensure success in implementing matrix management, a highly cooperative and mutually supportive environment must be created within the company. Managers and workers alike must be committed to the objectives of matrix management. A few key points are noted below:

1. Good communication channels (vertical and horizontal) must be established to allow for a free and continuing flow of information between projects and functional departments. Good communications must also be established from project to project.

2. Both project managers and functional department managers should participate in the initial establishment of company-wide and program-oriented objectives. Further, each must have an input and become involved in the planning process. The purpose is to help ensure the necessary commitment on both sides. Additionally, both project and functional managers must be willing to negotiate for resources.

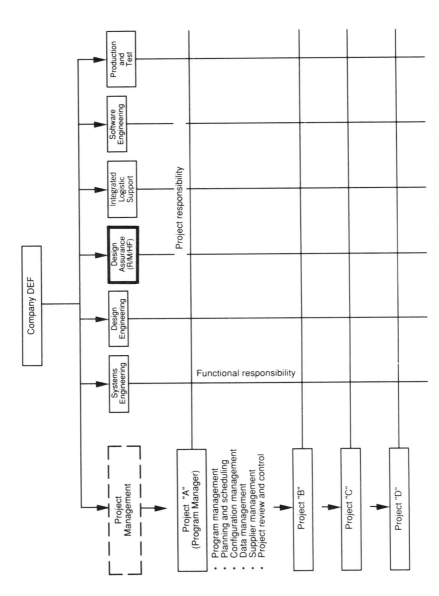

Figure 3.7. Pure matrix organization structure. (*Source:* Blanchard, B. S., *System Engineering Management*, John Wiley & Sons, N.Y., 1991, p. 264).

3. In the event of conflict, a quick and effective method for resolution must be established. A procedure must be developed with the participation and commitment from both project and functional managers.

4. For personnel representing the technical functions and assigned to a project, both the project manager and the functional department manager should agree on the duration of assignment, the tasks to be accomplished, and the basis upon which the individual(s) will be evaluated. The individual worker must know what is to be expected of him/her, the criteria for evaluation, and which manager will be conducting the performance review (or how the performance review will be conducted). Otherwise, a "two-boss" situation (each with different objectives) may develop and the employee will be caught in the middle!

The matrix structure provides the best of several worlds, that is, a composite of the pure project approach and the traditional functional approach. The major advantage pertains to the capability of providing the proper mix of technology and project-related activities. At the same time, a major disadvantage relates to the conflicts that arise on a continuing basis as a result of a power struggle between project and functional managers, changes in priorities, and so on. A few advantages and disadvantages are noted in Figure 3.8.

Organization for Maintainability. The material presented thus far has provided an overview of the major characteristics of the functional, project, and matrix organization structures. Some of the advantages and disadvantages of each are identified. It is important to thoroughly understand and have these characteristics in mind when developing an organizational approach involving maintainability. More specifically, when considering maintainability engineering objectives, the following points should be noted:

1. The function of maintainability must be oriented to the objective of "bringing a system into being" and have a "customer focus." Maintainability engineering should be involved in the initial establishment of system design requirements, and in the follow-on integration of design and support activities throughout system development, production, and operational use. In this regard, there is a natural close association with the "project" type of organizational structure.

2. The successful fulfillment of maintainability objectives requires a strong "technical thrust!" Not only must the maintainability engineer be knowledgeable about the latest methods and techniques associated with his/her discipline, but he/she must have some understanding of the design technologies associated with the system(s) in question. For the maintainability engineer to effectively *contribute* to the design process, he/she must be cognizant of design technologies and practices, and be able to "talk the designer's language." The transfer of knowledge from one project to another often facilitates this objective. Thus, the preferred organization structure should include selected "functional" elements in addition to the project orientation.

Advantages
1. The project manager can provide the necessary strong controls for the project while having ready access to the resources from many different functionally-oriented departments.
2. The functional organizations exist primarily as support for the projects. A strong technical capability can be developed and made available in response to project requirements in an expeditious manner.
3. Technical expertise can be exchanged between projects with a minimum of conflict. Knowledge is available for all projects on an equal basis.
4. Authority and responsibility for project task accomplishment are shared between the project manager and the functional manager. There is mutual commitment in fulfilling project requirements.
5. Key personnel can be shared and assigned to work on a variety of problems. From the company top-management perspective, a more effective utilization of technical personnel can be realized and program costs can be minimized as a result.

Disadvantages
1. Each project organization operates independently. In an attempt to maintain an identity, separate operating procedures are developed, separate personnel requirements are identified, and so on. Extreme care must be taken to guard against possible duplication of efforts.
2. From a company viewpoint, the matrix structure may be more costly in terms of administrative requirements. Both the project and the functional areas of activity require similar administrative controls.
3. The balance of power between the project and the functional organizations must be clearly defined initially and closely monitored thereafter. Depending on the strengths (and weaknesses) of the individual managers, the power and influence can shift to the detriment of the overall company organization.
4. From the perspective of the individual worker, there is often a split in the chain of command for reporting purposes. The individual is sometimes "pulled" between the project boss and the functional boss.

Figure 3.8. A matrix organization—advantages and disadvantages. (*Source:* Blanchard, B. S., *System Engineering Management,* John Wiley & Sons, N.Y., 1991, p. 266).

3. The nature of the maintainability engineering function, its objectives in terms of being one of the many disciplines within the design integration process, its many interfaces with other program activities, and so on, require the existence of good communication channels (both vertically and horizontally). Personnel within the maintainability engineering organization and assigned to a project must maintain effective communications with many other organizational elements, including systems engineering, reliability engineering, human factors, logistic support, the vari-

ous design disciplines involved in the system development effort, and the customer (refer to Section 2.4). The fulfillment of these requirements favors the "project" organizational approach.

Although the implementation of maintainability engineering requirements can actually be fulfilled through any one of a number of organizational structures, the preferred approach should respond to these three major considerations. It appears that the best organizational structure is constituted from a combination of "project" and "functional" requirements. Although a major *project* orientation is necessary in response to customer needs, a *functional* orientation is necessary to ensure consideration of the latest technology applications. The combined project–functional organization approach may vary somewhat depending on the size of the industrial firm. For a large firm, the organization structure illustrated in Figure 3.9 may be appropriate. Project activities are relatively large in scope (and in personnel loading) while, at the same time, supporting functional activities cover selected areas of expertise where centralization is justified. For smaller firms, the functional departments may be relatively large, and they provide support to individual projects on demand. This support is assigned on a task-by-task basis, and may include only a single individual performing maintainability functions on a part-time basis. In other words, there may be a number of small projects involving maintainability engineering tasks that, on an individual basis, cannot afford the assignment of a full-time engineer. In this instance, a maintainability engineer from the functional department may be assigned to work on several different projects concurrently. In practice, the degree of "project" emphasis and "functional" emphasis often shifts back and forth depending on both the size of the firm and the nature of the activity; that is, whether conceptual design, preliminary system design, or detailed design and development activities are in progress.

Project size varies not only with the type and nature of the system being developed, but with the specific stage of development. A large-scale system in the early stages of conceptual design may be represented by a small project organization, where maintainability requirements must be covered by the systems engineer, using a limited level of expertise from the functional maintainability organization as required. As system development progresses into the phases of preliminary system design and detail design and development, the organizational structure may shift somewhat, replicating the configuration in Figure 3.9. In other words, the characteristics and the structures of organizations are usually "dynamic" in nature. The organizational structure must be adapted to the needs of the project at that time, and these needs may shift as system development evolves.

3.1.3 Supplier Functions and Activities

Referring to Section 2.5, there may be many different categories of suppliers. Of particular interest are those suppliers involved in the design and development of new system components. Unless the component is large and complex (justifying a large *project* approach), a functional type of organization is likely to prevail within

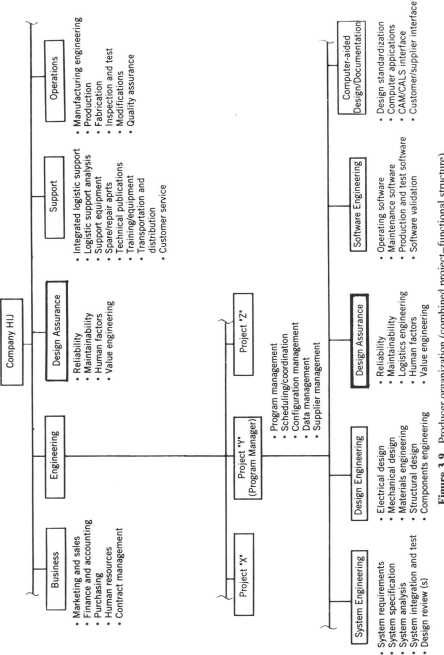

Figure 3.9. Producer organization (combined project–functional structure).

82

the supplier's overall structure. Maintainability functions will, in all probability, be limited to one or two design-related tasks tailored to meet a specific program need (e.g., the maintainability analysis tasks identified in Figure 3.3). Specific maintainability requirements must be included both within the technical specification (i.e., the Type "B" specification) and in the statement of work developed by the prime contractor for the supplier. As stated earlier, traceable requirements must be from the top down, and the appropriate level(s) of effort must be initiated to ensure that the requirements are met.

3.2   STAFFING THE ORGANIZATION

The requirements for staffing an organization initially stem from the results of the maintainability planning activity described in Chapter 2. Tasks are identified from both short- and long-range projections (refer to Section 2.3.3), and combined into work packages and the WBS. The work packages are then grouped and related to specific "position" requirements. The positions are, in turn, arranged within the organizational structure considered to be most appropriate for the need (refer to Figures 3.1–3.9).

With regard to specific position requirements for a maintainability organization, one should first have a good understanding of the basic functions of the organization. These are discussed in Chapter 2 and applied in the context of a "structure" in the earlier sections of this chapter. Review of the assigned tasks (Section 2.2), the nature and challenges of the organizational structure, and so on indicate that, in general, an entry-level "Maintainability Engineer" should have:

1. A basic formal education in some recognized field of engineering, that is, a baccalaureate degree in engineering or the equivalent.

2. An understanding of the overall design process (and the system life cycle) as it applies to the systems and products being developed by the company. For example, if the company is involved in the development of electrical/electronic systems, then it is desirable that the candidate have some knowledge of, and prior design experience in (if possible), electrical/electronic systems. A different type of experience would be required for aeronautical systems, mechanical systems, civil systems, and so on. The maintainability engineer, to be effective, must be able to "communicate" with the designer in the designer's "language."

3. An understanding of the system engineering process and the methods/tools that can be effectively utilized in bringing a system/product into being. For exple, the maintainability engineer must understand the process of requiremenition, functional analysis and allocation, synthesis and design optiand evaluation, and so on. Further, he or she must be familiar with involving the use of CAD, CAM, and continuous acquisition and (CALS), and must be able to interpret and understand design din varying formats throughout the system design process. Thneer must be able to effectively "contribute" to the design.

4. An understanding of the relationships among functions, including marketing, contract management, purchasing, system engineering, electrical engineering, mechanical engineering, reliability engineering, human factors, configuration management, production (manufacturing), quality control, customer and supplier operations, and so on. A basic understanding of the organizational interfaces described in Section 2.4 is desirable.

If an individual is to successfully implement the functions described in Chapter 2, then he/she should have some prior knowledge and experience in these areas. Additionally, the maintainability engineer must, of course, be knowledgeable about the methods/techniques utilized in the completion of the required maintainability tasks specified for a given project. For large program organizations, such as the one illustrated in Figure 3.3, maintainability functions may be broken down as follows:

1. *Planning and support activities* involving the initial development of plans and specifications, determination of supplier requirements, participation in design reviews, preparation of program reports, customer liaison, and so on.
2. *Design participation activities* involving the day-to-day participation in the system/product design process (the development of design criteria, evaluation of alternative design configurations, recommendations for design changes, etc.).
3. *Analysis activities* involving the accomplishment of various trade-off analyses, predictions, level of repair analyses, maintenance task analyses, life-cycle cost analyses, and so on.
4. *Test, evaluation, and verification activities* involving participation in system-level testing, the conductance of maintainability demonstrations, the assessment of systems/products in the field for maintainability considerations, and so on.

For large programs, where some degree of specialization can be supported within the maintainability organization, the requirement for additional entry-level skills may be necessary. For example, in the area of "maintainability analysis," it would be appropriate for the maintainability engineer to have some knowledge of statistical methods, selected operations research techniques (e.g., linear/dynamic programming, networking, queuing theory, control theory, and optimization), and data processing procedures. For the maintainability engineer assigned to the area of "test, evaluation, and verification," it would be appropriate for the individual to have had some "hands on" experience with the operation and support of hardware, software, and so on in the field (i.e., the consumer's environment).

With regard to smaller programs, it may not be possible, or economically feasible, to develop a large organization such as the one illustrated in Figure 3.3. In such instances, while the requirements for maintainability task completion still exist, it may be necessary to solicit the appropriate levels of expertise from outside sources (i.e., statistical services and data processing services). However, the maintainability engineer must still recognize the need for and have a good understanding of various analytical methods and their application to the day-to-day program tasks.

In staffing the organization, possible sources include (1) qualified personnel from within the company who are ready for promotion and (2) personnel from outside who are available through the open market. It is the responsibility of the maintainability department manager to work closely with the Human Resources Department in establishing the initial requirements for personnel, in developing position descriptions and advertising material, in recruiting prospects and conducting interviews, in the selection of qualified candidates, and in the final hiring of individuals for employment within the maintainability organization.

3.3 PERSONNEL TRAINING AND DEVELOPMENT

Nearly every engineer wants to know how he or she is doing on a day-to-day basis and what the opportunities are for growth. A response to the first part is derived through a combination of the "formal performance review," which is conducted on a regularly scheduled basis (either semiannually or annually), and the ongoing "informal communications process" with the boss. The engineer is given responsibility and seeks recognition and approval from the supervisor. As discussed earlier, there needs to be close communications, and the boss needs to provide some reinforcement the employee is doing a good job. Also, the employee needs to know as soon as possible when his/her work is unsatisfactory and improvement is desired. Waiting until the formal performance review is conducted to tell an employee that his/her work is unsatisfactory is poor practice and is demoralizing since, by virtue of not having heard any comments to the contrary, the employee has assumed that all is well! In a maintainability organization, where there are many interfaces (both internal and external), it is particularly important that an appropriately close level of communication is established from the beginning.

The answer to the second question pertaining to the opportunities for growth depends on (1) the climate provided within the organization and the actions of the manager which allow for individual development, and (2) the initiative on the part of the engineer to take advantage of the opportunities provided. Within a maintainability organization, it is *essential* that individual personal growth take place if the department is to function effectively, particularly because many of its functions are "service" oriented. The climate (or environment) must allow for individual development, and the individual maintainability engineer must seek opportunities accordingly. The maintainability department manager should work with each employee to prepare a tailored *development plan* for that employee. The plan, adapted to each person's specific needs, should allow for (and promote) personal development by making available a combination of the following:

1. Formal internal training designed to familiarize the engineer with the policies and procedures applicable to the overall company as a whole, as well as the detailed operating procedures of his/her own organization. This type of training should en-

able the individual to function more successfully within the framework of the total organization through familiarization with the many interfaces that he/she will encounter on the job (see Section 2.4).

2. On-the-job training through selective project assignments. Although the extensive shifting of personnel from job to job (or project to project) can be detrimental, it is sometimes appropriate to reassign an individual to work where he/she is likely to be more highly motivated! Every employee needs to acquire new skills, and occasional transfers may be beneficial, as long as the overall productivity of the organization does not suffer.

3. Formal technical education and training designed to upgrade the engineer relative to the application of new methods and techniques in his/her own field of expertise. This pertains to the necessity for the engineer to maintain currency (and avoid technical obsolescence) through a combination of (a) continuing education short courses, seminars, and workshops; (b) formal off-campus graduate engineering programs leading to an advanced degree and provided at the local level; and (c) long-term training involving opportunities for research and advanced education on some university or college campus. Currently, there are many opportunities available for an individual to progress through the pursuit of formal education and training.

4. A technical exchange of expertise with others in the field through participation in technical society activities, industry associations, symposia and congresses, and the like.

The maintainability department manager must recognize the need for the ongoing development of personnel in his/her organization, and should encourage each individual to seek a higher level of performance by offering not only challenging job assignments, but opportunities for growth through education and training. The long-term viability of any organization depends on personnel development. This, in turn, should enhance individual motivation and result in the fulfillment of maintainability engineering functions in a high-quality manner.

QUESTIONS AND PROBLEMS

1. Figure 2.3 illustrates major consumer–producer–supplier interfaces. Briefly describe how maintainability program requirements can be accomplished through this structure? What requisites must prevail to ensure that the results are effective and efficient?

2. The various types of organizational structure include the pure "functional," the "product line," the "project," and the "matrix." Briefly describe the structure and identify some of the advantages and disadvantages of each.

3. Refer to Question 2. Which type of organization structure is preferred from a maintainability engineering perspective? Why?

4. Describe the major organizational interface requirements (i.e., inputs and outputs) between maintainability engineering and—system engineering? design engineering (electrical, mechanical, etc.)? reliability engineering? human factors? logistic support? production or manufacturing?

5. Assume that you have been assigned responsibility for the design and development of a new system. Develop an organization chart (using any combination of structural approaches desired), identify the major elements contained therein (including the maintainability organization), and briefly describe some of the key interfaces.

6. In developing a maintainability organization for the conceptual design of a new system, what type of organization structure would you likely employ? Why? Would you utilize the same structure as the design evolves into preliminary system design, detail design and development, and so on? Why or why not?

7. Refer to Figure 3.3. The organization shown categorizes the maintainability functions into four basic areas. Describe and illustrate the interfaces that exist (by showing the flow of information).

8. Assume that you have just been assigned to establish and organize a maintainability engineering department for a newly formed company. What policies and procedures would you implement? What type of people would you need in terms of quantity, skill levels, individual backgrounds, and so on? What type of management style would you impose (autocratic, democratic, etc.)? Describe some of the characteristics.

9. Describe some of the major challenges associated with supplier management.

10. In terms of "organizational environment" for maintainability engineering, what factors need to be considered? Briefly describe the organizational environment that is appropriate for the successful implementation of maintainability engineering functions.

11. Based on your own perspective, describe the characteristics of a "Maintainability Engineer" (background, personal characteristics, and motivational factors).

12. As manager of the Maintainability Engineering Department, what steps would you take to ensure that your organization maintains a lead position relative to technical competency?

4

THE MEASURES OF MAINTAINABILITY

Maintainability requirements must be delineated initially in qualitative and quantitative terms. As the system design and development activity proceeds, the configuration defined must be evaluated against these specified requirements, and redesign efforts undertaken as needed to ensure effective results. This evaluative or feedback function is accomplished through a series of predictions, estimations, analyses, and demonstrations.

To facilitate the process of maintainability requirements definition, specification, allocation, and evaluation, appropriate quantitative measures of maintainability must be identified and defined. Multiple measures may be necessary to satisfactorily represent and model the relevant system design characteristics. Further, these measures may need to be adapted, depending upon the nature of the system being developed. The objective of this chapter is to introduce and address some of the more commonly used quantitative measures of system maintainability. Initially, a brief overview of selected reliability factors is presented. Further system effectiveness measures such as availability and dependability, which are impacted by maintainability, are also explored.

4.1 RELIABILITY FACTORS

Reliability can be defined as the probability that a system or product will perform in a satisfactory manner for a given period of time when operated under specified operating conditions. The frequency of maintenance of a given system is impacted by its reliability. In general, system reliability is inversely proportional to the frequency of corrective maintenance actions. The reliability function, $R(t)$, may be expressed as

$$R(t) = 1 - F(t) = \int_t^\infty f(u)du \qquad (4.1)$$

where $F(t)$ is the probability of system failure by time t, and t is a random variable with a density function $f(t)$. Assuming that the time variable is described by an exponential density function, the reliability function is transformed to [1]

$$R(t) = \int_t^\infty \frac{1}{\theta} e^{-u/\theta} du = e^{-t/\theta} \qquad (4.2)$$

where e is the natural logarithm base, t is the time interval of interest, and θ is the mean life. Mean life (θ) refers to the average lifetimes of all items under consideration and is equal to mean time between failure (MTBF) for the exponential time density function. The reciprocal of MTBF expresses the instantaneous failure rate λ. Therefore, the reliability function with an exponential time density can also be expressed as

$$R(t) = e^{-t/\text{MTBF}} = e^{-\lambda t} \qquad (4.3)$$

[handwritten annotation: IN TIME (t) you'll have of FAILURES]

Assuming an exponential density function for the time variable best represents a scenario where the failure rate is essentially constant over the useful system operating life. This is frequently assumed during the course of reliability analyses and predictions. While variable failure rates are experienced during the infant mortality and wearout periods of a system life cycle, the constant failure rate assumption often works well after the system attains a steady state of operation. The flat portion of the familiar bathtub curve, illustrated in Figure 4.1 and often used to describe system reliability characteristics over the life cycle, represent this steady operational state.

4.1.1 Failure Rate and Mean Time Between Failure (MTBF)

Failure rate and MTBF are the most commonly used reliability measures. Failure rate (λ) simply refers to the frequency of failures, or the rate at which failures occur over a unit interval of time. It may be defined in terms of number of failures per hour or per million hours, or of percent failures per 1000 hours. Failure rate (λ) is expressed as

$$\lambda = \frac{\text{number of failures}}{\text{total operating hours}} \qquad (4.4)$$

Failure rate as a reliability measure can be adapted to a particular system or mission scenario. It is often expressed in terms of a certain number of system operational cycles, mission phase, or distance. While determining the number of system

---

[1] This is equivalent to making the assumption that the number of failures occurring during a time interval are Poisson distributed. The two distributions are the same, except that the exponential distribution is continuous (e.g., time) and the Poisson distribution is discrete (e.g., number of failures).

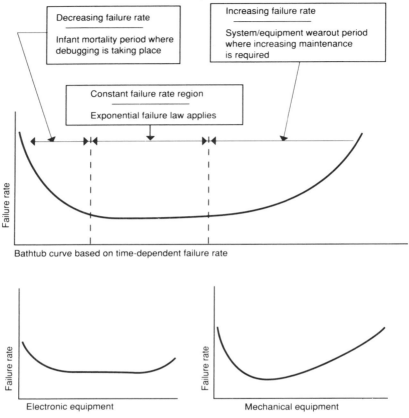

Figure 4.1. The bathtub system failure rate characteristic curve. (*Source:* Blanchard, B. S., *System Engineering Management,* John Wiley & Sons, N.Y., 1991, p. 79).

failures is usually not a problem, defining the time variable could be. The time variable is impacted by the nature of the experimental procedure, the component duty cycles, and mission scenarios. Consider the case where the system is repaired upon experiencing a failure and the test is continued to completion. The total time in this case is the product of the test time and the number of units undergoing test. Next, consider the case where failed units are not repaired and the time to failure in each instance is recorded. In this case, the total time is the sum of the individual test times of failed units and the product of the test duration and the number of units that successfully completed the test. As an example, suppose 10 units were tested under specified operating conditions where the test time is 600 hours, and assume that failed units are not repaired. Failures occur as follows:

- Unit 1 failed after 75 hours
- Unit 2 failed after 125 hours

- Unit 3 failed after 130 hours
- Unit 4 failed after 325 hours
- Unit 5 failed after 525 hours

Five units successfully completed the test cycle. The failure rate (λ), in number of failures per hour, can be expressed as

$$\lambda = \frac{5}{75 + 125 + 130 + 325 + 525 + 5(600)} = \frac{5}{4180} = 0.001196$$

As a second example, consider a given system with an operating cycle as depicted in Figure 4.2. The total operating time in this case is 152 hours and the system fails a total of six times as indicated. The failure rate per hour is

$$\lambda = \frac{\text{number of failures}}{\text{total operating time}} = \frac{6}{152} = 0.03947$$

Further, assuming an exponential time density function, the system mean life or the MTBF is

$$\text{MTBF} = \frac{1}{\lambda} = \frac{1}{0.03947} = 25.3357 \text{ hours}$$

In determining the frequency of corrective maintenance actions, care must be taken to consider the overall failure rate, including failures inherent in design (also called "primary" defects or "catastrophic" failures), "secondary" or dependent fail-

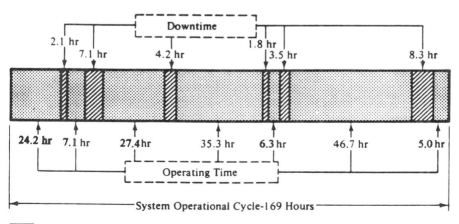

Figure 4.2. A typical system operational cycle.

TABLE 4.1   Combined System Failure Rate

Nature of System Failures	Failure Rate (number/hour)
Inherent reliability failure rate	0.000458
Dependent failure rate	0.000001
Manufacturing defects	0.000002
Wearout rate	0.000068
Operator-induced failure rate	0.000003
Maintenance-induced failure rate	0.000047
Equipment damage rate	0.000002
Total combined or operational failure rate	0.000581

ures, failures due to the introduction of manufacturing defects, and failures intro-duced by the operator or the maintenance personnel. The objective is to address all factors that may result in an inoperative system. This discussion leads to the concept of system *operational reliability* (R_o). The system operational reliability measure addresses not only the failures inherent in system design, but also secondary or dependent failures, and failures induced as a result of manufacturing defects and operator and maintenance personnel error. System operational reliability more closely tracks the failure rates actually experienced in the field. As an illustration, consider the overall or combined system failure rate presented in Table 4.1.

The concept of the reliability function and failure rates can be applied at the component level to compute reliability measures at the system level. Components within a system, depending upon their functionality, can be assumed to conform to a series network, a parallel network, or a combination thereof. This network be-comes obvious in a reliability block diagram. Networks with all components in series are the simplest to analyze. Failure of any one component causes a system level failure. Figure 4.3 depicts a series network.

For a pure series network, the system level failure rate is the sum total of all the individual component failure rates and can be expressed as

$$\lambda_s = \Sigma \lambda_i \tag{4.5}$$

where λ_s is the system-level failure rate, and λ_i is the individual component failure rate. In terms of MTBF, Equation (4.5) is transformed to

$$\text{MTBF}_s = \frac{1}{\Sigma 1/\text{MTBF}_1 + 1/\text{MTBF}_2 + \cdots + 1/\text{MTBF}_n} \tag{4.6}$$

Figure 4.3. A series component network.

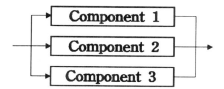

Figure 4.4. A parallel component network.

Further, the overall system reliability is

$$R_s = (R_1)(R_2)\ldots(R_n) \qquad (4.7)$$
$$= e^{-(\lambda_1 + \lambda_2 + \ldots + \lambda_n)t}$$

A pure parallel network, on the other hand, refers to a scenario where a number of similar components are in parallel and each one of them needs to fail in order to cause a system level failure. Figure 4.4 depicts a parallel network with three components.

System reliability for the arrangement in Figure 4.4 is expressed as

$$R(t) = 1 - (1 - R_1)(1 - R_2)(1 - R_3) \qquad (4.8)$$

In the case of n similar components, Equation (4.8) may be generalized to

$$R(t) = 1 - (1 - R)^n \qquad (4.9)$$

More often than not, series and parallel component relationships coexist in various combinations. The concepts discussed above can be applied to any combination. Consider the example illustrated in Figure 4.5. The reliability of the network as shown in Figure 4.5 is given by the equation

$$R(t) = [1 - (1 - R_a)(1 - R_b)(1 - R_c)][R_d][1 - (1 - R_e)(1 - R_f)(1 - R_g)] \qquad (4.10)$$

A convenient procedure to compute reliability for a combination network is to first analyze and reduce all the redundant or parallel elements to a unit reliability. System level reliability is then computed by finding the product of reliabilities of

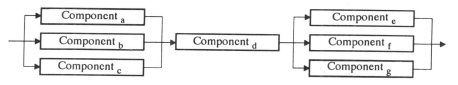

Figure 4.5. A combined series–parallel component network.

the units in series. Further, component redundancy can be classified into two broad types, active and passive (or standby). In the case of active redundancy, every redundant element is active and operational during system operation. Active redundancy can be classified further as full, partial, or conditional. In the case of passive or standby redundancy, however, redundant elements are activated only upon the failure of the operational element and with some external action taking place. Standby units may be similar to or different from the operational unit. For a more detailed coverage of this subject please refer to the bibliography in Appendix E.

4.1.2 Mean Time to Failure (MTTF) and Mean Time to First Failure (MTFF)

Mean time to system or component failure (MTTF) is often confused and used interchangeably with MTBF. While it can be used for repairable systems, it is a common reliability measure for nonrepairable systems and components such as light bulbs, transistors, and resistors. Mean time to failure is defined as the mean time to system failure measured from a particular reference point in time or mission state. It can be expressed as

$$\text{MTTF} = \frac{\text{total number of operating hours for an equipment population}}{\text{total number of population failures from a particular reference time}}$$

$$(4.11)$$

where the equipment items can be both repairable and nonrepairable. Mean time to first failure (MTFF) is another reliability measure often used for systems that exhibit a varying failure rate characteristic of redundant systems. It can be defined as the mean time to system failure measured from the reference point when the relevant systems are new and not used.

4.1.3 Probability of Survival

The probability of system survival or mission success (P_s) may be defined as the ratio of number of successful trials or missions to the total number of trials, including those that failed or were aborted. It may also be expressed as the probability of zero system failures during a certain time interval t, or

$$P_s = R(t) = e^{-\lambda t} \tag{4.12}$$

where $e^{-\lambda t}$ is the first term in a Poisson expansion series and represents the probability of zero failures or aborts during the time interval t. A constant failure rate is assumed in Equation (4.12). Extending the discussion to an operational system with a backup, the applicable equation is

0 + 1 FAILURE

$$P_s = e^{-\lambda t} + (\lambda t)e^{-\lambda t} \tag{4.13}$$

To further illustrate, assume that a system has a reliability of 0.778, and that the value of λt is 0.25. Also, the system is supported by two identical backups. The probability of system survival or mission success can now be expressed as

$$
\begin{aligned}
P_s &= e^{-\lambda t} + (\lambda t)e^{-\lambda t} + \frac{(\lambda t)^2 e^{-\lambda t}}{2!} \\
&= 0.778\left[1 + 0.25 + \frac{(0.25)^2}{(2)(1)}\right] \\
&= 0.778(1.2813) \\
&= 0.9968
\end{aligned}
\tag{4.14}
$$

This concept can be generalized to address a situation with x backups (or spares) and n operating units. The applicable general Poisson expression is

$$
f(x) = \frac{(n\lambda t)^x e^{-n\lambda t}}{x!}
\tag{4.15}
$$

4.1.4 Reliability Growth Curves

New systems and products often display a lower reliability during the early development phases. System reliability can be improved by analyzing and fixing some of the failure modes experienced. This concept is referred to as reliability growth and was formally analyzed for the first time in the mid-1960s by James T. Duane. Duane derived an empirical relationship based upon the MTBF improvement observed with respect to a range of aircraft components.

The concept of reliability growth is very relevant to maintainability analysis and can contribute to the overall effectiveness of the system support infrastructure. It impacts prediction of the frequency and types of system failures during the early design and development phases. The reliability growth curve in the Duane model can be expressed as

$$
\log(\text{MTBF}_c) = \log(\text{MTBF}_s) + \beta \, \log(T)
\tag{4.16}
$$

where MTBF_c and MTBF_s are the cumulative and starting mean time between failures, T is the total test or operating time, and β is the slope of the growth curve. The slope, β, indicates the effectiveness of the reliability growth program and has a strong correlation with the intensity of the effort. A given reliability growth curve can also be used to assess the test and evaluation time required to attain a target system reliability. Figure 4.6 depicts a typical reliability growth curve.

4.1.5 Software Reliability

Software has become an important facet of many modern diagnostic, test, and support equipment items. There is a certain amount of controversy in the literature

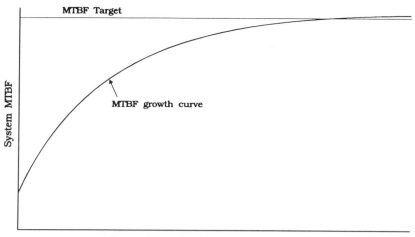

Figure 4.6. A typical reliability growth curve.

over the definition and the subsequent measurement of software reliability, and the associated failures and faults. A *software failure* is often defined as a characteristic of program behavior (e.g., a degradation in execution time or any other performance parameter), while a *fault* is an intrinsic part of the software program (e.g., an undeclared variable) and could be the cause of one or more failures. The occurrence of a software failure depends not only on the existence of a software fault, but also on the execution environment, for example, the operating system, the hardware platform, or the nature of the run or execution. Further, a software failure is defined as "an unacceptable departure of program operation from program requirements," and a software fault is "the software defect that causes a failure." [2]

The metrics of software failures more or less correspond to those utilized in the hardware world and include measures such as mean time between failure, failure rate, time to failure, and so on. Analogous to hardware, there are two general classifications of the time variable, *calendar time* and software *operating time* or duty cycle. Here, the software operating time or duty cycle is also referred to as the *execution time* or *clock time*. Obviously, the metrics of software reliability should consider time in terms of execution time or clock time.

Given the discussion above, one accepted definition of software reliability is "the probability of failure-free operation of a software component or system in a specified environment for a specified time." [2] For example, software reliability for a particular product may be expressed as being equal to 0.89 for an 11-hour time interval, when operated on a specified hardware platform and operating system, by

[2] Vick, C. R. and C. V. Ramamoorthy, *Handbook of Software Engineering,* Van Nostrand Reinhold Company, Inc., New York, 1984.

an operator with the specified skill level. Musa et al. define software failure intensity, another metric for software reliability, as "the rate of change of the mean value function or the number of failures per unit time."[3] This metric is expressed as a failure rate, for example, a software program may have a failure intensity of 0.01 failures/hour. In the two metrics defined above, higher modeling fidelity results if the time variable is expressed in terms of execution time and not calendar time.

4.2 MAINTAINABILITY FACTORS

Maintenance is performed on a system or component in the event of a failure, or as a preventive measure to preempt an expected failure. *Maintainability* on the other hand is a system design characteristic and addresses the ease, accuracy, timeliness, and economy of maintenance actions. Given that this characteristic is multidimensional, it has numerous measures. Maintainability may be measured in terms of a combination of elapsed times, labor hours and rates, maintenance cost and frequencies, and relevant logistic support factors. These measures facilitate the quantitative assessment of system maintainability. The objective is to influence design and produce/manufacture a system or unit that is effectively and efficiently supportable. Selected measures of maintainability are defined in this section.

4.2.1 Maintenance Elapsed-Time Factors

The objective of maintenance is to restore a system or unit to a satisfactory operational state, or to preempt expected failure and maintain a system's operational capability at a specified level. Maintenance can therefore be classified into two broad categories:

1. *Corrective maintenance*—the unscheduled actions initiated as a result of system failure (or perceived failure) that are necessary to restore a system to its expected or required level of performance. Corrective maintenance actions may include activities related to troubleshooting, disassembly, repair, remove and replace, reassembly, alignment, adjustment, checkout, and so on.
2. *Preventive maintenance*—the scheduled actions necessary to retain a system at a specified level of performance. Maintenance actions may include periodic inspections, calibration, condition monitoring, and/or replacement of critical items at designated time intervals.

Maintainability as a design characteristic is often measured in terms of the time required to perform maintenance. The easier and quicker it is to maintain a system, the better it is from a maintainability perspective. A schematic of the numerous

---

[3] Musa, J. D., A. Iannino, and K. Okumoto, *Software Reliability,* McGraw-Hill Book Company, Inc., New York, 1987.

maintenance-related time factors is depicted in Figure 4.7, while the more commonly used measures are defined below.

Mean Corrective Maintenance Time ($\overline{Mct}$). Corrective maintenance is performed primarily in response to an interruption of system operation or service caused by an unexpected failure. Since there is almost always a need for prompt operation restoration with minimum downtime, maximum emphasis is placed on corrective maintenance time reduction at the system level. A low diagnostics time coupled with a remove and replace maintenance concept can often effect a lower downtime at the operational level. Longer repair times can often be better tolerated at the "off-equipment" level. Some strategies for system-level corrective maintenance time reduction are discussed below.

1. *Fault recognition, location, and isolation.* Well-designed fault indicators that catch an operator's attention can contribute significantly to a lower corrective maintenance time. Historically, fault location and isolation are the most time-consuming elements within a corrective maintenance activity, particularly for electronic equipment. For mechanical items, the repair times are often the big contributor. Effective built-in test capability is an excellent way to further reduce maintenance times. Good maintenance procedures, well-trained maintenance personnel, and an unambiguous fault isolation capability all serve to reduce the time spent on this task.

2. *Accessibility.* There is a much higher degree of maintainability when high-failure-rate system elements are readily accessible. Whenever a model of a new system is available, it should be used to verify relative accessibility of system components. Consider the frustration of an automobile mechanic who must release engine mounts and jack up an engine to gain access to the spark plugs.

3. *Interchangeability.* Good functional and physical interchangeability facilitates the accomplishment of component removal and replacement, reduces maintenance downtime, and creates a positive impact on spares and inventory requirements.

4. *Redundancy.* It is very often possible to design-in redundant components such that they may be activated when needed and such that the system can continue to operate while the faulty unit is being repaired. Even though the overall maintenance work load may not be reduced, system downtime may be favorably impacted.

5. *Anthropometric considerations.* The human is an important element of the maintenance activity. Anthropometric considerations can also contribute to a higher system maintainability and include things such as the selection and proper placement of dials and indicators; readability and information processing aids; the size and placement of access gates; size, shape, and weight of components; readability of instructions; and so on. Refer to Section 4.3 for additional coverage of anthropometric and other human factor considerations.

A typical corrective maintenance cycle proceeds as depicted in Figure 4.8, and includes a series of steps such as: (1) failure detection, (2) fault isolation, (3) disassembly to gain access, (4) repair (or remove and replace), and (5) reassembly. The

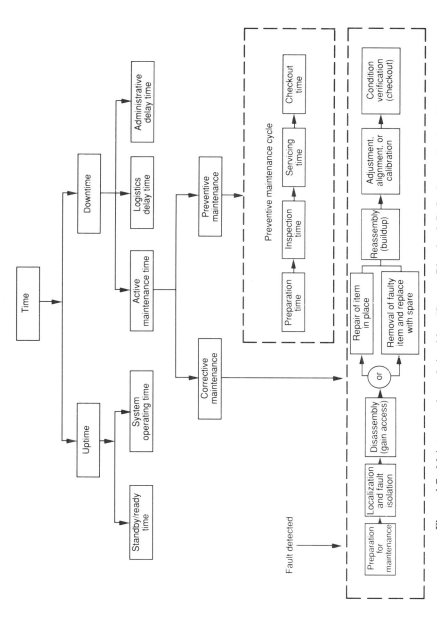

Figure 4.7. Maintenance time relationships. (*Source:* Blanchard, B. S., *System Engineering Management*, John Wiley & Sons, N.Y., 1991, p. 90).

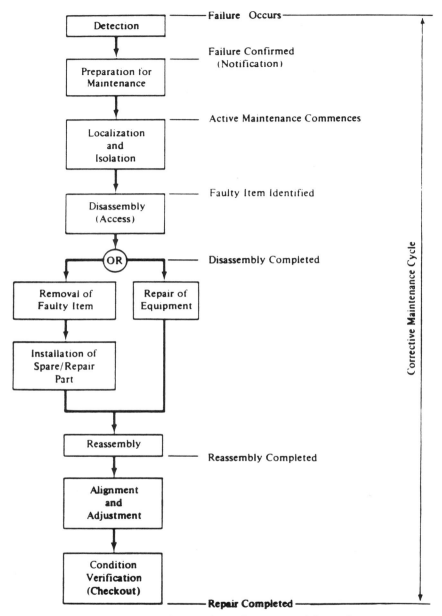

Figure 4.8. A general corrective maintenance activity cycle.

mean corrective maintenance time $(\overline{M}ct)$, or the mean time to repair (MTTR), is the composite value representing the arithmetic mean of individual maintenance cycle times for any number of individual maintenance actions over the system utilization phase.

The mean corrective maintenance time is expressed as

$$\overline{Mct} = \frac{\Sigma(\lambda_i)(Mct_i)}{\Sigma\lambda_i} \tag{4.17}$$

where λ_i, is the failure rate and Mct_i is the corrective maintenance time of the ith system element. Note that Equation (4.17) represents the "weighted" mean (using reliability factors) of the corrective maintenance tasks for the system being designed and is the preferred approach for computing $\overline{M}ct$. Note further that $\overline{M}ct$ considers only the downtime spent performing active maintenance, or the time spent working directly on the system. Logistics and administrative delay times are not addressed here.

Corrective maintenance time probability distributions usually fall into one of three common forms, as noted in Figure 4.9A:

1. *The normal* distribution applies mainly to mechanical or electromechanical hardware, usually with a remove and replace maintenance concept, where most individual repair tasks exhibit little variation around the mean.
2. *The exponential distribution* is sometimes assumed for electronic equipment with a good built-in test capability and a rapid remove and replace maintenance concept, particularly when applied for the purposes of reliability modeling. However, this assumption may lead to incorrect results, since most repair actions do require some minimal value of repair time.
3. *The log-normal distribution* applies mainly to electronic equipment without a built-in test capability and, as such, most tasks have unequal time durations. It can also apply to electromechanical equipment with widely variant individual repair times.

In the past, the maintenance time distribution for electronic equipment has often approximated a log-normal distribution. When the repair time is essentially constant, the repair rate and the designated time for maintenance can be applied as

$$P_r = 1 - P_{no\ repair} = 1 - e^{-\mu\tau} \tag{4.18}$$

where P_r is the probability of accomplishing a repair action within a designated time interval, $P_{no\ repair}$ is the probability of not accomplishing a repair action within the same designated time interval, μ is the repair rate (reciprocal of $\overline{M}ct$), and τ is the

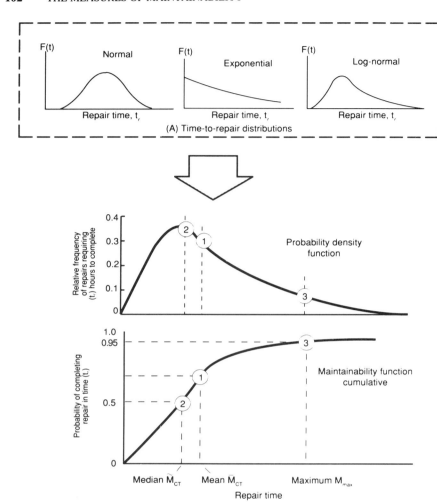

(B) Maintainability parameters related to the log-normal distribution

Figure 4.9. Maintainability distributions. (*Source:* Blanchard, B. S., *System Engineering Management,* John Wiley & Sons, N.Y., 1991, p. 92).

same designated time interval. Consider an example where the constant repair rate is 2 repairs per hour. The probability of repair in a time period equal to or less than an hour is given by

$$P_r = 1 - e^{-(2)(1)}$$
$$= 1 - 0.1353 = 0.8647$$

TABLE 4.2   Corrective Maintenance Times (minutes)

51	71	75	67	86	58	52	64	41	74
48	55	43	72	30	39	64	45	63	37
70	37	48	71	69	83	57	83	46	72
33	59	97	66	93	76	68	50	65	63
75	63	51	69	75	64	54	53	59	92

Further, the probability of repair in 30 minutes or less is given by

$$P_r = 1 - e^{-(2)(0.5)}$$
$$= 1 - 0.3679 = 0.6321$$

Note that Equation (4.18) represents the probability of effecting a single repair action. The probabilities of multiple repair-action completion are addressed in Section 4.6. Next, consider an example where data on a sample set of 50 corrective maintenance cycle times on a typical piece of equipment is available. The individual repair times (i.e., Mct_i) are listed in Table 4.2.

Based on the data in Table 4.2, a frequency distribution table, a frequency histogram, and a frequency polygon may be generated, as illustrated in Table 4.3 and Figures 4.10 and 4.11 respectively. The range of observations as listed in Table 4.2 is between 97 and 30 minutes, 67 minutes. Next, this range needs to be divided into a reasonable number of class intervals. The number of class intervals selected should be large enough to make explicit the shape of the data distribution and will likely depend upon the number and nature of the observations and their range. Ten class intervals are considered in this example and the resulting frequency distribution is as shown in Table 4.3.

Given the data in Table 4.3, one can plot a frequency histogram showing the

TABLE 4.3   Frequency Distribution

Class Interval	Frequency	Cumulative Frequency
29.5–36.5	2	2
36.5–43.5	5	7
43.5–50.5	5	13
50.5–57.5	6	19
57.5–64.5	9	28
64.5–71.5	9	37
71.5–78.5	7	44
78.5–85.5	2	46
85.5–92.5	2	48
92.5–99.5	2	50

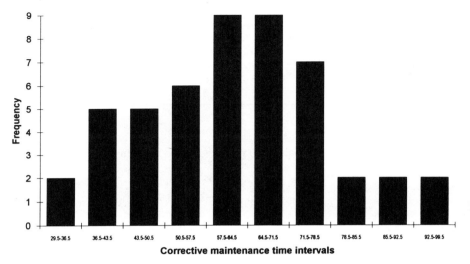

Figure 4.10. A frequency histogram.

maintenance times in minutes along the abscissa and the frequency of maintenance actions along the ordinate, as shown in Figure 4.10.

A frequency polygon can be derived from the histogram above by plotting a curve connecting the midpoints of each of the class interval blocks as shown in Figure 4.11. This indicates the approximate form of the probability distribution.

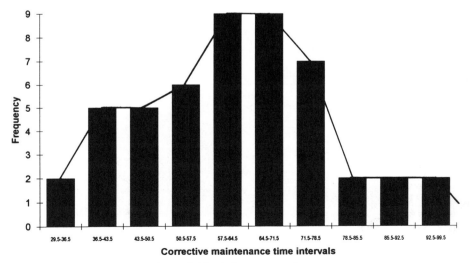

Figure 4.11. A frequency polygon.

Assuming a normal distribution for this example, the mean of the observations in Table 4.2 can be determined as

$$\overline{Mct}(\text{or MTTR}) = \frac{\Sigma Mct_i}{n} = \frac{3096}{50} = 61.92 \qquad (4.19)$$

where n is the sample size. Thus, the arithmetic mean or average corrective mainte-nance time for the sample data set is 62 minutes. Since normal distribution has been assumed, the arithmetic mean and the weighted mean of the observations is the same. This will change if the distribution appoximates the log-normal form instead, as shown in Figure 4.9B.

Standard deviation is a measure of dispersion of maintenance times about the mean. A larger standard deviation value indicates greater dispersion of observed values about the mean. The standard deviation for this example (i.e., normal distri-bution) can be calculated as

$$\sigma = \sqrt{\frac{\Sigma(Mct_i - \overline{Mct})^2}{n-1}} = \sqrt{\frac{12,138}{49}} = 15.74 \qquad (4.20)$$

For a normal distribution, 68% of the total population lies within one standard deviation on each side of the mean, 95% lies within two standard deviations on each side of the mean, and 99.7% lies within three standard deviations on each side of the mean. These characteristics are depicted in Figure 4.12.

To apply these concepts, assume that one wishes to determine the percentage of corrective maintenance times that lies between 52 and 72 minutes. This is graphi-cally represented in Figure 4.13. The objective is to find the percentage of total area

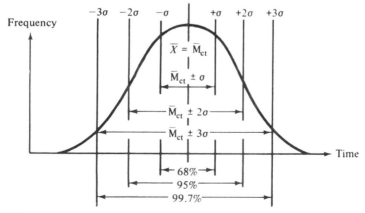

Figure 4.12. Normal distribution characteristics. (*Source:* Blanchard, One through Twelve in *Logistics Engineering and Management,* 4th ed., Blanchard ed., © 1992, Fig. 2-13, p. 42. Reprinted by permission of Prentice-Hall, Inc., Englewood Cliffs, N.J.).

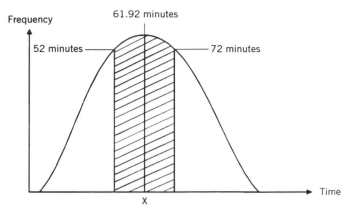

Figure 4.13. Normal distribution illustration.

represented by the shaded portion. This can be accomplished by first converting the 52- and 72-minute maintenance times into standard (Z) values, or the number of standard deviations above and below the mean of 62 minutes. This is accomplished as

$$Z\ (52\ \text{minutes}) = \frac{52 - 62}{16} = \frac{-10}{16} = -0.63$$

and

$$Z\ (72\ \text{minutes}) = \frac{72 - 62}{16} = \frac{10}{16} = 0.63$$

where a negative value for Z indicates a location below the mean, and a positive value indicates a point above the mean. From a normal distribution table, we can find the areas below the curve and to the left of the two points noted above. Point X_1 (52 minutes) represents an area equal to 0.2643%, and point X_2 (72 minutes) represents an area equal to 0.7357%. The shaded portion in Figure 4.13 is then the difference between these two percentage areas, or $X_2 - X_1 = 0.7357 - 0.2643 = 0.4714$, or 47.14%. In other words, 47.14% of all corrective maintenance action times fall between 52 and 72 minutes.

This discussion can be extended to determine confidence limits. In maintainability analysis only the upper confidence limit is addressed, since this is the most critical measure of concern. In practice, a specified confidence level or limit is assumed to be an upper limit. Once again referring to Table 4.1, since the observations listed were selected at random, they are statistically representative of the entire population. The upper limit, for a given confidence level, may be expressed as follows:

$$\text{upper limit} = \overline{M}ct + Z\left(\frac{\sigma}{\sqrt{N}}\right) \qquad (4.21)$$

where $\sigma/\sqrt{N}$ represents the standard error factor. Assuming a confidence limit of 90% (this is the same as saying that we are willing to be wrong 10% of the time), a standard (Z) value can be obtained. From a normal distribution table, $Z = 1.28$ for a confidence level of 89.97% and $Z = 1.29$ for a confidence level of 90.15%. Interpolating, $Z = 1.282$ for a confidence level of 90%. The upper limit can therefore be calculated as

$$\text{upper limit} = 61.92(1.282)\frac{15.74}{\sqrt{50}} = 64.77 \text{ minutes}$$

This is the same as stating that there is a 90% probability that the corrective maintenance times will be 64.77 minutes or less. Confidence limits for a log-normal distribution may be derived in a similar manner. The $\overline{Mct}$ and the σ must, however, be calculated using the corresponding equations for a log-normal distribution.

Mean Preventive Maintenance Time ($\overline{Mpt}$). Preventive maintenance constitutes activities to retain a system at a specified performance level and includes functions such as inspections, tuning, calibration, time/cycle replacements, and overhaul.

The overall objective of a preventive maintenance program is to postpone the wearout characteristics of an operational system. While a carefully tailored program can effect improved system performance and reduced downtime, a poorly designed preventive maintenance program can be costly and have a negative impact on an operating system. The mean preventive maintenance time ($\overline{Mpt}$) is given as

$$\overline{Mpt} = \frac{\Sigma(Mpt_i)(fpt_i)}{\Sigma fpt_i} \tag{4.22}$$

where Mpt_i is the elapsed time for the ith preventive maintenance task, and fpt_i is the frequency of the ith preventive maintenance task in actions per system operating hour. Here $\overline{Mpt}$ includes only the downtime expended as a result of accomplishing active maintenance. Administrative and logistics delay times are not included. Although some preventive maintenance tasks can be performed while the system is fully operational, a certain amount of system downtime results in most instances.

Median active corrective maintenance time ($\tilde{Mct}$). The median active corrective maintenance time is the fiftieth percentile of all repair time values and usually gives the best average location of the sample data. In the case of a normal distribution, the median, mode, and mean all have the same value, whereas in a log-normal distribution these values are displaced as indicated in Figure 4.9B. For a log-normal distribution, the median is calculated as follows:

$$\tilde{Mct} = \text{antilog}\frac{\displaystyle\sum_{i=1}^{n} \log Mct_i}{n} = \text{antilog}\frac{\Sigma(\lambda_i)(\log Mct_i)}{\Sigma\lambda_i} \tag{4.23}$$

where $\tilde{M}ct$ is the median in a log-normal distribution, which is equivalent to the geometric mean ($MTTR_g$).

Median Active Preventive Maintenance Time ($\tilde{M}pt$). The concepts applied to calculating the median active corrective maintenance time can be tailored to calculate the median active preventive maintenance time as well. The $\tilde{M}pt$ can be derived as

$$\tilde{M}pt = \text{antilog} \frac{\Sigma (fpt_i)(\log Mpt_i)}{\Sigma fpt_i} \qquad (4.24)$$

Maximum Active Corrective Maintenance Time (M_{max}). M_{max} represents a percentile below which a specified percentage of all corrective maintenance cycle times are expected to be accomplished. In other words, it specifies the upper limit on the percentage of tasks that can be allowed to exceed a given repair time duration. As an illustration, if the M_{max} is specified as two hours at the ninety-fifth percentile, then no more than 5% of the repair action times are expected to exceed 2 hours. In the case of a log-normal distribution, M_{max} is expressed as

$$M_{max} = \text{antilog}[\overline{\log Mct} + Z\sigma_{\log Mct_i}] \qquad (4.25)$$

where $\overline{\log Mct}$ is the mean of the logarithms of Mct_i, Z is the standard deviation value corresponding to the percentile value specified for M_{max}, and $\sigma_{\log Mct_i}$ is the standard deviation of the logarithms of the sample repair times, Mct_i, and is expressed as

$$\sigma_{\log Mcti} = \sqrt{\frac{\sum\limits^{N}(\log Mct_i)^2 - \left(\sum\limits_{i=1}^{N} \log Mct_i\right)^2 \Big/ N}{N-1}} \qquad (4.26)$$

When maintenance time distribution is approximated by a log-normal distribution, M_{max} cannot be derived directly from the observed maintenance time values. By taking the logarithm of each maintenance time duration, as indicated in Equation (4.25), the distribution is translated into a normal distribution. It is then convenient to use the Z standard deviation parameter as applied to a normal distribution.

Mean Active Maintenance Time ($\overline{M}$). $\overline{M}$ is the mean time required to perform preventive and corrective maintenance tasks. It is a function of the mean preventive and corrective maintenance times and their relative frequencies. $\overline{M}$ addresses only active maintenance times and does not consider administrative or logistics time delays and is expressed as

$$\overline{M} = \frac{(\lambda)(\overline{M}ct) + (fpt)(\overline{M}pt)}{\lambda + fpt} \qquad (4.27)$$

where λ is the corrective maintenance frequency and *fpt* is the preventive maintenance frequency.

***Logistics Delay Time* (LDT).** LDT is the time elapsed while waiting for some required logistics resource. This resource could be a spare part, a particular test and support equipment item, a facility, or a service or procedure. It may also include the time to perform a related administrative task such as completing a maintenance work order request. LDT does not include any portion of the active maintenance time duration, but very often contributes significantly to overall system downtime.

***Administrative Delay Time* (ADT).** ADT is the downtime as a result of some administrative priority or constraint, or any other cause not considered and included in computing LDT. Job assignment priorities and labor strikes are examples.

***Maintenance Downtime* (MDT).** MDT constitutes the sum total time required to either repair and restore a system to a specified performance level or to maintain and retain it at that level. It addresses not only the active maintenance times (both corrective and preventive), but also the administrative and logistics delay times, as depicted in Figure 4.7. MDT is usually expressed as a mean value and is a function of the mean preventive and corrective maintenance times, their relative frequencies, and the administrative and logistics delay times. It is expressed as

$$\text{MDT} = \overline{M} + \text{ADT} + \text{LDT} \tag{4.28}$$

Equipment downtime due to corrective maintenance is a function of corrective maintenance frequency, time to perform the individual maintenance tasks, and the time period of interest. Thus, the expected downtime may be computed as

$$T_D = \lambda t(\overline{M}ct) \tag{4.29}$$

where T_D is the equipment downtime, λ is the equipment failure rate, t is the time interval of interest, and $\overline{M}ct$ is the mean time to restore equipment to its full operational capability (equivalent to MTTR). As an illustration, consider a communications system with a projected failure rate equal to 14 failures per thousand hours of operation and an overall mean corrective time equal to 19 minutes. Assuming an operational profile spanning 1200 hours, the total system downtime is equal to

$$T_D = 1200(0.014)\,(0.317) = 5.32 \text{ hours}$$

In the event that the time of interest is restricted (e.g., a countdown sequence), a more rigorous form of Equation (4.29) is

$$T_D = t \sum_{i=1}^{n} (\lambda_i \overline{M}ct_i) \tag{4.30}$$

where λ_i and $\overline{M}ct_i$ are the failure rate and mean restoration time for each individual repairable or replaceable component within the equipment. Consider once again the communications system discussed above. Assume that this system is composed of three main assemblies, the receiver ($\lambda = 2$ failures/1000 operating hours; $\overline{M}ct = 14$ minutes), the transmitter ($\lambda = 6$ failures/1000 operating hours; $\overline{M}ct = 11$ minutes), and the controls ($\lambda = 6$ failures/1000 operating hours; $\overline{M}ct = 32$ minutes). Assuming the same operational profile of 1200 hours, Equation (4.30) may be applied as

$$T_D = 1200[(0.002)(0.233) + (0.006)(0.183) + (0.006)(0.533)]$$
$$= 5.72 \text{hours}$$

An even greater degree of sensitivity can be achieved by looking at not only the assemblies and subassemblies that constitute a system, but also the different modes of failure for each of the system elements. For scenarios where the percentage downtime is of interest (e.g., a monitoring system or a data processing system), the expression is

$$T_D = T_o\lambda(\overline{M}ct_m) \tag{4.31}$$

where T_o is the equipment operating time, which can be computed as the total mission or operational cycle time less equipment downtime as illustrated in Figure 4.2, or

$$T_o = T_m - T_D \tag{4.32}$$

where T_m is the total mission time. Equation (4.31) can therefore be transformed to

$$T_D = \lambda(\overline{M}ct_m)(T_m - T_D) = \frac{(\overline{M}ct_m)(T_m - T_D)}{\text{MTBF}} \tag{4.33}$$

Equation (4.33) can be further transposed to

$$T_D = \frac{T_m(\overline{M}ct_m)}{\text{MTBF} + (\overline{M}ct_m)} \tag{4.34}$$

Applying Equation (4.34) to the communications system discussed above, we have

$$T_D = \frac{1200(0.317)}{71.43 + 0.317} = 5.32 \text{ hours}$$

This result corresponds to the one obtained through the application of Equation (4.29). The probability that the equipment will be down (P_D) during any time period is expressed as

$$P_D = \frac{T_D}{T_m} = \frac{T_m(\overline{M}ct_m)}{T_m(MTBF + \overline{M}ct_m)} = \frac{\overline{M}ct_m}{MTBF + \overline{M}ct_m} \qquad (4.35)$$

Conversely, the inherent probability of the equipment being "up" or operational (P_U) is given as

$$P_U = 1 - P_D = \frac{MTBF}{MTBF + \overline{M}ct_m} \qquad (4.36)$$

Extending the discussion, the probability that two like units are down during the same time interval (a redundancy situation) is determined by squaring Equation (4.35) as

$$P_{D(\text{two units})} = \left(\frac{\overline{M}ct}{MTBF + \overline{M}ct}\right)^2 \qquad (4.37)$$

When more than two units in parallel are involved, the probability of the equipment being down is determined by changing the exponent of Equation (4.37) to the number of units in parallel. Conversely, the probability of being "up" can once again be computed as in Equation (4.36).

4.2.2   Maintenance Frequency Factors

Maintenance frequency, both corrective and preventive, is dependent upon the system reliability and wearout characteristics. The corrective maintenance frequency for a system is a function of the corresponding failure rate and MTBF. Preventive maintenance frequency for a system, on the other hand, depends, among other factors, on predicted wearout characteristics and trends of system components. Two of the more commonly used maintenance frequency factors are the mean time between maintenance MTBM and the mean time between replacement (MTBR).

Mean Time Between Maintenance (MTBM). The principle measure of maintenance frequency is MTBM, which is a function of both scheduled and unscheduled maintenance frequencies and is expressed as

$$MTBM = \frac{1}{1/MTBM_u + 1/MTBM_s} \qquad (4.38)$$

where $MTBM_u$ is the mean time between unscheduled (or corrective) maintenance and $MTBM_s$ is the mean time between scheduled (or preventive) maintenance. Here MTBM is considered to be both a reliability and maintainability parameter. The mean time between unscheduled (corrective) maintenance should closely approximate a composite system MTBF which addresses the inherent system failures, secondary or dependent failures, operator and maintenance-induced failures, and manufacturing defects. Operator and maintenance-induced failures must be minimized, if not eliminated. Further, the MTBM itself will approximate the MTBF in the absence of preventive maintenance actions.

Mean Time Between Replacement (**MTBR**). The MTBR is often used inter-changeably with the term mean time between demand (MTBD), which is also a factor of *MTBM*. Since all maintenance actions do not result in the need or demand for a replacement component, the MTBR factor is more than likely to be greater than MTBM. The MTBR measure is a significant input to spare part requirements analysis. This measure could apply to both preventive and corrective maintenance tasks, as long as a maintenance action requires item replacement. Further, such an activity generates a demand for spares and impacts the overall system logistic sup-port capability. An objective of maintainability in system design is to maximize MTBR.

A typical field experience relative to corrective maintenance is illustrated in Fig-ure 4.14. In this example, there are 200 corrective maintenance actions experienced at the system level. These were in response to actual or perceived system failures. This measure is addressed in the MTBM factor. Further, each maintenance task, whether in response to an actual failure or false alarm results in the consumption of resources, including maintenance personnel labor. This consumption of labor hours may be expressed in terms of mean maintenance man-hours per operating hour, or MMH/OH. Refer to Section 4.2.4 for a more detailed coverage of maintenance labor hour factors. In this example, 75% of all maintenance actions, or 150, result in a remove and replace activity to restore the system to satisfactory operation. The removal and replacement tasks are addressed in the MTBR measure and generate a

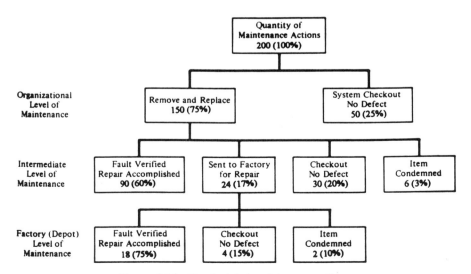

Figure 4.14. Unscheduled maintenance actions.

demand for spares. The remaining 25% of all maintenance tasks result in a system checkout without the positive identification of a fault.

The units removed from the system are shipped to the intermediate level shop for repair. Once again, only a certain percentage, in this case 80% or 120, of the units shipped to the intermediate level have a fault verified. These are either repaired at this level, condemned, or shipped to the next higher level for repair. The verified, or confirmed, failures are addressed in the system MTBF measure. In this illustration, only 108 of the 200 corrective maintenance actions result in a verified unit failure and subsequent repair.

4.2.3 Maintenance Cost Factors

Maintenance costs are generated as a result of corrective and preventive maintenance actions, and are based on the consumption of resources utilized in the performance of these maintenance actions. Such resources may include spare/repair parts and associated inventory, test and support equipment utilization, personnel, facilities, and data. The costs associated with the performance of maintenance are a major element of system life-cycle cost, and are included herein. Please refer to Section 1.2 and Figure 1.3 for more discussion of this topic. An objective of maintainability in design is to effect a reduction in system life-cycle cost through minimizing system support costs.

The following cost-related indices are examples of maintenance cost factors:

* Maintenance cost per system operating hour ($/OH)
* Maintenance cost per repair action
* Maintenance cost per month ($/month)
* Maintenance cost per mission or mission phase
* The ratio of maintenance cost to total system life-cycle cost
* Maintenance-related environmental clean-up cost

4.2.4 Maintenance Labor-Hour Factors

The maintainability factors considered in the previous sections relate to maintenance times and the cost associated with the performance of unscheduled and scheduled maintenance tasks. Often it is feasible to reduce maintenance times by increasing the number of maintenance personnel involved. This approach, however, may not be cost effective, and even less so if the skill-level requirements are high. A highly maintainable system is the result of a cost-effective trade-off between elapsed times, labor hours, personnel skills, and equipment and facility requirements.

Some of the commonly utilized labor-hour related maintainability measures are as follows:

* Maintenance man-hours per maintenance action (MMH/MA)
* Maintenance man-hours per month (MMH/month)

- Maintenance man-hours per operating hour (MMH/OH)
- Maintenance man-hours per mission or mission phase (MMH/mission).

4.3   HUMAN FACTORS

Very often in the development of maintainability requirements, emphasis is placed on the technical parameters related to maintenance frequencies and time duration, and the associated logistic support, while the human element of the process is ignored. To maximize maintainability effectiveness efforts, the human element and its interface with the equipment, both prime and support, needs to be addressed. The term "human factors," sometimes known as "ergonomics" or "human engineering," refers to the design of a system, product, process, or procedure for human use. When maintainability requirements are being established, particularly with regard to the human element, the relevant dimensions of the human body, its sensory abilities, the environmental stresses that are likely to impact its performance, and the often important emotional aspects, must be considered. For example, a high level of equipment vibration during maintenance, limited access to a critical item, temperature beyond the comfortable range of 55–75°F, noise outside the comfortable intensity level range of 50–80 dB, or confusing color codes and maintenance procedures are likely to increase the probability of maintenance-induced failures. Human engineering must consider factors such as:

1. *Anthropometric factors.* Anthropometric factors deal with the measurement of the dimensions and physical characteristics of the human body, including the standing height, sitting height, arm reach, hand size, weight, and so on. When establishing and implementing maintainability requirements, these parameters should be taken into account. These requirements are especially important in designing test and support facilities and equipment, influencing the design of the prime equipment itself for increased maintainability, delineating the maintenance procedures and activities that involve access to an embedded system component, application of force, the ability to lift weight, and so on. Both structural (i.e., when the body is static and in a fixed posture) and dynamic (i.e., when the body is in motion and involved in a certain physical activity) dimensions must be addressed. Also, the design engineer must consider dimensions relative to both males and females, and take into account the appropriate variability. For specific dimensions and design criteria and constraints, the reader should consult additional references.[4]

2. *Human sensory factors.* These factors relate more to the human sensory capacities, for example, sight, hearing, smell, feel or touch, and so on. In the design of test and diagnostic work centers, consoles, monitors, and read-out devices, the

---

[4] Anthropometry data are included in National Aeronautics and Space Administration and Space Administration (NASA); Volume 1, *Anthropometric Source Book;* Volume 2, *A Handbook of Anthropometric Data;* and Volume 3, *Annotated Bibliography;* NASA Reference Publication 1024, 1978. Additional references are included in Appendix E.

engineer must be cognizant of the human capabilities relative to sight as it pertains to horizontal and vertical fields of view, the detection of signals and objects from varying angles, and the detection of certain colors under varying degrees of brightness from varying angles. Sufficient illumination is important along with the field of view and color schemes for the satisfactory performance of tasks. Noise issues need to be addressed as well for increased operator comfort, productivity, and efficiency. Both noise level and its nature are relevant. For example, a human is likely to experience some physical sensation as the level approaches 120 dB, and pain at levels above 130 dB. The nature or characteristics of noise (e.g., whether it is steady or intermittent) is also a factor.

3. *Physiological factors.* It is important to recognize the impact of environmental stresses on human performance efficiency. Here, stress refers to an external or environmental situation that causes human productivity to decrease, and includes issues such as high and low temperature, humidity, high levels of vibration, high levels of noise, and large amounts of radiation or other toxic chemical effects in the vicinity where operational and maintenance activities are being conducted. The consequences of stresses such as these manifest themselves as reduced sensory capacities, slower motor response, and reduced mental alertness, leading to increased maintenance-induced error probability.

4. *Psychological factors.* These factors relate to the characteristics of the human mind, for example, emotional traits, attitudinal responses, and behavioral patterns relative to the performance of maintenance actions. While all the physical resources necessary for task completion may be available, the effectiveness of maintenance task completion is significantly impacted by personnel motivation, initiative, dependability, confidence, and so on. Maintenance policies and procedures which are perceived as being too difficult or complex, or a degrading management and supervisory style, can lead to frustration and a poor attitude. On the other hand, if the tasks are overly simple and routine, this could lead to a more casual approach to the job and a greater number of errors due to carelessness. Proper training for the personnel involved and a team approach to work accomplishment can preclude some of these problems.

4.4 LOGISTIC SUPPORT FACTORS

The implementation of maintainability requirements can impact system effectiveness and efficiency in two distinct ways. The first is by influencing the prime elements of the system from a "design for supportability" perspective. The second is by influencing the design of the support infrastructure. Further, the maintainability of the prime system elements can be significantly impacted by the effectiveness of the responding support capability. From a cost standpoint, it is important to plan for and implement a properly designed support structure to avoid the transition costs that result when a system must be "modified" after the fact to meet a given customer requirement. Selected logistic support factors are defined and discussed in this section.

4.4.1 Supply Support Factors

Supply support factors include considerations for spare and repair parts and the corresponding inventories. These factors directly influence the timeliness and effectiveness of both scheduled and unscheduled maintenance, and consequently the system downtime. The type and quantity of spare and repair part requirements need to be derived and implemented for all maintenance levels. Issues such as the number of spares to be procured and the procurement sources and frequencies also need to be adequately addressed. Most of these issues are a function of the expected spare and repair part demand rates and require consideration of:

1. *Spare and repair part requirements* in response to item replacement as a result of corrective and preventive maintenance. Here, spare parts are significant replaceable and repairable items, and repair parts are the smaller, nonrepairable items that are discarded upon failure. Further, the requirements and demand rates are a function of the item reliability, and the quantity of items used in the system and their corresponding duty cycles.

2. *The maintenance turnaround times* between the different levels of maintenance and the inventory storage facility. This includes items in the queue awaiting maintenance and the time that it takes to process an item through the maintenance cycle and back into the inventory ready for reuse as a spare. These times are impacted by the maintenance concept, test equipment and facilities, the transportation system in place, the personnel involved, and so on. Additional quantities of spares may often be required to prevent a stock-out condition.

3. *The procurement times involved relative to the expected item demand rates.* The item stock levels will be a function of the procurement lead times. Further, these levels may be different for spares as compared to repair parts (or discardables).

4. *Procurement sources and component obsolescence.* Very often, in the case where there is only a single procurement source, large spare and repair-item quantities may be purchased because these items will no longer be produced after a designated time prior to phaseout of the system, or if the procurement source is about to go out of business.

5. *Repairable item condemnation rates.* Additional spares may be stocked as a buffer against possible item condemnation. After an item has been cycled through the repair process a certain number of times, it is often not economically feasible to repair it any more.

6. *Component criticality.* In the event that a certain component in the system is critical to system operation or mission success, a higher stock level of spares may be recommended. This reduces the probability of not having a replacement available when needed.

7. *Spare and repair-item cost.* This factor plays a significant role in computing stock levels and the overall cost-effectiveness of the support infrastructure. Total cost should include not only the initial item procurement cost, but also the cost of maintaining the item in the inventory.

The initial maintenance concept drives the process of spare part requirements analysis. These requirements are later verified and validated through the maintenance task analysis described in Chapter 9. The planning for and implementation of an effective supply support capability is derived from a combination of equipment packaging schemes, level of repair analyses, test equipment studies, personnel skill level requirements, and so on. Maintainability plays a lead role in much of the support system planning and implementation effort. Maintainability objectives relative to supply factors include increased utilization of standard and interchangeable parts, and a reduction in spare level requirements. The consequence of such efforts will be reflected in a cost-effective support infrastructure and reduced life-cycle cost.

Concepts from reliability and probability theory are applied next to considerations related to supply support factors such as spare part quantity determination, the probability of success with spares availability considerations, and inventory considerations.

Spare Part Quantity Determination. Spare part quantity determination is a function of the probability of having a spare available when required, the reliability of the item in question, the quantity of items used in the system, and so on. A general expression, based on the Poisson distribution, often used in the determination of spare part quantity determination, is

$$P = \sum_{n=0}^{n=S} \frac{(R)[(-1)\ln R]^n}{n!} \tag{4.39}$$

where P is the probability of having a spare of a particular item available when required, S is the number of spare parts carried in stock, R is the composite reliability (probability of survival) $(R = e^{-K\lambda t})$, K is the quantity of parts used of a particular type, and $\ln R$ is the natural logarithm of R.

In Equation (4.39), the value of P directly impacts the quantities of the spare parts procured. Here P can also be referred to as the safety factor and reflects the level of protection desired in determining sparing requirements. The higher the safety factor, the greater the quantity of spares required. This also results in a higher cost for item procurement and inventory maintenance. The safety factor accounts for the risk relative to stock-out. The detemination of spare part quantities is addressed in more detail in Section 9.5.

When determining spare part quantities, one should consider system operational requirements (e.g., system effectiveness and availability) and establish the appropriate level at each location where corrective maintenance is accomplished. Different levels of corrective maintenance may be appropriate for different items. For instance, spares required to support prime equipment components which are critical to the success of a mission may be based on one factor; high-value or high-cost items may be handled differently than low-cost items; and so on. In any event, an optimum balance between stock levels and cost is imperative.

Probability of Success with Spares Availability Considerations. Assume that a single component is used in a unique system application along with a backup spare component. Determine the probability of system success having a spare available in time *t* (given that system failures are random and exponentially distributed). Note that this situation is analogous to an operating component and a standby parallel component (i.e., standby redundancy), as discussed in Section 4.1.1. The applicable expression in the case described above is

$$P = e^{-\lambda t} + (\lambda t)e^{-\lambda t} \tag{4.40}$$

With a component reliability of 0.8, the value of λt is 0.223. Substituting this value into Equation (4.40) gives a probability of success of

$$P = e^{-0.223} + (0.223)e^{-0.223}$$
$$= 0.8 + (0.223)(0.8) = 0.9784$$

This concept can be generalized and extended to address multiple items in a system backed by multiple spares, as discussed in Section 4.1.3.

Inventory Considerations. In progressing further, one needs not only to address the specific demand factors for spares, but to evaluate these factors in terms of the overall inventory requirements. Too much inventory may ideally respond to the demand for spares. However, this may not be a cost-effective solution and a significant amount of capital may be tied up in maintaining inventories. Further, losses may result from system design changes that render certain components obsolete and no longer required. Providing too little support, on the other hand, may result in frequent stock-outs, shortage costs, and possible penalties.

In general, it is necessary to implement a balance between the quantity of items in inventory at any given point in time, the frequency of purchase order transactions, item cost, and the quantity of items per purchase order. Figure 4.15 presents a graphical portrayal of an ideal and theoretical inventory cycle. A constant lead time and time demand rate (i.e., failure rate) is assumed. Stock depletions are represented by the sloping consumption line. As stock depletion attains a planned level, additional items are ordered (i.e., order point). Enough time is allowed for stock replenishment to avoid a stock-out situation.

Some of the relevant terms identified in Figure 4.15 are defined as follows:

1. *Operating level*—denotes the material item quantity required to support normal system operations in the interval between orders and successive material shipment arrivals.
2. *Safety stock*—refers to the additional stock required to compensate for unexpected demands, repair and recycle times, pipeline, procurement lead time, and unforeseen delays.
3. *Reorder cycle*—is the time interval between successive orders.
4. *Procurement lead time*—denotes the time span between the date of the material order to receipt of the shipment in the inventory. It includes (a) adminis-

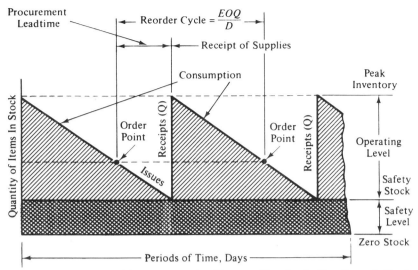

Figure 4.15. A theoretical inventory cycle. (*Source:* Blanchard, One through Twelve in *Logistics Engineering and Management,* 4th ed., Blanchard ed., Fig. 2-20, p. 61. Reprinted by permission of Prentice-Hall, Inc., Englewood Cliffs, N.J.).

trative lead time from the date that a decision is made to initiate an order to the receipt of the order at the supplier, (b) production lead time or the time from receipt of the order by supplier to completion of the manufacture of the item ordered, and (c) delivery lead time from completion of manufacture to receipt of the item in the inventory. The pipeline is included in the delivery lead time.

5. *Pipeline*—reflects the distance between the supplier and consumer, measured in days of supply. An increase in the demand rate may require more items in the pipeline.

6. *Order point (OP)*—is the point in time when orders are initiated for additional spare/repair parts. This point is often tied to a particular stock level and will likely be different for different items.

Figure 4.15 represents a puristic form of the inventory cycle; demands are often not constant and the reorder cycle changes with time in the real world. An actual inventory cycle may very well be significantly more disorderly, as depicted in Figure 4.16.

The total inventory system cost per period is the sum of the item, procurement, and holding costs for the period, and can be expressed as

$$\text{total cost} = \text{item cost} + \text{procurement cost} + \text{holding cost}$$

$$= C_i D + \frac{C_p D}{Q} + \frac{C_h Q}{2} \tag{4.41}$$

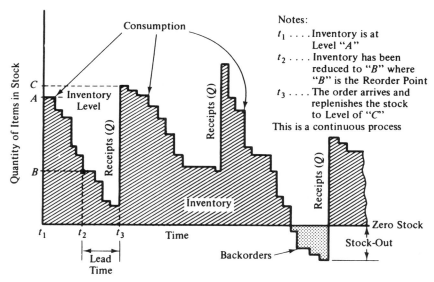

Figure 4.16. Illustration of an actual inventory cycle. (*Source:* Blanchard, One through Twelve in *Logistics Engineering and Management,* 4th ed., Blanchard ed., © 1992, Fig. 2-21, p. 62. Reprinted by permission of Prentice-Hall, Inc., Englewood Cliffs, N.J.).

where C_p is the average procurement cost, or the cost of ordering in dollars per order, C_h is the average holding cost, or the cost of carrying an item in inventory, C_i is the average item cost, D is the annual item demand (this is assumed constant), and Q is the item procurement quantity.

In general, the economic procurement principle equates the cost to order to the cost of hold, and the objective is to achieve the balance where the combined costs are at a minimum. This principle is graphically illustrated in Figure 4.17. The optimum procurement quantity Q^*, which results in a minimum total inventory system cost, can be computed by differentiating Equation (4.41) with respect to the item procurement quantity Q, and may be expressed as

$$Q^* = \text{EOQ} = \sqrt{\frac{2C_p D}{C_h}} \qquad (4.42)$$

where EOQ = Economic Order Quantity. Therefore, the minimum inventory system cost, TC^*, can be derived by substituting Equation (4.42) into Equation (4.41) as shown below:

$$TC^* = C_i D + \frac{C_p D}{\sqrt{2C_p D/C_h}} + \frac{C_h \sqrt{2C_p D/C_h}}{2} \qquad (4.43)$$

$$= C_i D + \sqrt{2C_p C_h D}$$

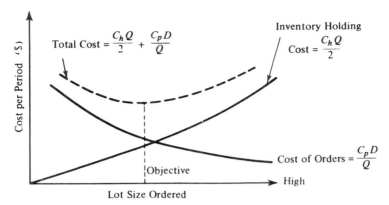

Figure 4.17. Inventory system cost trade-offs. (*Source:* Blanchard, One through Twelve in *Logistics Engineering and Management,* 4th ed., Blanchard ed., © 1992, Fig. 2-22, p. 63. Reprinted by permission of Prentice-Hall, Inc., Englewood Cliffs, N.J.).

Further, the number of purchase orders per year, N, is given as

$$N = \frac{D}{\text{EOQ}} \qquad (4.44)$$

The EOQ model as discussed in this section can generally be applied to address a scenario with relatively large quantities of spare and repair parts. Other methods of acquistion may become necessary for major high-value items and items considered critical to mission success.

4.4.2 Test and Support Equipment Factors

Test and support equipment is an important element in the overall logistic support infrastructure. This equipment must be provided in the proper quantity at the right location to ensure effective system maintenance. In determining specific test and support equipment requirements, it is important to define (1) the type and nature of items that are to be repaired and maintained at a given location and the frequency of test and support equipment utilization and mission times, and (2) the test functions that are to be performed along with the required performance parameter accuracy and tolerances that may need to be measured. Once again, the requirements are derived from the maintenance concept and may include items such as precision electronic and mechanical test equipment, material handling equipment, special jigs and fixtures, maintenance platforms and stands, and so on.

In accordance with the maintenance concept, test and support equipment may be deployed to numerous geographical locations throughout the country (or worldwide), within a well-designed system support organization. The objective is to provide the right item for the job intended, at the proper location, and in the correct quantity. While common measures such as availability, reliability, and maintainability apply to all types of test and support equipment; however, given their diverse nature, many other measures may have to be developed such that they are adapted to better represent test and support equipment effectiveness. These measures are likely to vary, depending on whether the test equipment in question is electronic or mechanical, the utilization profile, performance requirements, and so on. As an illustration, some "tailored" effectiveness measures may be: mean time to calibrate, mean time to diagnose fault, power requirements, weight, volume, tolerance of measurement, and response time. An objective is for the test and support equipment to positively impact prime equipment availability and functionality in a cost-effective manner.

A schematic of test and support equipment utilization and some associated effectiveness measures is depicted in Figure 4.18. Also, the same test equipment may have different effectiveness measures based on whether it was supporting on-site organization level maintenance or the "off-equipment" intermediate or depot-level maintenance. For example, test equipment weight and volume may assume a higher relative importance at the on-site equipment level.

While every piece of test and support equipment exists to satisfy an expected and justified need, some may be more critical than others. For example, test and support equipment at the intermediate and depot levels most likely respond to maintenance demands from numerous operational sites and, as such, the reliability and availability of test stations at these sites is of prime concern. Also, the criticality, type, and nature of repairable items arriving at these sites is likely to be significantly more diverse. This may generate the requirement for a well-thought-out system for turning around the items coming in for repair. A servicing policy, with priorities,

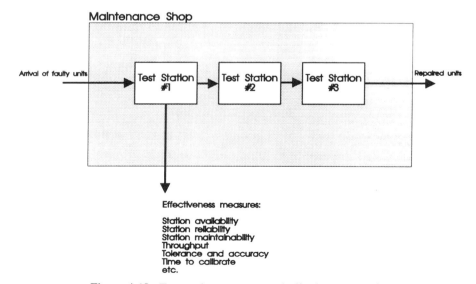

Figure 4.18. Test equipment usage and effectiveness metrics.

may need to be established. As repairable items come into the shop, they may be processed immediately or assigned to a queue depending upon their criticality and the availability of maintenance personnel, test and support equipment, and facilities.

In evaluating a test and support facility or process, the anticipated equipment utilization requirements need to be computed. This is a function of the maintenance concept, the relevant repairable item reliability and repair time distributions, equipment setup and process times, the quantity of equipment on site, and so on. Process times and quantity of test equipment will be impacted by the nature of the equipment itself, for example, whether it is manual, semiautomated, or fully automated. Further, the reliability and maintainability requirements for the test and support equipment and facilities need to be derived. This is, in turn, dependent on the utilization requirements. Very obviously, the test and support equipment will have to be more reliable than the equipment that it supports. The MTBM and MDT of this equipment is very relevant to the design of the support infrastructure and due attention needs to be focused on the logistic support requirements for the test and support equipment. In fact, if the complexity of the test equipment is high, as is very often the case today, these logistic support requirements may be extensive. As an illustration, consider the requirement to test and calibrate a piece of test equipment against a primary standard in a "clean-room" environment every so often. Thus, a host of factors and issues need to be addressed in planning for an effective allocation of test and support equipment within a cost-effective support organization.

It is a maintainability objective to promote the utilization of common and standard off-the-shelf tools, and test and support equipment. Such equipment is not only easy to maintain, but is also likely to be a more cost-effective alternative to developing special tools, jigs, fixtures, and other support equipment.

4.4.3 Personnel and Organizational Factors

Personnel factors are a function of the type, magnitude, and complexity of maintenance activity at a given location. These factors are an index of the degree of maintainability inherent in a particular system design (relevant maintainability measures were discussed in Section 4.2.4.). Further, these factors provide insight into the necessary organizational requirements that result from the defined maintenance concept. Some examples of relevant personnel factors are:

1. Direct labor time expended in the performance of maintenance. This needs to be derived for each personnel category and/or skill level. Maintenance labor times may be expressed by a number of measures, including:
   a. Maintenance man-hours per operating hour (MMH/OH)
   b. Maintenance man-hours per operating cycle or mission
   c. Maintenance man-hours per unit time period
   d. Maintenance man-hours per maintenance action (MMH/MA)

2. Indirect labor to support the prime maintenance activity (i.e., overhead).
3. Personnel attrition and turnover rates.
4. Anticipated maintenance training rates and times. This is impacted by personnel attrition and turnover.
5. Average maintenance work order processing times, or the frequency of work orders processed per unit of time.
6. The average administrative delay times anticipated or experienced, or the waiting time between an initial call for maintenance and the actual start of the maintenance task.

To ensure the effectiveness of the logistic support system, maintenance personnel must be treated as an important element. They must be available at the right time, at the correct location, and in the required quantities, and they must possess the specified skill levels.

A higher degree of inherent system maintainability will result in fewer organizational requirements. A maintainability objective is to effect simplicity in system design and the subsequent support and maintenance procedures. A consequence of such efforts is lower personnel skill requirements and minimal training.

4.4.4 Facility Factors

Facility factors directly impact the ease and economy of maintenance. The location of facilities is important along with the type, layout, and associated utilities provided. Facility factors address issues such as spare and repair part inventory and warehousing, related administrative personnel housing, utilities such as electricity and lighting, and layout of the test stations with respect to the material handing capability. The quantitative measures associated with a facility are likely to vary from one instance to another. Nonetheless, some of the more general and relevant factors are as follows:

1. *Facility utilization.* This refers to the percent utilization in terms of space occupancy. This may be a function of the type and frequency of corrective and preventive maintenance actions accomplished at a location.
2. *Energy utilization.* This can be expressed in terms of units of energy consumed per maintenance action, cost of energy consumed per maintenance action, and so on.
3. *Total facility cost for system support.* Total cost has two components, fixed and variable. Fixed cost may include factors such as the monthly rent or lease amount for the facility, insurance, and depreciation, while variable cost is likely to be a function of the facility utilization.
4. *Item process time or turnaround time (TAT).* This refers to the elapsed time required to process an item for maintenance and to restore it to an operational state.

4.4.5 Transportation and Handling Factors

The support system has to respond to maintenance demands from numerous operational sites spread over a wide geographical (often global) area. Transportation factors address the movement of human and material resources between various operational sites and maintenance facilities. Units requiring maintenance may be shipped from the operational or consumer site to the intermediate shop or depot, or to the manufacturer's facility for extensive repairs. Often, a team is dispatched along with test and support kits to the operational sites to accomplish the necessary maintenance. These decisions are derived from the maintenance concept developed during the early design process, leading to the more detailed activity of maintenance planning. Obviously, transportation and handling factors play a key role in maintainability and logistic support analyses.

The effectiveness of a transportation system is a function of the following factors:

1. *Transportation mode(s).* This includes modes such as rail, waterway, air, road, and pipeline.
2. *Transportation route(s).* This could be limited to within the United States, or global, in the event of off-shore and international operational sites. Issues such as customs requirements, policies, regulations, and cultural and legal factors need to be addressed.
3. *Transportation time.* This could be quantified and expressed in terms of the mean transportation time. Transportation time is of particular importance with regard to the delivery of spare and repair parts, turnaround times, and so on. It could directly impact spare and repair item inventory levels at the operational sites and maintenance facilities.
4. *Transportation cost.* This could be expressed in terms of measures such as cost per shipment, cost of transportation per carrier per mile, cost of packing and handling, transportation and material handling cost per month, and so on.

Transportation and handling factors are a significant input to associated analyses such as the maintenance engineering analysis, logistic support analysis, and life-cycle cost analysis. Once the relevant requirements have been delineated, they must be designed into the system support infrastructure to maximize its efficiency and cost-effectiveness.

4.4.6 Documentation Factors

Technical documentation is an important aspect of maintainability analysis. The proper maintenance procedures for both corrective and preventive maintenance must be stated and documented clearly. The relevant procedures could also be computerized in the form of interactive electronic maintenance manuals or tablets. Cost-effective computerization of complex maintenance procedures has been facilitated by the recent advances in computer technology. It is important that the documentation reflect any last-minute design changes for maximum effectiveness.

It is anticipated that with progressive implementation of the initiative, the preparation of technical data in a digital format will result in greater accuracy, elimination of redundancies, and higher quality. Further, data preparation and processing times will likely be reduced substantially, the ability to track last-minute design changes and modifications will be enhanced, and the appropriate data/information can be provided in a timely manner to support effective decision making.

Deciding on the level of detail for the relevant procedures is of special concern and will depend upon the level of maintenance, personnel skill levels, and the complexity of the prime and support equipment.

4.4.7 Computer Resource and Software Factors

Computer resource and software factors are an important element in the overall maintainability analysis and include all software, computer equipment, tapes/disks, databases, and related accessories necessary for the performance of system maintenance functions at all levels. Among other things, they relate to sophisticated condition monitoring equipment, modern electronic and computer-based test and diagnostic aids, and interactive electronic technical manuals and procedures. In many modern systems, software has evolved into a major element of support. This is particularly true of instances where computer applications and automation, digital databases, and so on, support maintenance and logistics functions. To be effective, logistic support resource requirements must address issues related to software and other computer resources. Resources necessary to support CALS requirements are also addressed.

There is substantial controversy in the literature over defining a metric for software reliability and robustness, and although software does not degrade or wear out in the same manner as hardware, its reliability is important and must be measured. For more discussion on software reliability please refer to Section 4.1.5.

Given the proliferation of higher-level languages, expert system shells, and so on, the number of lines of code required to implement a certain functionality has significantly decreased. This leads to a decrease in the number of faults or defects introduced during program development. Further, the popularity of modern software development methodologies (e.g., structured, modularized, and object-oriented programming) and the application of numerous systems engineering approaches and models (e.g., the waterfall model, the spiral model, the prototyping approach, and the evolutionary model approach) have also led to the development of more reliable and robust software programs.

4.5 AVAILABILITY FACTORS

System availability is expressed as the probability that the system will be in an operational state when called upon at a random point in time. For continuously operating systems, it is the probability that the system is operating at a random point in time. Availability is primarily a function of system reliability, maintainability, and supportability, and it may be expressed differently depending upon the nature of the system and/or the mission profile. Three of the more common measures of availability are defined below:

1. *Inherent availability* (A_i). Inherent system availability is the probability that the system will operate satisfactorily when called upon at any point in time under specified operating conditions and in an *ideal* logistic support environment. Here, ideal operating conditions refer to readily available maintenance personnel, spare and repair parts, test and support equipment, facilities, and so on. It does not consider any logistics or administrative time delays. Moreover, it excludes preventive or scheduled maintenance tasks. Inherent availability may be expressed as

$$A_i = \frac{\text{MTBF}}{\text{MTBF} + \overline{Mct}} \tag{4.45}$$

where MTBF is the mean time between failure and $\overline{Mct}$ is the mean corrective maintenance cycle time, as defined in Sections 4.1.1 and 4.2.1 respectively.

2. *Achieved availability* (A_a). Achieved system availability is the probability that the system will operate satisfactorily when called upon at any point in time under specified operating conditions and in an ideal logistic support environment. Once again, an ideal environment refers to a scenario with readily available maintenance personnel, spare and repair parts, test and support equipment, facilities, and so on. This availability metric does not consider any logistics or administrative delay times. The achieved system availability measure does address preventive maintenance tasks and may be expressed as

$$A_a = \frac{\text{MTBM}}{\text{MTBM} + \overline{M}} \tag{4.46}$$

where MTBM is the mean time between maintenance and $\overline{M}$ is the mean active maintenance time. Both MTBM and $\overline{M}$ are functions of corrective and preventive maintenance frequencies and time duration. These measures are discussed in detail in Sections 4.2.1 and 4.2.2.

3. *Operational availability* (A_o). Operational system availability is the probability that the system will operate satisfactorily when called upon at any point in time under specified operating conditions and in an actual logistic support environment. This availability measure is the closest to reality since it addresses not only the corrective and preventive maintenance tasks, but also the logistics and administrative time delays. It is expressed as

$$A_o = \frac{\text{MTBM}}{\text{MTBM} + \text{MDT}} \tag{4.47}$$

Since numerous measures for system availability exist, it is important that this metric be well defined and properly "tailored" to a given system or mission profile. As an illustration, consider the availability of a manufacturing process at a production facility. It may be expressed as

$$\text{availability } (A) = \frac{\text{loading time } - \text{ downtime}}{\text{loading time}} \tag{4.48}$$

where *loading time* refers to the time available per day (or per month) for manufacturing operations, and *downtime* is the time that the system is not operating because of equipment failures, overhaul, calibration and adjustment, exchange of dies and fixtures, and setup procedures.[5]

Inherent or achieved availability is often specified in equipment supplier contracts. This is because the supplier is likely to have little, if any, control over the effectiveness of the logistic support system and the associated time delays. Operational availability must however be considered when conducting a realistic evaluation of the overall system and its relevant support capability in a real world user environment.

Musa et al. define software availability as "the expected fraction of time during which a software component or system is functioning acceptably."[6] Further, while software maintainability refers to the ease, accuracy, and timeliness with which a software program can be corrected, it does not relate to software availability or system downtime in the traditional manner. According to Musa et al., this is because software maintenance and repairs very often do not idle the software component or system as is the case with hardware.

4.6 DEPENDABILITY FACTORS

Dependability is a measure of the system condition at one or more points during a mission, given that the system is operational and available at the start of the mission. It can be defined as the "probability that a system will complete its mission, given that the system was available at the start of the mission." Note that dependability for an unmanned system will be significantly impacted by its reliability. However, it can generally be improved through the incorporation of good maintainability and human engineering in the design of manned systems. System dependability (D) can be expressed as

$$D = R_o + M_o(1 - R_o) \tag{4.49}$$

[5]For a more detailed coverage of this material the reader is referred to any text on total productive maintenance, or TPM, which is a life-cycle approach to factory maintenance and support. Such references are included in Appendix H.
[6]Musa, J. D., A. Iannino, and K. Okumoto, *Software Reliability,* McGraw-Hill Book Company, New York, 1987.

where R_o is the operational reliability (i.e., the first term of the Poisson expansion series), M_o is the operational maintainability and is defined as the "probability that an item will be restored to a specified operational state or retained in that state within an allowable interval of time when maintenance is performed by appropriately trained personnel following the procedures delineated," and $(1 - R_o)$ is the system unreliability or the probability of failure. Equation (4.49) has several interesting characteristics.

Note that if M_o is zero, that is, there are no maintenance personnel involved, then the expression reduces simply to system reliability. On the other hand, if all maintenance actions can be accomplished within the specified time interval, then system dependability reflects an ideal situation, that is, a value equal to unity. In the real world, however, system dependability will have a value anywhere between 0 and 1.

Computation of M_o values for single repair actions is derived from the discussion in Section 4.2.1, and depends upon the repair-time probability distribution forms. For example, if M_{max} is equal to 34 minutes at the ninetieth percentile and the specified time interval for performing repairs is 34 minutes, then M_o is equal to 0.90. The method of computing M_{max} and M_o will, however, have to be adapted to the repair time distribution forms, as discussed in Section 4.2.1.

The approach above does not address the situation where there may be more than one system failure during the mission. If multiple system failures are likely to be significant, an alternate means for computing M_o must be utilized. In the case of multiple failures, M_o may be expressed as the ratio of failure rates, in all combinations, of those items that can be repaired in the specified time interval $(\lambda *)$, and the total failure rate of all combinations (λ_s), whether or not they are repairable in the allocated time interval. Thus,

$$M_o = \frac{\lambda *}{\lambda_s} \tag{4.50}$$

The advantages of a probabilistic approach to system reliability, maintainability, and dependability are many and include the ability to predict the degree of goodness of the figure of merit, the ability to work within time constraints, the flexibility of addressing any combination of systems and subsystems, the realism of including support resource and equipment considerations, and the provision of more realistic inputs to the sparing requirements analysis.

The dependability expression as in Equation (4.49) can be expanded further by considering the various combinations of possible failures. This is expressed as

$$D = R_o \left(1 + \lambda t M_1 + \frac{(\lambda t)^2}{2!} M_2 + \cdots + \frac{(\lambda t)^n}{n!} M_n \right) \tag{4.51}$$

where M_n is the probability of correcting n malfunctions, in all combinations of n, in the allowable time. For example, consider a system with reliability equal to 0.37 or 37%. In accordance with Equation (4.49), system dependability will also be

0.37 in the absence of a maintenance capability. However, this may violate certain operational requirements and therefore be unacceptable. In an attempt to increase dependability, assume that M_1, M_2, M_3, M_4, and M_5 are equal to 1.0, 0.9, 0.75, 0.5, and 0.33, respectively, for the same system. Applying Equation (4.51), the capability of correcting all single failures within the allowable time interval enhances system dependability to 0.736; the capability to effect complete repairs on 90% of all combinations of two failures increases dependability to 0.9013; and system dependability attains a value of 0.956 if 33% of all combinations of five failures or malfunctions can be addressed in the allotted time interval.

The marginal increase in dependability with an increase in the ability to restore a larger combination of system malfunctions is a function of the product of system failure rate (λ) and the time available for restoration (t). While the number of significant malfunction combinations is likely to be small for lower values of this product, a larger value of the product makes for the consideration of a larger combination of system failures or malfunctions.

The computation of the values of $M_{1, 2, \ldots, n}$ is an important step in the overall process of deriving system dependability. One approach, independent of the repair time distributions, is presented next. A truth table is initially developed for every combination of item failures feasible. This table is then examined to see if the combined maintenance task times and the available resources and facilities will permit system restoration in the allowable interval of time. The following example illustrates this process.

Consider a five-bay console with failure rates and maximum task times as shown in Table 4.4. Each bay is a "lowest replaceable item," and the failure of any one bay causes a system-level failure. The time available for performing maintenance is one hour (i.e., $t = 1$ hour), and the typical console mission completion time is 5 hours. Therefore, given the failure rates in Table 4.4, the console reliability can be expressed as

$$R_{console} = e^{-\lambda t} = e^{-(0.1)(5)} = 0.607$$

Adapting Equation (4.51) to the scenario above, M_1, or the probability of correcting

TABLE 4.4   Data for Five-Bay Console Example

System Element	Failure Rate	Maximum Task Time (minutes)
Bay #1	0.01	20
Bay #2	0.01	20
Bay #3	0.02	20
Bay #4	0.03	30
Bay #5	0.03	45
Console	0.10	

all single system malfunctions in the allowable time of 1 hour, can be expressed as

$$M_1 = \frac{\lambda_1 + \lambda_2 + \lambda_3 + \lambda_4 + \lambda_5}{\lambda_1 + \lambda_2 + \lambda_3 + \lambda_4 + \lambda_5} = \frac{0.10}{0.10} = 1.0$$

Consider next the situation where one of the five bays could not be repaired in the allowable time interval. This may be because of poor design, insufficient resources, or a reduction in the time allowed for system restoration. For example, for $t = 30$ minutes, M_1 will be

$$M_1 = \frac{\lambda_1 + \lambda_2 + \lambda_3 + \lambda_4}{\lambda_1 + \lambda_2 + \lambda_3 + \lambda_4 + \lambda_5} = \frac{0.07}{0.10} = 0.70$$

Further, given that only single malfunctions are being addressed, Equation (4.51) may be rewritten as

$$\text{FOR A SINGLE FAILURE} \quad D = e^{-\lambda t}(1 + \lambda^* t) \tag{4.52}$$

where λ^* is the sum of failure rates for those components that can be restored and is equal to 0.10. Next, the procedure is extended to address combinations of two system failures or malfunctions. A truth table is again developed for this case, as shown in Table 4.5.

On investigating Table 4.5, one realizes that all combinations of failures involving Bay 5 cannot be repaired in the allocated time interval. Given this, M_2 can be expressed as

TABLE 4.5   Truth Table for Five-Bay Console Example

Failure Combination	Task Times (minutes)	Total (minutes)
Bays 1 and 2	20 + 20	40
Bays 1 and 3	20 + 20	40
Bays 1 and 4	20 + 30	50
Bays 1 and 5	20 + 45	65
Bays 2 and 3	20 + 20	40
Bays 2 and 4	20 + 30	50
Bays 2 and 5	20 + 45	65
Bays 3 and 4	20 + 30	50
Bays 3 and 5	20 + 45	65
Bays 4 and 5	30 + 45	75
Bays 1 and 1	20 + 20	40
Bays 2 and 2	20 + 20	40
Bays 3 and 3	20 + 20	40
Bays 4 and 4	30 + 30	50
Bays 5 and 5	45 + 45	90

$$M_2 = \frac{\left[\frac{1}{2} \sum\limits_{i=1}^{4} \lambda^2{}_i + \sum\limits_{i,i=1,2}^{3,4} \lambda_i\lambda_j \right]}{\left[\frac{1}{2} \sum\limits_{i=1}^{5} \lambda^2{}_i + \sum\limits_{i,j=1,2}^{4,5} \lambda_i\lambda_j \right]} \qquad (4.53)$$

Further, Equation (4.50) for the situation above may be rewritten as

$$D = e^{-\lambda t} + \left(\sum_{i=1}^{5} \lambda_i \right) t e^{-\lambda t} + \left(\frac{1}{2} \sum_{i=1}^{4} \lambda^2{}_i + \sum_{\substack{i,j=1,2 \\ i \neq j}}^{3,4} \lambda_i\lambda_j \right) t^2 e^{-\lambda t} \qquad (4.54)$$

$$D = e^{-\lambda t} + \left[\sum_{i=1}^{5} \lambda_i \right] t e^{-\lambda t} + \frac{t^2}{2} [\lambda_1^2 + \lambda_2^2 + \lambda_3^2 + \lambda_4^2] e^{-\lambda t} \qquad (4.55)$$

$$+ t^2 [\lambda_1\lambda_2 + \lambda_1\lambda_3 + \lambda_1\lambda_4 + \lambda_2\lambda_3 + \lambda_2\lambda_4 + \lambda_3\lambda_4] e^{-\lambda t}$$

Equations (4.54) and (4.55) have been simplified since the dividend in Equation (4.53) equates to $\lambda^2/2$. In summary, Equation (4.54) may be stated as "system dependability is equal to the sum of the probability of no system malfunctions and the probability of the combinations of single or double malfunctions that do occur and can be repaired in the allocated time interval."

These concepts can be extended to the consideration of combinations of three, four, or more failure combinations. The dependability equation can be transformed to address each case. A generalized form of the dependability relation for combinations of up to four failures may be expressed as

$$D = e^{-\lambda t} \qquad (4.56)$$

$$+ te^{-\lambda t} \left[\sum_{i=1}^{n} \lambda_i^* \right]$$

$$+ t^2 e^{-\lambda t} \left[\left(\frac{1}{2} \sum_{i=1}^{n} \lambda_i^{*2} \right) + \left(\sum_{\substack{i,j=1-2 \\ i \neq h}}^{n-1,n} \lambda_i\lambda_j^* \right) \right]$$

$$+ t^3 e^{-\lambda t} \left[\left(\frac{1}{6} \sum_{i=1}^{n} \lambda_i^{*3} \right) + \left(\frac{1}{2} \sum_{i,j=1,2}^{n-1,n} \lambda_i^{*2}\lambda_j^* \right) + \left(\sum_{\substack{i,j,k=1,2,3 \\ j \neq j \neq k}}^{n-2,n-1,n} \lambda_i\lambda_j\lambda_k^* \right) \right]$$

$$+ t^4 e^{-\lambda \tau} \left[\left(\frac{1}{24} \sum_{n=1}^{n} \lambda_j^{*4} \right) + \left(\frac{1}{6} \sum_{\substack{i,j=1,2 \\ i \neq j}}^{n-1,n} \lambda_i^{*3} \lambda_j^* \right) + \left(\frac{1}{4} \sum_{\substack{i=1 \\ i \neq j}}^{n-1} \lambda_i^{*2} \sum_{j=2}^{n} \lambda_j^{*2} \right) \right.$$

$$\left. + \left(\frac{1}{2} \sum_{\substack{i=1 \\ i \neq j \neq k}}^{n} \lambda_i^{*2} \sum_{j,k=2,3}^{n-1,n} \lambda_j \lambda_k^* \right) + \left(\sum_{i,j,k,l=1,2,3,4}^{n-3,n-2,n-1,n} \lambda_i \lambda_j \lambda_k \lambda_l^* \right) \right]$$

System dependability with combinations involving more than four failures can be derived as well by using similar reasoning and extending Equation (4.56).

4.7 ECONOMIC FACTORS

The current economic trends of rising inflation and cost growth, combined with the reduction of available resources, have increased the importance of cost-effectiveness requirements for all evolving and existing systems. There is an enhanced awareness of the total system life-cycle cost in general, and system operation and maintenance costs in particular. In the past, little effort has been expended on making visible the total life-cycle cost of a system. This is reflected by the "iceberg" effect illustrated in Figure 1.3 (Chapter 1).

Decisions made during the early design and development phases usually have a significant impact on the costs incurred during the subsequent life-cycle phases. While a large percentage of the total system life-cycle cost can be attributed to the system operation and support phase, the greatest opportunity to influence this cost presents itself during the early design and development phases. Trends relative to life-cycle cost commitment and the opportunity to influence design are presented in Figure 1.4. As illustrated, even though a significant percentage of the life-cycle cost is not incurred until the production/construction and system utilization and support phases, early decisions pertaining to system configuration, technology application, materials selection, equipment packaging schemes, and so on, have a large affect on life-cycle cost. Further, the cost to make a system design change is higher and increasingly more difficult to implement during the latter phases of the system life cycle. The early involvement of maintainability engineers and analysts with the design process can have a positive influence on the total system cost. Objective and optimum decisions relative to maintenance and repair policies, level of repair concepts, preventive and corrective maintenance postures, spare and repair part allocations, and so on can significantly reduce the magnitude of the "downstream" sustaining support costs. The result will be the development and deployment of an effective and efficient supportable system.

For maximum benefit, when dealing with economic factors in design, a life-cycle approach is recommended. In keeping with this concept, the first step in performing an economic analysis is to define the life cycle as it applies to a particular system configuration along with the activities that constitute the evolving phases. The next step is to develop a cost breakdown structure (CBS) tailored to the specific system being evaluated. A CBS provides the framework to link functions and projected activities with resource requirements. A typical CBS is depicted in Figure 4.19. It constitutes a logical functional subdivision of total system cost to the level required to gain full visibility relative to system activities, operations, and support.[7] A more indepth presentation of a cost breakdown structure, along with a description of its cost categories, is included in Appendix B. As an input to the CBS, cost estimates are derived from a combination of historical data, experience with similar systems in the past, supplier cost proposals, predictions, parametric estimations, and/or engineering analysis. Further, cost estimates must consider interest rates and inflationary factors, learning curves, and so on.

Life-cycle cost profiles of alternative design configurations are presented to facilitate the economic evaluation process. Since factors such as reliability and maintainability, and decisions relative to specific logistic support requirements, are likely to vary from one alternative to another, costs will vary not only in magnitude but also in time of occurrence. Further, all alternatives must be reduced to an *equivalent basis* before decisions can be made on their relative desirability. In other words, the alternative profiles being considered must be converted to the point in time when the decision is being made. Thus, for a decision today, the profiles must be discounted to the present value amount.

Discounting a cost profile generates an evaluation metric or decision number that represents the cost streams for the alternatives as if all costs were incurred at the same point in time, rather than spread over a time interval. This facilitates a valid and equivalent relative comparison. Discounting is based on the concept that a dollar today is worth more than a dollar in the future because of the associated interest factors. Therefore, costs incurred at a future point in time cannot be equivalently compared with those incurred today without factoring in the time value of money. This concept can be extended to alternative cost profiles such as the ones depicted in Figure 4.20.

The single-payment present value relation can be used to translate a future cost into an equivalent present worth amount, and is expressed as

$$P = F \left[\frac{1}{(1+i)^n} \right] \tag{4.57}$$

where P is the present value, F is the cost incurred at a future point in time, i is the annual interest rate, and n is the relevant interest period. To illustrate the application

---

[7] For a more detailed treatise on the cost breakdown structure concept and its development, see Blanchard, B.S., *Logistics Engineering and Management*, 4th ed., Prentice-Hall, Inc., Englewood Cliffs, NJ, 1992; Fabrycky, W. J., and B. S. Blanchard, *Life-Cycle Cost and Economic Analysis*, Prentice-Hall, Inc., Englewood Cliffs, NJ, 1991.

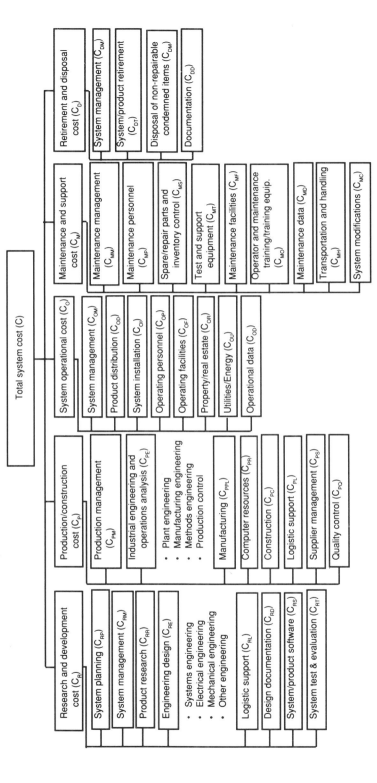

Figure 4.19. Cost breakdown structure (*Source:* Blanchard, B. S., *System Engineering Management*, John Wiley & Sons, N.Y., 1991, p. 132).

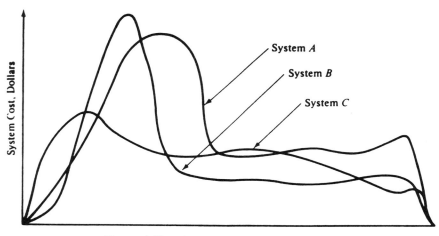

Figure 4.20. Alternative life-cycle cost profiles.

of Equation (4.57), consider the cost stream shown in Figure 4.21. Assuming an interest rate of 12%, the present value at time zero of the cost incurred at point A is

$$P = 1200\left[\frac{1}{(1+0.12)}\right] = \$1,071.43$$

For point B, the present value is

$$P = 3600\left[\frac{1}{(1+0.12)^2}\right] = 2869.90$$

The combined present value of the costs incurred at points A and B is the sum of the individual present values, or $1071.43 + 2869.90 = 3941.33. This concept is extended to include all costs over the system life cycle and to generate a single figure that represents the present equivalent life-cycle cost. The overall present value for the cost stream in the example above is equal to $14,941.60. Note, however, that the undiscounted system life cycle cost is $23,100. Present-value calculations can be simplified with the use of standard interest tables and the appropriate conversion factors are included in Appendix D, Table D.7.

Although the discussion above is limited to calculating the present value of costs

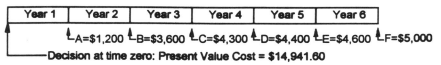

Figure 4.21. An example of a life-cycle cost stream.

incurred over time, very often an analyst may wish to evaluate alternatives with reference to some future point in time. Present value amounts may be translated to equivalent amounts at some point in the future using the single payment compound amount relation [8]

$$F = P[(1 + i)^n] \tag{4.58}$$

While this section has focused on life-cycle cost, revenues can be addressed simultaneously as well. Money has a time value and discounting is an effective mechanism to evaluate the relative merits of competing alternatives, and their associated revenue/cost profiles, in economic terms.

4.8 EFFECTIVENESS FACTORS

In measuring and/or assessing the overall value of a system, one should consider both the technical characteristics of the system and system cost (i.e., both sides of the balance illustrated in Figure 1.6). Numerous effectiveness factors can be expressed as figures of merit, representing the extent to which system effectively performs the functions intended in a cost-effective manner. These figures of merit need to be adapted to a particular system or mission scenario. Some of the more common and relevant terms are defined and discussed below.

1. *System effectiveness.* System effectiveness addresses the extent to which a system is capable of performing is intended functions. Depending upon the system characteristic, numerous figures of merit may be utilized to represent system effectiveness, to include the following:

a. *System technical and performance parameters,* and the extent to which the corresponding requirements are satisfied. Examples of system performance parameters are speed, range, accuracy, power output, throughout, weight, reliability, maintainability, supportability, and response time.

   As an illustration, when addressing a production capability (or manufacturing process), relevant performance parameters could include operating speed rate, quality rate, net operating speed rate, and performance effectiveness, defined as

$$\leftarrow \text{operating speed rate} = \frac{\text{theoretical cycle time}}{\text{actual cycle time}} \tag{4.59}$$

$$\longleftarrow \text{quality rate } (Q) = \frac{\text{processed amount} - \text{defect amount}}{\text{processed amount}} \tag{4.60}$$

---

[8] For an in-depth coverage of concepts relative to present value, future value, equal-payment series, interest tables, and so on, see Interest Factors for Annual Compounding in *Engineering Economy,* 8th ed., Thuesen, Fabrycky ed., © 1993, A10 (p. 677)–A19 (p. 686), A24 (p. 691), A25 (p. 692). Reprinted by permission of Prentice-Hall, Inc., Englewood Cliffs, N.J.

$$\longleftarrow \text{net operating rate} = \frac{\text{(processed amount)(actual cycle time)}}{\text{operating time}} \qquad (4.61)$$

$$\text{performance effectiveness} = \text{(operating speed rate)(net operating rate)} \qquad (4.62)$$

where *theoretical cycle time* represents the desired (or designed) time that it should take to process an item, as compared to the actual time; *processed amount* refers to the number of items processed per day; *operating time* is the difference between *loading time* and *downtime* (see Section 4.5); and *defect amount* represents the number of items rejected because of quality, and so on.

b. *Availability*, or the probability that the system will be available and capable of performing its intended function at any random point in time. As discussed in Section 4.5, availability is primarily a function of reliability, maintainability, and supportability.

c. *Dependability*, or the measure of the system operating condition at one or more points during the mission, given that the system is operational and available at the start of the mission. Dependability is also a function of system reliability, maintainability, and supportability.

From the discussions in this text, system effectiveness addresses the technical aspects of a system, as opposed to cost and other economic factors. Further, the effectiveness measure must be defined clearly and the relevant metrics adapted to a given situation. For example, in a particular application, system effectiveness may be expressed as:

$$\text{system effectiveness} = \text{(availability) (dependability) (performance)} \qquad (4.63)$$

To further illustrate the concept above, consider the effectiveness of a manufacturing process. In this case, the overall equipment effectiveness (OEE) is a function of the equipment availability, its performance effectiveness, and the corresponding quality rate. It is expressed as

$$\text{OEE} = \text{(availability)(performance effectiveness)(quality rate)} \qquad (4.64)$$

where availability, performance effectiveness, and quality rate are as defined in Equations 4.48, 4.60, and 4.62 respectively.[9] As is obvious from the discussion

---

[9]This approach to overall equipment effectiveness has been adapted from the concept of total productive maintenance (TPM), which is a total integrated maintenance management approach applied in a commercial factory environment and is being implemented with a measured degree of success by numerous Japanese and selected American companies. For a more detailed coverage of this subject, see Nakajima, S., *Total Productive Maintenance: An Introduction,* Productivity Press, Inc., Cambridge, MA, 1988; and Nakajima, S., Ed., *TPM Development Program: Implementing Total Productive Maintenance,* Productivity Press, Inc., Cambridge, MA, 1989.

above, maintainability aspects of system design significantly impact overall system effectiveness.

2. *Cost effectiveness.* A higher-level overarching objective in system design is to develop a system that not only satisfies all the necessary technical and performance-related requirements and constraints, but is also cost effective. The cost and economic facets of system design need to be balanced with the relative technical parameters to ensure cost-effective performance. The primary considerations and elements in a cost-effectiveness analysis are illustrated in Figure 4.22. This illustration presents not only the various factors that influence system cost effectiveness but also their relationships.

While the objective is to acquire the appropriate relationship between some measure of technical effectiveness and cost, trends in recent years have resulted in an imbalance between the two. The complexities of many systems have been increasing, primarily owing to the advent of new technologies in design. At the same time, with the ever-increasing emphasis on "performance" at the sacrifice of other key design parameters such as reliability and quality, the overall effectiveness of these systems has been decreasing and the costs have been going up. This is occurring at a time when competition is increasing, there is a greater degree of international cooperation and exchange, and the requirements for producing a well-integrated, cost-effective, high-quality system are even greater than in the past.

The true cost-effectiveness of a system is probably impossible to measure given the numerous intangibles that impact operations and support, and that cannot realistically be quantified. Nonetheless, the cost-effectiveness aspect of system design is often represented by numerous metrics and figures of merit (FOM), such as

$$\text{FOM} = \frac{\text{system effectiveness}}{\text{life} - \text{cycle cost}} \tag{4.65}$$

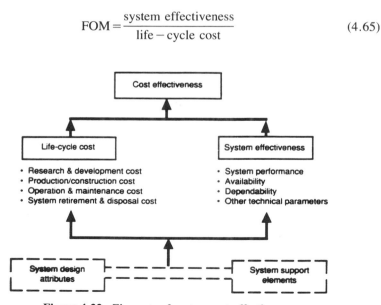

Figure 4.22. Elements of system cost effectiveness.

$$\text{FOM} = \frac{\text{system availability}}{\text{life} - \text{cycle cost}} \qquad (4.66)$$

$$\text{FOM} = \frac{\text{supply maintainability}}{\text{life} - \text{cycle cost}} \qquad (4.67)$$

Inherent design attributes impact both the technical and economic sides of the cost-effectiveness equation. In summary, numerous system characteristics interact, and the technical and economic consequences of these interactions need to be understood and evaluated.

QUESTIONS AND PROBLEMS

1. Define maintainability and maintenance. How do they relate?

2. Define reliability. What are its major characteristics? Define three common measures of reliability.

3. Broadly classify the common measures of system maintainability. Define two metrics from each classification.

4. A system accumulates 32 failures over a 10,000-hour operating time period. What is the system failure rate, MTBF, and reliability?

5. Ten units are tested to failure. Failures occur after 20, 40, 62, 78, 94, 112, 132, 142, 145, and 175 hours. What is the unit failure rate?

6. Twelve units are tested for 100 hours. Failed units are not repaired or replaced. Failures occur after 6, 14, 22, 48, 80, and 96 hours. What is the unit failure rate?

7. A system consists of 3 subassemblies connected in series. The individual subassembly reliabilities are

   Subassembly A = 0.97
   Subassembly B = 0.89
   Subassembly C = 0.92

   What is the overall system reliability?

8. What is the overall system reliability in Question 7 if subassemblies A, B, and C are connected in parallel?

9. Over a period of 10 years, a system successfully completes 985 missions and experiences only 15 aborts. What is the probability of mission success?

10. A system with a failure rate of 0.005 failures/hour is called upon for a mission involving 5 hours of operation. What is the probability of mission success? Is this the same as system reliability?

11. A system with an overall system level MTBF equal to 120 hours is to be operated for 2.6 hours. What is the system reliability?

12. If a system has reliability equal to 0.80 and is to be operated continuously for a time period of 1.5 hours, what is the expected failure rate?

13.

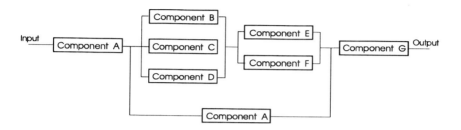

Determine the reliability of the network above given the following individual reliability values:

Component A = 0.90
Component B = 0.80
Component C = 0.70
Component D = 0.90
Component E = 0.80
Component F = 0.80
Component G = 0.90

14. Given that for a system the sum of the mean logistics and administrative time delays is 45 minutes and the mean maintenance downtime (MDT) is 1 hour, determine the mean active maintenance time $(\overline{M})$ for the system?

15. Given a MTTR of 30 minutes and an MTBF of 1000 hours, what is the inherent system availability?

16. With a specified inherent availability of 0.9990 and a calculated MTBF of 400 hours, what is the MTTR requirement?

17. The following corrective maintenance cycle times were observed:

Task Times (minutes)	Number of Tasks
10	1
12	2
14	3
16	4
18	6
20	10

(*continues*)

Task Times (minutes)	Number of Tasks
22	12
24	13
38	1
36	2
34	3
32	6
30	8
28	10
26	12

a. What is the range of the observations?

b. What is the appropriate class interval?

c. Plot the data in a frequency histogram and a frequency polygon. What form of repair time distribution is indicated?

d. What is the mean repair time?

e. Determine the standard deviation of the sample data in the table above.

f. The requirement for the mean repair time is 26 minutes with a 95% confidence level. Do the data above reveal compliance?

18. The following corrective maintenance cycle times were observed:

Task Times (minutes)	Number of Tasks
12	1
14	4
16	8
18	12
20	13
22	14
24	13
26	10
28	11
44	2
42	2
40	3
38	4
36	5
34	6
32	7
30	8

a. What is the range of observations in the table above?

b. Plot a frequency polygon for the observations above. What form of repair time distribution is indicated?

c. Determine the arithmetic and geometric means.

d. What is the standard deviation of the observations above?

e. What is the M_{max} value at the ninetieth percentile?

19. Calculate the achieved system availability given: (a) $\overline{Mct}$ = 30 minutes, (b) $\overline{Mpt}$ = 2 hours, (c) $MTBM_u$ = 250 hours, and (d) $MTBM_s$ = 1000 hours.

20. If an operational cycle takes 1000 hours for completion, and the total downtime during this cycle is 100 hours, determine the operational availability.

21. The following preventive maintenance durations and frequencies are anticipated:

a. 30 minutes every 25 operating hours

b. 30 minutes every 50 operating hours

c. 2 hours every calendar year (12 operating hours per month)

d. 5 hours every 5 calendar years

e. 10 hours every 10 calendar years

Determine the mean preventive maintenance time for the system above.

22. Refer to the sample data in Table 4.2. Given that the mean repair time is 62 minutes, what percentage of the repair tasks will be completed between 47 and 77 minutes?

23. Refer to the sample data in Table 4.2. Suppose the requirement is for the repairs to be accomplished in 64.5 minutes or less with an 85% confidence level. Do the observed values satisfy this requirement?

24. Discuss the relationship between the MTBM and MTBF measures for a system. When can one be greater than or equal to the other?

25. Given that the mean time between corrective and preventive maintenance actions is 150 and 250 hours, respectively, determine the system MTBM measure.

26. A data processing system has an MTTR of 20 minutes and a failure rate of 500 x 10^{-5} failures per hour. For an operating time of 1000 hours, what is the expected system downtime?

27. Calculate as many of the following system metrics as possible:

a. A_i

b. A_o

c. A_a

d. MTTR

e. M_{max}

f. MTBM

g. MTBF

h. $\overline{M}$

i. $MTTR_g$

Given the following information:

a. System failure rate $=$ 0.004 failures/hour
b. Total operating time $=$ 10,000 hours
c. Mean downtime $=$ 50 hours
d. Total number of maintenance actions $=$ 50
e. Mean preventive maintenance time $=$ 6 hours
f. Mean logistics and administrative time $=$ 30 hours

28. A five-bay console has failure rates and maximum task times as shown in the following table:

Bay No.	Failure Rate (failures/hour)	Maximum Task Times (minutes)
1	0.01	20
2	0.01	20
3	0.02	20
4	0.03	30
5	0.03	45
Total	0.10	

Each bay is a lowest replaceable item and failure of any one bay causes a console level failure. Given that the time available for maintenance is 1 hour, and the mission time is 5 hours, calculate:

a. Probability of no malfunctions during the mission time
b. Probability of mission success, if each bay is backed up by one spare
c. Probability of mission success, if each bay is backed up by two spares.

5

DEVELOPMENT OF THE
MAINTENANCE CONCEPT

The system life cycle encompasses a number of program phases, including advance planning and conceptual design, preliminary system design, detail design and development, production and/or construction, distribution, system operation, sustaining maintenance and support, and system retirement and material disposal. Maintainability, in one form or another, is an integral part of each phase of the system life cycle, as illustrated in Figure 1.5. Maintainability requirements are initially established as part of the overall system requirements definition process, included in "design to" criteria as goals for the system, and then incorporated through the day-to-day system design and development activity. Finally, maintainability characteristics in design are assessed through a combination of predictions, analyses, and test and evaluation activities.

Chapters 1–4 covered an introduction to maintainability terms and definitions, maintainability planning and organizations, and some of the measures of maintainability. Chapters 5–12 describe various maintainability activities as they are accomplished throughout the system life cycle. These activities commence with the identification of a consumer need and the accomplishment of feasibility studies, the definition of system operational requirements, and the development of the system maintenance concept. This chapter addresses these early activities, from which maintainability requirements evolve.

5.1 CONSUMER NEEDS ANALYSIS AND FEASIBILITY STUDY

The system life cycle generally commences with the identification of a "want" or "desire" for some item(s) arising out of a perceived deficiency. An individual and/or organization defines a deficiency in an existing capability, for example, inadequate system performance. As a result, the "need" for a new system is defined along with the priority, the date when the new system is required for operational use, and an estimate of the resources required for investing in the new capability. Where possible, the need should be defined in specific quantitative terms so that one can completely define the problem before proceeding with the design and development of the system. Defining the statement of need in complete and unambiguous terms is extremely important if one is to ensure maximum customer satisfaction in the end. Frequently, there are ambiguities, descriptors omitted, and sometimes incorrect statements in a customer-originated statement of need! Careful research is necessary to verify these anamolies and correct them as part of the marketing strategy. A preliminary maintenance concept, expressing the approach planned or preferred by the customer, is a part of this statement of need. This, in turn, requires close ongoing communications between the customer and the prime contractor/producer from the beginning![1]

Given a complete statement of need, it is appropriate to accomplish a feasibility analysis, the scope of which will vary depending on the type and complexity of the requirement. The feasibility analysis leads to the description of a preferred *technical approach* for system design and development. It is necessary to (1) identify all possible alternatives that will fulfill the requirement; (2) screen and evaluate the most likely candidates in terms of performance, effectiveness, logistic support requirements, environmental impact, and economic criteria; and (3) select a preferred approach.

Included within this process is the identification of possible technology applications, the availability of such, the potential cost, and the associated risks. Examples of technology applications may include the use of fiber optics for communications, the use of artificial intelligence for the accomplishment of procedural functions/ tasks, the utilization of digital design techniques, the selection of specific materials for structural purposes, and so on. At this stage in conceptual design, maintainability characteristics must be considered as part of the technology selection process, particularly since these early decisions have a large impact on the ultimate life-cycle cost of a system. Not only are the reliability factors important, but it is also essential that maintainability and supportability criteria be included in the overall evaluation of the candidate alternatives.

A feasibility analysis is often conducted as part of, or an extension to, a preliminary market analysis and leads to (1) definition of system operational requirements, (2) development of a system maintenance concept, and (3) description of a system configuration that is feasible within the constraints of available technology and resources (i.e., money, human resources, equipment, materials, or a combination thereof). In the event that current technology applications are limited, it may be necessary to initiate research activity with the objective of developing new methods/ techniques for specific applications. The accomplishment of such may be directed not only to the fulfillment of basic performance requirements, but to maintainability and supportability applications and to the various elements of logistic support. The application of new materials, new diagnostic techniques, new material handling

---

[1] Quite often the use of the Quality Function Deployment (QFD) technique serves as an excellent tool in promoting the necessary communications and in assisting the customer in the specification of requirements.

methods, new design and manufacturing processes, and so on may have significant implications from a maintainability perspective.

5.2   SYSTEM OPERATIONAL REQUIREMENTS

The system maintenance concept begins with a series of statements and illustrations defining the input criteria to which the system should be designed. It is then further developed into a description of the planned levels of maintenance, major functions to be accomplished at each level and the attendant organizational responsiblities, "on-equipment" versus "off-equipment" maintenance, basic support policies, design criteria associated with the elements of support (e.g., built-in test versus external test, personnel skill levels to which to design), effectiveness factors (i.e., the applicable measures described in Chapter 4, such as MTBM, MDT, and $\overline{M}ct$), and anticipated maintenance environmental requirements.

To define the maintenance concept, one needs to first establish the overall configuration for the system in terms of its projected *operational requirements*. At this point, the following questions should be addressed:

1. What are the basic performance requirements for the system, and how is the system to be utilized?
2. Where will the system (and its components) be utilized in terms of consumer locations?
3. How long will the system be utilized, that is, what is the planned life cycle?

The answer to these and comparable questions leads to the definition of system operating characteristics, the maintenance and support concept for the system, and the identification of specific design guidelines or criteria. The general sequencing of activities is illustrated in Figure 5.1.

Referring to block 3 of the figure, the operational concept (as defined in this text) includes the following information:

1. *Mission definition:* identification of the prime and alternative (or secondary) functions that the system is to perform. What is the system to accomplish (i.e., design objective)? The mission may be defined through one or a set of scenarios, or operational profiles, as illustrated in Figure 5.2. While each and every operational scenario may be difficult (if not impossible) to define, it is important that the "dynamics" of the system operating characteristics be identified. The operating on–off sequences, anticipated duty cycles, and so on must be addressed in order to design for the appropriate stress levels and, as a result, for the proper levels of system reliability and maintainability.

2. *Performance and physical parameters:* definition of specific operating characteristics that the system must incorporate in order to accomplish its intended functions, for example, size, weight, shape of components, range, accuracy, bits, ca-

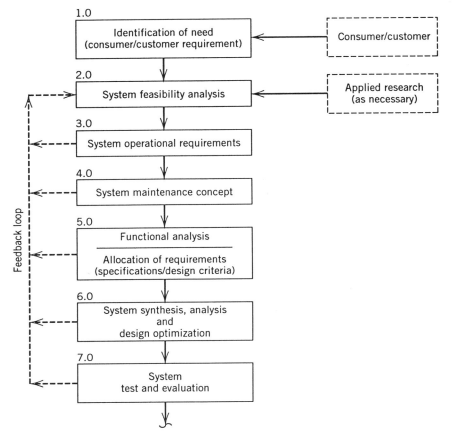

Figure 5.1. An abbreviated sequence of design activities.

pacity, flow rate, transmit, and receive. What are the "critical" system performance parameters?

3. *Operational deployment and distribution:* identification of the anticipated number and geographical location of customer sites. What equipment, personnel, software, facilities, data, and so on are to be distributed, to what location, and when? When does the system become fully operational? System operational requirements may involve a worldwide distribution, such as illustrated in Figure 5.3.

4. *Operational life cycle (planned horizon):* anticipated time that the system will be in operational use. Who will be operating the system and for what period of time? While the planned life cycle period may change, a "baseline" needs to be established from the beginning.

5. *Effectiveness factors:* system requirements specified as figures of merit to define cost/system effectiveness, operational availability, readiness rate, depend-

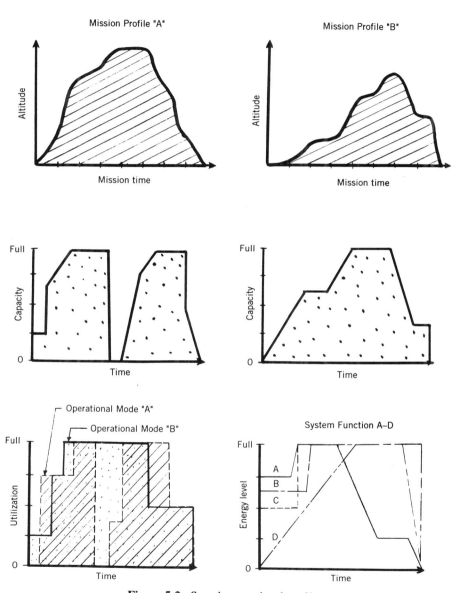

Figure 5.2. Sample operational profiles.

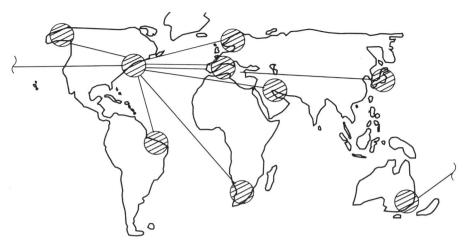

Figure 5.3. System operational requirements (geographical distribution). (*Source:* Blanchard, B. S., *System Engineering Management,* John Wiley & Sons, N.Y., 1991, p. 24).

ability, MTBM, facility utilization, operator skill levels and personnel efficiency, and so on. Given that the system will perform, how effectively or efficiently will it accomplish its mission? The appropriate metrics must be specified and related to a given mission scenario, or operational profile.

6. *Environment:* definition of the environment in which the system is expected to operate (e.g., flat or mountainous terrain, arctic or tropics, airborne, ground, shipboard, temperature and humidity extremes, and vibration/shock requirements). This should include a range of values (as applicable), and should cover all transportation, handling, and storage modes. Not only must the operating environment be defined, but how will the system (and its components) be handled in transit? What will the system be subjected to during operational use, and for how long? A complete environmental profile should be developed.

The establishment of operational requirements forms the basis for system design, and the subsequent definition of support requirements. Although conditions may change, some initial assumptions are necessary. For example, system components will be utilized differently at various consumer locations; the distribution of system components may vary as the need changes; and/or the length of the life cycle may change as a result of obsolescence or the effects of competition. Nevertheless, the information presented above needs to be developed, with a "baseline" established, in order to proceed with system design.

From a historical perspective, the operational requirements for many new systems were either not developed very well, or were developed by some outside organization (a consultant, a marketing group, or some equivalent organizational entity), the results were placed in a file while awaiting a decision to proceed with preliminary system design, and they were then forgotten when subsequent design activity did resume. At that point, with the need for this type of information readily apparent but not available, individual design groups generated information based on their own assumptions. Also, not all of the design groups were referencing the same baseline, and conflicting requirements evolved. This, in turn, led to systems being developed that did not meet consumer requirements, and the initiation of corrective action through downstream costly modifications occurred. In other words, if the applicable operational requirements are not well defined and integrated into the design process from the beginning, the results later on could turn out to be quite costly.

The definition of system operational requirements is a necessary first step in the establishment of criteria for the design of a specific system configuration. Included within these requirements are the appropriate cost-effectiveness measures, which address the necessary considerations for reliability, maintainability, supportability, and related characteristics. System operational requirements lead directly into the further definition of the system maintenance concept.

5.3 LEVELS OF MAINTENANCE

Maintenance levels pertain to the division of functions and tasks for each area where maintenance is performed. Maintenance tasks may be accomplished (1) on the prime elements of the system (e.g., equipment, facilities, and software at the consumer's site where these elements are normally operating; (2) off-line at a nearby location in a mobile van or a fixed shop on those components that have been removed from the operational site for maintenance; (3) at a remotely located overhaul facility or manufacturer's plant; and/or (4) at a specific supplier's facility. Task complexity, personnel skill-level requirements, frequency of occurrence, special facility needs, political factors, economic and environmental criteria, and so on dictate to a great extent the specific functions that will be accomplished at each level.

The establishment of maintenance levels evolves from the definition of system operational requirements, that is, the geographical distribution of system components, performance factors, technology applications, and system effectiveness requirements and the anticipated frequencies of maintenance. Figure 5.4 illustrates the flow of activities and materials, where maintenance activities are represented by the flow from the consumer's operational site back to the supplier/depot level of maintenance and the provision of logistic support resources (e.g., spare parts) is represented by the flow from the supplier to the operational site.

While the number of maintenance levels (and the depth of activity accomplished at each level) may vary from one system application to another, the "organizational," "intermediate," and "depot" or "factory" levels are assumed as a baseline for the purposes of further discussion.

1. *Organizational maintenance.* Organizational (on-system/equipment) maintenance is accomplished on the prime elements of the system at the consumer's operational site (e.g., airplane, vehicle, communication facility, or manufacturing line).

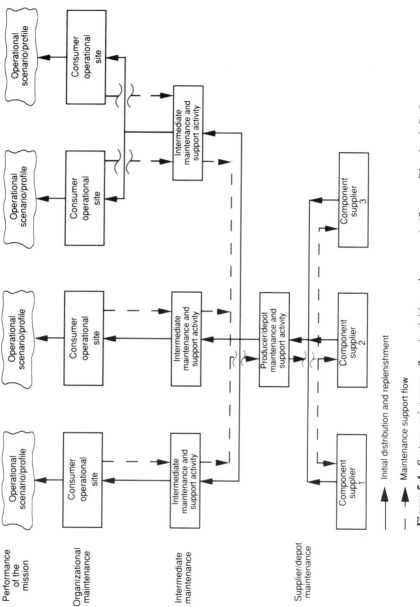

Figure 5.4. System maintenance flow (activities and resources). (*Source:* Blanchard, B. S., *System Engineering Management*, John Wiley & Sons, N.Y., 1991, p. 28).

Generally it includes tasks performed by the using organization on its own equipment by personnel usually involved with the operation and use of that equipment, and having minimum time available for detailed system maintenance. Maintenance at this level normally is limited to periodic checks of equipment performance, visual inspections, cleaning of equipment, some servicing, external adjustments, and the removal and replacement of components. Personnel assigned to this level generally do not repair the removed components, but forward them to the intermediate level. From the maintenance perspective, the least-skilled personnel are assigned, and the design of system components must take this fact into consideration (i.e., design for simplicity).

2. *Intermediate maintenance.* Intermediate maintenance tasks are performed by mobile, semimobile, and/or fixed specialized organizations and installations. At this level, end items removed from the operating system may be repaired through the removal and replacement of major modules, assemblies, and/or piece parts. Scheduled maintenance requiring equipment disassembly may also be accomplished. Available maintenance personnel are usually more skilled and better equipped than those at the organizational level, and are responsible for accomplishing more detailed maintenance.

Mobile or semimobile units are often assigned to provide close support to deployed operational elements of the system. These units may include vans, trucks, or portable shelters containing some test and support equipment and spare parts. The objective is to provide on-site maintenance (beyond that accomplished by organizational-level personnel) to facilitate the return of the system to its full operational status on an expedited basis. A mobile unit may be used to support more than one operational site. A good example is the roving maintenance vehicle that is deployed from the airport hangar to an airplane parked at a commercial airline terminal gate and needing extended maintenance.

Fixed installations (permanent shops) are generally established to support both organizational-level tasks and the mobile or semimobile units. Maintenance tasks that cannot be performed by the lower levels, because of limited personnel skills and test equipment, are performed here. High personnel skills, additional test and support equipment, more spares, and better facilities often enable equipment repair to the module and piece-part level. Fixed shops are usually located near the operational system, but within specified geographical areas (e.g., at the same airport where there are aircraft at terminal gates or in the same building where there is manufacturing equipment producing a product). Rapid maintenance turnaround times are usually not as imperative here as at the lower levels of maintenance.

3. *Supplier/manufacturer/depot maintenance.* The supplier or depot level constitutes the highest type of maintenance, and supports the accomplishment of tasks above and beyond the capabilities available at the intermediate level. Physically, this may be a specialized repair facility supporting a large number of systems/equipments/software in the inventory, or it may constitute the manufacturer's main plant. Depot facilities are "fixed" and mobility is not a problem. Complex and bulky capital equipment, large quantities of spares, environmental control provisions, and so on can readily be made available if required. The high-volume potential at this

level fosters the use of assembly-line techniques which, in turn, permits the use of relatively unskilled labor for a large portion of the workload, with a concentration of highly skilled specialists in selected areas where high-precision tasks are performed.

The supplier/manufacturer/depot level of maintenance includes the complete overhauling, rebuilding, and calibration of equipment, as well as the accomplishment of other highly complex maintenance actions. This facility is generally remotely located and provides support for a number of product lines throughout a wide geographical area, including assuming the role of inventory manager for spare and repair parts.

A brief comparison of the three levels of maintenance described herein is presented in Figure 5.5.

5.4 BASIC REPAIR POLICIES

Within the overall structure illustrated in Figure 5.4, there are a number of possible repair policies. A repair policy specifies the anticipated extent to which the repair of a system (or its components) will be accomplished (if at all). In the development of the initial maintenance concept, basic repair policies must be considered in order to provide input design criteria for equipment packaging and diagnostic routines, as well as the elements of logistic support. The repair policy may dictate that an item should be designed to be nonrepairable, partially repairable, or fully repairable. A typical repair policy is illustrated in Figure 5.6.

1. *Nonrepairable item.* A nonrepairable item is generally modular in construction with a relatively low replacement and disposal cost, and is discarded when a failure occurs. No repair is accomplished and the item is replaced by a spare. The residue is then dispositioned as a "throwaway" or is reclaimed and recycled for other uses. Referring to Figure 5.6, the policy may indicate that either Unit "A," Unit "B," or Unit "C" are discarded at the organizational level when the applicable item fails. No intermediate-level activity is required, except for supplying a spare unit.

If this policy is selected, system design criteria should be established to promote a positive built-in self-test capability (high self-test thoroughness) to ensure that a failure has actually been confirmed prior to discarding the applicable unit. If, on the other hand, a failure is suspected but not confirmed, there is a good possibility of discarding a good unit, which can be costly!

The system should be designed such that the units are easily removable (i.e., of a "plug-in" variety that do not require a lot of external fasteners). Since the unit is to be discarded at failure, the outside package can be hermetically sealed to improve reliability and add protection against humidity and corrosion. Further, there is no need for internal accessibility, test points, plug-in assemblies, modularization, and so on, which may result in a lighter-weight, more reliable unit with a lower total production cost.

Criteria	Organizational Maintenance	Intermediate Maintenance		Supplier/Manufacturer/Depot Maintenance
		Mobile or semimobile units	Fixed units	
		Truck, van, portable shelter, or equivalent	Fixed field shop	
Done where?	At the operational site or wherever the prime equipment is located			Supplier/Manufacturer/depot facility — — — — Specialized repair activity, or manufacturer's plant
Done by whom?	System/equipment operating personnel (low maint. skills)	Personnel assigned to mobile, semimobile, or fixed units (intermediate maintenance skills)		Depot facility personnel or manufacturer's production personnel (mix of intermediate fabrications skills and high maintenance skills)
On whose equipment?	Using organization's equipment	Equipment owned by using organization		
Type of work accomplished?	Visual inspection Operational checkout Minor servicing External adjustments Removal and replacement of some components	Detailed inspection and system checkout Major servicing Major equipment repair and modifications Complicated adjustments Limited calibration Overload from organizational level of maintenance		Complicated factory adjustments Complex equipments repairs and modifications Overhaul and rebuild Detailed calibration Supply support Overload from intermediate level of maintenance

Figure 5.5. A simple comparison of the major levels of maintenance.

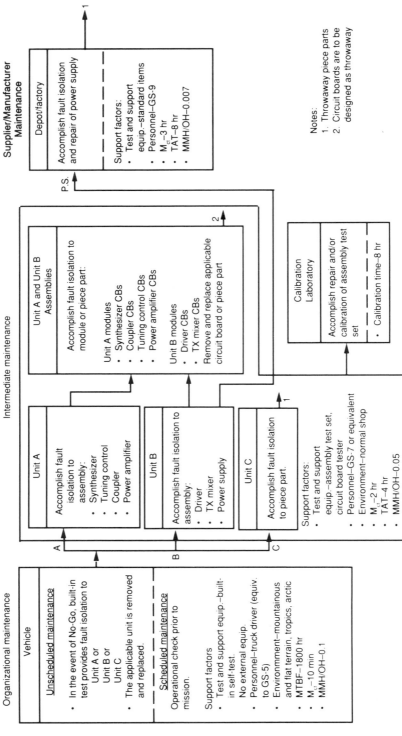

Figure 5.6. System maintenance concept flow (repair policy). (*Source:* Blanchard, B. S., *System Engineering Management*, John Wiley & Sons, N.Y., 1991, p. 32).

Logistic support requirements are minimal. Spare units must be stocked at each intermediate-level facility, or at a location close to the point of need. No lower-level spares are required. Test and support equipment may be necessary to initially check units as they enter the inventory; however, no detail maintenance test equipment is required. Low personnel skills will suffice since maintenance is limited to a "remove and replace" function. Maintenance procedures are considerably simplified since there is no need to include coverage of unit maintenance. The objective is to weigh the cost of spares and unit disposal against the requirements for logistic support if the unit were repaired, and it is important to address some of these factors at this point in the system life cycle.

2. *Partially repairable system.* A partially repairable system may assume various forms. Referring to Figure 5.6, the policy illustrated indicates that unit repair is accomplished when a failure occurs. Unit repair constitutes the removal and replacement of assemblies, and the assemblies are repaired through the removal and replacement of modules and/or circuit boards. These, in turn, are discarded at failure.

The selection of a specific repair policy is highly dependent on operational requirements. For example, the system operational availability (A_o) may dictate a mean downtime requirement of such short duration that it can be met only by providing for quick repair capability at the organizational level. Since the personnel skills and available equipment at the organizational level are limited, a need exists to design the equipment for easy and positive failure identification and for the rapid removal and replacement of the applicable item once the failure has been confirmed. Thus, design criteria should cover built-in self-test features, modularization (plug-in units), and accessibility to the unit level.

At the intermediate level, a different requirement exists. The activity here is designed to support organizational maintenance needs necessary to meet operational mission objectives. Referring to Figure 5.6, internal test provisions (either as an extension of the system built-in self-test or as a set of separate test points) are necessary to enable fault isolation to the module/circuit board level. The modules/circuit boards should be readily accessible and easily removable from the various assemblies. Further, these modules and/or circuit boards should be designed such that it becomes economical for discard at failure.

Alternative policies may dictate that the units are repairable and the assemblies are nonrepairable; or the units are repairable, the synthesizer and driver assemblies are repairable, and the other assemblies are nonrepairable; or any combination of similar factors. The repair policy establishes goals for equipment/component design in terms of what is repairable and what is not, and the level of maintenance at which repair is accomplished. The support policy must consider *all* applicable levels of maintenance since a decision at one level impacts on the other levels, and one must support system operational requirements.

Logistic support needs can be identified on a preliminary basis for each repair policy being considered. For the policy illustrated in Figure 5.6, spare units, assemblies, modules, and circuit boards are stocked at the intermediate maintenance facility. While no external test equipment is required at the organizational level, an assembly test set and circuit board tester are required at the intermediate shop. Personnel skill levels are specified together with effectiveness factors. These and other requirements are evaluated in terms of defining an optimal repair policy for the system.

3. *Fully repairable system.* A fully repairable system promotes a degree of maintenance beyond the repair policy illustrated in Figure 5.6. Individual modules and circuit boards within Unit "A" and Unit "B" assemblies are classified as being repairable. In this instance, specific design criteria must cover the module/circuit board down to the piece-part level. This policy reflects a requirement for the greatest amount of logistic support in terms of test and support equipment, spares/repair parts, personnel and training, technical data coverage, and facilities. Again, the overall policy selection must consider life-cycle cost in the decision-making process.

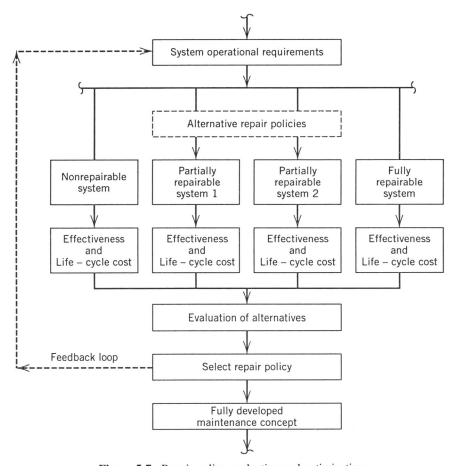

Figure 5.7. Repair policy evaluation and optimization.

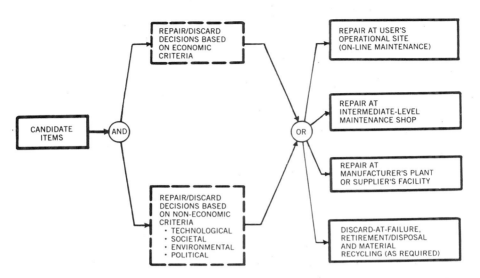

Figure 5.8. Level of repair analysis decision factors.

In developing the maintenance concept, one must evaluate system operational requirements and identify repair policies that will support these requirements. It is quite possible that there will be as many policy variations as one can imagine. The intent is to narrow the field down to one or two feasible approaches.

At this stage in the life cycle not too much is known about the details of system design; thus, one must assume such alternatives as those shown in Figure 5.7. Each option, which reflects different characteristics of system design and support, is evaluated in terms of an appropriate effectiveness figure of merit and life-cycle cost. Data input factors are based on historical experience, early predictions, and so on. The policy selected will be based on the relative merits of each when compared on an equivalent basis. If two policies are considered to be acceptable at this point, the maintenance concept will include the consideration of the two policies until enough detailed data are available to enable the accomplishment of a more in-depth comparative analysis. Figure 5.7 reflects the basic trade-off evaluation approach. Further, a preliminary level of repair analysis may be accomplished in supporting early decisions pertaining to individual candidate items such as Unit "A," Assembly "1," Module "XYZ," and so on. The level of repair analysis is discussed further in Chapter 6. The approach is illustrated in Figure 5.8, where candidate items are evaluated in terms of repair versus discard decisions on the basis of economical, technical, and other criteria.

5.5   LOGISTIC SUPPORT REQUIREMENTS

As part of the initial maintenance concept, design criteria must be established relating to the various elements of logistic support. Referring to Section 1.3.4, these elements include supply support (spares and repair parts, associated inventories, etc.), test and support equipment, personnel and training, transportation and handling equipment, facilities, data, and computer resources. Such criteria, as an input to design, may cover self-test provisions, built-in versus external test requirements, packaging and standardization factors, personnel quantities and skill levels, transportation and handling constraints, and so on.

Referring to Figure 5.4, the various elements of logistic support are initially addressed for each of the maintenance levels shown. There will be spare part requirements, there will be test equipment needs, and there will be personnel requirements. From a maintainability perspective, some guidelines should be provided to ensure good design for supportability, for example, design for maximum standardization to reduce the variety and quantity of spares, design to ensure compatibility between the prime equipment and the elements of external test equipment, design for minimum personnel skills, and design for ease of packaging and handling.

5.6   EFFECTIVENESS REQUIREMENTS

The maintenance concept should include some quantitative "design-to" effectiveness factors applied to the system support capability. In the supply support area, for example, this may include a spare part demand rate, the probability of a spare part being available when required, the probability of mission success given a designated quantity of spares, or the economic order quantity as related to inventory procurement. For test equipment, the length of the queue while waiting for test, the test station process time, and the test equipment reliability or availability are key factors. In transportation, transportation rates, transportation times, the reliability of transportation, and transportation costs are of significance. For personnel and training, one should be interested in personnel quantities and skill levels, training rates, and training equipment reliability. In software, the number of errors per line of code may be an important measure.

Many of these factors (described in Section 4.4) can be specified as part of the maintenance concept and applied with respect to the activities identified for each level of maintenance, as shown in Figure 5.6. Additionally, they must be complementary to and supportive of the effectiveness factors specified under operational requirements for the prime elements of the system. It is meaningless to specify a tight quantitative requirement applicable to the repair of a prime equipment item when it takes 6 months to acquire a needed spare part. The effectiveness requirements applicable to the support capability must be well integrated into the requirements for the system overall.

5.7 ENVIRONMENTAL FACTORS

Within the context of the maintenance concept, one needs to address environmental issues, in terms of both the impact of external environmental factors on the performance of maintenance and the accomplishment of maintenance activities and their impact on the environment. Often, environmental issues are adequately addressed with respect to operational requirements (refer to Section 5.2), but not well addressed when considering maintenance and support activities.

Environmental requirements from a maintenance perspective stem from system operational requirements, for example, how the system is to be utilized, where the components of the system are to be located, and so on, as reflected by the activities identified in Figure 5.4. Maintenance tasks are accomplished at each level, and there are transportation and and handling functions that are necessary to support task completion. Environmental factors include temperature, humidity, vibration and shock, noise, and the like. Whether maintenance tasks are to be accomplished in the tropics or in the arctic, in a mountainous versus a flat terrain, on shipboard or in an aircraft, or outside or in a sheltered environment is a major consideration relative to ensuring that the ultimate system is maintained effectively and efficiently. If the appropriate environmental impacts are not adequately considered in the accomplishment of maintenance tasks, significant system degradation could result.

On the other end of the spectrum, one needs to assess the possible impacts on the environment from the completion of maintenance activities—the discarding of nonrepairable items, the storage of certain materials over a long period of time, and the use of chemical fluids for cleaning certain areas, which could result in the discharge of toxic wastes. Within the context of the initial maintenance concept, possible environmental impacts must be addressed, and "design to" criteria must be specified to ensure that the accomplishment of system maintenance activities will not lead to environmental degradation in any form.

5.8 THE RESPONSIBILITIES FOR MAINTENANCE

The accomplishment of maintenance tasks may be the responsibility of the consumer, the producer (or supplier), a third party, or a combination thereof! As indicated in Figure 5.5, system operating personnel may perform a limited number of maintenance functions at the organizational level, personnel representing the consumer organization or the producer may provide intermediate-level maintenance support, and the producer or a supplier may provide the necessary depot-level maintenance and support. Additionally, a "third party" (not consumer, producer, or a supplier of system components) may be involved in the accomplishment of "contracted" maintenance at any level. The responsibilities for maintenance may vary, not only with different components of the system, but as one progresses in time through the system operational utilization and sustaining support phase.

Decisions pertaining to organizational responsibilities may impact system design, particularly with regard to the inclusion of diagnostic provisions, the type of packaging, and test and evaluation schemes. Also, repair policies may be heavily influenced by organizational decisions. Contractual warranty provisions may dictate that repair must be accomplished only by company "A." The complexity of an item may be such that repair can only be accomplished in facility "Y." For political or ecological reasons, repair can only be accomplished at location "Z." There may be any number of options; however, the major issues need to be addressed and included as considerations in the development of the maintenance concept.

5.9   DEVELOPMENT OF THE MAINTENANCE CONCEPT

In developing the maintenance concept, one must first analyze the operational requirements for the system and then define a maintenance concept that will support these requirements. This definition process is iterative and usually includes the accomplishment of a number of different trade-off studies. For example, one may wish to evaluate the benefits of a two-level versus a three-level maintenance approach; a repair versus discard decision (i.e., the accomplishment of a preliminary level of repair analysis); a reliability/maintainability/availability analysis involving alternative methods of support; a built-in versus external test capability; an automatic versus manual approach to maintenance; an alternative organizational approach to maintenance; and so on. An example of a "two versus three levels of maintenance" evaluation, along with some of the factors that need to be considered, is presented for illustrative purposes.

Assume that you are in the process of defining the maintenance concept early in the conceptual design of System "X," and that you wish to evaluate the feasibility of either designing the system for two levels of maintenance support, or for three levels of maintenance support.

In the design for *two levels* (i.e., organizational and supplier maintenance), the system packaging scheme includes six assemblies, operating in series, and integrated into the overall system housing. In the event that corrective maintenance is required, faults are isolated to a specific assembly through a built-in test capability; the applicable assembly is removed and replaced with a spare; and the faulty item is sent back to the supplier for repair.

In the design for *three levels* (i.e., organizational, intermediate, and supplier maintenance), the system packaging scheme includes three assemblies operating in series within Unit "A," and three assemblies operating in series within Unit "B." The units are integrated into the overall system housing. In the event that corrective maintenance is required, faults are isolated to the unit level through a built-in test capability; the applicable unit is removed and replaced with a spare; and the faulty unit is returned to the intermediate maintenance shop for repair. Unit repair includes isolation to a faulty assembly using external test equipment; the assembly is removed and replaced with a spare; and the malfunctioned assembly is returned to the supplier for repair. The two configurations are illustrated in Figure 5.9.

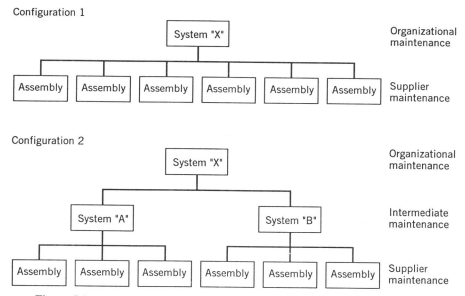

Figure 5.9. Comparison of alternatives—two versus three levels of maintenance.

The objective is to determine the most feasible approach given the following information (i.e., Configuration 1 or Configuration 2?):

1. System "X" is expected to operate 2000 hours per year, at a single customer installation, and the system acquisition cost (i.e., design and production) for Configuration 1 is $250,000, and for Configuration 2 it is $175,000. The difference is primarily attributed to the more extensive built-in test capability required for Configuration 1.

2. Assume that all assemblies are equivalent in terms of reliability, and that the average assembly failure rate is 0.001 failures per hour. Also, assume that all repair times are equivalent.

3. The average personnel cost per maintenance action (MA) is $100 at the organizational level, $200 at the intermediate level, and $300 at the supplier level.

4. For Configuration 1, three spare assemblies are required for organizational maintenance support, and the cost per assembly is $20,000. For Configuration 2, two spare units are required for organizational maintenance support and two spare assemblies are required for the intermediate level of maintenance support. The cost of a spare unit is $50,000, and the cost of a spare assembly is $15,000. These figures include the costs of provisioning and inventory maintenance.

5. The cost of external test equipment to support unit-level repair is $75,000, and assembly-level repair is $50,000. These figures include the costs of operating and maintaining the support equipment.

6. The cost of facility utilization at the intermediate level of maintenance is $75/MA, and for the supplier level of maintenance is $30/MA.
7. Transportation costs associated with unit-level maintenance is $100/MA, and for assembly-level maintenance it is $75/MA. Assume that the cost of maintenance data is $25/MA for Configuration 1 and $40/MA for Configuration 2.

The results from the evaluation are included in the following table:

Evaluation Factor	Configuration 1 (Cost $)	Configuration 2 (Cost $)
System "X" acquisition cost (design and production)	250,000	175,000
Personnel cost (cost per maintenance action at 12 MAs/year)	4,800	7,200
Spare part cost (unit and assembly spares)	60,000	130,000
Test equipment	50,000	125,000
Facilities	360	1,260
Transportation	900	2,100
Data	300	480
Total	366,360	441,040

Configuration 1 is preferred.

In summary, the maintenance concept provides the basis for the establishment of maintainability and supportability requirements in system design. In the development of the maintenance concept, criteria are established for system test and diagnostic routines, equipment packaging schemes, accessibility provisions, the standardization and interchangeability of components, and so on. Not only do these criteria impact the prime mission-oriented segments of the system, but they should also provide guidance in the design and/or procurement of the necessary elements of logistic support. After the definition of system operational requirements, the maintenance concept development task is one of the first steps in the overall requirements definition process for any category of systems whether a transportation system, a communications system, an airborne or shipboard system, an information processing system, or a manufacturing system. Additionally, the maintenance concept forms a baseline for the development of a detailed maintenance plan, prepared during the detail design and development phase. As a final check, relative to the completeness of the maintenance concept, the following questions should be addressed:

1. Have the levels of maintenance (i.e., organizational, intermediate, supplier, manufacturer, depot) been specified and defined?

2. Have the basic maintenance functions been identified for each level of maintenance?

3. Have the organizational responsibilities for each level of maintenance been established (i.e., consumer, producer, supplier, outside source, or combination thereof)?

4. Have the level of repair policies been established (repair versus discard, repair at intermediate/depot level)? Have the criteria for level of repair decisions been adequately defined?

5. Have the appropriate quantitative effectiveness factors been established for maintenance frequencies, maintenance task times, maintenance labor hours, maintenance cost, turnaround times, transportation times, test and support equipment availability and utilization, spare part demands and associated inventory factors, pipeline times, facility utilization, software reliability, and so on? These factors, as applicable, should be applied to ALL maintenance levels, and not just for organizational maintenance, since the interaction effects are many (i.e., the impact of activities at one level on the other levels).

6. Have the appropriate design criteria been established for the different elements of logistic support at each level of maintenance, that is, spares and repair parts, test and support equipment, personnel and training, transportation and handling, facilities, data, and computer resources?

7. Have the appropriate environmental factors (i.e., constraints) been established for the maintenance activities at each level?

These, and many other questions of a similar nature, may be helpful in the development of the system maintenance concept.

QUESTIONS AND PROBLEMS

1. Why is the definition of system operational requirements important? What information is included?

2. Why is it important to define specific mission scenarios (or operational profiles) within the context of the system operational requirements?

3. What information should be included in the system maintenance concept?

4. When in the system life cycle should the maintenance concept be developed? Why?

5. How do system operational requirements influence the maintenance concept (be specific)?

6. How does the maintenance concept affect system/product design?

7. When evaluating alternative repair policies, what criteria would you use? Why?

8. Select a system of your choice and develop the operational requirements for that system. Based on the results, develop a maintenance concept from the established operational requirements. Construct a maintenance flow, identify repair policies, and apply quantitative effectiveness factors as appropriate.

9. Develop a maintenance concept flow for a major home appliance.

10. Develop a maintenance concept flow for a manufacturing capability.

11. What factors should be considered in determining which maintenance functions should be accomplished at the organizational level, at the intermediate level, and at the supplier/manufacturer/depot level?

12. Refer to Figure 5.6. If the reliability MTBF is less than the specified 1800 hours, what are the possible impacts on the overall maintenance concept? Describe some of the interrelationships. Describe some of the impacts on system design as a result of the $\overline{M}ct$ of 10 minutes. What is the impact of logistics pipeline times and facility turnaround times on the total logistic support capability?

13. Refer to Figure 5.4. What factors must be considered in determining spare/repair part requirements at the intermediate-level maintenance shop?

14. When developing the maintenance concept, *all* applicable levels of maintenance must be considered on an integrated basis. Why?

15. Describe how the maintenance concept can lead into the development of maintainability criteria for design. Provide some specific examples.

16. Describe how the assignment of maintenance responsibilities can influence system design for maintainability.

17. Refer to Figure 5.9 (and the illustrated problem). Assume that the failure rate is 10 times the specified value and calculate the results for each of the two alternatives. What are the results if the failure rate is one-tenth of the specified value?

18. How can the accomplishment of maintenance tasks influence the environment? Provide some examples.

19. How does the maintenance concept relate to the maintenance plan?

6

MAINTAINABILITY ANALYSIS

Within the broad context of "Maintainability Analysis" (as compared to "maintenance engineering analysis," discussed in Chapter 9), a series of functions/tasks are
directed toward the *design for maintainability*. These include the initial definition
of system requirements, the accomplishment of functional analysis, the top-down
allocation of requirements to the various elements of the system, system synthesis
and analysis, and design optimization. More specifically, this area of activity
evolves through the definition of system operational requirements and the maintenance concept discussed in Chapter 5, and leads into the definition of maintenance
functions (as part of the functional analysis), maintainability allocation, and the
conductance of various trade-off studies, the results of which impact maintainability
characteristics in design. The basic process is illustrated in Figure 6.1. This chapter
covers the activities within the shaded block, that is, functional analysis, allocation,
and some typical trade-off studies that are often inherent within a maintainability
program.

6.1 FUNCTIONAL ANALYSIS[1]

The system functional analysis is a significant activity during the preliminary design
phase. The objective of the functional analysis activity is to translate system level
requirements, both qualitative and quantitative, into design requirements at different levels in the system hierarchy. This translation is an iterative process and is
accomplished through the employment of a graphical modeling notation/language
such as functional flow diagrams.

A functional approach assures that the total system definition fully recognizes
and addresses all the involved elements and aspects of the system over its intended
life cycle. The possibility of ignoring an essential system component or requirement
is reduced. A comprehensive functional analysis involves analyzing every function

---

[1]The source for much of the material in Sections 6.1–6.3 is Blanchard, B. S., *System Engineering
Management*, John Wiley & Sons, Inc., New York, 1991.

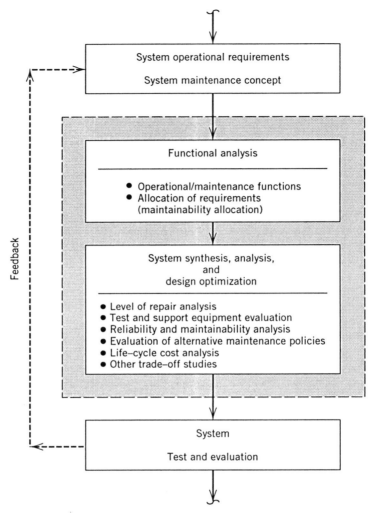

Figure 6.1. Maintainability analysis in the system development process.

for the subfunctions that need to be completed for its accomplishment, the resources required, the inputs intended, and the outputs expected. Functions and subfunctions, along with the corresponding inputs and outputs, are depicted in functional flow diagrams as a hierarchy. This depiction allows an explicit delineation of the functional relationships (series or parallel) that may exist, and also facilitates the definition of interface functions and requirements.

The system functional analysis and decomposition is conducted to a level deemed necessary to support the subsequent allocation, design, analysis, and evaluation activities. In summary, the system functional analysis is a logical, systematic, and comprehensive approach to system design and development which involves the consideration of not only the operational functions but also those that are necessary to support and sustain the system. Moreover, this approach facilitates the traceability between top-level system requirements and the detailed design requirements allocated down the system hierarchy to the constituting modules and components.

Graphical modeling languages such as functional flow diagrams, behavior diagrams, and Integrated DEFinition language (IDEF) models are developed primarily to support and facilitate the process of translating system requirements into design requirements for the constituting elements and their interfaces. They are a pictorial depiction of system requirements in functional terms. It is important to emphasize that a functional analysis and the functional flow diagrams are helpful for understanding a system from the perspective of what needs to be accomplished. In other words, the objective and purpose of this activity is not to focus on the *hows* but the *whats*.

The functional definition of the system is subsequently linked to a physical system design configuration through the allocation of individual functions and corresponding requirements to hardware, software, firmware, facilities, or people components. Acquisition of any and all of the constituting system elements should be first justified through a functional analysis. Any proposed system element not so justified is an automatic candidate for elimination.

6.1.1 Functional Flow Diagrams

Functional flow diagrams are a graphical portrayal of the intended system's functionality. System level requirements are depicted as functions in serial and parallel relationships, and in a top-down hierarchy, with the most abstract and general functions at the top. The top-level functions are successively broken down to lower levels, depicting greater and greater detail, thus leading to the development of a top-down functional hierarchy.

The evolution of functional flow diagrams proceeds along two lines. First, an increasingly detailed definition of the top-level functions leads to the development of the functional tree or the functional hierarchy. Second, information relative to function inputs, outputs, and resource requirements, portrayed in conjunction with the functional flow diagrams, is proportional to the progress of the design and development activity and the state of available information. In other words, given that the early design phase is characterized by a lack of information, an early version of the system functional flow diagrams may portray only gross level functions that need to be accomplished in order to satisfy the system level requirements. This represents the genesis of the translation of system level requirements into detailed design requirements and ensures (1) correlation between system requirements and the subsequent engineering design effort, (2) documentation of critical system formulation decisions, (3) consideration of the "total" system requirements (i.e., operational and logistic support) during the initial system formulation phase, and (4) explicit delineation of the interface requirements and conditions. Information relative to input, output, and resource requirements can be added to this graphical portrayal as and when available.

It is important for the functional analysis and corresponding functional flow diagrams to address both operational and support activities as they are expected to occur over the intended system's life cycle. Maintenance planning should be initiated at this stage to assure the development of a system that is not only technologically feasible and satisfactory but also easily and efficiently supportable.

In summary, some key features of the overall functional flow diagrams are:

1. All activities throughout the system life cycle should be considered. Further, the method of presentation should reflect the proper functional sequences and interface interrelationships.
2. Attention should be focused on *what* is required rather than *how* to accomplish a function.
3. The graphical portrayal should be flexible enough to incorporate additional information as and when it becomes available, and also to allow for information reduction if too much detail has been presented. The objective is to iteratively and progressively work down to a level of detail where resources can be identified relative to how a function should be accomplished.

A standard symbology for the functional flow diagrams will ensure better understanding and communications between the groups and individuals involved in the design effort. It is recommended that certain basic practices and symbols be used in the physical layout of the functional diagrams. The paragraphs below provide some guidance to facilitate achieving this objective.

1. *Function block.* Each separate function in a functional flow diagram should be presented in a single box enclosed by a solid line. Blocks used for reference to other flows (i.e., at a higher level, in a separate functional flow diagram, or a repetition within the same functional flow diagram) should be indicated as partially enclosed boxes labeled "REF." Here function refers to a definite, finite, and discrete action accomplished by the system (hardware and/or software), the associated personnel and/or facilities, or a combination thereof. Questionable or tentative functions should be depicted in dotted blocks. Further, it is important to address both operational and maintenance functions.

2. *Function numbering.* Functions identified on the functional flow diagrams at every level should be numbered in a manner that preserves the continuity of functions and facilitates traceability between the gross top-level functions and the detailed functions at the lowest level in the functional hierarchy. This facility indirectly allows traceability between system level requirements and the detailed allocated design requirements at the system component level.

Functions on the top-level functional diagram should be numbered 1.0, 2.0, 3.0, and so on. Functions that further indenture these top functions should contain the same parent identifier and should be coded at the next decimal level for each indenture. For example, the first indenture of function 4.0 would be 4.1, the second 4.1.1, the third 4.1.1.1, and so on. For expansion of a higher-level function within a particular level of indenture, a numerical sequence should be used to preserve the continuity and operational sequence of the functions. For example, if more than one function is required to amplify function 4.0, the first level of indenture, the sequence should be 4.1, 4.2, 4.3, . . . , 4.n. For expansion of function 4.3 at the second level, the numbering shall be 4.3.1, 4.3.2, . . . , 4.3.n. Where several levels of indenture appear on a single functional diagram, the same pattern should be maintained. While the basic ground rule should be to maintain a minimum level of indentures on any one particular flow, it may become necessary to include several levels to support and preserve the continuity of functions and to minimize the number of flows required to functionally depict the system.

3. *Functional reference*. Each functional diagram should contain a reference to its next higher functional diagram through the use of a reference block. For example, function 4.3 should be shown as a reference block in the case where functions 4.3.1, 4.3.2, . . . , 4.3.n are being used to expand function 4.3. Reference blocks should also be used to indicate interface functions as appropriate.

4. *Flow connections*. Lines connecting functions should indicate only the functional flow and should not represent either a lapse in time or any intermediate activity. Vertical and horizontal lines between blocks should indicate that all functions so interrelated must be performed in either a parallel or series sequence. Diagonal lines may be used to indicate alternative sequences (cases where alternative paths lead to the next function in the sequence).

5. *Flow directions*. Functional diagrams should be laid out so that the functional flow is generally from left to right and the reverse flow, in the case of a feedback functional loop, from right to left. Primary input lines should enter the function block from the left side; the primary and desired output (or GO line) should exit from the right; and the undesired output (or NO-GO line) should exit from the bottom of the box.

6. *Summing gates*. A circle should be used to depict a summing gate. As in the case of functional blocks, lines should enter and/or exit the summing gate as appropriate. The summing gate is used to indicate convergent, divergent, parallel, or alternative functional paths and is annotated with the term AND or OR. The term AND is used to indicate that parallel functions leading into the gate must be accomplished before proceeding to the next function, or that paths emerging from the AND gate must be accomplished after the preceding function(s) has been completed. The term OR is used to indicate that any of several alternative paths (alternative and mutually exclusive functions) converge to, or diverge from, the OR gate. The OR gate thus indicates that alternative paths may lead to or follow a particular function.

7. *Go and No-Go paths*. The symbols G and $\overline{G}$ are used to indicate GO and NO-GO paths, respectively. The symbols are entered adjacent to the lines, leaving a particular function to indicate alternative functional paths.

8. *Numbering procedures for changes to functional diagrams.* Addition of functions to existing functional flow diagrams can be accomplished by locating the new function(s) in its correct position without regard to the numbering sequence. The new function(s) should be numbered using the first unused number at the level of indenture appropriate for the new function.

9. *Function inputs, outputs, and resources requirements.* Every function identified on the functional flow diagram is associated with a desired output(s), input(s), and a set of resource requirements necessary for its accomplishment. This information is identified and tabulated in conjunction with the flow diagrams.

Using this guide to a standard graphical symbology, the maintenance functions, wherever and whenever appropriate, evolve directly from the operational functions identified. In other words, functions to be addressed in a completed functional flow diagram can be classified into two groups, operational functions and maintenance functions. The generic graphical symbology discussed earlier is illustrated in Figure 6.2.

Functional flow diagrams are designated as top level, first level, second level, and so on. The top-level functional flow diagram shows only the gross-level operational functions. The first-, second-, and third-level functional flow diagrams, and so on, represent progressive expansion of individual functions from the top-level functional flow at the preceding level. These detailed functions can represent both operational and maintenance activities. The indenture relationships of functions by level are illustrated further in Figure 6.3. The functional inputs, outputs, and resource requirements should be tabulated as shown in Table 6.1.

6.1.2   Operational Functions

Operational functions refer to activities that are performed in direct response to the accomplishment of mission requirements. This may include a description of multiple mission phases and modes of system operation and utilization. For example, typical operating functions may include (1) "prepare aircraft for flight" (as part of a commercial aircraft mission scenario), (2) "initiate communications between the producer and the consumer" (as part of a communications system scenario), (3) "transport material from the factory to the warehouse" (as part of a product distribution scenario), (4) "manufacture 15 products in a seven-day time frame" (as part of a production scenario), and (5) "process 'Y' data to eight company distribution outlets" (as part of a data processing scenario). System functions necessary to describe the various modes of operation are described and presented in a block diagram format. Note also that the first word in the functional block is always an "action" verb.

Through the progressive expansion of functional activities, directed to defining the "whats" (versus the "hows"), one can evolve from the mission profile illustrated in Figure 6.4 down to a specific commercial aircraft capability such as "communications." A communications subsystem is identified, trade-off studies are conducted, and a detailed design approach is selected. Specific resources that are necessary to

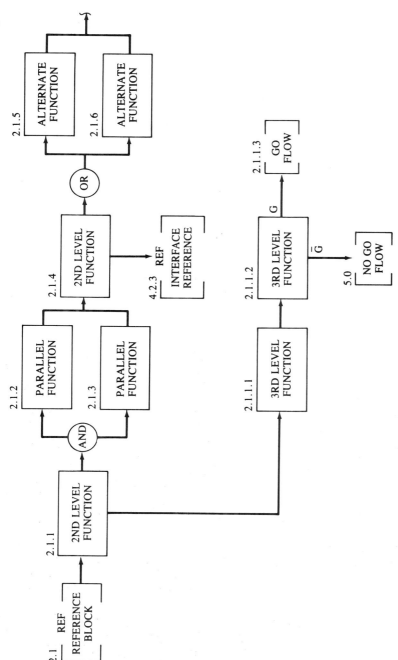

Figure 6.2. General format for developing functional flow diagrams. (*Source:* Systems Engineering and Analysis in *Systems Engineering and Analysis 2/e,* Blanchard/Fabrycky, ed., © 1990, Figure A.1. Reprinted by permission of Prentice-Hall, Inc., Englewood Cliffs, N.J.).

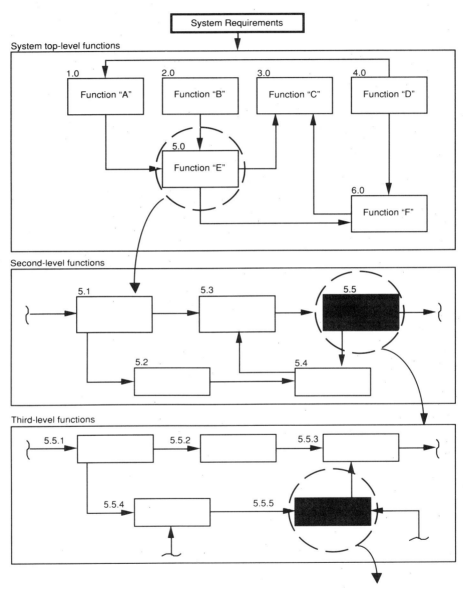

Figure 6.3. System functional breakdown. (*Source:* Blanchard, B. S., *System Engineering Management,* John Wiley & Sons, N.Y., 1991, p. 35).

TABLE 6.1 Table of Functional Inputs, Outputs, and Resource Requirements

CONCEPTUAL SYSTEM ENGINEERING DESIGN PROCESS

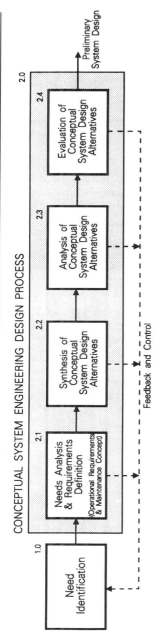

Activity Number	Activity Description	Required Inputs	Expected Outputs	Tools and Techniques or "Best Practices"
1.0	Need Identification	Customer surveys; marketing inputs; shipping and servicing department logs; market niche studies; competitive product research	A specific qualitative and quantitative needs statement responding to a current deficiency. Care must be taken to state this need in functional terms.	Benchmarking; statistical analyses of data (i.e., data collected as a result of surveys and consolidated from shipping and servicing logs, etc.)
2.1	Needs Analysis and Requirements Definition	A specific qualitative and quantitative needs statement expressed in functional terms.	Qualitative and quantitative factors pertaining to system performance levels, geographical distribution of products, expected utilization profiles, user/consumer environment; operational life-cycle, effectiveness requirements, the levels of maintenance and support, consideration of the applicable elements of logistic support, the support environment, and so on.	Quality Function Deployment (QFD); input/output matrix; checklists; value engineering; statistical data analysis; trend analysis; matrix analysis; parametric analysis, various categories of analytical models and tools for simulation studies, trade-offs, etc.
2.2	Synthesis of Conceptual System Design Alternatives	Results from needs analysis and requirements definition process; technology research studies; supplier information	Identification and description of candidate conceptual system design alternatives and technology applications.	Pugh's concept generation approach; brain storming; analogy; checklists.
2.3	Analysis of Conceptual System Design Alternatives	Candidate conceptual solutions and technologies; results from the needs analysis and requirements definition process	Approximation of the "goodness" of each feasible conceptual solution relative to the pertinent parameters, both direct and indirect. This goodness could be expressed as a numeric rating, probabilistic measure or fuzzy measure.	Indirect system experimentation (e.g., mathematical modeling and simulation); parametric analyses; risk analyses
2.4	Evaluation of Conceptual System Design Alternatives	Results from the analysis task in the form of a set of feasible conceptual system design alternatives	A single or shortlisted set of preferred conceptual system designs. Further, a "feel" for how much better the preferred approach(es) is relative to all other feasible alternatives.	Design dependent parameter approach; generation of hybrid numbers to represent candidate solution "goodness"; conceptual system design evaluation display.

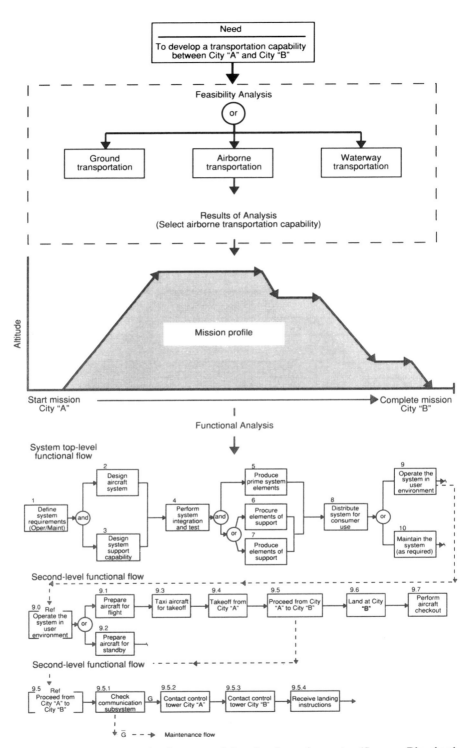

Figure 6.4 Evolutionary development of functional requirements. (*Source:* Blanchard, B. S., *System Engineering Management,* John Wiley & Sons, N.Y., 1991, p. 36).

respond to the stated functional requirements can be identified and presented in a format similar to that presented in Table 6.1. In other words, one can drive "downward" from the system level to identify the resources needed to perform certain functions (i.e., equipment, facilities, human resources, software, and data). Also, given a specific equipment requirement, one can progress "upward" for the *justification* of that requirement. The functional analysis provides the mechanism for "updown" traceability, and can be applied to any category of system, large or small.

6.1.3 Maintenance Functions

System maintenance functions evolve from operational functions, and are presented within the context of the maintenance concept described in Chapter 5. For instance, given a specified set of inputs, there are desired outcomes in terms of performance expectations associated with each function. A check of the applicable requirements will indicate whether or not the particular function is performing satisfactorily. This check will lead to either a GO or a NO-GO decision. A GO decision leads to the activation and subsequent check of the succeeding function. A NO-GO decision provides a starting point for the development of detailed maintenance functional flow diagrams and logic troubleshooting flow diagrams (in the event of corrective maintenance). These functional diagrams are used to further describe the activities for each level of maintenance specified in the system maintenance concept. Maintenance functions identified at this stage should address and reflect maintainability and supportability requirements delineated at the system level. Further, these functions are an index to the nature and quantity of logistic support resources needed during the system utilization phase.

Figure 6.5 represents an abbreviated maintenance functional flow diagram for the commercial aircraft communications subsystem, and is an extension of the operational flow in Figure 6.4. Note that the maintenance functions evolve directly from the operational functions, and that "traceability" is provided through the block numbering system. Individual functions in this diagram can be expanded downward to the depth necessary to allow for determination of the "hows" and identification of resources. Table 6.2 summarizes the specific inputs/outputs (and associated resources) for several of the functions described in Figure 6.5.

Referring to Table 6.2, sample inputs, outputs, and resource requirements are identified for three of the maintenance functions shown in Figure 6.5. Given this basic definition of functions, the next step is to apply some quantitative effectiveness factors for each functional block. This may include specific numerical performance measures, elapsed times and labor-hour requirements, probabilities of success, costs goals, and other measures as appropriate. The objective is to allocate the applicable factors specified in the maintenance concept, presented in Figure 5.6 for example, to the functional requirements illustrated in Figure 6.5. This, in turn, leads to the specification of "design-to" criteria for the resources that will be needed for system maintenance and support. This process is iterative, leading to a more in-depth definition of system design.

The overall evolutionary process defined herein for the communications subsys-

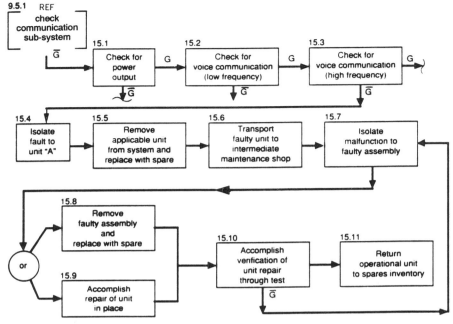

Figure 6.5. Maintenance functional flow diagram.

tem of the commercial aircraft is applicable to any type of system, large or small. To illustrate, Figure 6.6 is included to cover a simple home lawn mower requirement, and Figures 6.7 and 6.8 include the operational and maintenance flow diagram development process for a manufacturing capability.

6.1.4 Alternative Graphical Modeling Languages

The objectives of the functional analysis process are accomplished by analyzing a system from a functional perspective. This is facilitated by developing functional flow diagrams, as discussed in Section 6.1.1. Notwithstanding its widespread usage, functional flow diagramming is only one of the more popular graphical modeling languages in use today to analyze, organize, and document a system's functionality. Two alternative graphical modeling languages, "*N*-squared" charts and "IDEF" models, are discussed briefly in this section.

N-Squared Charts. *N*-squared charts focus attention on data flows between functions. Input and output relationships between a system's functions are explicitly addressed. Functions are depicted as blocks and placed along the diagonal of an $n \times n$ matrix, where n is the total number of functions being addressed. As may be obvious, some of the functional relationships (parallel versus serial) are obscured

TABLE 6.2 Functional Inputs, Outputs, and Resources–Communications Subsystem[a]

Function Number	Function Description	Inputs	Outputs	Resource Requirements	General Comments
15.4	Isolate fault to Unit "A"	Provide test signal to Unit "A"	No voice modulated signal from Unit "A"	Communications subsystem built-in test capability, operator personnel	Verification that Unit "A" is faulty
15.5	Remove applicable unit from the system and replace with a spare	Faulty Unit "A" (on aircraft), spare unit from stock	Good Unit "A" installed in prime system (on aircraft)	Standard tool kit, organization maintenance personnel with appropriate skills, spare Unit "A," built-in test for checkout	Accomplished on the commercial aircraft at the terminal gate
15.6	Transport faulty unit to the intermediate maintenance shop	Faulty Unit "A" at organization level	Faulty Unit "A" at intermediate level, failure report	Container for transportation of Unit "A," transportation vehicle, maintenance personnel with proper skills	Unit "A" is at intermediate shop ready for corrective maintenance

[a] Refer to Figure 6.5.

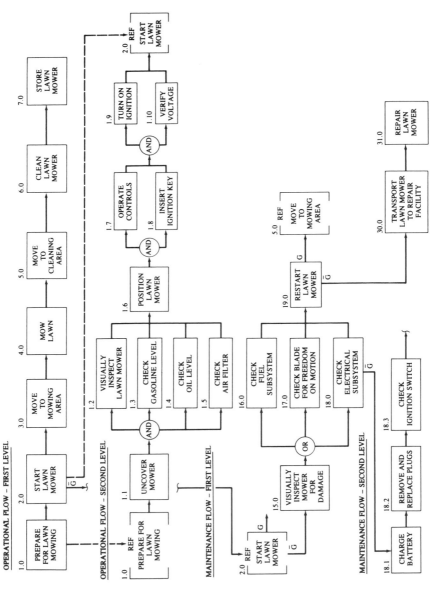

Figure 6.6. Operational and maintenance functional flows—lawn mowing capability. (*Source:* Systems Engineering and Analysis in *Systems Engineering and Analysis* 2/e, Blanchard/Fabrycky, ed., © 1990, Figure A.3. Reprinted by permission of Prentice-Hall,

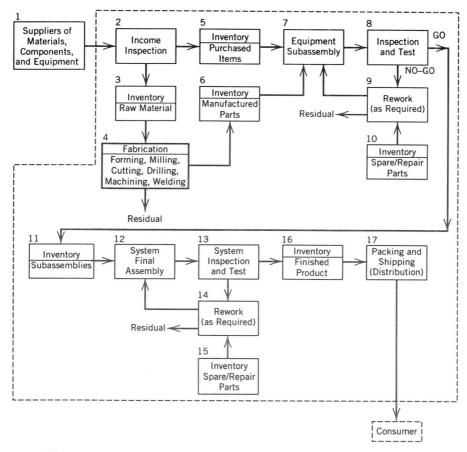

Figure 6.7. Simplified operational flow diagram—manufacturing capability.

when utilizing this modeling format. Data flows between any two functions are presented in blocks where horizontal and vertical lines from the two functions in question intersect. In other words, there are two blocks for every two functions where data flow items between these two functions are listed. Data items that flow down the diagonal (i.e., from a function at a higher level on the diagonal to a function at a lower level) are listed in the intersection block above the diagonal, whereas data items that flow up the diagonal (e.g., feedback or control) are listed in the intersection block below the diagonal. This concept is further illustrated in Figure 6.9, which is a partial depiction of the scenario addressed in Figure 6.5 using functional flow diagrams. Note that functions 15.8 and 15.9 are mutually exclusive as depicted in Figure 6.5, but this relationship is not being made explicit in the N-squared chart.

182

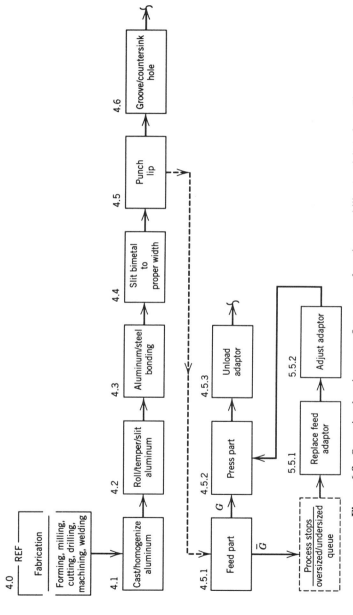

Figure 6.8. Operational-maintenance flow—manufacturing capability (partial) (refer to Figure 6.7).

An n-squared chart with functions on the diagonal and data flow items in the off-diagonal cells:

1	2	3	4	5	6	7
Isolate fault to Unit A	Positive Identification of fault in Unit A					
	Remove applicable unit from system and replace with spare		Unit A fault classification			
		Transport faulty unit to intermediate maintenance shop				
			Isolate malfunction to faulty assembly	Positive identification of fault in relevant assembly	Positive identification of fault in relevant assembly	
				Remove faulty assembly and replace with spare		Assembly replacement confirmation
					Accomplish repair of unit in place	Repair completion confirmation
				Verification results	Verification results	**Accomplish verification of unit repair through test**

Data flow Item placement convention

Figure 6.9. An *n*-squared chart (example).

183

Integrated Definition Language. ICAM **DEF**inition language (IDEF) has its genesis in the US Air Force sponsored Integrated Computer-Aided Manufacturing (ICAM) program initiated by MANTECH (Manufacturing Technology) over a decade ago. While numerous IDEF techniques exist to facilitate the modeling of functions and activities (e.g., $IDEF_0$), data (e.g., $IDEF_{1x}$), and simulation (e.g., $IDEF_2$), only $IDEF_0$, developed to model system functionality, is addressed in this section. $IDEF_0$ is yet another graphical modeling language used to model functions and activities along with the data flows that link these functions. An $IDEF_0$ model is made up of two basic components:

1. A *square block* is used to depict functions or activities.
2. *Arrows* are used to represent functional inputs, outputs, control or constraints, and mechanism data or information. These arrows are also called ICOMS (**I**nputs, **C**ontrols, **O**utputs, and **M**echani**S**ms). Further, inputs always enter a function block from the left and an output always exits a function block from the right. The controls and constraints for a function are shown as arrows entering a functional block from the top, whereas the necessary mechanisms, data, or information required to accomplish a function are depicted as entering a function block from below. Every ICOM is labeled. This generic and accepted modeling symbology is illustrated in Figure 6.10.

An $IDEF_0$ model can be decomposed to show increasing detailed information in a manner similar to functional flow diagrams. This decomposition is depicted in Figure 6.11. To further illustrated the potential usage of $IDEF_0$ models, Figure 6.5 has been redrawn using the $IDEF_0$ modeling technique shown in Figure 6.12.

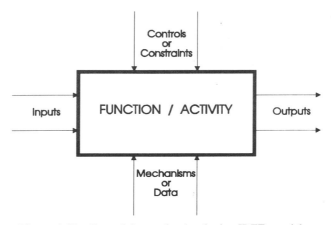

Figure 6.10. General format for developing $IDEF_0$ models.

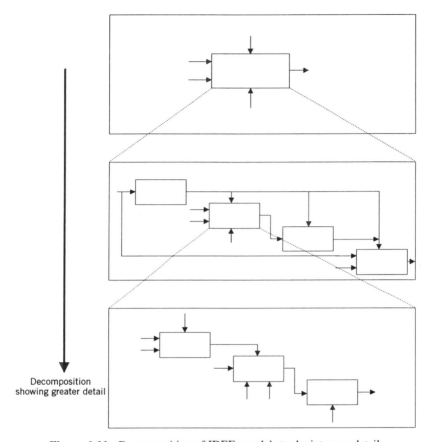

Figure 6.11. Decomposition of IDEF$_o$ models to depict more detail.

6.2   ALLOCATION OF REQUIREMENTS

The developing of system operational requirements and the maintenance concept, described in Chapter 5, results in the definition of specific design criteria at the *system* level. This is the level where the overall requirements for maintainability in design are specified. The functional analysis (Section 6.1) defines the system and its major components in functional terms, progressing downward to the point where the system configuration begins to take shape and specific resource requirements are identified. While the initial objective is to define the "whats," it is now appropriate to respond in terms of the "hows;" that is, how should a given function be accomplished? This, in turn, leads to the identification and evaluation of various possible alternatives, and the system begins to assume a certain degree of structure.

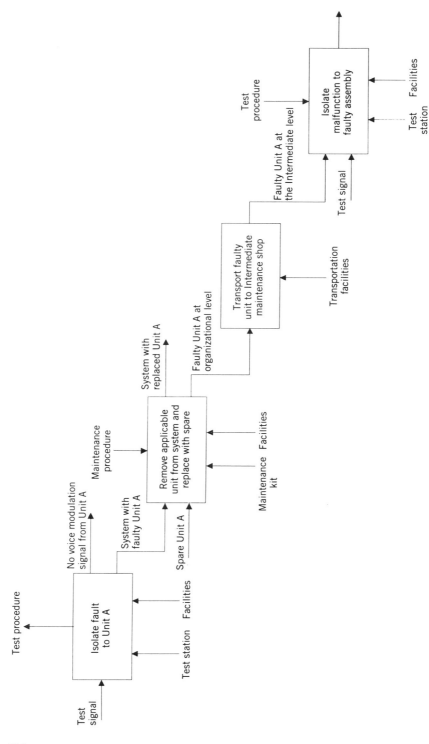

Figure 6.12. An IDEF₀ model (example).

In some instances, a function might best be accomplished through the utilization of equipment. In other cases, the combination of a computer and software might be appropriate. In any event, the system may be broken down into components such as those illustrated in Figure 6.13. However, the identification of these components must evolve through the process of functional definition, the analysis of inputs/outputs/resource requirements for each function, and the packaging of functions into units of equipment, software, and so on. Figure 6.14 conveys the concept, in an abbreviated form, leading from system definition at the left to the identification of three equipment items at the right—Units "A," "B," and "C." The characteristic of "functional packaging" is the goal here and it is particularly important in the design for maintainability.

With the system structure now defined, the question is—what requirements should be specified for the various system components (e.g., Units "A," "B," and "C" in Figure 6.14) such that, when combined, they will fulfill the requirements for the overall system? As overall system performance is dependent on the individual performance of each of its components, it is necessary to establish requirements at the component level to ensure that the system requirements will be met. The alternative is to procure and/or design the necessary system components (with characteristics selected by individual designers and usually based on performance only—not

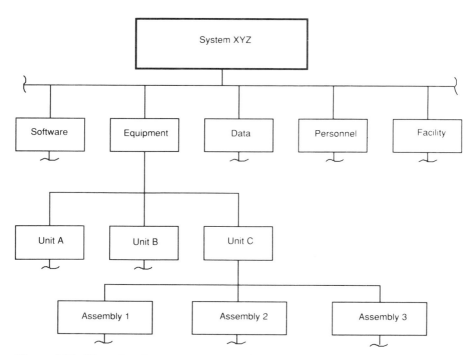

Figure 6.13. Hierarchy of system components (examples). (*Source:* Blanchard, B. S., *System Engineering Management,* John Wiley & Sons, N.Y., 1991, p. 42).

reliability, maintainability, supportability, etc.), combine these into a system structure, and *hope* that the results will be compatible in terms of meeting consumer requirements.

In other words, there are two basic approaches! The first is a "top-down" approach, working from the system level down to the extent required to provide some input controls on design. The second is a "bottom-up" approach, selecting and combining "favored" components, and *hoping* that the product output will fulfill the need in an effective and efficient manner. Historically, the bottom-up approach has been employed in many instances and the results have not been very satisfactory, particularly with regard to the lack of consideration of design characteristics such as reliability, maintainability, and so on.

As the prime objective here is to design and develop a system that will meet the specified requirements in an effective manner, it is important that a methodical and organized top-down approach be assumed from the beginning. Leaving things to "chance" often results in costly "downstream" modifications to get the system to perform in the manner originally intended. This top-down approach involves the allocation, or apportionment, of requirements at the system level down to the various applicable components of the system. These requirements, stated both qualitatively and quantitatively, are then included in second-tier specifications used in the procurement and/or acquisition of those components, that is, the development, process, product, and/or material specifications noted in Figures 2.1 and 2.16 (Chapter 2). The basic question is—what requirements should be included in the specification for the development of Unit "B" of System XYZ (refer to Figure 6.14)?

In the allocation of requirements, it is necessary to perform a reliability allocation first and then a maintainability allocation, since the latter really depends on reliability factors.

6.2.1 Reliability Allocation [2]

After an acceptable reliability factor has been established for the system, it must be allocated (or broken down) among the various subsystems, units, assemblies, and so on. The allocation commences with the generation of a reliability block diagram, a simplified version of which is presented in Figure 6.15. The block diagram is an extension of the functional analysis, and employs some of the component series–parallel relationships illustrated in Figures 4.3–4.5. The intent is to develop a reasonable approximation, early in the life cycle, of those elements or items that must function for the successful operation of the system.

Referring to the figure, there is a progressive expansion from the system level down. The reliability requirement for the system (e.g., R, MTBF, λ) is specified for the entire network identified at Level 1, and an individual requirement is specified for each individual block in the network. For instance, the reliability of block

---

[2] An in-depth coverage of reliability allocation is not intended, nor practical, within the confines of this text. A review of additional literature is recommended. Refer to Appendix H.

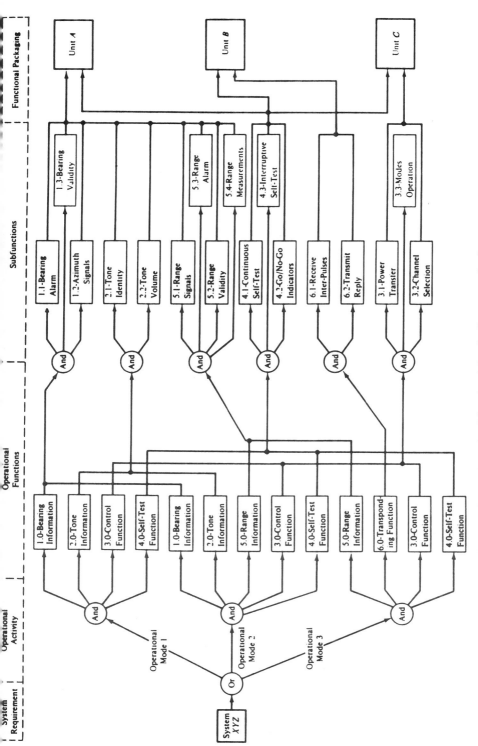

Figure 6.14. An abbreviated functional flow leading to a packaging concept.

189

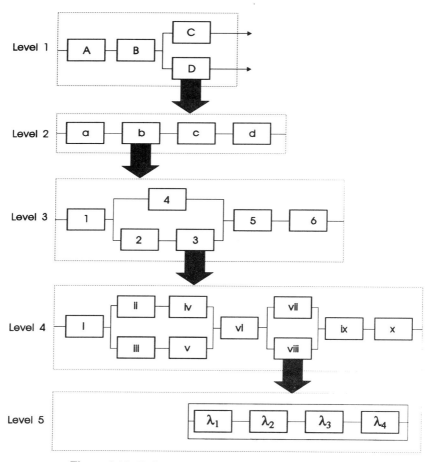

Figure 6.15. Reliability block diagram approach (*N* levels).

D, function x, may be expressed as the probability of survival of 0.95 for a given 4-hour mission profile. Similar requirements are specified for blocks A, B, and C. These, when combined, will indicate the reliability for the system.

Block diagrams are generated to cover each of the major functions identified in Figure 6.14. Success criteria (GO/NO-GO parameters) are established and failure rates (λs) are applied to all of the blocks, the combining of which provides an overall factor for a series of blocks constituting a function or a subfunction. Depending on the function, one or more of these diagrams can be related to a physical entity such as Unit "A," or an assembly of Unit "A." The failure-rate information provided at the unit/assembly level represents a reliability design goal which, in turn, represents the anticipated frequency of corrective maintenance, which is utilized in the determination of maintainability design requirements.

The approach used in determining failure rates may vary depending on the maturity of system definition. Failure rates may be derived from direct field and/or testing experience covering similar items, prediction reports from earlier configurations that are similar in nature, and/or engineering estimates based on judgment. In some instances, weighting factors are used to compensate for system complexity and environmental stresses. When accomplishing reliability allocation, the following steps are considered appropriate:

1. Evaluate the system function flow diagram(s) and identify areas where design is known and failure-rate information is available or can be readily assessed. Assign the appropriate factors and determine their contribution to the top-level system reliability requirement. The difference constitutes the portion of the reliability requirement which can be allocated to the other areas.

2. Identify the areas that are new and where design information is not available. Assign complexity weighting factors to each functional block. Complexity factors may be based on an estimate of the number and relationship of parts, the equipment duty cycle, the mode of operation and criticality of the path, whether an item will be subjected to temperature extremes, and so on. That portion of the system reliability requirement which is not already allocated to the areas of known design is allocated using the assigned weighting factors.

The end result should constitute a series of lower-level values which can be combined to reflect the system reliability initially specified as part of the operational requirements (see Chapter 5).

A reliability mathematical model is developed to relate the individual "block" reliability to the reliabilities of its constituent blocks or elements. The procedure simply consists of developing a mathematical expression (or series of expressions) that represents the probability of survival for a small portion of the proposed configuration. Multiple applications of this process will eventually reduce a relatively complex system to an equivalent serial configuration. It is then possible to represent the system with a single probability statement. Some of the mathematical relationships used in this process are described in Section 4.1.

When allocating a system-level requirement, one should construct a simplified functional breakdown, as illustrated in Figure 6.16. This evolves from Figure 6.14, and it is assumed in this example that a MTBF of 450 hours is required for System "XYZ." Initially, failure rates are identified for items of known design and are deducted from the overall system requirement. A complexity factor may be assumed for each of the remaining items. These factors are then used to apportion failure rates to the next lower level and on down. As a check, failure rates at the assembly level are totaled to obtain the unit failure rate, and the unit failure rates support the system failure rate (note that Units "A," "B," and "C" represent a series operation). The MTBF is usually assumed to be the reciprocal of the failure rate, and the reliability (R) of the system can be determined from Equation (4.3).

Referring to Figure 6.16, a reliability block diagram showing the functional rela-

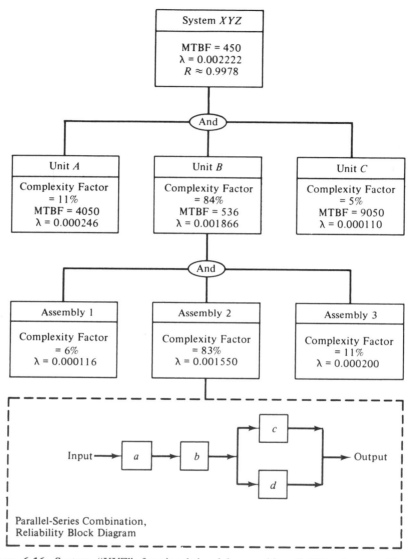

Figure 6.16. System "XYZ" functional breakdown with reliability factors. (*Source:* Blanchard, One through Twelve in *Logistics Engineering and Management,* 4th ed., Blanchard ed., © 1992, Fig. 5-11, p. 141. Reprinted by permission of Prentice-Hall, Inc., Englewood Cliffs, N.J.).

tionships of four elements (*a, b, c,* and *d*) illustrates the makeup of Assembly 2. The mathematical expression is

$$R_{\text{assy.2}} = (R_a)(R_b)[R_c + R_d - (R_c)(R_d)] \qquad (6.1)$$

Using the general expression in Equation (4.3), reliability factors can be determined and, in this instance, the total failure rate for Assembly 2 should not exceed 0.001550.

Review of the equipment breakdown configuration illustrated in Figure 6.16 indicates a top-level system requirement supported by factors established at the unit level and on down. Unless otherwise specified, the requirements at the unit level may serve as design criteria. For instance, the engineer responsible for Unit "B" shall design Unit "B" such that the failure rate shall not exceed 0.001866. This is the value that will be included in the development specification, or in the product specification if the item is to be purchased "off-the-shelf!" As the design progresses, reliability predictions are accomplished and the predicted value is compared against the requirement of 0.001866. If the predicted value does not meet the requirement (i.e., higher failure rate or lower MTBF), the design configuration must be reviewed for reliability improvement and design changes are implemented as appropriate. The allocated factors not only provide the designer with a reliability criterion goal, but serve as an indicator of the expected number of corrective maintenance actions resulting from anticipated equipment failure.

6.2.2. Maintainability Allocation

The process of translating system maintainability requirements (e.g., MTBM, $\overline{Mct}$, $\overline{Mpt}$, and MMH/OH) into lower-level design criteria is accomplished through maintainability allocation. The allocation requires the development of a simplified functional breakdown of system elements as illustrated in Figure 6.16. This hierarchial breakdown is based on the maintenance concept, functional analysis data, and a description of the basic repair policy—the levels of maintenance and whether a system is to be repaired through the replacement of a unit, an assembly, or a part.

For the purposes of illustration, it is assumed that System XYZ must be designed to meet an inherent availability (A_i) requirement of 0.9989, a MTBF of 450 hours, and a MMH/OH (for corrective maintenance) of 0.2, and a need exists to allocate $\overline{Mct}$ and MMH/OH to the unit and assembly levels. The $\overline{Mct}$ value is

$$\overline{Mct} = \frac{\text{MTBF}(1 - A_i)}{A_i} \qquad (6.2)$$

or

$$\overline{Mct} = \frac{450(1 - 0.9989)}{0.9989} = 0.5$$

Thus, the system's $\overline{Mct}$ requirement is 0.5 hour, and this requirement must be allocated to Units "A," "B," and "C," and the assemblies within each unit. The allocation process is facilitated through the use of a format similar to that illustrated in Table 6.3.

TABLE 6.3   System XYZ Maintainability Allocation (Unit Level)

1	2	3	4	5	6	7
Item	Quantity of Items per System (Q)	Failure Rate $(\lambda) \times 1000$ hr	Contribution of Total Failures $C_f = (Q)(\lambda)$	Percent Contribution $C_p = C_f/\Sigma C_f \times 100$	Estimated Corrective Maint. Time Mct_i (hr)	Contribution of Total Corrective Maint. Time $C_t = (C_f)(Mct)$
Unit "A"	1	0.246	0.246	11	0.9	0.221
Unit "B"	1	1.866	1.866	84	0.4	0.746
Unit "C"	1	0.110	0.110	5	1.0	0.110
Total			$\Sigma C_f = 2.222$	100		$\Sigma C_t = 1.077$

$\overline{M}ct$ for System "XYZ" $= \dfrac{\Sigma C_t}{\Sigma C_f} = \dfrac{1.077}{2.222} = 0.485$ hour (requirement: 0.5 hour)

Referring to Table 6.3, each item category and the quantity (Q) of like items per system are indicated. Allocated reliability factors (from Figure 6.16) are specified in column 3, and the degree to which the failure rate of each unit contributes to the overall failure rate (represented by C_f) is entered in column 4. The average corrective maintenance time for each unit is estimated in column 6. These factors represent the composite of the average times that it takes to repair the unit, and are often based on the use of complexity factors since the inherent characteristics of equipment design are not usually known at this early point in the life cycle. As a goal, the item that contributes the highest percentage to the expected total number of failures (Unit "B" in this case) should require a low $\overline{M}ct$, and those items that are not expected to fail very often may require a higher $\overline{M}ct$. On certain occasions, however, the design costs associated with obtaining a low $\overline{M}ct$ for a complex item may lead to a modified approach, which is feasible as long as the end result (i.e., $\overline{M}ct$ at the system level) falls within the overall quantitative requirement. Note that there are many possible solutions. The analyst should attempt to analyze each possible solution, and choose that which is most cost-effective![3]

The estimated value for C_t for each unit is entered in column 7, and the sum of the contributions for all units can be used to determine the overall system $\overline{M}ct$ as

$$\overline{M}ct = \frac{\Sigma C_t}{\Sigma C_f} = \frac{1.077}{2.222} = 0.485 \qquad (6.3)$$

In Table 6.3, the calculated $\overline{M}ct$ for the system is within the requirement of 0.5 hour.

Once the allocation is accomplished at the unit level, the resultant $\overline{M}ct$ values can be allocated to the next lower equipment indenture item. For instance, the 0.4 hour $\overline{M}ct$ value for Unit B can be allocated to Assemblies "1"–"3," and the procedure for allocation is the same. An example of the allocated values for the assemblies of Unit "B" is included in Table 6.4.

TABLE 6.4   Unit "B" Maintainability Allocation (Assembly Level)

1	2	3	4	5	6	7
Assembly 1	1	0.116	0.116	6%	0.5	0.058
Assembly 2	1	1.550	1.550	83%	0.4	0.620
Assembly 3	1	0.200	0.200	11%	0.3	0.060
Total			1.866	100%		0.738

$$\overline{M}ct \text{ for Unit "B"} = \frac{\Sigma C_t}{\Sigma C_f} = \frac{0.738}{1.866} = 0.395 \text{ hour (Requirement: 0.4 hour)}$$

[3] In any event, the maintainability parameters are dependent upon reliability parameters. Also, it will frequently occur that reliability allocations are incompatible with maintainability allocations (or vice versa). Hence, a close feedback relationship between these activities is mandatory.

The $\overline{M}ct$ value covers the aspect of "elapsed" or "clock" time for restoration actions. Sometimes this factor, when combined with a reliability requirement, is sufficient with regard to establishing the necessary maintainability characteristics in design. On other occasions, specifying $\overline{M}ct$ by itself is not adequate since a number of different design approaches may meet the $\overline{M}ct$ requirement, but not necessarily in a cost-effective manner. Meeting a $\overline{M}ct$ requirement may result in an increase in the skill levels of personnel accomplishing maintenance actions, increasing the quantity of personnel in the performance of maintenance functions, and/or incorporating some form of automation for manually performed activities. In each instance, there are costs! Thus, one may wish to impose additional constraints on the design, such as the skill levels of personnel at each maintenance level (e.g., "the design shall be such that all maintenance actions at the organization level can be performed by a high school graduate with basic entry skills"), and the maintenance man-hours per operating hour (MMH/OH) for significant equipment items. In other words, a

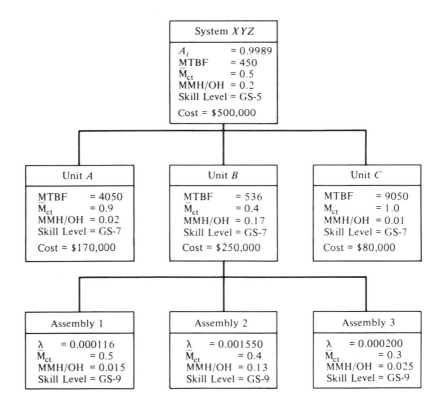

Figure 6.17. The allocation of System "XYZ" requirements. (*Source:* Blanchard, One through Twelve in *Logistics Engineering and Management,* 4th ed., Blanchard ed., © 1992, Fig. 5-12, p. 145. Reprinted by permission of Prentice-Hall, Inc., Englewood Cliffs, N.J.).

requirement may dictate that an item be designed such that it can be repaired within a specified elapsed time with a given quantity of personnel possessing skills of a designated level. This will influence the design in terms of accessibility, packaging schemes, handling requirements, diagnostic provisions, and so on—that is, the design for maintainability.

The maintainability measure of MMH/OH is a function of the frequency of maintenance (corrective and/or preventive maintenance) and the complexity of tasks in the performance of maintenance. The system-level measure is allocated on the basis of system operational requirements (i.e., system operating hours), the anticipated quantity of maintenance actions, the elapsed time per maintenance action, and the number of personnel assigned in each instance. Experience data are used where possible. Following the completion of the allocation process for each applicable indenture level of the system, all values are included in a functional breakdown, as illustrated in Figure 6.17.

The allocation of system requirements must be represented within the context of the defined maintenance concept described in Chapter 5. In addition to considering the availability of personnel, one must also consider the availability of spare parts, test equipment, facilities, and so on. This is particularly true for systems with a specified quantitative operational availability (A_o) requirement or equivalent. Reliability and maintainability factors need to be related to those factors pertaining to the elements of logistic support. For example, inherent maintenance times ($\overline{M}ct$ values) may be extended somewhat, providing that there is an available spare part, and maintenance times can be reduced through the utilization of automatic test equipment. Basically, the establishment of maintainability requirements for design must consider reliability factors, human factors, logistic support factors, economic factors, and so on.

6.3 DESIGN TRADE-OFF TECHNIQUES

With the basic requirements for the various components of the system having been specified through the allocation process, it is now appropriate to identify possible alternative design approaches for meeting these requirements. Initially, it is desirable to list *ALL* candidate approaches to ensure against inadvertent omissions, and then eliminate those candidates that are clearly unattractive for one reason or another, leaving only a few for evaluation. Those few candidates are then analyzed through the accomplishment of a trade-off study, with the objective of selecting a preferred approach.

In the accomplishment of a trade-off study, there is an overall process that is intuitive! The analyst needs to first define the problem in terms of clarifying the objectives of the analysis, identify the ground rules and constraints, describe the feasible alternatives, and define an approach for problem resolution. Next, the analyst needs to select the appropriate evaluation criteria (define and prioritize the measures to be applied in the evaluation—refer to the evaluation parameters in Figure 1.6), describe the input data needs, identify the potential areas of risk and uncertainty, and so on. The next step is to select the appropriate methods or techniques to facilitate the evaluation process (the analytical approach and the model to be used). With the analytical approach and the tool(s) defined, the analyst must proceed with the data collecting process, using a combination of allocation results, predictions, forecasts, and/or historical data from an existing data bank. The applicable candidate approaches are evaluated, sensitivity analyses are accomplished, areas of risk are defined, and a recommended approach is proposed. This overall process, which is an inherent aspect of system engineering analysis, is illustrated in Figure 6.18.[4]

Referring to the figure, a key factor in the performance of any trade-off study is the selection and application of the appropriate analytical techniques and the model(s) to be used in the evaluation process. The objective is to select a model (or a series of models) that will enable the analyst to develop a simplified representation of the real world, abstracting the features of the situation relative to the problem being addressed. The model is a tool employed by the analyst to assess the likely consequences of the various courses of action being examined. It must be adapted to the problem at hand, and the output must be oriented to the selected evaluation criteria. The model, in itself, is not the decision maker, but a tool that provides the necessary data in a timely manner in support of the decision-making process.

Throughout the system design and development process, a variety of problems need to be addressed. For example, trade-off studies are accomplished during the

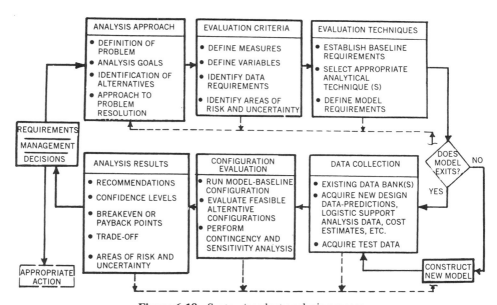

Figure 6.18. System/product analysis process.

[4] A more in-depth description of the system engineering analysis process is included in Blanchard, B. S., *System Engineering Management*, John Wiley & Sons, 1991.

conceptual design phase in the evaluation and selection of alternative technology applications. Further, the analyst may wish to look at alternative system utilization profiles, equipment distribution schemes, and the two-level versus three-level maintenance approach and alternative repair policies described in Chapter 5. As the design progresses, it is appropriate to evaluate alternative equipment/software packaging schemes, the degree of built-in versus external test provisions, and the selection and application of different components from alternative suppliers. Additionally, one may wish to evaluate alternative approaches to production—different manufacturing processes, automation versus the accomplishment of tasks manually, and so on.

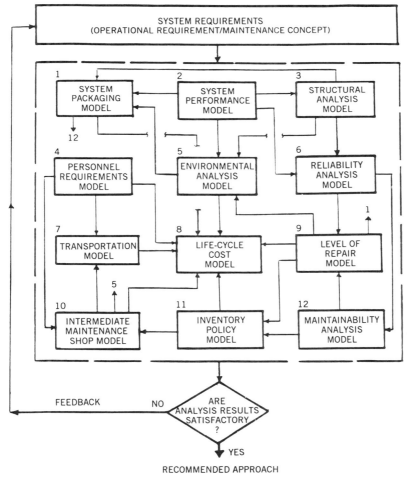

Figure 6.19. The integration of analytical models (example).

In any event, many different categories of problems need to be addressed and, as a result, the process of selecting tools to facilitate the evaluation process may vary from one situation to the next. Whatever the case, however, the utilization of these tools (or models) must be considered on an "integrated" basis. An integrated database must be established such that the output from one analysis (i.e., modeling exercise) can be fed directly as an input to another analysis effort. In other words, the models must "talk" to each other as reflected in Figure 6.19.

Regardless of the nature of the design trade-off study, maintainability must be a consideration in the process. Although many different factors are involved in the evaluation of alternatives, maintainability is an important characteristic of design and often has a significant impact on the results. As a means of illustrating how maintainability impacts the process, five abbreviated examples are presented:

1. *Level of repair analysis (Example 1)*—compares the options of designing an item for repair at the intermediate level of maintenance, for repair at the manufacturer/supplier/depot level, or for discard at failure. This type of analysis is initially accomplished to help in defining repair policies as part of the maintenance concept (refer to Chapter 5), and is used later in defining specific logistic support resource requirements.

2. *Evaluation of alternative design configurations using multiple criteria (Example 2)*—involves the comparison of three alternative design configurations using multiple evaluation criteria, weighting factors to establish levels of importance, and so on.

3. *Reliability and maintainability analysis (Example 3)*—compares three alternatives in terms of reliability and maintainability, given a specific system availability requirement.

4. *Evaluation of alternative maintenance policies in a manufacturing environment (Example 4)*—compares an in-house maintenance policy, a policy supported by one supplier, and a policy supported by two suppliers.

5. *Life-cycle cost analysis (Example 5)*—compares two alternative system design configurations, illustrating the life-cycle cost analysis process. This includes the steps of problem definition, definition of operational requirements and maintenance concept, cost breakdown structure (CBS), cost estimation, generation of cost summaries, break-even analysis, and so on. Application of the life-cycle cost analysis process is appropriate throughout all phases of system design and development.

6.3.1   Level of Repair Analysis (Example 1)

Referring to Section 5.4, defining the system maintenance and support concept requires the early establishment of repair policies. Given a three-level maintenance concept, one of the decision factors relates to the question of whether an item should be designed to be "repairable" or to be "discarded" in the event of failure? If it is designed to be repairable, at what level of maintenance should the repair be accomplished—at the intermediate level or at the manufacturer or depot level? Although these questions can be applied to any component of the system (e.g., equipment, unit, assembly, subassembly, module, and/or an element of software), this case study applies to the design of Assembly "A-1". This assembly is one of 15 assemblies in Unit "B" of System "XYZ." The objective is to evaluate design alternatives for the assembly on the basis of life-cycle economic criteria.

1. *The analysis process.* The accomplishment of a level of repair analysis requires that the item being evaluated be presented in terms of a system operational requirement, a maintenance concept, and a program plan. In this instance, it is assumed that System "XYZ" is installed in an aircraft. When a maintenance action is required, there is a built-in test capability within the aircraft that allows one to isolate the fault to Unit "A," Unit "B," or Unit "C." The applicable unit is removed, replaced with a spare, and the faulty item is transported to the intermediate-level maintenance shop for corrective maintenance. In the maintenance shop, fault isolation is accomplished within the unit to the assembly level. The faulty assembly is removed, replaced with a spare, and the unit is checked out and returned to the inventory as an operational spare. The basic question pertains to the disposition of the assembly.

In approaching this problem, the first step is to accomplish a level of repair analysis on Assembly "A-1" as an individual entity. Subsequently, the results of this part of the analysis need to be viewed in the context of the whole, that is, the results of similar analyses involving Assembly "A-2," "A-3," . . . , and "A-15," and the applicable assemblies of Unit "A" and Unit "C." There is usually a feedback effect between the individual assembly analysis, the unit-level analysis, and the overall maintenance concept for the system as a whole.

In completing the level of repair analysis Assembly "A-1," the following information is provided:

a. System "XYZ" is installed in each of 60 aircraft which are distributed equally at five operating sites over an eight-year time period. System utilization is on the average of 4 hours per day, and the total operating time for all systems is 452,600 hours.

b. As stated above, System "XYZ" includes three units—Unit "A," Unit "B," and Unit "C." Unit "B" includes 15 assemblies, one of which is Assembly "A-1." The estimated acquisition cost for Assembly "A-1" (to include design and development cost and production cost) is $1700 each if the assembly is designed to be repairable, and $1600 each if the assembly is to be designed to be discarded at failure. The design for repairability considers the incorporation of diagnostic provisions, accessibility, internal labeling, and so on, which is apt to cost more in terms of design and production costs.

c. The estimated failure rate (or corrective maintenance rate) of Assembly "A-1" is 0.00045 failures per hour of system operation. When failures occur, repair is accomplished by a single technician who is assigned for the duration of the allocated active maintenance time. The estimated $\overline{M}ct$ is 3 hours. The loaded labor rate is $20 per labor hour for intermediate-level maintenance and $30 per labor hour for depot-level maintenance.

d. Supply support includes three categories of cost—the cost of spare assemblies in inventory, the cost of spare components to enable the repair of faulty assemblies, and the cost of inventory management and maintenance. Assume that 5 spare assemblies will be required in inventory when maintenance is accomplished at the intermediate level, and that 10 spare assemblies will be required when maintenance is accomplished at the depot level. For component spares, assume that the average cost of material consumed per maintenance action is $50. The estimated cost of inventory maintenance is assumed to be 20% of the inventory value (the summation of the costs for assembly and component spares).

e. When assembly repair is accomplished, special test and support equipment is required for fault diagnosis and assembly checkout. The cost per test station is $12,000, which includes acquisition cost and amortized maintenance cost. This cost is that part of the total cost that is attributed to the maintenance requirement for Assembly "A-1," and there are five test stations required for intermediate-level maintenance.

f. Transportation and handling cost is considered as being negligible when maintenance is accomplished at the intermediate level. However, assembly maintenance accomplished at the depot level will involve an extensive amount of transportation. For depot maintenance, assume $150 per 100 pounds per one-way trip (independent of distance), and that the packaged assembly weighs 20 pounds.

g. The allocation for Assembly "A-1" relative to maintenance facility cost is categorized in terms of an initial fixed cost, and a sustaining recurring cost proportional to facility utilization requirements. The initial fixed cost is $1000 per installation, and the assumed usage cost allocation is $1.00 per direct maintenance labor hour at the intermediate level and $1.50 per direct labor hour at the depot level.

h. Technical data and maintenance software requirements constitute the maintenance instructions to be included in the technical manuals to support assembly repair activities, and the failure reporting and maintenance data covering each maintenance action in the field. Assume that the cost for preparing and distributing maintenance instructions (and supporting computer software) is $1000, and that the cost for field maintenance data is $25 per maintenance action.

i. There will be some initial formal training costs associated with maintenance personnel when considering the assembly repair option. Assume 30 student-days of formal training for the intermediate level of maintenance (for the five sites in total) and 6 student-days for depot level maintenance. The cost of training is $150 per student-day. The requirement for replenishment training as a result of attrition or turnover is considered as being negligible.

j. As a result of maintenance, there will be a requirement for disposal and/or the recycling of material. The assumed disposal cost is $20 per assembly and $2 per component. Possible revenues gained from recycling are not considered herein.

The objective is to evaluate Assembly "A-1" based on the information provided. Should Assembly "A-1" be designed for (1) repair at the intermediate level of maintenance, (2) repair at the depot level of maintenance, or (3) discard at failure?

2. *The analysis results.* Figure 6.20 presents a worksheet with the results from the evaluation of Assembly "A-1." Based on the information shown, it is recommended that the assembly be *repaired at the depot level of maintenance.*

Prior to making a final decision, however, one should review the data in Figure 6.20 in terms of "high-cost" contributors and the sensitivities of various input factors. Some of the initial assumptions may have a great impact on the analysis results and, perhaps, should be challenged. Also, the analyst may wish to review the source of prediction data covering reliability, maintainability, and some of the input cost factors.

Given that the repair policy decision for Assembly "A-1" is verified in terms of its evaluation in an "isolated" sense (i.e., a decision has been made relative to the results of the individual analysis in Figure 6.20), it is essential that this decision be reviewed in context with other assemblies of System "XYZ" and with the maintenance concept. Figure 6.21 reflects the results of individual level of repair analyses accomplished for each of the major assemblies in Unit "B." The same approach used for Assembly "A-1" is used for the evaluation of Assemblies "A-2" through "A-15."

Referring to Figure 6.21, there are several major choices: (1) adopt the individual repair policy for each assembly (i.e., an overall "mixed" policy), or (2) adopt a uniform overall policy for *all* assemblies based on the lowest total policy cost (i.e., repair at depot). In this instance, it appears that the "mixed" policy ($1,793,741) is preferred, based on the assumptions made. However, before making a final decision, each option must be reviewed in terms of possible "feedback" effects that may occur, life-cycle cost implications, and associated risks. For example, if the "repair-at-intermediate," "repair-at-depot," or "discard at failure" policy is assumed (for technical, political, social, environmental, or other reasons), the analyst needs to challenge some of the input assumptions leading to the data in Figure 6.20. The initially assumed test equipment configuration will likely be significantly different if a "discard at failure" decision is made, versus one of the repair options.

Figure 6.22 illustrates the basic process that is applicable in the performance of a level of repair analysis, whether the analysis is accomplished in conceptual design or during the detail design and development phase. The only difference is the depth and availability of the input data. The results of such an effort will have a great impact on the maintainability characteristics in design, that is, the proposed system/

Evaluation Criteria	Repair at Intermediate Cost ($)	Repair at Depot Cost ($)	Discard at Failure Cost ($)	Description and Justification
1. Estimated acquisition cost for Assembly A-1 (to include design and development, production cost)	1,700/Assembly or 102,000 (47.8%)	1,700/Assembly or 102,000 (54.7%)	1,600/Assembly or 96,000 (19.5%)	Acquisition costs based on 60 systems. Assembly design and production costs are less in the discard case (simplified configuration).
2. Maintenance labor cost	12,240 (5.7%)	18,360 (9.8%)	Not applicable	Based on 452,600 hours of operation and a maintenance rate of 0.00045, the estimated quantity of maintenance actions is 204. When repair is accomplished, one (1) technician is assigned on a full-time basis. The Mct is 3 hours. The labor rate is $20/hour for intermediate and $30/hour for depot.
3. Supply support – spare assemblies	8,500 (4%)	17,000 (9.1%)	326,400 (66.4%)	For intermediate maintenance, 5 spare assemblies are required to compensate for turnaround time, the maintenance que, etc. 10 spares are required for depot maintenance. 100% spares are required for the discard case.
4. Supply support – spare components	10,200 (4.8%)	10,200 (5.5%)	Not applicable	Assume $50 per maintenance action
5. Supply support – inventory maintenance	3,740 (1.8%)	5,440 (2.9%)	65,280 (13.3%)	Assume 20% of the inventory value (spare assemblies and spare components)
6. Special test and support equipment	60,000 (28.1%)	12,000 (6.4%)	Not applicable	Special test equipment is required in the repair case. The acquisition cost is $12,000 per installation. There are five (5) installations at intermediate and one(1) at depot.
7. Transportation and handling	Negligible	12,240 (6.6%)	Not applicable	Transportation costs at the intermediate level are negligible. For depot maintenance, assume 408 one-way trips at $150/100 pounds. One assembly weighs 20 pounds.
8. Maintenance training	4,500 (2.1%)	900 (0.5%)	Not applicable	Assume 10 students for 3 days at $150 per student day for intermediate, and 2 students for 3 days at $150 per student day for depot.
9. Maintenance facilities	5,612 (2.6%)	1,918 (1%)	Not applicable	Assume $1.00 per direct maintenance manhour for intermediate, and $1.50 per direct manhour for depot. Also, assume an initial fixed cost of $1,000 per installation.
10. Technical data	6,100 (2.9%)	6,100 (3.3%)	Not applicable	For repair case, assume $1,000 for the cost of preparation of maintenance instructions. Also, assume $25 per maintenance action for maintenance data.
11. Disposal	408)0.2%)	408 (0.2%)	4,080 (0.8%)	Assume $20 per assembly and $2 per component as the cost of disposal.
Total Estimated Cost	$213,300	$186,566	$491,760	

Figure 6.20. Repair versus discard evaluation (Assembly "A-1").

| Assembly Number | Repair Policy | | | | Decision |
	Repair at Intermediate	Repair at Depot	Discard at Failure	"Mixed" Policy	
A-1	$213,300	$186,566	$491,760	$186,566	Repair-Depot
A-2	130,800	82,622	75,440	75,440	Discard
A-3	215,611	210,420	382,452	210,420	Repair-Depot
A-4	141,633	162,912	238,601	141,633	Repair-Intermediate
A-5	132,319	98,122	121,112	98,122	Repair-Depot
A-6	112,189	96,938	89,226	89,226	Discard
A-7	125,611	142,206	157,982	125,611	Repair-Intermediate
A-8	99,812	131,413	145,662	99,812	Repair-Intermediate
A-9	128,460	79,007	66,080	66,080	Discard
A-10	167,400	141,788	314,560	141,788	Repair-Depot
A-11	185,850	142,372	136,740	136,740	Discard
A-12	135,611	122,453	111,502	111,502	Discard
A-13	105,667	113,775	133,492	105,667	Repair-Intermediate
A-14	111,523	89,411	99,223	89,411	Repair-Depot
A-15	142,119	120,813	115,723	115,723	Discard
Policy Cost	$2,147,905	$1,920,808	$2,679,555	$1,793,741	Repair-Depot

Figure 6.21. Summary of repair level decisions (all Unit "B" assemblies).

equipment/software packaging schemes or extent of the degree of modularization incorporated, the level of diagnostics or the degrees of built-in versus external test required, the personnel quantities and skill-level requirements, and so on. Also, the maintenance frequency, elapsed time, and MMH/OH factors will be influenced significantly by the decisions evolving from the level of repair analysis. While the decision in this example was based primarily on economic factors, "noneconomic" screening factors may prevail in some level of repair analyses.

6.3.2. Evaluation of Alternative Design Configurations Using Multiple Criteria (Example 2)

Company ABC is responsible for the design and development of a large system which, in turn, is comprised of a number of subsystems. Subsystem XYZ is to be procured from an outside supplier, and there are three different configurations being

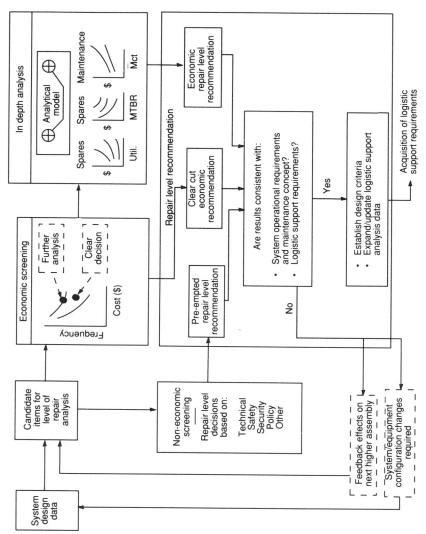

Figure 6.22. Level of repair analysis process. (*Source:* Blanchard, B. S., *System Engineering Management*, John Wiley & Sons, N.Y., 1991, p. 334).

Item	Evaluation parameter	Weighting factor	Configuration A		Configuration B		Configuration C	
			Base rate	Score	Base rate	Score	Base rate	Score
1	Performance – input, output, accuracy, range, compatibility	14	6	84	9	126	3	42
2	Operability – simplicity and ease of operation	4	10	40	7	28	4	16
3	Effectiveness – Ao, MTBM, Mct, Mpt, MDT, MMH/OH	12	5	60	8	96	7	84
4	Design characteristics – reliability, maintainability, human factors, supportability, producibility, interchange-ability	9	8	72	6	54	3	27
5	Design data – design drawings, specifications, logistics data, operating and maintenance procedures	2	6	12	8	16	5	10
6	Test aids – common and standard test equipment, calibration standards, maintenance and diagnostic computer programs	3	5	15	8	24	3	9
7	Facilities and utilities – space, weight, volume, environment, power, heat, water, air conditioning	5	7	35	8	40	4	20
8	Spare/repair parts – part type and quantity, standard parts, procurement time	6	9	54	7	42	5	30
9	Flexibility/growth potential – for reconfiguration, design change acceptability	3	4	12	8	24	6	18
10	Schedule – research and development, production	17	7	119	8	136	9	153
11	Cost – life cycle (R & D, investment, O & M)	25	10	250	9	225	5	125
Subtotal				753		811		534
Derating factor (development risk)				113 15%		81 10%		197 20%
Grand Total		100		640		730		427

Figure 6.23. Evaluation summary (three alternatives). (*Source:* Blanchard, B. S., *System Engineering Management,* John Wiley & Sons, N.Y., 1991, p. 341).

evaluated for selection. Each of the candidate configurations represents an existing design, with some redesign necessary for compatibility with the requirements of the new system. The evaluation criteria include such parameters as performance, operability, effectiveness, design characteristics, schedule, and cost. Both qualitative and quantitative considerations are covered in the evaluation process, and maintainability characteristics in design are inherent within several of the categories noted. Figure 6.23 identifies the evaluation parameters, with items 3, 4, 6–9, and 11 being of particular interest for maintainability.

1. *The analysis process.* The analyst commences with the development of a list of evaluation parameters, as depicted in Figure 6.23. In this instance, there is no single parameter (or figure of merit) that is appropriate by itself, but there are 11 factors that must be considered on an integrated basis. Given the evaluation parameters, the next step is to determine the level of importance of each. Quantitative weighting factors from zero to 100 are assigned to each parameter in accordance with the degree of importance. The Delphi method, or some equivalent evaluation technique, may be used to establish the weighting factors. The sum of all weighting factors is 100.

For each of the 11 parameters identified in Figure 6.23, the analyst may wish to develop a special checklist, including criteria against which to evaluate the three proposed configurations. For instance, a sample checklist supporting Item 11 ("cost–life cycle") is presented in Figure 6.24. Using this and similar checklists for each factor, the three supplier proposals are evaluated independently. Base rating values from zero to 10 are applied according to the degree of compatibility with the

Rating (Points)	Evaluation Criteria – Item II – Cost-life cycle*
20–25	The supplier has justified his design on the basis of life-cycle cost, and has included a complete life-cycle cost analysis in his proposal (i.e., cost breakdown structure, cost profile, etc.).
15–19	The supplier has justified his design on the basis of life-cycle cost, bud did not include a complete life-cycle cost analysis in his proposal.
10–14	The supplier's design has not been based on life-cycle cost; however, he plans to accomplish a complete life-cycle cost analysis and has described the approach, model, etc., that he proposes to use in the analysis process.
5–9	The supplier's design has not been based on life-cycle cost, but he intends to accomplish a life-cycle cost analysis in the future. No description of approach, model, etc., was included in his proposal.
0–4	The subject of life-cycle cost (and its application) was not addressed at all in the supplier's proposal.

*Refer to Figure 6.23 for individual criteria

Figure 6.24. Checklist of evaluation criteria for supplier proposals (sample).

desired goals. If a "highly desirable" evaluation is realized, a rating of 10 is assigned.

The base rate values are multiplied by the weighting factors to obtain a score. The total score is then determined by adding the individual scores for each configuration. Since some redesign is required in each instance, a special derating factor is applied to cover the risk associated with the failure to meet a given requirement. The resultant values from the evaluation are summarized in Figure 6.23.

2. *The analysis results.* Referring to Figure 6.23, Configuration B represents the preferred approach based on the highest total score of 730 points. This configuration is recommended on the basis of its inherent features relating to performance, operability, effectiveness, design characteristics, design data, and so on.

6.3.3. Reliability and Maintainability Analysis (Example 3)

Referring to the hierarchy of system evaluation factors in Figure 1.6 (Chapter 1), maintainability is just one of the many design parameters that must be considered. Another, and closely related, parameter is reliability. These two parameters are often "traded-off" in order to meet a higher-level requirement such as "availability" (refer to Section 4.5, Equation 4.45). In this instance, there is a need to replace an equipment item as part of System "XYZ," and there are three alternatives being considered on the basis of reliability and maintainability, with the objective of meeting an overall inherent availability requirement.

1. *The analysis process.* There is a requirement to replace an existing equipment in the inventory (with a new item) for the purpose of improving operational effectiveness. The current need specifies that the equipment must operate 8 hours per day, 360 days per year, for 10 years. The existing equipment meets an availability of 0.961, a MTBF of 125 hours, and a $\overline{M}ct$ of 5 hours. The new system must meet an availability of 0.990, a MTBF greater than 300 hours, and a $\overline{M}ct$ not to exceed 5.0 hours.

An anticipated quantity of 200 equipments is to be procured. Three different alternative design configurations are being considered to satisfy the requirement, and each configuration constitutes a modification of the existing equipment.

Figure 6.25 graphically shows the relationships between inherent availability (A_i), MTBF, and $\overline{M}ct$, and illustrates the allowable area for trade-off. The selected configuration must reflect the reliability and maintainability characteristics represented by the shaded area. Obviously, the existing design is not compatible with the new requirement.

Further, the figure indicates that three alternative design configurations are being considered. Each configuration meets the availability requirement, with configuration A having the highest estimated reliability MTBF and configuration C reflecting the best maintainability characteristics with the lowest $\overline{M}ct$ value. The objective is to select the best of the three configurations on the basis of cost.

When considering cost, there are costs associated with research and development (R&D) activity, investment or manufacturing costs, and operation and maintenance

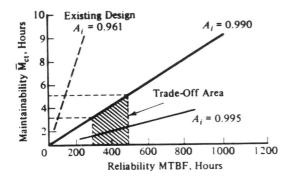

Conf.	A_i	MTBF	$\bar{M}_{ct}$
Existing	0.961	125	5.0
Alt. A	0.991	450	4.0
Alt. B	0.990	375	3.5
Alt. C	0.991	320	2.8

Other systems are eligible for consideration as long as the effectiveness parameters fall within the trade-off areas

Figure 6.25. Reliability—maintainability trade-off.

(O&M) costs. For instance, improving reliability and/or maintainability characteristics in design will result in an increase in R&D and investment (manufacturing) cost. In addition, experience has indicated that such improvements will usually result in lower O&M cost, particularly in the areas of maintenance personnel and support cost and the cost of spare/repair parts. Thus, initially the analyst looks only at these categories. If the final decision is close, it may be appropriate to investigate other categories. A summary of partial cost data is presented in Figure 6.26.

Referring to the figure, the delta costs associated with the three alternative equipment configurations are included for R&D and investment. Maintenance personnel and support costs, included as part of O&M cost, are based on estimated operating time for the 200 equipments throughout the required 10-year period of use (i.e., 200 equipments operating 8 hours per day, 360 days per year, for 10 years) and the reliability MTBF factor. Assuming that the average cost per maintenance action is $100, maintenance personnel and support costs are determined by multiplying this factor by the estimated quantity of maintenance actions, which is determined from total operating time divided by the MTBF value.

2. *The analysis results.* Configuration A satisfies the system availability, reliability, and maintenance requirements with the least life-cycle cost.

Category	Conf. A	Conf. B	Conf. C	Remarks
R&D cost				High-reliability parts, packaging, accessibility
•Reliability design	$ 17,120	$ 15,227	$ 12,110	
•Maintainability design	2,109	4,898	7,115	
Investment cost manufacturing (200 systems)	$3,422,400	$3,258,400	$3,022,200	$17,112/Equipment A; $16,292/Equipment B; $15,111/Equipment C
O&M cost				
•Maintenance personnel and support	$1,280,000	$1,536,000	$1,800,000	12,800 Maint. actions/Equipment A; 15,360 Maint. actions/Equipment B;
•Spare/repair parts	342,240	325,840	302,220	18,000 Maint. actions/Equipment C; 10% of manufacturing cost for spares
Total cost	$5,063,869	$5,140,365	$5,143,645	

Figure 6.26. Comparison of costs (three configurations).

6.3.4 Evaluation of Alternative Maintenance Policies in a Manufacturing Environment (Example 4)

As the new plant manager for Company "DEF," you are responsible for a factory full of capital equipment and you have been directed to implement a new TPM program. As an initial task, you have decided to evaluate three different maintenance-policy approaches. The intention is to select one of these approaches (as a baseline) for application through the next 10 years, followed by the implementation of a continuous process improvement effort directed toward productivity improvement. For the purposes of establishing a frame of reference, it is assumed that normal plant operation is 8 hours per day, 40 hours per week, and 52 weeks per year. Three candidates are being evaluated on an equivalent basis using economic criteria (the assumed discount rate is 12%).

1. *The analysis process.* The three alternative approaches are:
   a. Alternative "A" includes the development of an in-house maintenance capability. Based on past experience, an estimate has been made covering the costs associated with the accomplishment of both corrective maintenance and preventive maintenance for the next 10 years. The figures are:

Year 1—$70,000	Year 5—$105,000	Year 8—$135,000
Year 2—$82,500	Year 6—$115,000	Year 9—$153,000
Year 3—$91,000	Year 7—$120,000	Year 10—$165,000
Year 4—$95,500		

   b. Alternative "B" includes the accomplishment of all corrective and preventive maintenance activities by Supplier "GHI." It is estimated that the cost of corrective maintenance will be $2500 per unscheduled maintenance action (assume that $MTBM_u = 104$). The cost of accomplishing all preventive maintenance requirements will be $50,000 per year for the next 10 years.
   c. Alternative "C" includes the accomplishment of all corrective maintenance activities by Supplier "JKL" and all preventive maintenance activities by Supplier "MNO." From a proposal submitted by Supplier "JKL," the cost for corrective maintenance will be $2200 per unscheduled maintenance action (assume that $MTBM_u = 80$). For preventive maintenance, Supplier "MNO" estimated that it would cost $40,000 per year for the next 10 years.

   Which alternative would you select based on a 10-year life-cycle cost?

2. *The analysis results.* Referring to Figure 6.27, it appears as though the preferred approach is to select Alternative "C" based purely on the economic factors presented. However, the cost difference between Alternative "C" and the next best alternative (i.e., Alternative "B") is small, and before making a final decision, the analyst may wish to review some of the assumptions upon which the analysis is based. For instance, Alternative "C" involves two different suppliers—one responsible for corrective maintenance and the other responsible for preventive maintenance. If preventive maintenance is not performed in a high-quality manner, the result may be an increase in corrective maintenance requirements. The interaction

Year	UNDISCOUNTED COST ($)			Discount Factor (12%)[a]	DISCOUNTED COST ($)		
	Alternative "A"	Alternative "B"	Alternative "C"		Alternative "A"	Alternative "B"	Alternative "C"
1	70,000	100,000	97,200	0.8929	62,503	89,290	86,790
2	82,500	100,000	97,200	0.7972	65,769	79,720	77,488
3	91,100	100,000	97,200	0.7118	64,774	71,180	69,187
4	95,500	100,000	97,200	0.6355	60,690	63,550	61,771
5	105,000	100,000	97,200	0.5674	59,577	56,740	55,151
6	115,000	100,000	97,200	0.5066	58,259	50,660	49,242
7	120,500	100,000	97,200	0.4524	54,514	45,240	43,973
8	135,000	100,000	97,200	0.4039	54,527	40,390	39,259
9	153,000	100,000	97,200	0.3606	55,172	36,060	35,050
10	165,000	100,000	97,200	0.3220	53,130	32,200	31,298
Total	1,132,000	1,000,000	972,000	5.6503	588,915	565,030	549,209

[a]Refer to the interest tables in Appendix D for the appropriate discount factors.

Figure 6.27. Economic evaluation of alternative factory maintenance policies.

effects are many! Also, as more experience is gained, it may be desirable to accomplish some trade-off studies between the performance of corrective maintenance and preventive maintenance, obtaining a balance as illustrated in Figure 6.28. This can be accomplished more efficiently when dealing with a single supplier (versus trying to work with two different suppliers). Thus, when undertaking a sensitivity analysis (i.e., by evaluating the underlying assumptions and input data in terms of possible impact on the output results), Alternative "B" might be more appropriate.

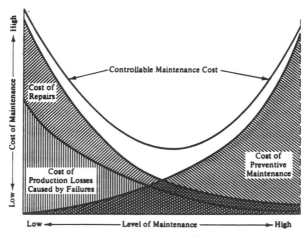

Figure 6.28. Corrective versus preventive maintenance cost considerations.

Given that Alternative "B" has been selected, the next step is to determine (through data collection and analysis) the various "high-cost" contributors in the maintenance area. What functions, equipment, processes, and so on constitute the largest contributors to the cost of maintenance? Are there any reliability and/or maintainability improvements that can be made to reduce the quantity of maintenance actions, reduce maintenance costs, and improve overall productivity in the factory? In essence, an analysis of this type can be effectively applied with the objective of improving the maintainability of an existing system capability.

6.3.5 Life-Cycle Cost Analysis (Example 5)

Life-cycle cost analyses may be accomplished to varying levels of depth throughout the system development process, in the evaluation of alternatives, to ensure that the ultimate design configuration selected reflects the appropriate economic considerations (refer to Sections 1.2, 1.3.15, and 1.3.16). The proper balance between the cost side of the spectrum and the system effectiveness side, illustrated in Figure 1.6, must be attained.

The completion of a typical life-cycle cost analysis generally requires that one follow certain steps, such as those presented in Figure 6.29. Referring to the figure, one needs to progress from the problem definition stage, through the definition of system requirements, the development of a CBS, the estimation of costs and the development of cost profiles, the accomplishment of a break-even analysis and a sensitivity analysis, the identification of areas of risk, and so on. While this process basically follows the steps illustrated in Figure 6.18, the important consideration here is to ensure that economic issues are addressed in terms of the entire system life cycle. An example of the steps employed in the accomplishment of a life-cycle cost analysis is presented in depth in Appendix D. However, to illustrate some of the steps involved, an abbreviated example is presented here. Maintainability is one of the many characteristics in design that can have a significant impact on life-cycle cost; therefore, it is important at this point to become familiar with some of the key steps in the process.

1. *The analysis process.* A ground vehicle currently in the development phase requires the incorporation of a radio communication equipment. A decision is needed as to the type of equipment deemed most feasible from the standpoint of performance, reliability, maintainability, and life-cycle cost. Budget limitations suggest that the equipment unit cost (based on life-cycle cost) should not exceed $20,000.

   a. The accomplishment of a cost-effectiveness evaluation requires further expansion of the problem definition. A description of system operational requirements and the maintenance concept (Chapter 5) are essential. In addition, one needs to know the program time frame. The communication equipment is to be installed in a light vehicle. The equipment shall enable communication with other vehicles at a range of 200 miles, overhead aircraft at an altitude of 10,000 feet or less, and a centralized area communication facility. The system must have a reliability MTBF of 450 hours, a $\overline{M}ct$ of 30

Step	Activity
1	Define the problem.
2	Identify feasible alternatives – configurations to be evaluated through the LCC analysis.
3	Project the alternatives in terms of system requirements: a. Define operational requirements b. Define the system maintenance concept. c. Identify and categorize life-cycle activities.
4	Develop a cost breakdown structure (CBS) – cost categories and cost estimating relationships.
5	Develop a cost model (or select one "off-the-shelf") that is sensitive to the problem at hand, and that can be effectively utilized to facilitate the analysis process.
6	Estimate the appropriate costs for each activity, and for each year in the projected life cycle – known cost factors, analogous cost factors, and parametric cost estimating relationships. Include the effects of inflation, learning curves, etc.
7	Develop a cost profile (inflated costs) for each alternative being evaluated.
8	Develop a cost summary (discounted present value costs) for each alternative, and compare the results in terms of preference.
9	Accomplish a breakeven analysis showing the points in time a given alternative assumes a preferred position.
10	Identify the "high-cost" contributors, and determine the cause-and-effect relationships.
11	Accomplish a sensitivity analysis.
12	Accomplish a risk analysis, and identify the potential areas of high risk.
13	Recommend a preferred approach – select the most desirable alternative.

Figure 6.29. The basic steps in a life-cycle cost analysis. (*Source:* Blanchard, B. S., *System Engineering Management,* John Wiley & Sons, N.Y., 1991, p. 134).

minutes, and a MMH/OH requirement of 0.2. The operational and maintenance concepts and program time frame are illustrated in Figure 6.30.

b. Review of all possible supplier sources indicates that two design configurations appear (based on preliminary design data) to meet the specified requirements. Each is evaluated on an equivalent basis in terms of reliability MTBF and total life-cycle cost.

c. The next step is to identify data needs and to structure the analytical model for use in the evaluation process. For each configuration the analyst needs:

• A reliability allocation or prediction providing estimated component failure rates and a system MTBF. The system MTBF must be 450 hours or greater.

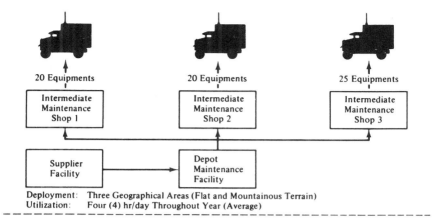

Deployment: Three Geographical Areas (Flat and Mountainous Terrain)
Utilization: Four (4) hr/day Throughout Year (Average)

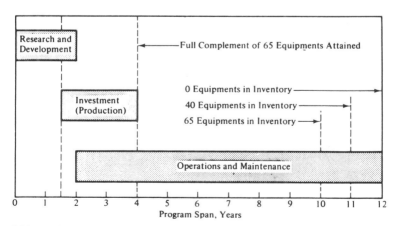

Figure 6.30. Basic system concepts. (*Source:* Blanchard, One through Twelve in *Logistics Engineering and Management,* 4th ed., Blanchard ed., © 1992, Fig. 6-7, p. 170. Reprinted by permission of Prentice-Hall, Inc., Englewood Cliffs, N.J.).

- A maintainability allocation or prediction providing $\overline{M}ct$ and MMH/OH factors for the equipment and its components.
- A gross-level maintenance analysis describing maintenance tasks, task times and frequencies, and basic logistic support requirements (test and support equipment, spare/repair parts, personnel and training, technical data, facilities, transportation and handling). The maintenance analysis is discussed in Chapter 9.
- A life-cycle cost analysis involving a definition of cost categories plus input cost factors. The details of a cost analysis are presented in Appendix D.
- An analytical model structured on the basis of problem definition, the evaluation criteria, and the data available to the analyst.

2. *The analysis results.* The problem is to select the best among two alternatives on the basis of reliability, maintenance, and life-cycle cost. A comparison of reliability and life-cycle cost data for each of the two configurations is illustrated in Figure 6.31. In this instance, configuration *A* is the preferred alternative with the highest reliability and lowest life-cycle cost.

A breakdown of life-cycle cost is presented in Figure 6.32. Note that the acquisition cost (R&D and Investment) is higher for configuration *A* ($478,033 versus $384,131). This is due partially to a better design using more reliable components. Although the initial cost is higher, the overall life-cycle cost is lower because of a reduction in maintenance actions resulting in lower O&M costs. These characteristics in equipment design have a tremendous effect on life-cycle cost.

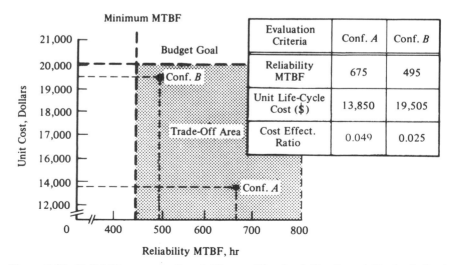

Figure 6.31. Reliability versus unit cost. (*Source:* Blanchard, One through Twelve in *Logistics Engineering and Management,* 4th ed., Blanchard ed., © 1992, Fig. 6-8, p. 171. Reprinted by permission of Prentice-Hall, Inc., Englewood Cliffs, N.J.).

Figure 6.33 projects the life-cycle cost in the form of a profile. The anticipated cost for each year is indicated. These figures can be applied directly for budgeting purposes when inflated, or for comparison purposes when discounted. As stated earlier, configuration A is preferred on the basis of total life-cycle cost. Prior to a final decision, however, the analyst should perform a break-even analysis to determine the point in time where configuration A becomes more effective than configuration B. Figure 6.34 illustrates a payback point that is 6 years and 5 months, or a little more than 2½ years after a full complement has been acquired in the operational inventory. This point is early enough in the life cycle to support the decision. On the other hand, if the payback point were much further out in time, the decision might be questioned.

Cost Category	Configuration A		Configuration B	
	P.V. Cost	% of Total	P.V. Cost	% of Total
1. Research and development (C_R)	$70,219	7.8	$53,246	4.2
(a) Program management (C_{RM})	9,374	1.1	9,252	0.8
(b) Advanced R & D (C_{RR})	4,152	0.5	4,150	0.4
(c) Engineering design (C_{RE})	41,400	4.5	24,581	1.9
(d) Equipment development and test (C_{RT})	12,176	1.4	12,153	0.9
(e) Engineering data (C_{RD})	3,117	0.3	3,110	0.2
2. Investment (C_I)	407,814	45.3	330,885	26.1
(a) Manufacturing (C_{IM})	333,994	37.1	262,504	20.8
(b) Construction (C_{IC})	45,553	5.1	43,227	3.4
(c) Initial logistic support (C_{IL})	28,267	3.1	25,154	1.9
3. Operations and maintenance (C_O)	422,217	46.9	883,629	69.7
(a) Operations (C_{OO})	37,811	4.2	39,301	3.1
(b) Maintenance (C_{OM})	384,406	42.7	844,328	66.6
• Maintenance personnel and support (C_{OMM})	210,659	23.4	407,219	32.2
• Spare/repair parts (C_{OMX})	103,520	11.5	228,926	18.1
• Test and support equipment maintenance (C_{OMS})	47,713	5.3	131,747	10.4
• Transportation and handling (C_{OMT})	14,404	1.6	51,838	4.1
• Maintenance training (C_{OMP})	1,808	0.2	2,125	Neg.
• Maintenance facilities (C_{OMF})	900	0.1	1,021	Neg.
• Technical data (C_{OMD})	5,402	0.6	21,452	1.7
(c) System/equipment modifications (C_{ON})	. . .	. . .	. . .	. . .
(d) System phaseout and disposal (C_{OP})	. . .	. . .	. . .	. . .
Grand total*	$900,250	100%	$1,267,760	100%

*The cost values presented are hypothetical but realistically derived. A 10% discount factor was used in determining present-value costs.

Figure 6.32. Life-cycle cost analysis breakdown.

| Category | Program Year | | | | | | | | | | | | |
	1	2	3	4	5	6	7	8	9	10	11	12	Total
Research and Development	32,119	38,100	—	—	—	—	—	—	—	—	—	—	70,219
Investment	—	94,110	156,852	156,852	—	—	—	—	—	—	—	—	407,814
Operations and Maintenance	—	—	12,180	32,480	60,492	57,472	53,480	50,484	50,470	50,494	37,480	17,185	422,217
Total	32,119	132,210	169,032	189,332	60,492	57,472	53,480	50,484	50,470	50,494	37,480	17,185	$900,250

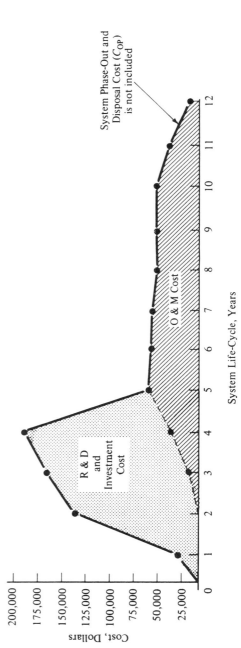

Figure 6.33. Cost profile for configuration A. (*Source:* Blanchard, One through Twelve in *Logistics Engineering and Management*, 4th ed., Blanchard ed., © 1992, Fig. 6-9, p. 173. Reprinted by permission of Prentice-Hall, Inc., Englewood Cliffs, N.J.).

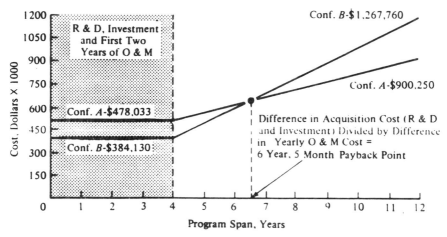

Figure 6.34. Investment payback (break-down analysis).

Referring to Figure 6.32, the analyst can readily pick out the high contributors (those that contribute more than 10% of the total cost). These are the areas where a more refined analysis is required and greater emphasis is needed in providing valid input data. For instance, maintenance personnel and support cost (C_{OMM}) and spare/repair parts cost (C_{COMX}) contribute 23.4% and 11.5%, respectively, of the total cost for configuration A. This leads the analyst to reevaluate the design in terms of impact on personnel support and spares; the prediction methods used in determining maintenance frequencies and inventory requirements; the analytical model to ensure that the proper parameter relationships are established; and cost factors such as personnel labor cost, spares material costs, inventory holding cost, and so on. If the analyst wishes to determine the sensitivity of these areas to input variations, he or she may perform a sensitivity analysis. In this instance, it is appropriate to vary MTBF as a function of maintenance personnel and support cost (C_{OMM}) and spare/repair parts cost (C_{OMX}). Figure 6.35 presents the results.

The analyst or decision maker should review the break-even analysis in Figure 6.34 and determine how far out in time he or she is willing to go and remain with configuration A. Assuming that the selected maximum payback point is 7 years, the difference in alternatives is equivalent to approximately $65,000 (the present value difference between the two configurations at the 7-year point). This indicates the range of input variations allowed. For instance, if the design configuration A changes or if the reliability prediction is in error resulting in a MTBF as low as 450 hours (the specified system requirement), the maintenance personnel and support cost (C_{OMM}) will increase to approximately $324,000, an increase of about $113,340 above the baseline value. Thus, although the system reliability is within the specified requirements, the cost increase due to the input MTBF variation causes a decision shift in favor of configuration B. The analyst must assess the sensitivity of significant input parameters and determine their impact on the ultimate decision.

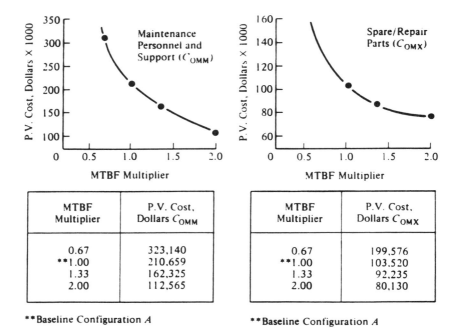

Figure 6.35. Sensitivity analysis.

6.4   MAINTAINABILITY DESIGN CRITERIA

Chapter 5 covers the steps necessary to define the specific qualitative and quantitative requirements at the system level, and Sections 6.1 and 6.2 cover the allocation of these requirements to the subsystem level and below. With this information, it is now possible to generate detail design criteria. Such criteria may constitute specific requirements in areas of equipment packaging and modularization, standardization, interchangeability, the mounting of components, accessibility, the degree of self-test features and the placement of test points for external test, the extent of automation versus manual provisions, repair versus discard levels, safety features, labeling, and so on. These criteria may be stated qualitatively or quantitatively, and are employed as guidelines for the design engineer. Qualitative criteria must support the quantitative goals developed through allocation. The criteria thus established have a direct impact on maintainability and on system/product design overall. In regard to the development of such criteria, a few examples are provided for illustrative purposes:

1. Through the level of repair analysis, it is possible to determine whether an item should be designed for "repair" or for "discard at failure." If the reliability of an item is high enough (e.g., one anticipated failure in 50,000 hours of system

operation) and the unit cost is low enough, it may not be economically feasible to establish a capability (with the required test equipment, personnel, facilities, data, etc.) to enable the repair of that item when a failure occurs. Thus, the item is discarded at failure and there is no need to incorporate provisions for accessibility, test points, modular packaging, and so on within that item. This, of course, can have a significant impact on spare parts, test equipment requirements (i.e., the level of diagnostics incorporated within the overall system), personnel training, and the extent of coverage in the maintenance manual. In any event, maintainability criteria should support the "design for discard" decision.

2. Referring to Unit "B" of System "XYZ" in Figure 6.17, the allocated $\overline{M}ct$ value of 0.4 hour means that in the event of malfunction, the maintenance technician must be able to complete the corrective maintenance cycle (refer to Figure 4.8) in 24 minutes. Based on experience data for similar equipment, about 60% of the total corrective maintenance (on the average) is consumed in diagnostics through the accomplishment of localization and fault-isolation tasks. Assuming that this percentage is valid for System "XYZ," an estimate of allowable localization and fault-isolation time for Unit "B" would be 14 minutes. Complying with a 14-minute goal in design would necessitate the availability of a few readily accessible test points or readout devices to allow for positive fault isolation to any one of the three assemblies within Unit "B." If the time allocated for diagnostics is less (e.g., 7 minutes), it may be necessary to incorporate a built-in automatic test capability down to the assembly level. Given the 14-minute diagnostic time requirement, the maintenance technician must accomplish the necessary disassembly, remove and replace (or repair in place), reassembly, and checkout or verification tasks in the remaining 10 minutes. This, of course, infers that the three assemblies of Unit "B" must be directly accessible and must not require the removal of another assembly in order to gain access. In addition, each assembly should be modular in construction with plug-in and/or quick-release features, and should be completely interchangeable so as to minimize alignment and adjustment requirements after the spare item is installed. In any event, specific maintainability criteria must include both qualitative and quantitative factors dealing with functional packaging, interchangeability, accessibility, the type and level of diagnostic test capability, and so on.

3. The allocated skill-level and maintenance man-hour requirements, identified in Figure 6.17, dictate that the design of the system components (where human interaction exists) be constrained to the extent that anticipated maintenance tasks can be adequately and effectively accomplished within the prescribed limits. This infers that the design will incorporate simple readout devices for rapid decision making, the proper layout of front panel displays, automation of complex operating functions, standardization of components, labeling, and other provisions that will facilitate ease and simplicity in the accomplishment of maintenance functions. The proper consideration of these provisions is facilitated through the application of good human engineering principles.

The establishment of design criteria (i.e., design guidelines) must be consistent

<u>M</u> Criteria

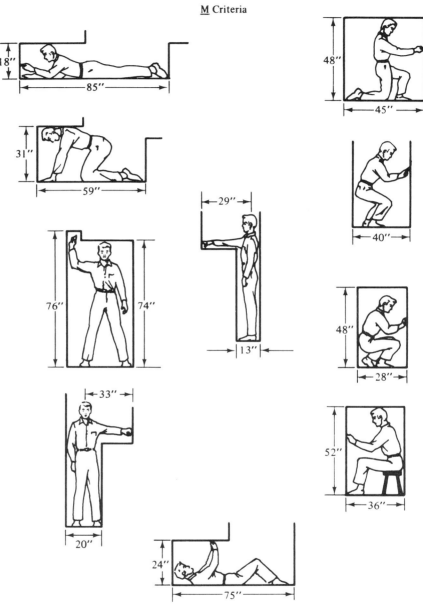

Figure 6.36. Limiting clearances required for various body positions. (*Source:* Blanchard, B., *Logistics Engineering and Management,* 4th ed., Prentice-Hall, Englewood Cliffs, N.J., 1991, p. 210).

MINIMUM OPENINGS FOR USING COMMON HAND TOOLS

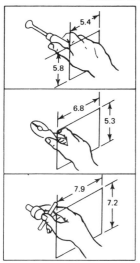

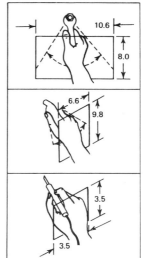

SPACE REQUIRED FOR USING COMMON HAND TOOLS

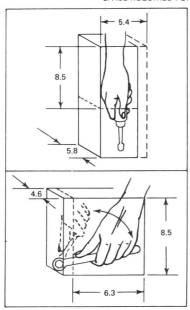

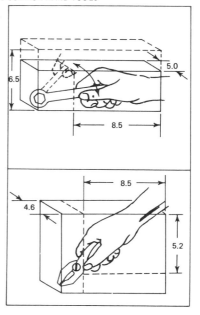

Figure 6.37. General criteria (accessibility). (*Source:* AMCP 706-134, *Engineering Design Handbook: Maintainability Guide for Design,* U.S. Army, Washington, D.C., p. 144).

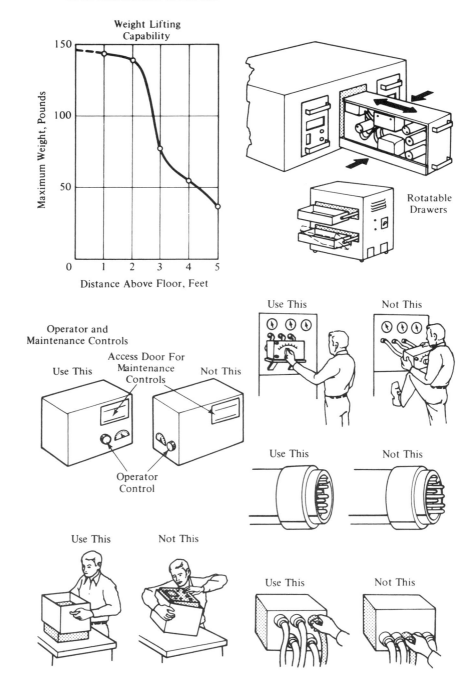

Figure 6.38. General criteria. (*Source:* Blanchard, B., *Logistics Engineering and Management,* 4th ed., Prentice-Hall, Englewood Cliffs, N.J., 1991, p. 213).

Mounting Method	Maintainability Considerations	
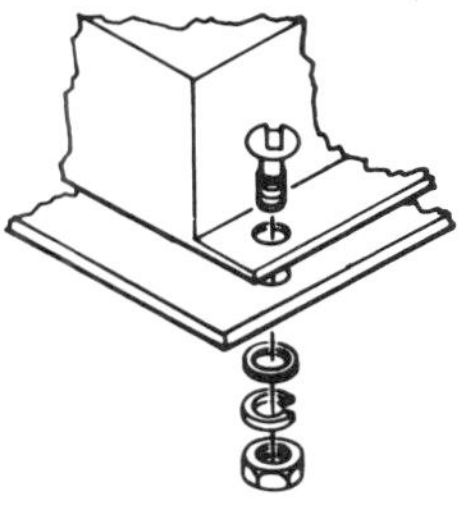	Description:	Horizontal surface mounted. Screws through clearance holes with nuts, flat washers, and lock washers.
	Advantages:	No significant advantages.
	Disadvantages:	Two-handed operation. Possible loss of nuts, washers, or screws. Alignment of holes difficult.
	Tools Required:	Screwdriver and wrench or "spintite,"
	Operating Time:	Approximately 0.8 minute per fastener.
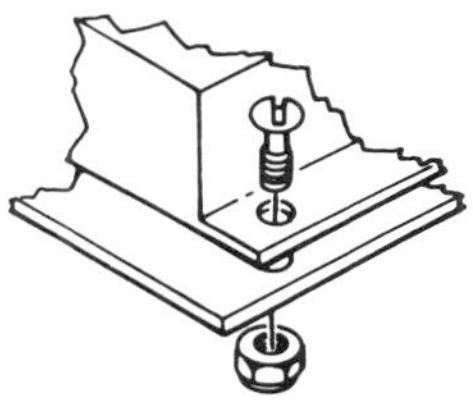	Description:	Horizontal surface mounted. Screws with stop nuts through clearance holes.
	Advantage:	No washer required.
	Disadvantages:	Stop nut difficult to turn. Two-handed operation. Alignment of holes difficult. Possible loss of nuts.
	Tools Required:	Screwdriver and wrench or "spintite."
	Operating Time:	Approximately 1.4 minutes per fastener.
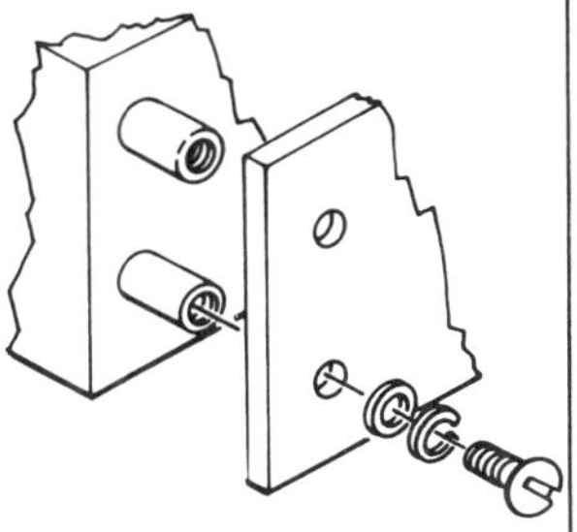	Description:	Vertical surface mounted. Screws with flat washers and lock washers into tapped holes.
	Advantage:	One-handed operation if part is supported.
	Disadvantages:	Parts must be supported to start nuts or screws. Alignment of holes difficult. Possible loss of washers or screws.
	Tools Required:	Screwdriver.
	Operating Time:	Approximately 0.8 minute per fastener.

Figure 6.39. Comparison of component mounting methods. (*Source:* AMCP 706-134, *Engineering Design Handbook: Maintainability Guide for Design,* U.S. Army, Washington, D.C., pp. 149, 150).

with system operational requirements, the maintenance concept, and the factors defined through allocation. This definition is further extended through the accomplishment of early trade-off studies such as those illustrated in Section 6.3. The objective is to provide some specific guidance to the designer, the consideration and incorporation of such which will lead to a highly maintainable product. Figures 6.36–6.40 are included to illustrate, in a simplistic manner, a few representative areas that directly pertain to maintainability in design.

Mounting Method	Maintainability Considerations	
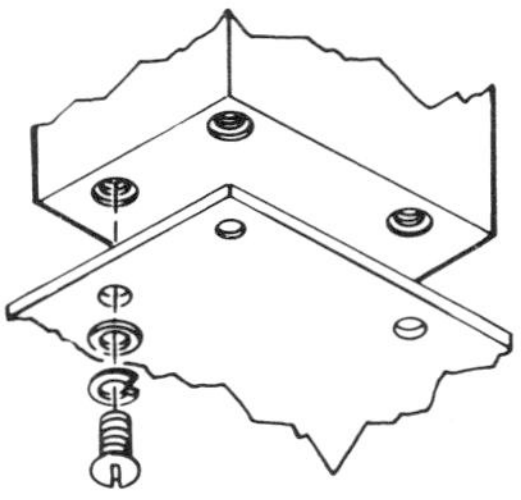	Description:	Horizontal surface mounted. Screws with flat washers, and lock washers into tapped hole in part.
	Advantage:	Can usually be performed with one hand if necessary.
	Disadvantages:	Possible loss of washers or screws. Alignment of holes difficult.
	Tools Required:	Screwdriver
	Operating Time:	Approximately 0.8 minute per fastener.
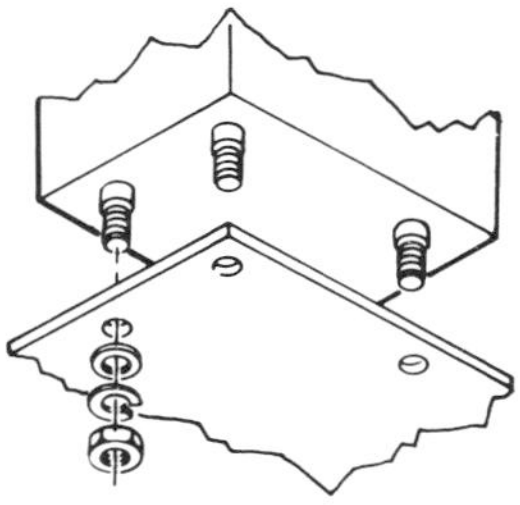	Description:	Horizontal surface mounted. Studs through clearance holes with nuts, flat washers, and lock washers.
	Advantage:	One-handed operation. Studs act as locating pins.
	Disadvantages:	Possible loss of nuts or washers. Part must be lifted.
	Tools Required:	Wrench or "spintite."
	Operating Time:	Approximately 0.7 minute per fastener.
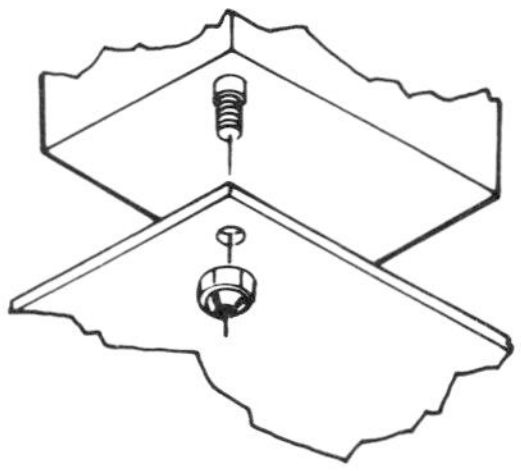	Description:	Horizontal surface mounted. Stop nuts on studs.
	Advantages:	No washer required. One-handed operation. Studs act as locating pins.
	Disadvantages:	Possible loss of nuts. Part must be lifted.
	Tools Required:	Wrench or "spintite."
	Operating Time:	Approximately 1.3 minutes per fastener.

Figure 6.39. (*Continued*)

6.5   DEVELOPMENT OF DESIGN SPECIFICATIONS

Referring to Figure 6.13, the system is broken down into its components (i.e., equipment, software, personnel, data, and/or facility) through the functional analysis process described in Sections 6.1 and 6.2. System-level requirements are then allocated and design criteria are developed for each component. It is now necessary to ensure that these various system components are developed and/or purchased to meet the established criteria. This is accomplished through the development of design specifications and the inclusion of the applicable maintainability criteria therein.

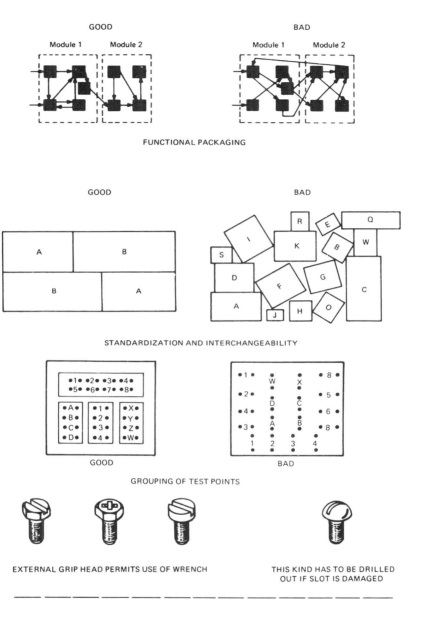

FUNCTIONAL PACKAGING

STANDARDIZATION AND INTERCHANGEABILITY

GROUPING OF TEST POINTS

EXTERNAL GRIP HEAD PERMITS USE OF WRENCH

THIS KIND HAS TO BE DRILLED
OUT IF SLOT IS DAMAGED

U-TYPE LUGS FACILITATE REPAIRS

Figure 6.40. General criteria. (*Source:* AMCP 706-134, *Engineering Design Handbook; Maintainability Guide for Design,* U.S. Army, Washington, D.C., pp. 145, 147, 148).

Figure 2.16 (Chapter 2) illustrates a sample "specification tree." With regard to System "XYZ," it is assumed that Unit "B" must be newly designed. Therefore, the design requirements shown for Unit "B" in Figure 6.17 must be included in the Type "B" Development Specification for that item (i.e., MTBF = 536, $\overline{M}ct = 0.4$, MMH/OH = 0.17). Next, assume that Unit "C" represents a standard inventory item. In this case, the requirements for this unit must be included in the Type "C" Product Specification (i.e., MTBF = 9050, $\overline{M}ct = 1.0$, MMH/OH = 0.01). These "specific" criteria may then be supplemented with some "generic" factors, such as those illustrated in Figures 6.36–6.40. In any event, each of the major system components must be covered by maintainability goals and objectives included in the applicable design specifications.

QUESTIONS AND PROBLEMS

1. What is meant by "functional analysis?" What purpose does it serve? Why is it important? Can a functional analysis be accomplished for any system?

2. Assume that you have identified a need for a new transportation capability (e.g., transportation between your home and your place of employment). Describe the current "deficiency" (justifying the need), both in qualitative and quantitative terms; select a mode of transportation; define the operational requirements (to include a system utilization profile, the direct application of technical performance measures, etc.); describe the maintenance concept; accomplish a functional analysis (to include an operational functional flow diagram to at least four levels and a maintenance functional flow diagram to at least three levels); and allocate system-level requirements to the depth necessary to establish design criteria as an input to design. Referring to Tables 6.1 and 6.2, identify functional inputs, outputs, and resource requirements. Show the traceability of requirements from the top down.

3. Select a system of your choice and construct a functional diagram showing three levels of operational functions and four levels of maintenance functions. Referring to Tables 6.1 and 6.2, identify functional inputs, outputs, and resource requirements. How do the operational functions and maintenance functions relate? (provide an illustration).

4. Select a segment of the functional flow diagrams from either Question 2 or Question 3, and construct an "N-squared chart."

5. Briefly describe what is meant by an IDEF$_0$ model? Provide an illustration.

6. What is the purpose of the block numbering practice in the development of functional flow diagrams?

7. In equipment design, what are the benefits of functional packaging?

8. How does the functional analysis impact maintainability (or vice versa)?

9. What is the purpose of allocation? How does it impact system design? To what depth should allocation be accomplished?

10. Briefly describe the steps in reliability allocation, maintainability allocation, the allocation of logistic support factors, and the allocation of cost factors.

11. What is the relationship between reliability allocation and maintainability allocation? (provide an example).

12. Select a system of your choice and assign top-level requirements. Accomplish a reliability allocation and a maintainability allocation to the second indenture level.

13. Referring to Figure 6.17, System "XYZ" has the following requirements: MTBF = 650, $\overline{M}ct$ = 0.6, MMH/OH = 0.7, and unit LCC = $100,000. Allocate these requirements to Units "A"—"C" and to Assemblies 1–3 of Unit "B."

14. In Figure 6.41, allocate the quantitative factors to the unit level as indicated.

15. What is a "model?" List some of the basic characteristics desired.

16. Refer to Figure 6.18. Describe some of the interrelationships (data input/output factors) for the various models shown.

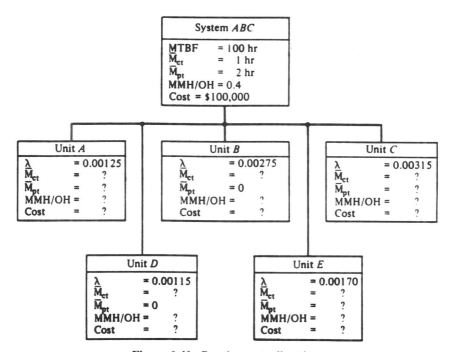

Figure 6.41. Requirements allocation.

17. What is a "sensitivity analysis?" What are the benefits of a sensitivity analysis?

18. Refer to the level of repair analysis problem in Section 6.3.1. Assume that the reliability failure rate (λ) of Assembly "A-1" is 10 times the value specified for the baseline case. What option would you select? What option would you select if the failure rate (λ) is 0.1 of the specified value?

19. Refer to Figure 6.23. Explain how each of the 11 evaluation factors relates to or has an impact on maintainability.

20. What steps would you perform in accomplishing a life-cycle cost analysis involving two or more alternatives?

21. Refer to Figure 6.32. What do the figures tell you? How do these figures relate to equipment design? What actions (if any) would you take to improve the system design for maintainability?

22. What is a "break-even analysis" (in the context discussed herein)? Of what value is it?

23. Refer to the life-cycle cost problem in Section 6.3.5. What would the results be if a 5% discount factor is used? if a 15% discount rate is used? How will the utilization of these factors impact the break-even analysis in Figure 6.34?

24. What is meant by "design criteria?" How are design criteria developed, and how are they specified? (provide some examples).

7
MAINTAINABILITY IN DESIGN

A major objective is, of course, to ensure that maintainability characteristics are included in system/product design. Specific qualitative and quantitative requirements are identified through the needs analysis, the accomplishment of feasibility studies, and the development of system operational requirements and the maintenance concept described in Chapter 5. These requirements are addressed through the implementation of program planning activities and the organizational tasks identified in Chapters 2 and 3, respectively.

Of particular significance is the day-to-day design participation process and the program tasks that are directed to facilitate the incorporation of "maintainability in design." As the responsible "design engineer" (or design team) progresses toward the definition of a specific design configuration, he or she must consider several different design factors, acquiring the proper balance between maintainability and the many other factors that must be addressed to meet the consumer needs. This consideration is best accomplished through the representation of a maintainability specialist as part of the design team. This individual (or individuals) can assist the design process by developing design criteria, as discussed in Sections 6.4 and 6.5, and working with other members of the design team as appropriate. Not only must the maintainability specialist be knowledgeable in the specification of maintainability requirements, but he/she must be able to "communicate" with the designer and "contribute" to the overall design effort.

The success in meeting this day-to-day objective is highly dependent on having the appropriate tools available for accomplishing the necessary design analysis and evaluation activities. The utilization of models for the purpose of requirements allocation, the availability of various design analysis methods to help in the design definition process, and the use of tools for system/product evaluation are key areas where the maintainability specialist can contribute positively to the ultimate design output. This chapter covers some of these key tools, technologies, and aids.

7.1 DESIGN REQUIREMENTS

Referring to Figure 1.5 (Chapter 1), maintainability functions are performed throughout the system/product life cycle. In the early conceptual design phase, major trade-off studies are completed as part of the overall system definition process. A level of repair analysis may be accomplished in defining the maintenance concept, a reliability versus maintainability trade-off may be necessary in defining a higher-level system availability requirement, a life-cycle cost analysis may be used in the evaluation of possible alternative technology applications, and so on (refer to Section 6.3 and Appendix B).

In the preliminary system design phase, system-level requirements are defined in functional terms and allocated downward to the various components of the system (refer to Section 6.2). Reliability and maintainability models are developed to facilitate this process. As the design progresses, a fault tree analysis (FTA) and a failure mode, effects, and criticality analysis (FMECA) are accomplished to ensure that reliability and maintainability characteristics are incorporated into the configuration being developed. Reliability and maintainability predictions are accomplished periodically to assess whether the design configuration (at that time) meets the initially specified consumer requirements. A human factors analysis is often accomplished to evaluate the human-machine interface requirements and the personnel quantities and skill levels necessary for the operation and maintenance of the system. A maintenance task analysis is accomplished to (1) evaluate the system relative to its inherent maintainability characteristics, and (2) determine the specific system support resources required in terms of spare parts, test equipment, facilities, data, and so on.

In the detail design and development phase, as the design configuration becomes somewhat fixed and engineering and prototype models are being developed, reliability tests and maintainability demonstrations are accomplished to provide an actual assessment of the system/product characteristics. The maintenance task analysis is expanded, and a logistic support analysis is accomplished to assess overall system supportability.

Throughout this overall system/product design process, numerous tools can be applied to assist the maintainability specialist in accomplishing his (or her) functions. Depending on the phase and the degree of design definition, any one (or a combination) of these tools can be utilized effectively at any time. The objective is to become familiar with some of these tools, methods, and/or aids. Section 7.2 and Appendix C describe some of the major tools used in design analysis and evaluation.

7.2 DESIGN ANALYSIS AND EVALUATION TOOLS

7.2.1 Reliability Analysis and Modeling

Throughout the system design process, reliability models are used to aid in the accomplishment of reliability allocation, the evaluation of alternative configura-

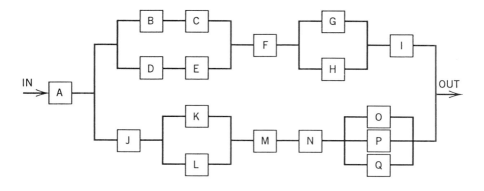

Component	Reliability	Component	Reliability	Component	Reliability
A	0.84	G	0.87	M	0.83
B	0.86	H	0.88	N	0.85
C	0.89	I	0.89	O	0.84
D	0.86	J	0.86	P	0.89
E	0.87	K	0.85	Q	0.89
F	0.82	L	0.86		

Figure 7.1. A simplified reliability block diagram.

tions, the accomplishment of reliability prediction, the conductance of stress–strength analysis, and for the purposes of reliability assessment downstream. As discussed in Section 4.1, there are series models, series-parallel models, parallel redundancy models, and so on. Models are developed through the generation of block diagrams, evolving from the functional analysis described in Section 6.1 and into a specific reliability block diagram, as illustrated in Figure 6.15. This illustration shows a breakout of the system by level, and the blocks at each level can be described quantitatively using the mathematical relationships discussed in Section 4.1.

For the purposes of illustration, an abbreviated reliability block diagram is shown in Figure 7.1. The objective is to "model" the system, both functionally and in terms of its physical components. Series and parallel relationships are shown, and individual quantitative measures of reliability can be assigned for each block. Initially, these factors can be derived through the process of allocation. Later, as the design definition process evolves and more information is available covering the various components of the system, failure rate information from historical data-bases or from component supplier sources can be applied. In the example presented in Figure 7.1, it is assumed that the overall system requirement specifies a reliability of 70%, and that the allocated values for each of the blocks are as indicated in the figure. The task is to determine whether the system configuration modeled meets the overall requirement. Applying the mathematical relationships in Section 4.1, the system reliability is:

$R_{BC} = (0.86)(0.89) = 0.7654$

$R_{DE} = (0.86)(0.87) = 0.7482$

$R_{BCDE} = 0.7654 + 0.7482 - (0.7654)(0.7482) = 0.9409$

$R_{GH} = 0.87 + 0.88 - (0.87)(0.88) = 0.9844$

$R_{BCDEFGHI} = (0.9409)(0.82)(0.9844)(0.89) = 0.6759$

$R_{KL} = 0.85 + 0.86 - (0.85)(0.86) = 0.9790$

$R_{OPQ} = [(1 - 0.84)(1 - 0.89)(1 - 0.89)] = 0.9981$

$R_{JKLMNOPQ} = (0.86)(0.9790)(0.83)(0.85)(0.9981) = 0.5929$

$R_{BCDEFGHIJKLMNOPQ} = 0.6759 + 0.5929 - (0.6759)(0.5929) = 0.8681$

$R_S = (R_A)(R_{BCDEFGHIJKLMNOPQ}) = (0.84)(0.8681) = \underline{0.7292}$

With a calculated reliability of 72%, the configuration shown does meet the overall requirement.

The model in Figure 7.1 is relatively simple. Often, when constructing a reliability block diagram for a large system, there are hundreds of components to consider, the notation becomes quite complex, and the use of a computerized model is required to facilitate the calculations. With an appropriate computer program, one can expeditiously:

1. Evaluate alternative design approaches where different combinations of series-parallel relationships are being considered.
2. Evaluate the reliability of the system (or various elements of the system) when different component parts are being considered for a specific application. This applies to evaluating component stress–strength characteristics, the effects of component derating, and so on.

Of major concern in the design for system reliability are the stress–strength characteristics of its components. Component parts are designed and manufactured to operate in a specific manner when utilized under normal conditions. If additional stresses are imposed owing to electrical loads, temperature variations, vibration, shock, humidity, and the like, then unexpected failures will occur and the reliability of the system will be less than anticipated. Also, if materials are utilized in a such manner that nominal strength characteristics are exceeded, fatigue occurs and the materials may fail much earlier than expected. In any event, overstress conditions will result in reliability degradation, causing an increase in maintenance requirements, and understress conditions may be costly as a result of overdesign; that is, incorporating more than what is actually necessary to do the job.

A stress–strength analysis is often undertaken to evaluate the probability of identifying a situation where the value of stress is much larger than (or the strength is much less than) the nominal value. Such an analysis may be accomplished through the following steps:

1. For selected components, determine nominal stresses as a function of loads, temperature, vibration, shock, physical properties, time, and so on.

2. Identity factors affecting maximum stress, such as stress concentration factors, static and dynamic load factors, stresses as a result of manufacturing and heat treating, environmental stress factors, and so on.

3. Identify critical stress components and calculate critical mean stresses (e.g., maximum tensile stress and shear stress).

4. Determine critical stress distributions for the specified useful life of the component. Analyze the distribution parameters and identify component safety margins. Applicable distributions may assume a normal, Poisson, Gamma, Weibull, log-normal, or other pattern.

5. For those components that are critical and where design safety margins are inadequate, corrective action must be initiated. This may consist of component-part substitution, the addition of some redundancy, or a complete redesign of the system element in question.

Reliability computerized models may be used to facilitate the accomplishment of a stress–strength analysis. Different reliability factors, or a range of factors with a specific distribution, may be applied to each element in the reliability block diagram. Cause and effects are evaluated, and individual component failures rates (λ) may be adjusted as appropriate to reflect the effects of the stresses on the components involved.

The reliability model serves as an excellent tool for use in the early allocation of requirements, the conductance of stress analysis, reliability prediction, and the ultimate assessment of a given system configuration. The results from these various activities provide a key input necessary in the design for maintainability. The results from reliability allocation are used in accomplishing a maintainability allocation; the stress–strength analysis helps to identify some of the weak spots in the system, where greater emphasis is needed in terms of maintenance and support; reliability predictions are a required input for maintainability predictions and for the maintenance task analysis; and reliability assessment data are required in the development of both maintenance task analysis data and logistic support analysis data. In essence, the maintainability engineer must be familiar with the reliability analysis and modeling task, as the results will have a significant impact on system design for maintainability.

7.2.2 Failure Mode, Effects, and Criticality Analysis

The failure mode, effects, and criticality analysis (FMECA) is a design technique to systematically identify and investigate potential system (product or process) weaknesses. It consists of a methodology for examining all the ways in which a system failure can occur, potential effect(s) of failures on system performance and safety, and the seriousness of these effects. The FMECA consists of two distinct analyses—the failure mode and effects analysis (FMEA), which is then extended to analyze failure mode criticality, called criticality analysis (CA). Over and above the more obvious benefits of identifying actions or changes that could eliminate or reduce the chance of potential failures and evaluating product/process risk, the

FMECA also enhances knowledge of the system and provides increased insight into its expected behavior. Further, output from a FMECA conducted in a timely manner provides invaluable input to the development of a cost-effective preventive maintenance program and to the tailoring of a focused control plan.

The FMECA is a useful technique utilized during the conceptual and preliminary design phases, and evolves through the detail design and development phase. For greater effectiveness, it is "tailored" in terms of emphasis and detail to the system under study, the organization, and the system development phase. While the analysis is best used to affect "before-the-fact" enhancements to system design and the corresponding support infrastructure, it can also be used as an "after-the-fact" tool to evaluate and continuously improve existing systems. In either case, the objective is to increase system effectiveness, reduce system maintenance and support costs, increase productivity, and increase overall international competitiveness.

A system-level failure mode, effects and criticality analysis, depending upon its emphasis and orientation, can generally be classified as either a "product-oriented" design FMECA or a "process-oriented" process FMECA. While a design FMECA addresses issues directly related to the primary equipment performance, upkeep, and safety, a process FMECA focuses on the processes used for raw material transformation and the subsequent production and assembly of the primary equipment, and how variations and failures in these processes could impact equipment functionality overall.

To further clarify the difference between a process and a design FMECA, consider an engine gasket in an automobile, as depicted in Figure 7.2. While conducting a FMECA on the automobile from a design perspective, failure modes associated with every component in the automobile, including the gasket, are considered along with their impact on automobile operation. To illustrate the point, during the course of operation the gasket might "crack," resulting in increased engine oil consumption, which in turn may result in higher operating expenses, and so on. The focus in conducting this FMECA is on the automobile, its lower-level assemblies and the components thereof, and the associated failure modes and effects. On the other hand, a process FMECA would focus on the material transformation processes utilized to actually produce the different components of the automobile, including the engine gasket. For example, assume that it requires three processes to manufacture the gasket—pressing, stamping, and finishing. Variations in each of these processes could result in the production of gaskets that are substandard in terms of material composition, tolerances, and so on. In many cases these variations are detected as part of the overall production quality assurance programs; however, some lower-quality gaskets are likely to be installed in the engine, thereby causing problems during the automobile operational phase. The emphasis in a process FMECA is to analyze the process used to produce the components that make up the prime equipment from the perspective of how these processes could result in a compromise of component and, therefore, equipment quality. It should be obvious from the discussion above that the design and process perspectives impact each other, and there is a definite overlap in the scopes of the two FMECA orientations, as depicted in Figure 7.2.

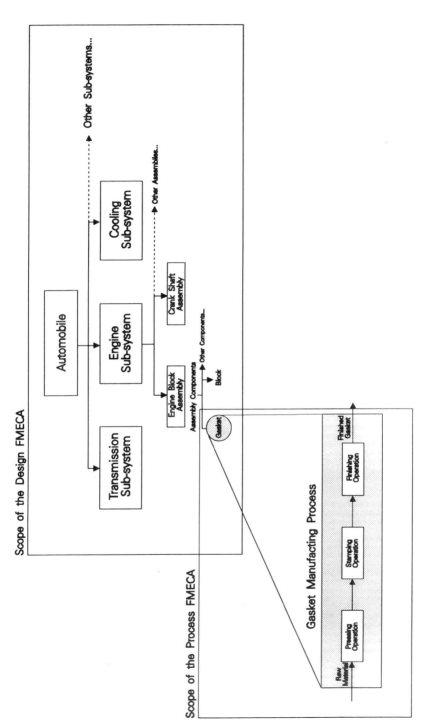

Figure 7.2. Design and process FMECA focus and scope.

The FMECA can also be conducted using either the top-down functional approach, the bottoms-up hardware approach, or a combination of these two approaches. The technique used is a function of the amount of data and information available to the designer or analyst, and the state of the design and development activity. Often, and very early in the design and development phase, the analysis involves investigating the system from a functional perspective. This functional approach to conducting a FMECA usually evolves into the more comprehensive hardware/software approach during the detailed design phase, when information with respect to component parts and their characteristics becomes available to the analyst.

A FMECA is best begun during the conceptual and preliminary design phases when the system is analyzed more from a functional perspective. For maximum effectiveness, however, the analysis should evolve as additional information becomes available to the analyst. It should also reflect all design changes and their impact on the overall system. Over and above the more obvious benefits that accrue from a FMECA, it can make significant contributions to system feasibility studies during the conceptual and preliminary design phases through the delineation of functional conflicts, incompatibilities, and/or bottlenecks early in the design phase. Figure 7.3 outlines the steps involved in completing a failure modes and effects analysis.

While the general approach to conducting a FMECA remains the same, as is illustrated in Figure 7.3, there are slight variations, depending upon the focus and orientation of the analysis, and the extent of information available to the analyst. The steps that constitute this general approach are discussed next.

1. *Define system (product or process) requirements.* For each product or process, it is important to address not only the desired but also the undesired outcomes or outputs. What is the product/process to accomplish? Further, these requirements need to be traced back to the need initially identified and the customer requirements. All relevant performance and effectiveness factors need to be addressed.

2. *Accomplish functional analysis.* This involves defining the system in functional terms. System functionality is clearly delineated using a symbolic representation such as a functional flow diagram. The functional representation of a system is often complemented with the associated data flow diagram in a format such as N-squared charts for increased insight into a system's behavioral characteristics. Functional flow analysis, along with some selected symbolic languages, are discussed in detail in Section 6.1.

3. *Accomplish requirements allocation.* This is a top-down breakout of system-level requirements to each functional entity in the system (product or process) functional hierarchy. It is important to identify the applicable performance, effectiveness, input/output, throughput, speed, and other factors for each functional block. Allocation or apportionment of system-level requirements is discussed in Section 6.2.

4. *Identify failure modes.* In the context of this analysis, a failure mode is the manner in which the system element fails to accomplish its function. For example,

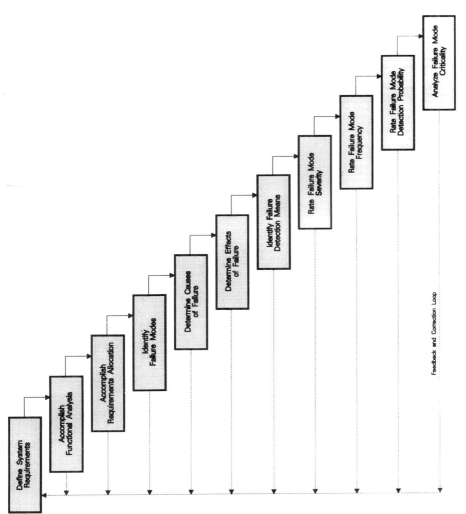

Figure 7.3. General approach to conducting a FMECA.

The figure contains the following process boxes:

- Define System Requirements
- Accomplish Functional Analysis
- Accomplish Requirements Allocation
- Identify Failure Modes
- Determine Causes of Failure
- Determine Effects of Failure
- Identify Failure Detection Means
- Rate Failure Mode Severity
- Rate Failure Mode Frequency
- Rate Failure Mode Detection Probability
- Analyze Failure Mode Criticality

Feedback and Correction Loop

switch fails in "open" position; pipe "ruptures;" "shear" due to stress; and so on. This is true whether the focus of the analysis is a process or a piece of equipment. As a caution, when describing a failure mode, it is recommended that the analyst be rather specific. Delineating a generic failure mode such as "degraded performance" or "low output" may cause some confusion while determining causes.

5. *Determine the causes of failure.* This involves analyzing the process or product in order to delineate the cause(s) responsible for the occurrence of any particular failure. A team approach to conducting a FMECA facilitates the identification of a more complete set of potential causes. While experience with similar systems is a definite "plus" in accomplishing this step in the analysis process, techniques such as Ishikawa's *cause and effect diagram,* also called the "fishbone diagram," can prove to be highly effective in delineating potential causes responsible for a failure.[1] A cause and effect diagram is a systematic and graphical portrayal of the various factors that impact the successful functioning of a product or process. Given that for a complex system or subsystem, the number of relevant factors or variables is likely to be fairly extensive, Juran and Gryna recommend the delineation of the more significant or "major variables" which could represent clusters of "satellites" or minor variables.[2] To illustrate the utilization of this technique, consider a manufacturing process to stamp a gasket. While analyzing this process for potential failure-mode causes, some of the major variables could be (a) the equipment, (b) the manufacturing environment, (c) manufacturing controls and procedures, and (d) the raw material. These major variables could in turn represent a host of minor variables. For example, the manufacturing environment could be broken down into more detailed minor variables such as operating temperature, magnetic field, dust, and vibration. This scenario is reflected through a cause and effect diagram in Figure 7.4. Such a graphical portrayal of factors that impact the effective functioning of a product or process provides valuable insight to the analyst while delineating potential failure mode causes.

6. *Determine the effects of failure.* Failures impact, often in multiple ways, the performance and effectiveness of not only the associated functional element but also the overall system. When conducting a FMECA, it is important to consider the effects of failures on the next higher-level functional entity along with the impacts on the overall system. On the other hand, while analyzing a process it is important to address failure effects on subsequent processes and to the end customer or consumer.

7. *Identify failure detection means.* In the context of a process-oriented FMECA, this refers to the current process controls, which may detect the occurrence of failures or defects. However, when the FMECA has a design focus, this refers to the existence of any design features, aids, gauges, readout devices, or verification procedures that will result in the detection of potential failure modes.

---

[1] Ishikawa, Kaoru, *Guide to Quality Control,* 2nd rev. English ed., Asian Productivity Organization, Tokyo, 1982.
[2] Juran, J. M. and F. M. Gryna, Jr., *Quality Planning and Analysis,* 2nd ed., McGraw-Hill Book Company, New York, 1980.

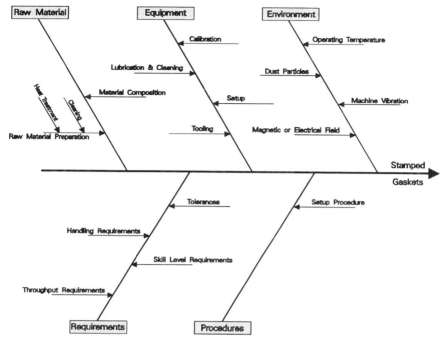

Figure 7.4. The Ishikawa cause and effect (fishbone) diagram.

8. *Rate the severity of a failure mode.* Severity, in the context of this analysis, refers to the seriousness of the effect or impact of a particular failure mode. The military standard, MIL-STD-1629A, proposes the classification of failure effects into the following four severity classes:[3]

a. *Catastrophic:* a failure that can result in the loss of personal life and/or a complete system loss.

b. *Critical:* a failure that can potentially cause serious personal injury and/or significant system damage and loss of system functionality.

c. *Marginal:* a failure that can result in minor injury to personnel, minor system damage, and/or degraded system functionality.

d. *Minor:* a failure not severe enough to cause any personal injury or system damage, but resulting in the requirement for some corrective maintenance.

On the other hand, the FMECA as utilized in the automotive industry generally involves rating failure mode severity on a scale of 1–10, as depicted in Table 7.1a.[4]

---

[3] MIL-STD-1629A, *Procedures for Performing a Failure Mode, Effects, and Criticality Analysis,* Department of Defense, Washington, DC, 1984.
[4] *Potential Failure Mode and Effects Analysis* (FMEA), Reference Manual, FMEA-1, AIAG, 1993. Developed by FMEA teams at Ford Motor Company, General Motors, Chrysler, Goodyear, Bosch, and Kelsey-Hayes, under the auspices of ASQC and AIAG.

TABLE 7.1   Rating Failure Mode Severity, Frequency, and Probability of Detection

a. Severity of Effects	Rating
Minor: Unreasonable to expect that the minor nature of this failure would have any real effect on the system performance. Customer will probably not even notice the failure.	1
Low: Low severity ranking owing to nature of failure causing only a slight customer annoyance. Customer will probably only notice a slight deterioration of the system performance.	2 3
Moderate: Moderate ranking because the failure causes some customer dissatisfaction. Customer is made uncomfortable or is annoyed by the failure. Customer will notice some subsystem performance deterioration.	4 5 6
High: High degree of customer dissatisfaction due to the nature of the failure such as an inoperable system. Does not, however, violate system safety or cause noncompliance with government regulations.	7 8
Very high: Very high severity ranking when a potential failure mode affects safe system function or causes noncompliance with government regulations.	9 10

b. Failure Mode Occurrence Frequency	Rating	Failure Probability
Remote: Failure is unlikely.	1	<1 in 10^6
Low: Relatively few failures.	2	1 in 20,000
	3	1 in 4,000
Moderate: Occasional failures.	4	1 in 1,000
	5	1 in 400
	6	1 in 80
High: Repeated failures.	7	1 in 40
	8	1 in 20
Very high: Failure is almost inevitable.	9	1 in 8
	10	1 in 2

c. Probability of Detection	Rating
Very high: Design verification (DV) or current process controls (PC) will almost certainly detect a potential failure mode.	1 2
High: DV or current PCs have a good chance of detecting a potential failure mode.	3 4
Moderate: DV or current PCs may detect a potential failure mode.	5 6
Low: DV or current PCs will not likely detect a potential failure mode.	7 8

c. Probability of Detection	*Rating*
Very low: DV or current PCs will probably not detect a potential failure mode.	9
Absolute certainty of nondetection: DV or current PCs will/cannot detect a potential failure mode.	10

Source: Adapted from *Potential Failure Mode and Effects Analysis* (FMEA) Reference Manual, FMEA-1, AIAG, 1993.

9. *Rate the frequency of occurrence of a failure mode.* Given that a function or a physical component within a system is likely to fail in multiple ways, this step addresses the frequency of occurrence of each individual failure mode, also called the modal failure frequency. Obviously, the sum of all modal failure frequencies for a system element must equal its failure rate. The military standard, MIL-STD-1629A, proposes two approaches, one qualitative and the other quantitative, to address the frequency of occurrence of a failure mode. The qualitative approach is recommended only in the event that specific failure-rate information is not available to the analyst. Table 7.2 lists the qualitative rating of failure mode frequencies. The FMECA, as performed within the automotive industry, involves rating failure mode frequencies on a scale of 1–10 as depicted in Table 7.1*b*.

TABLE 7.2   Qualitative Ranking of Failure Probabilities

Rating	Name and Explanation
A	*Frequent:* A high probability of occurrence during item operation. High probability is defined as a single failure mode probability greater than 0.20 of the overall item probability of failure.
B	*Reasonably probable:* A moderate probability of occurrence during item operation. Probable in this context is defined as a single failure mode probability greater than 0.10 but less than 0.20 of the overall probability of item failure.
C	*Occasional:* An occasional probability of occurrence during item operation. Occasional probability is defined as a single failure mode probability greater than 0.01 of the overall probability of item failure.
D	*Remote:* An unlikely probability of occurrence during item operation. Remote probability is defined as a single failure mode probability greater than 0.001 but less than 0.01 of the overall probability of item failure.
E	*Extremely unlikely:* A failure whose probability of occurrence is negligible during item operation. Extremely unlikely is defined as a single failure mode probability less than 0.001 of the overall probability of item failure.

Source: Adapted from *Procedures for Performing a Failure Mode, Effects, and Criticality Analysis,* MIL-STD-1629A, 1984.

In the event that adequate failure-rate information and data are available on the various system elements, the quantitative approach is recommended.[5] This approach requires part or functional failure rates, operating times and duty cycles, failure mode ratio (i.e., the ratio of modal failure frequency to total functional or item failure frequency), and the failure effect probabilities as input.

10. *Rate the probability that a failure will be detected.* This refers to the probability that the design features/aids and verification procedures will detect potential failure modes in time to prevent a system-level failure. When this analysis has a process orientation, this refers to the probability that a set of process controls currently in place will be in a position to detect and isolate a failure before it gets transferred to the subsequent processes or to the end customer/consumer. This probability is once again rated on a scale of 1–10 as generally practiced within the automotive industry.[6]

11. *Analyzing failure mode criticality.* The objective of this step is to consolidate the information generated thus far in an effort to delineate the more critical aspects of the system design. Criticality in the context of this analysis is a function of the frequency of occurrence of a failure mode, its severity, and the probability that it will be detected in time to preclude its impact at the system level.

On the commercial side of the spectrum, primarily the automotive industry, use is made of a metric called the *risk priority number* or RPN, which can be expressed as:

$$\text{RPN} = (\text{severity rating})(\text{frequency rating})(\text{probability of detection rating}) \qquad (7.1)$$

The RPN reflects failure-mode criticality. Obviously, a failure mode with a high frequency of occurrence, with significant impact on system performance, and which is difficult to detect is likely to have a very high RPN, that is, a high "criticality." Conducting a Pareto analysis (as part of the FMECA) helps in visualizing the more critical failure modes, which require greater attention on part of the design team and management. Such a visual portrayal helps in differentiating the few critical failure modes from the insignificant many.

While it is important to get a "feel" for the more critical failure modes, it is failure causes that need to be resolved if improvements to product and process designs are to be affected. Your authors recommend a failure-cause Pareto analysis in conjunction with a failure-mode Pareto analysis. This is because very often the same cause is likely to "trigger" multiple failures, and conducting a causal Pareto analysis will provide insight into the most damaging failure causes. Examples of Pareto analyses focused on failure modes and causes are depicted in Figures 7.5*a* and *b,* respectively.

---

[5] This approach is consistent with the quantitative approach proposed in MIL-STD-1629A, *Procedures for Performing a Failure Mode, Effects, and Criticality Analysis,* Department of Defense, Washington, D.C., 1984.

[6] *Potential Failure Mode and Effects Analysis* (FMEA), An Instructional Manual, Ford Motor Company, 1988; *Failure Mode and Effects Analysis,* Saturn Quality System, Saturn Corporation.

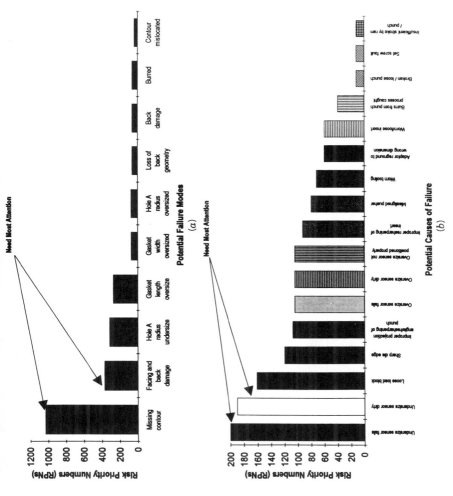

Figure 7.5. Example failure mode and cause Pareto analyses.

245

PROCESS FAILURE MODE, EFFECTS, AND CRITICALITY ANALYSIS

Part or Process Name/No. Gasket/Stamping _____
Design/Production Responsibility _____

Model No. _____
Design Release Date 03/11/XX
Production State Date 05/22/XX

Prepared by __Dinesh V__
FMECA Date 04/14/XX
Approved by __B. Blanchard__

Reference Number	Process Description	Potential Failure Mode	Potential Causes of Failure	Potential Effects of Failure on Process *	Potential Effects of Failure on Customer *	Current Controls	OCCURRENCE	SEVERITY (process)	SEVERITY (customer)	DETECTION	RPN	Recommended Action Items	Responsibility
(4)	Press Punch Holder	a) Missing Contour	a) Broken/Loose Punch	a) Jam Finishing Operation.	a) Blocked Engine Oil Flow and Engine Seizure.	a) 100% Autodetect.	4	4	10	1	160		
		b) Contour Mislocated	b) Insufficient Stroke by Ram/Punch	a) Jam Finishing Operation.	a) Blocked Engine Oil Flow and Engine Seizure.	a) 100% Autodetect.	3	4	10	1	120		

Figure 7.6. Sample FMECA worksheet.

* Given the process orientation of this illustrative FMECA, your authors recommend delineating failure effects into two types: (a) effects that are internal to the manufacturing process, and (b) effects that are likely to impact the customer or consumer. This will ensure an increased customer focus during the course of a FMECA.

The next step in the analysis is to identify compensatory provisions, or to redesign system components in order to address critical failure modes and eliminate the more serious failure causes. The subsequent evolution of the FMECA needs to reflect these changes and point to other areas of concern. A partially completed FMECA worksheet with a process orientation is depicted in Figure 7.6.

The military standard, MIL-STD-1629A, proposes two different approaches, one qualitative and the other quantitative, to address failure mode criticality. The qualitative approach involves locating failure modes in a criticality matrix, as shown in Figure 7.7. This location is a function of the severity classification, in accordance with the classification discussed in Step 8, and frequency of occurrence, as listed in Table 7.2.[7]

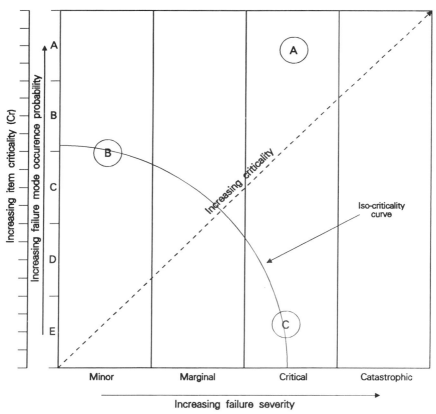

Figure 7.7. A criticality matrix.

[7]MIL-STD-1629A does not explicitly address and include the probability of detecting a failure in the criticality analysis of failure modes. Your authors recommend addressing this issue in the overall criticality analysis.

The quantitative approach involves the calculation of failure mode criticality numbers, given as

$$C_{im} = \beta_i \alpha_i \lambda_p t \tag{7.2}$$

where C_{im} is the ith failure mode criticality number. It is a function of β_i, the conditional probability that the ith failure mode will have a particular severity classification; α_i, the failure mode ratio (the ratio of the frequency of occurrence of the ith failure mode to the overall item failure frequency); and t, the operating time.

Subsequent to the calculation of failure-mode criticality numbers, item criticality numbers are computed by taking the sum of all of relevant failure mode criticality numbers which have a particular severity classification. This is given as

$$C_{pr} = \sum_{i=1}^{n} \beta_i \alpha_i \lambda_p t \tag{7.3}$$

where C_{pr} is the item criticality number for the pth item with n failure modes under a specific severity classification. These item criticality numbers are next located on a criticality matrix, as shown in Figure 7.7.

In the context of the criticality matrix, the criticality of a particular item is a function of its distance from the origin. The greater this distance, the higher the item criticality. For example, item A is more critical than item B because it is further away from the origin of the matrix. The readers may note that indifference curves, or isocriticality curves, exist on this matrix, as depicted in Figure 7.7. For example, the criticalities of items B and C are more or less equal. In other words, since item criticality is a function of frequency of occurrence and severity classification, it is possible to have the same criticality for a variety of combinations of failure frequencies and severity classifications.

7.2.3 Fault Tree Analysis

During the very early stages of the system design process, and in the absence of information required to complete a FMECA (as discussed in Section 7.2.2), a fault tree analysis (FTA) is often conducted to gain insight into critical aspects of the feasible and selected system design concepts and technology applications.

Fault tree analysis is a deductive approach involving the graphical enumeration and analysis of the different ways in which a particular system failure can occur, and the probability of its occurrence. A separate fault tree is developed for every critical failure mode, or undesired top-level event. Attention is focused on this top-level event and the first-tier causes associated with it. Each of these causes is next investigated for *its* causes, and so on. This "top-down" causal hierarchy, and the associated probabilities, is called a fault tree.

One of the outputs from a FTA is the probability of occurrence of the top-level event or failure. In the event that this probability is unacceptable, the causal hierarchy developed provides system engineers with insight into aspects of the system to which redesign efforts may be directed or compensatory provisions provided. The FTA can have the most impact if initiated during the conceptual and preliminary design phases, when design and configuration changes can be most easily and cost effectively implemented.

The logic used in developing and analyzing a fault tree has its foundations in boolean algebra. Axioms from boolean algebra are used to collapse the initial version of the fault tree to an equivalent reduced tree with the objective of deriving the minimum cut sets. Minimum cut sets are unique combinations of basic failure events that can cause the undesired top-level event to occur. These minimum cut sets are necessary to evaluate a fault tree from a qualitative and quantitative perspective.

The steps necessary to conduct a FTA are depicted in Figure 7.8 and discussed next.

1. *Identify top-level event.* The first and most important step is to identify and define the top-level event. It is necessary for the analyst to be rather specific in defining this event. A generic and nonspecific definition is likely to result in a broad-based fault tree with too wide a scope and lacking in focus.

2. *Develop the fault tree.* Once the top-level event has been satisfactorily defined, the next step is to construct the initial causal hiearchy in the form of a fault tree. Once again, a technique such as Ishikawa's *cause and effect diagram,* discussed in Section 7.2.2, can prove beneficial. While developing the fault tree, all hidden failures (and the relevant combinations thereof) must be considered and incorporated.

For the sake of consistency and communication, a standard symbology to develop the fault tree is recommended. Table 7.3 depicts and defines the symbology to comprehensively represent the causal hierarchy and interconnects associated with a particular top-level event. Further, an illustrative fault tree is depicted in Figure 7.9. The symbols OR1 and OR2 represent the two OR logic gates, and1–and8 represent eight AND logic gates, I-1 through I-8 represent eight intermediate fault events, b1–b5 represent five basic events, and u1 and u2 represent two undeveloped

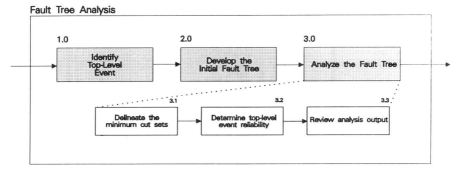

Figure 7.8. Steps to complete a FTA.

TABLE 7.3   Fault Tree Construction Symbology

Fault Tree Symbol	Discussion
	The ellipse represents the *top-level event*. Obviously, the ellipse always appears at the very top of a fault tree.
	The rectangle represents an *intermediate fault event*. A rectangle can appear anywhere in a tree except at the lowest level in the hierarchy.
	A circle represents the *lowest level failure event,* also called a *basic event*. Basic events are likely to appear at the lowest level in a fault tree.
	The diamond represents an *undeveloped event*. Undeveloped events could be further broken down, but are not for the sake of simplicity. Very often, complex undeveloped events are analyzed through a separate fault tree. Undeveloped events appear at the lowest level in a fault tree.
	This symbol, sometimes called the house, represents an *input event*. An input event refers to a signal or input that could cause a system failure.
	This symbol represents the *AND logic gate*. In this case, the output is realized only after all the associated inputs have been received.
	This symbol represents the *OR logic gate*. In this case, any one or more of the inputs need to be received for the output to be realized.
	This symbol represents the *ORDERED AND logic gate*. In this case, the output is realized only after all the associated inputs have been received in a particular predetermined order.
	This symbol represents the *EXCLUSIVE OR logic gate*. In this case, one and only one of the associated inputs needs to be received for the input to be realized.

failure events. While constructing a fault tree, it is important to break every branch down to a reasonable and consistent level of detail.

3. *Analyze the fault tree.* The third step in conducting the FTA is to analyze the initial fault tree developed. A comprehensive analysis of a fault tree involves both a quantitative and a qualitative perspective. The important steps in completing the analysis of a fault tree are:

   a. *Delineate the minimum cut sets.* As part of the analysis process, the minimum cut sets in the initial fault tree are first delineated. These are necessary to evaluate a fault tree from a qualitative and/or quantitative perspective. The objective of this step is to reduce the initial tree to a simpler equivalent reduced fault tree. The minimum cut sets can be derived using two different

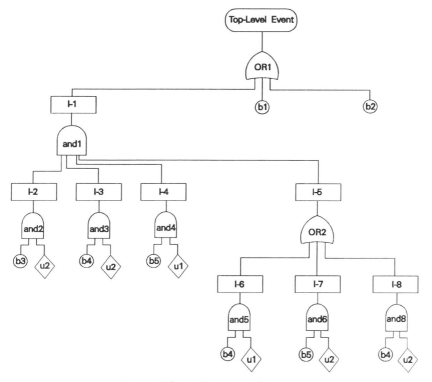

Figure 7.9. An illustrative fault tree.

approaches. The first approach involves a graphical analysis of the initial tree, an enumeration of all the cut sets, and the subsequent delineation of the minimal cut sets. The second approach, on the other hand, involves translating the graphical fault tree into an equivalent boolean expression. This boolean expression is then reduced to a simpler equivalent expression by eliminating all the redundancies and so on. As an example, the fault tree depicted in Figure 7.9 can be translated into a simpler and equivalent fault tree, through boolean reduction, as depicted in Figure 7.10.

  b. *Determine the reliability of the top-level event.* This is accomplished by first determining the probabilities of all relevant input events, and the subsequent consolidation of these probabilities in accordance with the underlying logic of the tree. The reliability of the top-level event is computed by taking the product of the reliabilities of the individual minimum cut sets.[8]

---

[8] *Fault Tree Analysis Application Guide,* prepared by the Reliability Analysis Center (RAC), Rome Air Development Center, is an excellent "how-to" source for the application of FTA depicting numerous case studies.

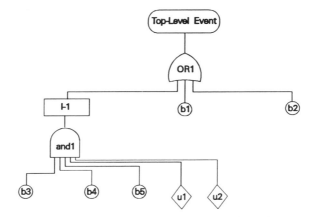

Figure 7.10. A reduced equivalent fault tree (refer to figure 7.9).

c. *Review analysis output.* If the derived top-level probability is unacceptable, necessary redesign or compensation efforts will need to be initiated. The development of the fault tree and subsequent delineation of minimum cut sets provides engineers and analysts with the kind of foundation needed for making sound decisions.

Over and above the more obvious benefits, a FTA provides invaluable input to subsequent analyses such as system testability analysis, system diagnostics, and troubleshooting. A FTA is most beneficial if conducted, not in isolation, but as part of an integrated system analysis process. Moreover, it needs to evolve and reflect all the design changes implemented.

7.2.4 Reliability Centered Maintenance

Reliability centered maintenance (RCM) is a systematic approach to develop a focused, effective, and cost-efficient preventive maintenance program and control plan for a product or process. This technique is best initiated during the early system design process and evolves as the system design, development, and deployment activities progress. The technique, however, can also be used to evaluate preventive maintenance programs for existing systems with the objective of continuous improvement.

The RCM technique was developed in the 1960s primarily through the efforts of the commercial airline industry.[9] The essence of this technique is a structured deci-

---

[9] A maintenance steering group (MSG) was formed in the 1960s which undertook the development of this technique. The result was a document called "747 Maintenance Steering Group Handbook: Maintenance Evaluation and Program Development (MSG-1)" published in 1968. This effort, focused toward a particular aircraft, was next generalized and published in 1970 as "Airline/Manufacturer Maintenance Program

sion tree which leads the analyst through a "tailored" logic in order to delineate the most applicable preventive maintenance tasks (their nature and frequency). The overall process involved in implementing the RCM technique is depicted in Figure 7.11. The major steps are discussed next.

1. *Identification of critical system functions and/or components.* The first step in this analysis is to identify critical system functions and/or components; for example, airplane wings, car engine, printer head, video head, and so on. Criticality in terms of this analysis is a function of the failure frequency, the failure effect severity, and the probability of detection of the relevant failure modes. The concept of criticality is discussed in more detail in Section 7.2.2. This step is facilitated through outputs from the system functional analysis (see Section 6.1.1) and the failure mode, effects, and criticality analysis (see Section 7.2.2). This is also depicted in Figure 7.11, Blocks 1.0–4.0.

2. *Application of the RCM decision logic and PM program development.* The critical system elements are next subjected to the tailored RCM decision logic. The objective here is to better understand the nature of failures associated with the critical system functions or components. In each case, and whenever feasible, this knowledge is translated into a set of preventive maintenance tasks, or a set of redesign requirements. An illustrative RCM decision logic is depicted in Figure 7.12. Numerous decision logics, with slight variations to the original MSG-3 logic and tailored to better address certain types of systems, have been developed and are currently being utilized.[10]

These slight variations notwithstanding (as illustrated in Figure 7.12), the first concern is whether *a failure is evident or hidden.* A failure could become evident through the aid of certain color-coded, visual gauges and/or alarms. It may also become evident if it has a perceptible impact on system operation and performance. On the other hand, a failure may not be evident (i.e., hidden) in the absence of an appropriate alarm, and even more so if it does not have an immediate or direct impact on system performance. For example, a leaking engine basket is not likely to reflect an immediate and evident change in the automobile's operation, but it may in time and, after most of the engine oil has leaked, cause engine seizure.

In the event that a failure is not immediately evident, it may be necessary to either institute a specific fault-finding task as part of the overall PM program or design in an alarm that signals a failure (or pending failure).

Planning Document-MSG2." The MSG-2 approach was further developed and published in 1978 as "Reliability Centered Maintenance," Report Number A066-579, prepared by United Airlines, and in 1980 as "Airline/Manufacturer Maintenance Program Planning Document-MSG3." The MSG-3 report has been revised and is currently available as "Airline/Manufacturer Maintenance Program Development Document (MSG-3), 1993." These reports are available from the Air Transport Association.

[10]RCM decision logics, with some variations, have also been proposed in: (a) MIL-STD-2173(AS)— *Reliability-Centered Maintenance Requirements for Naval Aircraft, Weapons Systems, and Support Equipment;* (b) AMC-P-750-2—*Guide to Reliability-Centered Maintenance;* and (c) Moubray, John, *Reliability-Centered Maintenance,* Butterworth-Heinemann, London, United Kingdom, 1991.

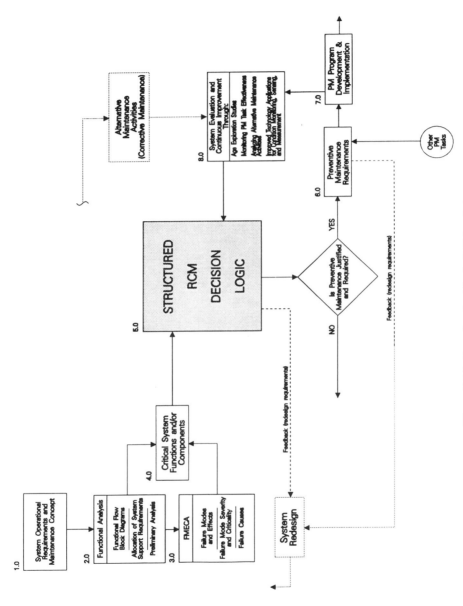

Figure 7.11. Steps involved in the RCM technique.

The next concern is whether the failure is likely to compromise personal safety or system functionality. Queries exist in the decision logic to clarify this and other likely impacts of failures. This step in the overall process can be facilitated by the results of the FMECA (Section 7.2.2). The objective is to better understand the basic nature of the failure being studied. Is the failure likely to compromise the system or personnel safety? Does it have an operational or economic impact? For example, a failure of an aircraft wing may be safety related, whereas a certain failure in the case of an automobile engine may result in increased oil consumption without any operational degradation, and will therefore have an economic impact. In another case, a failed printer head may result in a complete loss of printing capability and is said to have an operational impact, and so on.

Once the failure has been identified as a certain type, it is then subjected to another set of questions. However, in order to answer this next set of questions adequately, the analyst must thoroughly understand the nature of the failure from a *physics of failure* perspective. For example, in the event of a crack in the airplane wing, how fast is this crack likely to propagate? How long before such a crack causes a functional failure?

These questions have an underlying objective of delineating a feasible set of compensatory provisions or preventive maintenance tasks. Is a lubrication or servicing task applicable and effective and, if so, what is the most cost-effective and efficient frequency? Will a periodic check help preclude the failure, and at what frequency? Periodic inspections or checkouts are likely to be most applicable in situations where a failure is unlikely to occur immediately, but is likely to develop at a certain rate over a period of time. The frequency of inspections can vary from very infrequently to continuously, as in the case of condition monitoring (see Section 7.3.4). Some of the more specific queries are presented in Figure 7.12. In each case, the analyst must not only respond with a "yes" or "no," but he or she should also give specific reasons for each response. Why would lubrication either make, or not make, any difference? Why would periodic inspection be a *value-added* task? It could be that the component's wearout characteristics have a predictable trend, in which case inspections at predetermined intervals could preclude corrective maintenance. Would it be effective to discard and replace certain system elements in order to upgrade the overall inherent reliability? And if so, at what intervals or after how many hours of system operation (e.g., changing the engine oil after 3000 miles of driving)? Further, in each case a trade-off study, in terms of the benefit/cost and overall impact on the system, needs to be accomplished between performing a task and not performing it.[11]

In the event that a set of applicable and effective preventive maintenance requirements are delineated, they are input to the preventive maintenance program development process and subsequently implemented as shown in Figure 7.11, Blocks

---

[11] MIL-STD-2173(AS)—*Reliability-Centered Maintenance Requirements for Naval Aircraft, Weapons Systems, and Support Equipment,* January 1986, is an excellent source on the thought process behind responding to the queries of the RCM decision logic along with the issues that need to be considered.

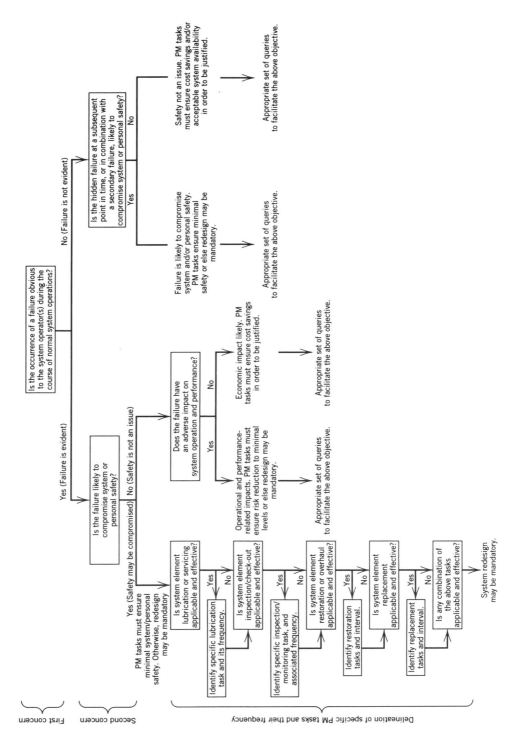

256

5.0–7.0. If no feasible and cost-effective provisions or preventive maintenance tasks can be identified, a redesign effort may need to be initiated.

3. *PM program implementation and evaluation.* Very often, the PM program initially delineated and implemented is likely to have failed to consider certain aspects of the system, delineated a very conservative set of PM tasks, or both. Continuous monitoring and evaluation of preventive maintenance tasks along with all other (corrective) maintenance actions is imperative in order to realize a cost-effective preventive maintenance program. This is depicted in Figure 7.11, Block 8.0. Further, given the continuously improving technology applications in the field of condition monitoring, sensing, and measurement, PM tasks need to be reevaluated and modified whenever necessary.

Often, when the RCM technique is conducted in the early phases of the system design and development process, decisions are made in the absence of ample data. These decisions may need to be verified and modified, whenever justified, as part of the overall PM evaluation and continuous improvement program. Age exploration studies are often conducted to facilitate this process. Tests are conducted on samples of unique system elements or components with the objective of better understanding their reliability and wear-out characteristics under actual operating conditions. Such studies can aid the evaluation of applicable PM tasks, and help delineate any dominant failure modes associated with the component being monitored and/or any correlation between component age and reliability characteristics, as depicted in Figure 7.13. If any significant correlation between age and reliability is

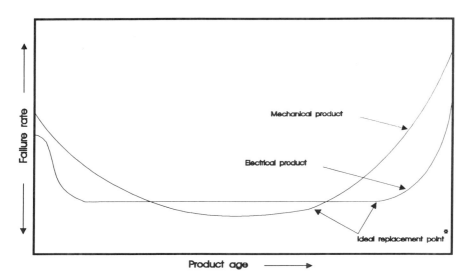

Figure 7.13. Correlation between product age and reliability.

noticed and verified, the associated PM tasks and their frequency may be modified and adapted for greater effectiveness. Also, redesign efforts may be initiated to account for some, if any, of the dominant component failure modes.

The preventive maintenance requirements identified during the RCM analysis are subsequently translated into a set of specific preventive maintenance tasks along with suggested frequencies. The extent and scope of these requirements and tasks reflect upon the overall maintainability of a product or process. Extensive preventive maintenance requirements are likely to negatively impact the system life-cycle cost, and may indicate an inadequate system design from a maintainability perspective.

The preventive maintenance tasks and their frequencies, identified during this analysis, become an input to the system maintainability prediction process, discussed in Chapter 8. This information is necessary to assess the system mean time to repair metric. Further, the results from the RCM analysis are incorporated into the overall system maintenance task analysis, discussed in Chapter 9, to identify the resources (facilities, test equipment, tools, maintenance personnel skill level requirements, etc.) that may be required.

7.2.5   Critical Useful Life Analysis

A critical useful life item is one which, because of its short life, is incapable of satisfying the functional requirements imposed by its application unless corrective or preventive maintenance is accomplished. The identification of these "short-life" items is accomplished during the preliminary and detail design phases, using the results of the FMECA and the RCM analysis as a basis for justification. The critical items are listed along with their expected life in terms of system operating time, operating cycles, or calendar time. This listing significantly influences the requirements for maintenance, personnel support, tools and test equipment, and spares or repair parts. In the interest of system/product design for maintainability, all short-life critical items should be eliminated if at all possible.

The critical-useful-life analysis is often accomplished as a task within a reliability program; however, while it is envisioned that the reliability of a system can be extended through the accomplishment of scheduled replacements (even if the items replaced are on the "short-life" list), the impacts on maintainability and supportability can be extremely negative!

7.2.6   Reliability Prediction

Reliability prediction is an evolving check on the system design and development activity with respect to the reliability requirements initially delineated during the system requirements definition stage. It is continuously refined as increasing amounts of engineering data become available to the analyst. The predicted system reliability values, expressed as failure rate (λ), R, MTBF, or MTBM, are compared against the corresponding requirements initially allocated down the system hiear-

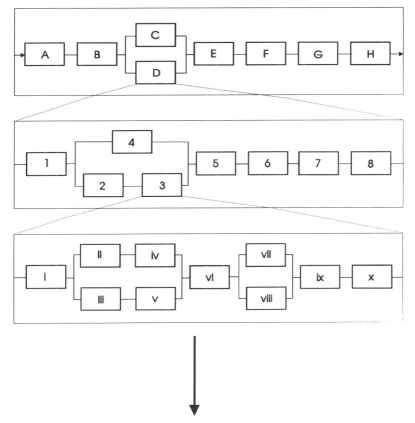

Figure 7.14. Sample reliability block diagram expansion.

chy. Subsequent to this comparison, attempts are made to identify specific design improvement opportunities as required.

System-level reliability requirements are allocated down the system hierarchy through the utilization of reliability block diagrams. The reliability block diagram, as illustrated in Figure 7.1, is an extension of the functional flow diagrams discussed in Section 6.1.1. The objective in developing a reliability block diagram is to provide a reasonable approximation of system elements which must successfully accomplish their corresponding functions in order for the overall system to operate effectively. To the extent feasible, each of these functional entities (represented by blocks in the reliability block diagram), should be functionally independent of other functional entities at the same level within the overall system hiearchy. Minimizing the number of interface functions and information flows simplifies system modeling and the overall analysis process. Figure 7.14 depicts another reliability block diagram, along with its progressive expansion as the system design and development progresses to a more detailed state.

The determination of system-level reliability requirements is accomplished during the requirements definition and conceptual design phase. This system-level reliability requirement is allocated down to the block level during the preliminary design phase (refer to Section 6.2). The process of reliability prediction during the preliminary and detailed design phases, and the subsequent comparison with the allocated reliability factors, provides a continuously evolving check on the design process.

The approach used to predict and determine block failure rates may vary, depending upon the extent of information available to the analyst and the maturity of system definition. Failure rates may be derived from direct field and/or test experience with similar subsystems or components, parametric analyses, reliability prediction reports covering components and assemblies which are of a similar type and functionality, research using stress analysis conducted on similar components, and/or engineering estimates based upon judgment and experience with similar components and assemblies in the past. In many instances, weighting factors are used to adjust component reliability metrics in order to compensate for changing operational environments, operational stresses, and varying functional complexities. Some of the basic reliability prediction techniques are summarized as follows:

1. *Prediction may be based upon the analysis of similar equipment.* This technique is particularly useful and recommended only when the absence of relevant data precludes the utilization of more sophisticated techniques. The prediction approach uses MTBF or failure-rate values for similar components and assemblies, with matching functionality and similar reliability characteristics, and near identical levels of complexity. The reliability of the system element that is part of the current design is assumed to be equal to that of equipment most comparable in terms of performance and complexity. This technique does not address environmental issues, stresses, part quantities or types and, even though convenient to accomplish, is not very accurate.

2. *System reliability predictions may be based upon an estimate of active element groups (AEG).*[12] An AEG is the smallest functional building block that controls or converts energy. An AEG includes one active element (e.g., relay, transistor, pump, or machine) and a number of supporting passive elements to form a group such as a relay circuit or an amplifier circuit. By estimating the number of AEGs, and using a complexity chart, one can predict system level reliability in terms of failure rates or MTBF.

3. *System reliability prediction may be accomplished based upon an equipment parts count.* Depending upon the data source utilized, the number of component part-type categories or groups, assumed stress levels, and operating environments, a variety of methods are used to predict system-level reliability based on this general approach.[13] A system design parts/material list is used and component parts are first

---

[12] A good source for more details on this approach to reliability prediction is NAVORD OD 44622— *Reliability Data Analysis and Interpretation.*

[13] One popular data source to facilitate this approach is MIL-HDBK-217—*Reliability Prediction of Electronic Equipment* (latest revision).

classified in accordance with certain designated part-type categories. Thereafter, failure rates are assigned to the parts and consolidated in accordance with the underlying component part interdependencies and relationships, as discussed in Section 7.2.1. This consolidation leads to the derivation of a predicted reliability metric at the system level. Interdependencies between system elements can be better visualized through the utilization of reliability block diagrams.

The "goodness" of this prediction is a function of the extent and accuracy of the historical data available. An example illustrating this approach is presented in Table 7.4.

4. *System reliability prediction may be based upon competent stress analysis.* In the presence of greater engineering and component part operational information, more sophisticated reliability prediction techniques may be applied. When conducting a prediction in accordance with this approach, part types and quantities are determined, failure rates are applied, and stress ratios and issues relating to operating environments are addressed and incorporated. Further, interaction effects between components are also taken into account.

The basic approach involves adjusting component base failure rates to account for different application stresses, component quality levels, operational environments, and so on as

$$\lambda_{actual} = \lambda_{base}(\pi_{application})(\pi_{environment})(\pi_{quality})\dots \tag{7.4}$$

Some data sources that support this approach involve a multiplication factor to address the component burn-in phase.[14] While most of the data sources that support

TABLE 7.4 Reliability Prediction Data Summary

Component Part	Failure Rate/Part (%/1000 Hours)	Quantity of Parts	(Failure Rate/Part) (Quantity)
Part A	0.161	10	1.610
Part B	0.102	130	13.260
Part C	0.021	72	1.512
Part D	0.084	91	7.644
Part E	0.452	53	23.956
Part F	0.191	3	0.573
Part G	0.022	20	0.440
Failure rate (λ) = 48.995%/1000 hours			Σ = 48.995%
$MTBF = \dfrac{1000}{0.48995} = 2041$ hours			

[14] Two of the more popular data sources that support this approach are (a) Bell Communications Research (Bellcore) Report, TR-NWT-000331, September, 1990 and (b) MIL-HDBK-217—*Reliability Prediction of Electronic Equipment.* While these two sources are generic and available for use by anyone, many companies supplement these sources with their own proprietory data reflecting company experience with certain component types in unique applications.

this approach have traditionally concentrated on electronic component types, research is currently under way to develop a similar set of models to address mechanical component reliabilities as well.

5. *Reliability prediction based upon component physics of failure.* This approach is based on a detailed analysis of each component failure mechanism from the perspective of established engineering materials, physics, and chemistry principles. This approach is likely to be most useful when dealing with unique component types. While this prediction methodology is highly sophisticated and may require greater investment in comparison with the other approaches, it is recommended whenever feasible.

The output from system reliability allocation and prediction studies provides a direct input to other related analyses including FMECA and FTA (Sections 7.2.2 and 7.2.3), RCM (Section 7.2.4), reliability growth modeling (Section 7.2.7), maintenance task analysis and the logistic support analysis (Chapter 9), and so on. System reliability provides an index to the frequency of maintenance, its nature, and the quantity of maintenance actions anticipated throughout the useful operational life of the equipment. It is, therefore, imperative that reliability prediction results be accurate since this will impact the effectiveness and accuracy of subsequent analyses.

7.2.7 Reliability Growth Modeling

The concept of reliability growth and reliability growth curves was discussed earlier in Section 4.1.4. Reliability growth refers to the positive improvement in a reliability parameter over a period of time due to changes in product design or in the manufacturing process. The concept of reliability growth and growth curves was addressed for the first time by James T. Duane in the mid-1960s. Numerous models for projecting system reliability growth have been postulated since. Two such models, the *Duane Model* and the *Gompertz Model,* are briefly outlined and discussed in this section.[15]

To fully realize the benefits of system reliability growth, it must be properly managed. This involves systematic planning for reliability achievement as a function of time and the application of resources, and controlling the ongoing rate of achievement by reallocation of resources based on comparisons between planned and assessed system reliability values. The reliability growth management and planning process is depicted in Figure 7.15.

Furthermore, the rate at which system reliability "grows" is a function of how rapidly activities in the R-growth management process can be accomplished, how real the identified problems are, and how well the redesign effort solves the identified problems without introducing new problems.

R-growth modeling involves the assessment of system reliability trends through

---

[15] These two models are discussed in more detail in Kececioglu, D., *Reliability Engineering Handbook,* Vol. 2, Prentice Hall, Englewood Cliffs, NJ, 1991.

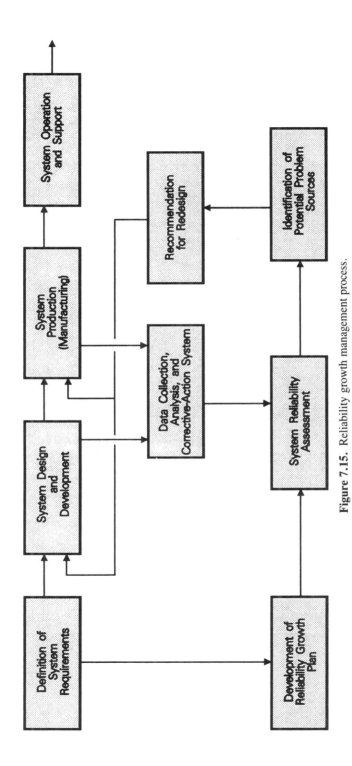

Figure 7.15. Reliability growth management process.

reliability predictions; test, analyze, and fix (TAAF) procedures; developmental testing; and so on. These trends from previous programs are utilized in the development of idealized R-growth curves, as depicted in Figure 7.16. An appropriate idealized curve is selected for application, depending upon the nature of the design and development program. This curve then serves as an evolving baseline for the current design and development program. Attempts are made to assess current system reliability. These assessments are next extrapolated, in accordance with the idealized R-growth curve selected, to the conclusion of the system design process, as depicted in Figure 7.16. The objective is to predict if the system, given its current state, will qualify from a reliability requirements perspective, and to delineate activities and actions that will facilitate this process.

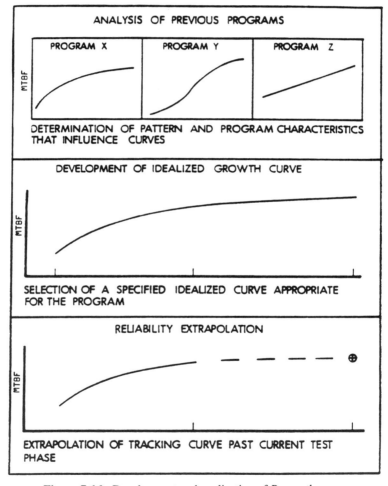

Figure 7.16. Development and application of *R*-growth curves.

While there is an obvious benefit of reliability growth modeling from the perspective of system reliability requirements and analysis, this concept is very applicable to the maintainability posture of a system as well. The projected system reliability growth impacts system reliability and maintainability trade-off studies, the overall system availability, and the support infrastructure effectiveness. It impacts prediction of the frequency and types of system failures during the early design and development phases and through to the period of program time where the mature reliability is expected to be achieved.

Duane derived an empirical relationship based upon the MTBF improvement observed with respect to a range of aircraft components. Figure 4.6 depicts an illustrative R-growth curve.

The reliability growth curve in the Duane model can be expressed as

$$\mathrm{MTBF_c = MTBF_s}(T)^\beta \tag{7.5}$$

where $\mathrm{MTBF_c}$ and $\mathrm{MTBF_s}$ are the cumulative mean time between failures at a certain future projected point in the design and development process, and at the start of the testing and observation; T is the total test or operating time; and β is the slope of the growth curve. The slope indicates the effectiveness of the reliability growth program and has a strong correlation with the intensity of the effort. Equation (7.5) reduces to a linear relationship between the current and future cumulative system reliabilities, for a given reliability growth parameter and time, in a log-log field, as

$$\log(\mathrm{MTBF_c}) = \log(\mathrm{MTBF_s}) + \beta \, \log(T) \tag{7.6}$$

Equation (7.5) expresses reliability as cumulative mean time between failures. It can also be written with reliability expressed as cummulative system failure rate as

$$\lambda_c = \lambda_s T^{-\beta} \tag{7.7}$$

where λ_c and λ_s are the cumulative system failure rates at a certain future projected point in the design and development process, and at the start of the testing and observation; T is the total test or operating time; and β is the slope of the growth curve. Equation (7.7) can also be reduced to a linear logarithmic form as

$$\log \lambda_c = \log \lambda_s - \beta \, \log T \tag{7.8}$$

A given reliability growth curve can also be used to assess the test and evaluation time required to attain a target system reliability. The Duane Model for projecting reliability growth was postulated by James T. Duane in a paper in 1964.[16] Since

---

[16] Duane, J. T., "Learning Curve Approach to Reliability Monitoring", *IEEE Trans. Aerospace,* **2,** (2), 1964.

then, numerous other empirical models have been proposed and used, including the Gompertz Model.[17] The form of this model is

$$R_r = (\alpha)R_\infty^{\beta\tau} \tag{7.9}$$

where R_T is the system reliability at some future projected point in time T, R_∞ is the upper reliability limit that the system is capable of achieving, $(\alpha)R_\infty$ is the system reliability at the beginning of the reliability growth test and analysis activity or the observed system reliability at $T = 0$, and β is the reliability growth parameter. Further, both α and β have values in the range between 0 and 1.

7.2.8 Environmental Stress Screening

Environmental stress screening (ESS) is an ongoing process in which environmental stimuli, such as rapid temperature cycling and random vibration, are applied to system components (usually electronic items) to precipitate latent defects to early failure. A latent defect (or a flaw) refers to some irregularity due to manufacturing processes or materials which will advance to the failure state when exposed to environmental or other stimuli. For example, a cold solder joint on a circuit board represent a flaw. After vibration, shock, and/or thermal cycling, the joint will likely crack and the board will fail. Some flaws can be stimulated by temperature cycling, some by vibration, some by voltage cycling, some by shock, and/or some by another stimulus. Often, a failure will not occur until a combination of such stimuli are applied.

An environmental stress screening program is often included as part of a formal reliability and/or quality program, and is tailored to a particular system requirement. Individual "screens" (or tests) are conducted with the objective of "driving out" potential failures prior to the delivery of equipment to the consumer. Of particular interest relative to maintainability is the likelihood of possible equipment failure (due to manufacturing and related processes) after the system is distributed and in operational use. The maintainability engineer should become familiar with ESS methods and their applications, with possible system maintenance and support requirements in mind.[18]

7.2.9 Reliability Qualification Testing

Reliability testing may include a combination of system qualification testing, component life or longevity testing, growth testing, environmental stress screening,

---

[17] For a more detailed discussion of the Gompertz Reliability Growth Model and the Modified Gompertz Model the reader is referred to Kececioglu, D., *Reliability Engineering Handbook,* Vol. 2, Prentice Hall, Englewood Cliffs, NJ, 1991.

[18] The ESS methods are described in DOD-HDBK-344, *Military Handbook,* "Environmental Stress Screening (ESS) Of Electronic Equipment," Department of Defense, Washington, DC. Additionally, the Institute of Environmental Sciences (IES) published "Environmental Stress Screening Guidelines" in 1981.

production sampling, and acceptance testing. The emphasis here is on reliability qualification testing, conducted during the detail design and development phase to provide an assessment of system reliability (MTBF) in terms of the initially specified requirements.

Reliability qualification tests are usually conducted using preproduction prototype models of the system, representative of the same configuration that the consumer will utilize in the field. The objective is to operate the system (or selected components thereof) in a simulated consumer environment, progressing through a series of operational scenarios, and measuring successful time/cycles completed against system failures. The operational scenarios selected should evolve from the operational functional block diagrams (a major segment of the system functional analysis) and should simulate, to the extent possible, how the system is expected to be utilized later on. When it is impossible to realistically simulate a utilitarian profile in a representative operational environment, an assumed duty cycle may be used in reliability testing. Figure 7.17 illustrates an example of a test duty cycle. The system (or applicable elements) may be installed in an environmental test chamber, or an equivalent facility, and operated following the illustrated profile.

Reliability qualification testing is generally accomplished following a specific plan, with associated producer and consumer risk factors, that has previously been negotiated and agreed upon by the contractor and the customer. Figure 7.18 illustrates a typical reliability sequential test plan that is representative of what is implemented on many programs.

Referring to the figure, the objective is to operate the system in accordance with a specified environmental scenario, accumulating successful operating time and progressing along the abscissa to the right toward the "Accept" region. As time accumulates without the advent of a failure, a high system MTBF may be assessed early, and testing may be discontinued given that the system has met the initially specified requirements. On the other hand, as failures occur during testing, one

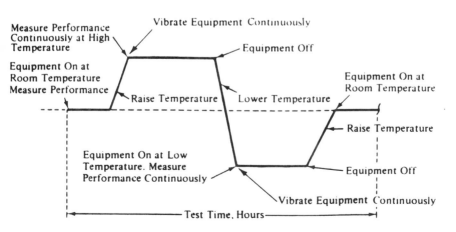

Figure 7.17. Program test duty cycle.

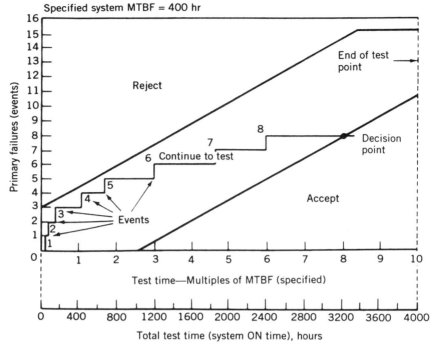

Figure 7.18. Reliability test plan.

progresses upward along the ordinate toward the "Reject" region. If the number of failures is high in a relatively short period of time, the system will be rejected in terms of meeting the specification requirement. In the event that the design is marginal (in terms of meeting a MTBF requirement), reliability testing continues until an "Accept–Reject" decision is made. Accept–reject criteria will vary with the program, the type and complexity of the system, negotiated consumer-producer risk factors, and so on.[19]

Reliability qualification testing constitutes as excellent source for the collection of maintenance-related data and the evaluation of maintainability characteristics in design. As failures occur, the follow-on repair actions should be accomplished in accordance with the proper procedures, using the recommended elements of logistic

---

[19]The requirements associated with reliability testing can be quite extensive, and are far beyond the rather cursory level of coverage in this text. The purpose here is to convey enough information to allow the maintainability engineer to gain some insight to reliability testing and the information that can be derived from such! For more in-depth coverage, refer to (a) MIL-STD-781D, *Reliability Design Qualification and Production Acceptance Tests: Exponential Distribution,* Department of Defense, Washington, DC and (b) Lloyd, D. K. and M. Lipow, *Reliability: Management, Methods, and Mathematics,* 2nd ed., published by the authors Defense and Space Systems Group, TRW Systems and Energy, Redondo Beach, CA, 1977.

support as specified for the field (i.e., recommended test and support equipment, spares/repair parts, facilities, data, etc., based on the results of the maintenance task analysis). Maintenance frequencies, task sequences, and task times can be recorded for each repair action. As the test progresses, one can calculate MTBF, MTBM, $\overline{M}ct$, and MMH/OH, and compare these values with the initially specified requirements for the system. Given that the information acquired from reliability testing is valid, it may be possible to reduce the data requirements for maintainability demonstration, that is, use the results from reliability tests as a substitute for maintainability test requirements. In any event, the maintainability engineer should be familiar with the requirements, concepts, and methods associated with reliability testing since it provides an excellent source for the acquisition of maintainability information.

7.2.10 Failure Reporting, Analysis, and Corrective Action

A major task in a typical reliability program is the establishment of a "failure reporting, analysis, and corrective-action system" (FRACAS). The objective is to develop and implement a closed-loop failure reporting system and the procedures for analysis, for determining the causes of failures, and for documenting the required corrective action initiated. A reliability analysis is accomplished in response to each failure to determine the mode of failure (i.e., the way in which a component failed—rupture, break, a switch in an "open" position, a stuck relay, etc.) and the effects of the failure on other components and on the overall system. Failures must be traced from the initial "symptom" at the system level, through the diagnostics, and down to the removal and replacement of the faulty item. The maintenance procedures used and the resources required in repairing the system should be recorded.

A Failure Review Board (FRB) is generally established to review critical failures, failure trends, and corrective-action status, and to ensure that the appropriate action is being taken in a timely manner to resolve any outstanding problems. The results of FRB actively should be reported periodically to other interested project personnel.

The FRACAS and FRB activity are critical not only from a reliability perspective, but from a maintainability viewpoint as well! First, the results have an obvious impact on system support requirements, and the maintenance task analysis must be updated to reflect any changes in this area. Second, any proposed corrective action for the purposes of solving a reliability problem may impact maintainability. Thus, prior to arriving at a final solution to a given problem, one must ensure that the maintainability characteristics in design have not been degraded as a result and that the impact on life-cycle cost does not reflect an upward trend.

7.2.11 Maintainability Analysis and Modeling

Within the context of "maintainability analysis" (as compared to "maintenance task analysis"), there are a series of functions/tasks that are directed toward the *design for maintainability*. These include the definition of system requirements for maintainability, the accomplishment of functional analysis and the development of maintenance functional flow diagrams, the top-down allocation of maintainability requirements to the various elements of the system, system synthesis, analysis and trade-off studies, and design optimization. These activities, accomplished by the maintainability engineer as part of the design team, evolve from the system maintenance concept defined during the conceptual design phase and extend into the preliminary and detail design and development phases. The nature of the functions/tasks in this area are described in Chapter 6.

7.2.12   Maintainability Prediction

As the design progresses, it is desirable to periodically assess the system configuration, as it exists at a given point in time, in terms of whether the initially specified maintainability requirements are being met. Assume that a system requirement includes a $\overline{M}ct$ of 30 minutes. It is appropriate to evaluate the design configuration, based on available design data (drawings, part/material lists, trade-off study reports, electronic data), and predict the anticipated $\overline{M}ct$. If the predicted value is greater than the specified value, corrective action will be required, resulting in a possible redesign effort. Referring to Figure 2.1, maintainability predictions are often accomplished when design baselines are established and for the purposes of review as part of the formal design review process. Maintainability prediction is discussed further in Chapter 8.

7.2.13   Maintenance Task Analysis

During the conceptual, preliminary, and detailed design and development phases, a "maintenance task analysis" (MTA) is accomplished by analyzing a given design configuration to (1) provide an assessment of that configuration in terms of the maintainability and supportability characteristics in design, and (2) identify the resources required for the sustaining maintenance and support of that system configuration throughout its planned life cycle. Such resource requirements may include the identification of the anticipated tasks for all categories of maintenance, maintenance personnel and training requirements (quantities and skill levels), test and support equipment, spares/repair parts and associated inventories, transportation and handling requirements, facilities, technical data, and computer resources.

Accomplishment of the maintenance task analysis (MTA) depends on the results of the reliability analysis and prediction, the FMECA, the RCM, reliability testing, maintainability analysis and modeling, and maintainability prediction. Chapter 9 includes extensive coverage of the MTA.

7.2.14   Maintainability Test and Demonstration

In the early conceptual and preliminary design phases of a program, the evaluation of maintainability characteristics in design is accomplished using various analytical methods/tools (allocation, prediction and analysis, etc.). As preproduction proto-type models of the system (and its components) become available during the detail design and development phase, it is appropriate to utilize these models in accomplishing a maintainability demonstration. A qualified system is installed in a simulated consumer operational environment; failures are induced; maintenance activities commence and system repair is accomplished using the logistic support resources described in the MTA. At the same time, data are collected for the purposes of evaluating maintenance tasks, task sequences and times, and the effectiveness of the proposed support capability. As the number of demonstrated tasks increases, one can calculate $\overline{M}ct$, $\overline{M}pt$, MMH/OH, and related measures of maintainability. Maintainability demonstration is described in detail in Chapter 11.

7.2.15 Human Factors Analysis

Human factors, sometimes known as "ergonomics" or "human engineering," refers to the *design of a system/product with the human being in mind,* that is, the design for operability, the design for ease of maintenance, and so on. Considerations in design must include anthropometric factors (the physical dimensions of the human body), human sensory factors (sight or vision, hearing, feel or touch, smell), physiological factors (the effects of environmental stress on the body), and psychological factors (factors pertaining to the human mind—emotions, traits, attitudinal responses, and behavioral patterns as they relate to job performance). These characteristics are discussed further in Section 4.3.[20]

Human factor requirements initially stem from the functional analysis described in Section 6.1. Operational and maintenance functions are described (i.e., the WHAT requirements), and resources are identified in response to HOW the functions are to be accomplished. A functional requirement may specify the need for a piece of equipment, an item of software, a facility, data, a human being, or a combination of these items. Those functions that are allocated to the human being are of particular interest here. Operational functions may require a human operator and the accomplishment of maintenance functions may require a maintenance technician. In each instance, the design must consider the human being and his/her interface with equipment, software, data, and so on. The development of human factor requirements is illustrated in Figure 7.19.

Of particular interest within the context of a typical human factors program is the accomplishment of a *detailed operator task analysis.* The major operational functions involving the human are broken down into:

---

[20]The objective is to provide a short introduction to human factors (i.e., those activities that are of particular significance to the maintainability engineer), and not to cover the subject in depth. However, for more information, three good references are (a) Meister, D., *Behavioral Analysis and Measurement Methods,* John Wiley & Sons, Inc., New York, 1985; (b) Sanders, M. S. and E. J. McCormick, *Human Factors In Engineering Design,* 6th ed., McGraw-Hill Co., New York, 1987; and (c) Woodson, W. E., *Human Factors Design Handbook,* McGraw-Hill Co., New York, 1981. Additional references are included in Appendix E of this text.

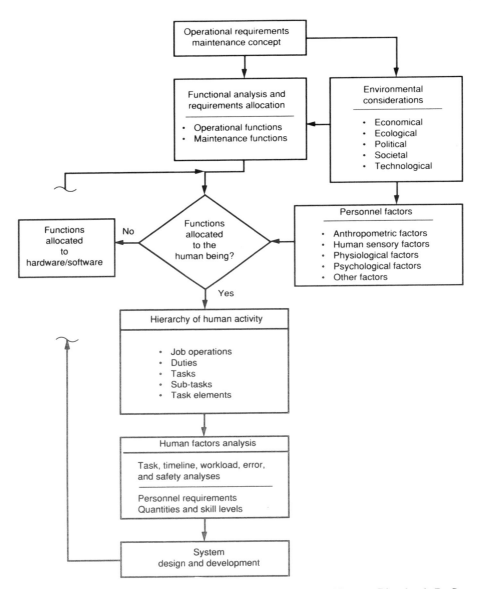

Figure 7.19. The development of human factor requirements. (*Source:* Blanchard, B. S., *System Engineering Management*, John Wiley & Sons, N.Y., 1991, p. 105).

1. *Job operation*—combination of duties and tasks (e.g., "operate a motor vehicle").
2. *Duty*—set of related tasks (e.g., "drive the motor vehicle in traffic on a daily basis").
3. *Task*—composite of related specific activities (e.g. "apply appropriate pressure on accelerator to maintain desired vehicle speed," or "shift gears in order to maintain speed").
4. *Subtask*—an element or "breakout" of a task (e.g., "shift gear from first to second").
5. *Task element*—the smallest logical definable facet of activity that requires individual behavioral responses (e.g., "a decision pertaining to a single physical action").

The detailed operator task analysis constitutes an evaluation of a given design configuration (based on drawings, design data, reports, part lists, etc.). Functions, tasks, and subtasks are identified. For each subtask, the required action and action stimulus, information feedback, subtask classification in terms of workstation assignment, subtask time, and personnel skill level are recorded. The human factors analyst, in the evaluation effort, must be knowledgeable about human psychology, behavioral responses in decision making, and the skills and knowledge required for subtask/task performance. These factors are combined for all subtasks and grouped into duties and job operations; the results lead to an identification of operator personnel quantities and skill-level requirements. Additionally, potential problems in design are noted (e.g., levels of difficulty or the possibility of operator-induced problems).

The identification of operator personnel quantities and skill levels for the newly designed system leads to the development of training program requirements. A comparison of the results of the operator task analysis with the current makeup of the consumer's organization (i.e., the "user") will lead to a description of the extent and depth of the training necessary to bring the skills of the anticipated consumer operator(s) up to a level specified for the new system.

A second significant design task in a human factors program is the development of an *operational sequence diagram* (OSD). The OSD evolves from the functional analysis, is based on an analysis of available design data, and reflects a sequential step-by-step record of each action performed by the operator, describing the interface requirements with major equipment items. Figure 7.20 presents an example of an OSD, where a communications sequence between operators and workstations is illustrated. Through a symbolic presentation, different actions are shown that, in turn, lead to the identification of specific design requirements, for example, the addition of an appropriate visual and/or audio indicator, the application of the proper format for information transmission and retention, accessibility provisions for inspection and condition monitoring, and so on. The OSD constitutes an extension of the detailed operator task analysis, with emphasis on the interface requirements between the operator and his/her workstation design configuration.

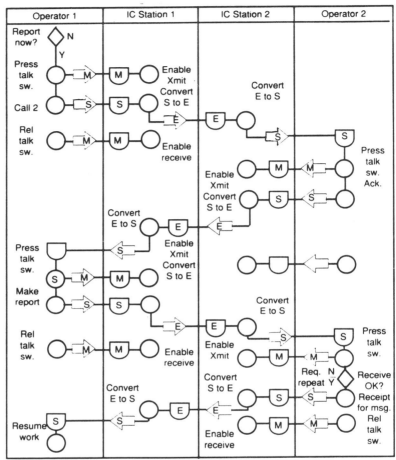

Notes on operational sequence diagram

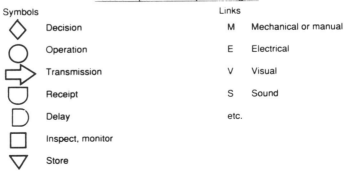

Stations or subsystems are shown by columns; sequential time progresses down the page.

Source: MIL-H-46855, Military Specification, "Human Engineering Requirements For Military Systems, Equipment and Facilities," Department of Defense, Washington, D.C.

Figure 7.20. Operational sequence diagram (example).

A third key area of activity is the aspect of *personnel test and evaluation*. The purpose is to demonstrate selected human-activity sequences to verify the adequacy of the personnel resources provided (based on recommendations from the detailed operator task analysis), the adequacy of operating procedures, and the effectiveness of the proposed training program. Demonstrations are conducted using a combination of analytical computer simulations, physical mockups (wooden, metal, cardboard, and/or combinations thereof), and preproduction prototype systems and equipment. Computer simulations may include the insertion of a fifth percentile female or a ninety-fifth percentile male into the work space, in a sitting or standing position, in order to evaluate activity sequences and space requirements. A great deal of information can be acquired through utilization of the appropriate computer graphics, employing a database which presents the human being in a three-dimensional configuration.

As the design progresses and a preproduction prototype model of the system (and its components) becomes available, testing may include the utilization of personnel, trained as recommended for the consumer, in the performance of selected operator and/or maintenance task sequences accomplished in accordance with approved procedures. These activities are monitored with the objective of identifying areas of task difficulty and the introduction of human errors. The conductance of such tests should not only allow for the evaluation of critical human-machine interfaces, but should provide reliability information pertaining to operator functions, maintainability data when maintenance tasks are performed, verification and validation of information in formal technical manuals/procedures, verification of the adequacy of the training program for operator and maintenance personnel, and so on.

In summary, a typical human factors program applicable to the design and development of a new system may commence with the preparation of a human factors program plan, at a time in the program concurrent with the development of a reliability program plan and a maintainability program plan. Program task requirements will include (1) the accomplishment of a system/function/task analysis, (2) the development of operational sequence diagrams, (3) a workload and timeline analysis, (4) an error analysis, (5) a safety and hazard analysis, (6) the development of mockups, (7) personnel training requirements and the development and implementation of a training program, (8) personnel test and evaluation, (9) the implementation and monitoring of supplier activities, (10) participation in design reviews, and (11) many other related program tasks. While a formal program effort may be quite comprehensive, only three tasks are mentioned herein.

With regard to the detailed task analysis and the development of operational sequence diagrams, the emphasis herein is primarily oriented to "operator" functions, while "maintenance" functions and tasks are not often addressed. Nevertheless, the same techniques and human factors criteria can be applied in determining the personnel and training requirements for maintenance. One of the requirements of the MTA, described in Chapter 9, is to determine personnel quantity and skill-level requirements for maintenance activities. In accomplishing such, it would be appropriate for the maintainability engineer to become familiar with human factor methods and design criteria.

Relative to the personnel test and evaluation task, the assessment of the "operator" and the interfaces with his/her workstation is a prime objective in a typical human factors program. However, equally important is the evaluation of the maintenance technician in the performance of corrective and preventive maintenance tasks. Again, many of the techniques employed here for personnel test and evaluation can also be applied to the maintainability test and demonstration requirements described in Chapter 11.

7.2.16 Safety/Hazard Analysis

"Safety" is a system design characteristic. The selection of certain materials in the design and construction of a system component could produce harmful toxic effects on the human; the placement and mounting of components could cause injuries to the operator and/or the maintainer; the use of certain fuels, hydraulic fluids, and/or cleansing liquids could result in an explosive environment; the location of certain electronic components close together may cause the generation of an electrical hazard; the performance of a series of strenuous tasks during the operation or maintenance of the system could cause personal injury; and so on.

Safety is important, both from the standpoint of the human operator and/or maintainer, and from the standpoint of the equipment and other elements of the system. Through faulty design, one can create problems that could result in human injury. Also, problems can be created that result in damage to other elements of the system. In other words, the design concerns deal with both *personal* and *equipment* safety.

In the design and development of a new system, one may evaluate the system with regard to the incorporation of safety characteristics by using a combination of methods. The FTA, described in Section 7.2.3, is a top-down analytical process, using deductive analysis and Boolean methods, for determining system events that will, in turn, cause undesirable events, or hazards. The FTA can be effectively applied early in design when limited data are available. The FMECA, described in Section 7.2.2, covers anticipated failure modes and their effects on other elements of the system. Through the utilization of this design tool, one can identify possible hazards, from both personal and equipment safety perspectives. A third tool is the *hazard analysis,* usually accomplished as a task in most safety programs. Like the FMECA, the objective is to evaluate the design and determine possible events that could result in hazards at the system level. By simulating failures, critical activities, and so on, at the component level, one can (through a "cause-and-effect" analysis) identify possible hazards, anticipate the frequency of occurrence, and classify in terms of criticality. Criticality levels include "catastrophic," "critical," "marginal," and "negligible." Recommendations for design change are made where appropriate, with a high-level of attention given to catastrophic and critical areas.[21]

---

[21] For more in-depth coverage of system safety, three good references are (a) Hammer, W., *Occupational Safety Management and Engineering,* 4th ed., Prentice-Hall, Inc., Englewood Cliffs, NJ, 1989; (b) Roland, H.E. and B. Moriarty, *System Safety Engineering and Management,* John Wiley & Sons, Inc., New York, 1983; and (c) MIL-STD-882, Military Standard, "System Safety Program Requirements," Department of Defense, Washington, DC (latest edition).

7.3   UTILIZATION OF DESIGN TECHNOLOGIES AND AIDS

Throughout the early design and development process, there are many instances where the maintainability engineer, in accomplishing his or her objective, may wish to utilize certain technologies and/or aids to facilitate the process. For instance, numerous techniques from the field of operations research can be used in the development and implementation of maintainability models. Artificial intelligence (i.e., expert systems) can be applied in the development of maintenance procedures and decision-making processes. Computer-aided tools and methodologies such as CAD, CAM, and CALS are being used in the design process to an increasing degree, and the maintainability engineer must be knowledgeable about these methods if he or she is to effectively contribute to the design. Condition monitoring and testing methods may be recommended for an accurate assessment of the system and its components, and the maintainability engineer must be cognizant of possibilities in this area. Mockups and equivalent models may be developed to demonstrate certain maintainability principles. In any event, the maintainability engineer has available to him/her a number of aids/tools which, if properly applied, can produce highly beneficial results. Some of these tools are discussed in this section.

7.3.1   Operations Research Methods and Models

Early stages of the system design and development process are characterized by a scarcity of relevant knowledge and information about the system. The designer(s), on the other hand, is required to make decisions and commitments during these early stages which are likely to have a significant impact on most "downstream" issues relating to production/construction, deployment, utilization/operation, sustaining support and upkeep, and phase-out or retirement. Development and utilization of models using methods from the field of operations research can, to some extent, bridge this "gap" between the designer's state of knowledge and the extent of commitments made during the earlier stages of the system design process.[22]

Mathematical models offer designers an opportunity to conduct indirect experimentations with tailored abstractions of reality through simulation and other "what-if" analyses. The result is increased insight, facilitating a more enlightened and informed decision-making process. This argument applies very well to system maintainability issues. The analyst is better able to respond to such questions as what is the impact on spare/repair parts if the equipment reliability is degraded? What is the impact on support equipment if the prime equipment packaging design and concept changes? What is the impact on overall system maintenance if the equipment is utilized to a greater extent than initially planned?

Numerous analytical techniques and methods are applicable and may be invoked

---

[22]For more details on applicable models and techniques from the field of operations research, see (a) Fabrycky, W. J., P. M. Ghare, and P. E. Torgersen, *Applied Operations Research and Management Science,* Prentice Hall, Inc., Englewood Cliffs, NJ, 1984 and (b) Hillier, F. S. and G. J. Liberman, *Introduction to Operations Research,* McGraw-Hill Publishing Company, New York, 1990.

to address a host of decision-making situations in the domain of system maintainability engineering and management. The technique or method most applicable is a function of the extent of knowledge available to the analyst, the nature of the problem area, and constraints relative to time and resources.

Very often simulation and other dynamic verification techniques are utilized in the early design phases as an effective form of indirect experimentation to assess system maintainability, along with other technical parameters. For instance, the Monte Carlo simulation technique could be applied, given a predetermined set of constraints or conditions, to study and assess system maintainability. Dynamic verification techniques may be applied to system functional models in order to identify functional bottlenecks, incompatibilities, or conflicts. It is important to mention here that very often it is not the optimal point solution that is important during the course of these analyses, but the stability of the design over a range of potential values. This stability is often referred to as the "robustness" of the design alternative.

Given certain assumptions, reliability and maintainability models can be developed by assigning probability distributions to system reliability and maintainability. These models can subsequently be used to predict system failure rates and the quantity of maintenance actions, and to estimate the system inherent availability metric. Models of this type can range from the simplistic to the very sophisticated. Linking reliability and maintainability models to spare and repair parts inventory, allocation, and optimization models in a realistic manner can be a complicated exercise, but the effort could yield substantial insight into the effectiveness of the overall support infrastructure. For example, reliability and maintainability distributions can be incorporated into models from the domain of queuing theory to predict and assess spare and repair part inventories.[23]

Network models are often useful in evaluating aspects of the system which involve the movement of material over time, and may be employed in evaluating the movement of spare/repair parts and other repairables between the different levels of repair facilities, the storage facilities, and the deployment sites during the course of preventive or corrective maintenance actions. Once again, probability distributions may be invoked to determine the preferred path, given a set of alternatives (e.g., PERT, discussed in Section 2.3.2). In fact, network models could be a part of an interconnected and interdependent set of models utilized to optimize spare and repair part levels and allocations.

Optimization techniques, such as linear programming, goal programming, nonlinear programming, and dynamic programming, are often a fundamental part of many modeling exercises. Linear programming is a modeling technique which involves the optimization of an effectiveness function in the face of certain constraints. Both the effectiveness function and the constraints have a linear formulation. Such a technique could be applied to optimize the utilization of maintenance

---

[23] An excellent reference for more details on the development of reliability, maintainability, and supportability models is Knezevic, J., *Reliability, Maintainability, and Supportability—A Probabilistic Approach,* McGraw-Hill Book Company, London, 1993.

facilities and/or equipment with limited capacity, to the allocation of system maintainability, and so on. Special classes of linear programming problems address situations from the domain of transportation. Such formulations may find application in certain spares allocation and location problems. Linear programming can be generalized to address and simultaneously optimize multiple objectives. This generalized form of linear programming is called goal programming. Goal programming is an effective and popular multiattribute optimization technique and can be applied when optimizing two or more, often conflicting, system design parameters. For instance, this technique can be applied in the simultaneous allocation of system maintainability and reliability, in the face of constraints relative to system reliability, maintainability and availability requirements, mission profile requirements, life-cycle cost and other budgetary limits, and so on. Dynamic programming, on the other hand, is a technique utilized if the solution involves a discrete set of possible alternatives with sequentially related constraints. It is particularly useful when the analysis involves a large number of variables, such as spares modeling.

Accounting models are used when the functions involved are, generally speaking, of an additive nature. Such models, along with their variations, are often used in the estimation of system maintenance cost and life-cycle cost, where individual costs or resources linked to functional activity areas are summed up while adjusting them for interest/inflation rates and learning curve impacts. Benefit–cost ratios are often used to justify preventive maintenance tasks and their frequencies while conducting a reliability-centered maintenance analysis, as discussed in Section 7.2.4. Further, accounting models are used to determine economic order quantities in inventory modeling.

As may be obvious from this discussion, a number of quantitative methods are available for use in the development of mathematical models, for the purpose of indirect experimentation. Further, a model often involves the integrated application of a number of methods or techniques. For instance, optimization is oftentimes conducted in conjunction with linear and nonlinear programming, queuing techniques, probabilistic functions, accounting methods, and so on, in a given model. It is important to note here that modeling often involves a trade-off between model complexity and model integrity, or the desire to represent a design or decision situation in a very realistic manner. The effectiveness and utility of a model often depends upon this trade-off. The field of heuristics has evolved as a result of this very important aspect of modeling. Heuristics developed as result of extensive experimentation and observation frequently yield near optimal solutions for a fraction of the time and resources that may otherwise be required.

When developing models, certain cautions must be exercised. First, a single, complex, all-inclusive model cannot offer some of the advantages, in terms of simplicity and flexibility, that are available through the use of multiple models designed to function as an integrated set. Large models often exceed available computation and storage capabilities, require additional capabilities, or require extensive computing times. Thus, when developing models, a modular approach to the problem is recommended.

Second, when developing models, there is always the danger of making assumptions and grossly oversimplifying the design or decision situation. In many instances, the model ultimately developed may reflect poor judgment and a loss of investment in terms of time and resources. The analyst must guard against such a situation.

Finally, there is the danger that the analyst, after working with the model long enough, begins to believe in it beyond the limits of reason. It is important to understand that a mathematical model can only facilitate the decision-making process *in partnership* with the analyst. After all, a mathematical model, like any other model, is just a tailored abstraction of reality, and not reality itself!

7.3.2 Fuzzy Logic-Based Methods and Models

The system design and development process is continuously iterative and evolving toward greater resolution. The only constant denominator is the uncertainty that has to be addressed, and even this is constantly changing in character, and needs to be considered and addressed using different analytical techniques depending upon the state of evolution of the system design process. For instance, uncertainty during the conceptual design phase may result from (1) imprecise "capture" of the consumer/customer need, (2) imprecise translation of this need into qualitative and quantitative system-level requirements, (3) imprecise and vague prioritization of system-level requirements, (4) uncertain user profile, and (5) uncertain market and technology trends.

Uncertainty in system design can be classified broadly into two distinct categories, stochastic uncertainty and design imprecision. For instance, consider system maintainability. Design imprecision reflects itself through the uncertainty relative to the system maintainability requirement *(Is it precisely 30 minutes MTTR? At what level above 30 minutes would the design be unacceptable?)*. While this requirement should be translated from market/user needs analysis, benchmarking, and so on, a "crisp" and precise assignment of the maintainability requirement may not be very realistic.

On the other hand, the random or stochastic component of the overall design uncertainty is reflected through the very nature of maintainability and its dependence on a host of factors *(inherent, environmental, and operational)* that are stochastic in nature. Further, while probability theory has been used to address stochastic uncertainty very effectively, the theory of fuzzy subsets and fuzzy logic offers an alternate method to explicitly analyze uncertainty due to imprecision in a graceful and formal manner. The theory of fuzzy subsets and fuzzy logic can be utilized to complement the usage of probability theory in the modeling of complex system behavior. The two approaches are compatible and often allow a more realistic and robust modeling of system functionality.

The concept of fuzzy sets was defined for the first time in the mid-1960s by Lotfi Zadeh as a mechanism for dealing with situations where uncertainty is as a result of the absence of "crisp" criteria for membership in a set, rather than the presence of random variables.[24] For example, let X be the field of reference encompassing a

[24]Zadeh, L. A., "Fuzzy Sets," *Information and Control*, 8 (1965).

definite range of objects (i.e., the relevant universe). Consider a subset $\tilde{A}$ where the transition between membership and nonmembership is gradual rather than abrupt. In other words, this fuzzy subset of the relevant universal set obviously has no well-defined boundaries. For example, assume $\tilde{A}$ to be the set of all tall children in a classroom X. There will obviously exist children in the classroom who are definite members of this set, and those who are definitely not. But then there also exist children who are the so-called *borderline* cases.

Traditionally, the grade of membership one is assigned to elements that completely belong to $\tilde{A}$ (in the example above, that would be the children who are definitely tall); conversely elements that absolutely do not belong to $\tilde{A}$ are assigned a membership value zero. Quite naturally, the grades of membership for the *borderline* cases fall in the range between zero and one. Further, the more an element or object x belongs to $\tilde{A}$, the closer to one is its grade of membership, $\mu_{\tilde{A}}(x)$. Use of a numerical scale, such as the interval [*zero, one*], allows a convenient representation of gradation in membership. Precise membership values do not exist by themselves; they are tendency indices that are subjectively assigned by an individual or a group.

Subsequent to developing the concept of fuzzy sets, Zadeh proposed the concept of linguistic variables and the extension principle in 1973 and 1975.[25,26] These two concepts served as a catalyst in the application of fuzzy logic, and can be involved and extended to address a host of situations in the domain of maintainability analysis and maintenance.

Linguistic variables have words or sentences, rather than numbers, in natural language as their values. The totality of values that a linguistic variable can assume constitutes its *term set*. A *term set* could conceptually have an infinite number of elements. For example, the term set of the linguistic variable *Reliability* might read as follows:

T(Reliability) = reliable + not reliable + very reliable + not very reliable + very very reliable + . . . + unreliable + not unreliable + very unreliable + not very unreliable + . . .

where a plus sign denotes a union.

Further, in the case of the linguistic variable *Reliability,* the numerical variable *reliability,* whose values are real numbers in the range between 0 and 1, constitutes the base variable for *Reliability*. In terms of this variable, a linguistic value such as *reliable* or *unreliable* may be interpreted as a label for a fuzzy restriction on the values of the base variable. This fuzzy restriction constitutes the meaning of *reliable* or *unreliable*.

A fuzzy restriction on the values of the base variable is characterized by a *membership function* (similar to a characteristic function), which associates with each value of the base variable a number in the interval [0,1], which in turn represents

[25] Zadeh, L. A., "Outline of a New Approach to the Analysis of Complex Systems and Decision Processes," *IEEE Trans. Systems, Man, Cybern.*, SMC-1, 1973.

[26] Zadeh, L. A., "The Concept of a Linguistic Variable and its Application to Approximate Reasoning I, II, and III," *Inf. Sci.*, **8, 9** (1975).

its compatibility with the fuzzy restriction. For example, the compatibility of the numerical reliabilities 0.98, 0.87, and 0.62 with the fuzzy restriction labeled *reliable* might be 1.0, 0.75, 0.15, respectively. The meaning of reliable could then be represented by a graph, as in Figure 7.21a, which is a plot of the membership function of *reliable* with respect to the base variable *reliability*.

Here it is important to understand the distinction between the notions of compatibility and probability. As an illustration, the statement that the compatibility of 0.87 with *reliable* is 0.75 has no relation to the probability of the reliability value being equal to 0.87. The correct interpretation of the compatibility value of 0.75 is that it is merely a subjective indication of the extent to which the reliability value 0.87 fits one's conception of the label *reliable*.

Linguistic hedges and connectives are linguistic operators that modify the meaning of the operand in a specified fashion. Thus, if the meaning of *reliable* is defined by the membership function in Figure 7.21a, then the meaning of *very reliable* could be obtained by squaring the membership function of *reliable*, while that of *not reliable* would be given by subtracting the membership function of *reliable* from unity. These modifications are shown in Figure 7.21b.

The concept of a linguistic variable includes within itself the concept of a fuzzy variable. In other words, the values of a linguistic variable (elements of its *term set*)

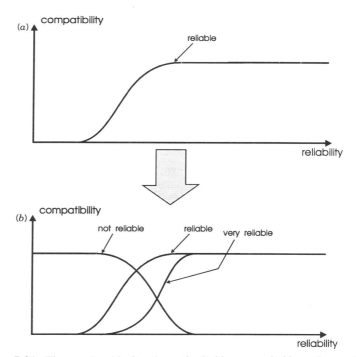

Figure 7.21. The membership functions of *reliable, very reliable,* and *not reliable.*

are themselves fuzzy variables. Therefore, the linguistic variable *Reliability* can take on values such as *reliable, unreliable,* and soon, which are all fuzzy variables. In this context, a fuzzy variable, such as reliable, can be completely defined by a triple given as

$$[\tilde{A}, X, R(\tilde{A};x_i)]$$

where $\tilde{A}$ is the fuzzy variable name, such as *reliable;* X is the universe of discourse, such as $0-1$; and $R(\tilde{A};x_i)$ is a fuzzy subset of X and represents the fuzzy restriction on values of x_i imposed by the fuzzy variable $\tilde{A}$.

The objective of this brief introduction to the concept of fuzzy and linguistic variables is to convey their use in the assignment of imprecise maintainability requirements, in the estimation of imprecise and subjective performance levels, and in the assignment of relative priorities. More specifically, along with other applications, linguistic variables have been applied to better address the subjective and imprecise issues concerning failure severity, detectability, and occurrence (during the very early phases of the design process) while conducting a criticality analysis as part of the failure mode and effects analysis.[27]

The extension principle allows the generalization of crisp mathematical concepts to the fuzzy framework. It provides the means for any function f that maps points $x_1, x_2, \ldots, x_n$ in the crisp set X to the crisp set Y to be generalized such that it maps fuzzy subsets of X to Y. Formally, given a function f mapping points in set X to points in set Y and any fuzzy set $\tilde{A}\epsilon\tilde{P}(X)$, where

$$\tilde{A} = \mu_1/x_i + \mu_2/x_2 + \cdots + \mu_n/x_n$$

and $\tilde{P}(X)$ is the power set of the universe of discourse, or the set of all possible fuzzy subsets of $X,$ the extension principle states that

$$f(\tilde{A}) = f(\mu_1/x_1 + \mu_2/x_2 \cdots + \mu_3/x_n)$$
$$= \mu_1/f(x_1) + \mu_2/f(x_2) + \cdots + \mu_n/f(x_n)$$

To illustrate the principle, consider X *(universe)* to be the set of integers. Then a fuzzy subset, *small,* of X may be given as

$$small = 1/1 + 0.9/2 + 0.8/3 + 0.7/4 + 0.6/5 + 0.5/6 + 0.4/7 + 0.3/8 + 0.2/9 + 0.1/10$$

Further, let the function f by a cubing operation or $f(small) = small^3$; then, according to the extension principle,

$$small^3 = 1/1 + 0.9/8 + 0.8/27 + + 0.7/64 + 0.6/125 + 0.5/216 + 0.4/343 + 0.3/512 + 0.2/729 + 0.1/1000$$

[27]Pelaez, C. E. and J. B. Bowles, *Using Fuzzy Logic for System Criticality Analysis,* Proceedings, Reliability and Maintainability Symposium, Anaheim, CA, 1994.

If the fuzzy set $\bar{A}$ is not discrete but continuous, and of the form

$$\bar{A} = \int_x \mu_{\bar{A}}(x_i)/x_i$$

then the extension principle may be applied as

$$f(\bar{A}) = f\left[\int_x \mu_{\bar{A}}(x_i)/x_i\right] = \int_y \mu_{\bar{A}}(x_i)/f(x_i)$$

The extension principle affords a means for rigorous manipulation of fuzzy sets, in general, and fuzzy numbers, in particular (fuzzy variables that are convex, normal, and subsets of the set of real numbers are called fuzzy numbers). An extensive presentation of the theoretical underpinnings, and potential applications, of fuzzy sets and fuzzy logic can be found in selected references included in Appendix E.

Concepts from the field of fuzzy set theory and fuzzy logic have been applied to address a host of problems in reliability and maintainability analysis.[28] These concepts have been extended for application in the field of failure diagnosis, troubleshooting, and fault detection, where one may have to synthesize and analyze a number of imprecise inputs or symptoms (visual, audio, or otherwise). Fuzzy algebra has been applied to hazard detection in combinatorial switching circuits and the detection of static hazards in combinatorial switching circuits.[29] The development of expert systems for the purposes of fault isolation offers another application opportunity. Fuzzy linguistic variables may be used to more realistically model the experts' knowledge and reasoning process in situations characterized by imprecise and vague data. Further, in the development of such expert systems, fuzzy logic has also been utilized in the storage and retrieval of imprecise and incomplete data.[30]

7.3.3 Artificial Intelligence (Expert Systems)

A major objective of maintainability is to simplify the accomplishment of maintenance tasks by reducing task times and improving human reliability in the performance of these tasks. Complex tasks need to be completed quickly, accurately, reliably, and with a minimum of personnel training. This is particularly true for modern systems characterized by vastly increasing scope and complexity. The question is—what techniques can be applied to facilitate the accomplishment of both corrective and preventive maintenance tasks in an effective and efficient manner?

---

[28] Verma, D. and J. Knezevic, *Application of Fuzzy Logic in the Assurance Sciences*, Proceedings, Reliability and Maintainability Symposium, Anaheim, CA, 1994.

[29] Hughes, J. S and A. Kandel, "Applications of Fuzzy Algebra to Hazard Detection in Combinatorial Switching Circuits," *Int. Comput. Int. Sci.*, **6** (1977).

[30] Gaines, B. R., *Logical Foundations for Database Systems*, Fuzzy Reasoning and its Applications, E. H. Mamdani and B. R. Gaines (Eds.), Academic Press, London, 1981.

Artificial intelligence in general, and its application in the development of expert systems in particular, offers one such technique.

Artificial intelligence (AI) can be defined as the science that attempts to replicate human behavior and intelligence through computer programs. Webster's Dictionary defines *intelligence* as *the ability to learn or understand or to deal with new or trying situations, the skilled use of reason, the ability to apply knowledge to manipulate one's environment or to think abstractly as measured by objective criteria, the ability to foresee and solve problems, to perceive one's environment and to adapt to it and deal with it effectively.* It may never be possible to instill a computer with all of the characteristics that imply intelligence, but attempts to implement subsets of the list of characteristics given above are practical and have been successful.

Research in the field of AI has led to the development of several branches with specific objectives and applications. Some of the more popular research fields are (1) natural language processing, (2) expert systems, (3) robotics, (4) voice recognition, (5) computer vision and visualization systems, and (6) computer learning.

Developments in these disciplines complement each other in a host of different ways. Of all the research areas mentioned above, advancement in the domain of expert systems has been the most significant and provides an insight into the potential of the science of AI. Expert systems represent the first commercial application of research conducted in AI. According to Edmunds, expert systems can be defined as follows: [31]

An expert system is a computer-based system designed to assimilate the knowledge of human experts, making that knowledge conveniently available to other people in a useful way.

Adapting this definition to maintainability and system maintenance, an expert system can be defined as a system that "captures" the expertise of one or more maintainability design or system maintenance experts, is replicable and portable, and makes this captured knowledge and experience of experts available to non-experts in a logical user-friendly manner through an interactive computer-based tool. Generally speaking, an expert system is composed of the following six components:

1. *A knowledge base.* A knowledge base is where an expert maintenance technician's knowledge is stored. One mechanism of storing this knowledge is in the form of rules. This knowledge may include not only facts, but also heuristics and beliefs which result from extensive experience of experts in a particular field.
2. *A working memory.* Facts relative to any immediate maintenance or diagnostic problem are input by the user and placed on the working memory of the

[31] Edmunds, R. A., *The Prentice Hall Guide to Expert Systems,* Prentice Hall, Englewood Cliffs, NJ, 1988.

expert system. This is also where any intermediate or final conclusions are posted by the expert system.

3. *An inference engine.* The inference engine correlates the known facts about the current maintenance, diagnostic, or troubleshooting requirement with the expert knowledge in the knowledge base and attempts to arrive at a solution. This "matching" or correlation is in accordance with any of the rule priorities which may be inherent in the expert system's knowledge base.

4. *A knowledge acquisition capability.* An important aspect of an expert system is its knowledge acquisition facility. It is essential that an expert system be amenable to change, refinement, and additions as more information becomes available to further impact maintainability and maintenance downtime. In fact, in a "good" expert system this is usually easy to do.

5. *An explanation capability.* An explanation capability is often necessary for the expert system to delineate and give the reasoning behind offering a particular suggestion to a nonexpert maintenance technician conducting troubleshooting on a complex system or attempting to isolate a failure.

6. *A friendly user interface and input/output facility.* Users of the expert system interact with it, and the knowledge internal to it, through the user interface or the input/output facility.

A schematic showing the interrelationship between the components of an expert system is depicted in Figure 7.22.[32]

Expert systems are a successful reality today. One of the more popular application domains for expert systems is system maintenance in general, and system diagnostics, troubleshooting, testability, and fault isolation (alarm investigation) in particular. Success stories abound, resulting in significantly reduced system downtimes and costs, decreased reaction times, significant reduction in the number of false alarms, and a much more effective utilization of the scarce expert maintenance personnel resource.

One of the many expert systems developed and implemented at Texas Instruments, Inc., called the Intelligent Machine Prognosticator (IMP) is claimed to have decreased the mean time to repair a highly sophisticated machine used in the production of semiconductors by some 36%. Similarly, one of the many expert systems being used by the Campbell Soup Company in Camden, New Jersey, aids the fault diagnosis process relative to large and complex soup-sterilizing equipment. The company realized that certain problems could only be diagnosed and resolved by a scarce number of expert maintenance personnel, nearing retirement. An expert system was developed to address this problem and has been implemented within eight

---

[32] This schematic of an expert system has been adapted from Maher, M. L., D. Sriram, and S. J. Fenves, *Tools and Techniques for Knowledge-Based Expert Systems for Engineering Design,* Technical Report Number DRC-12-22-84, Design Research Center, Carnegie Mellon University, Pittsburgh, PA, December 1984.

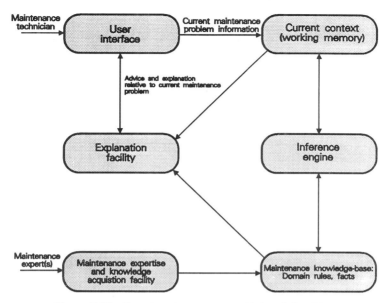

Figure 7.22. Expert system component interrelationships.

of their plants in the United States and Canada. These and other successful applications are briefly outlined in Table 7.5.[33]

Within the domain of artificial intelligence, and with the specific purpose of applying information from a variety of sources to solve certain problems, such as failure diagnosis, condition monitoring, and troubleshooting of complex systems, the development of the blackboard architecture holds special promise. In simple terms, the blackboard architecture consists of a blackboard, or a central database, and a set of knowledge sources, as depicted in Figure 7.23. All information relative to a given problem is contained on the blackboard and the various knowledge sources address different, but related, aspects of this problem and together, in an organized manner, attempt to resolve it. The different knowledge sources communicate with each other only indirectly, through the central blackboard. Based on the current state of the problem, each knowledge source proposes the next action. A *scheduler* pools all proposed actions, evaluates them, and selects the most appropriate action. Once an action has been taken, the knowledge sources, based on the current state of the problem and the incremental solution, propose a fresh set of proposals. This process continues until the problem is solved.

The blackboard architecture can facilitate the development of more powerful

[33] The Texas Instruments and Campbell Soup Company expert system application case studies are adapted from Herrod, R. A., "AI: Promises Start to Pay Off," *Manuf. Eng.*, March 1988.

TABLE 7.5 Successful Expert Systems Applied to Diagnostics and Maintenance

Expert System Name	Application
DELTA	Diesel electric locomotive troubleshooting aid (DELTA) was the first commercially available expert system applied to diagnostics and maintenance. It was initially named CATS-1 and was developed by General Electric Company in 1983.
DEFT	Diagnostic expert final test (DEFT) is an expert system developed and applied by the IBM Corp. as a diagnostic aid for mainframe computer disk drives beginning in 1986.
PRIDE	Pulse radar intelligent diagnostic environment (PRIDE) is an expert system developed by the United States Army to aid in the diagnostics of the pulse acquisition radar of the Hawk missile system.
FOREST	An expert system to support equipment maintenance through fault diagnosis and isolation in automatic test equipment.
IMP	The intelligent machine prognosticator (IMP) is an expert system developed by Texas Instruments, Inc. and used for fault diagnosis in highly complex semiconductor production equipment. Contains approximately 1000 rules addressing about 90% of all known problems with the equipment and has resulted in a 36% reduction in MTTR.
ET	Expert technician (ET) is another expert system developed by Texas Instruments, Inc., to aid junior- and senior-level technicians with the troubleshooting of a complex electronic device.
HH	Hotline helper (HH) is yet another expert system developed by Texas Instruments, Inc. to aid nonexperts diagnose printer problems over the telephone.

expert systems for the purpose of failure diagnosis and troubleshooting, specifically in the case of complex systems where responses from multiple sensors may have to be coordinated while analyzing symptoms and navigating toward the underlying causes.[34]

In summary, the advantages to be gained through the thoughtful application of the expert systems technology include:

---

[34] There are numerous references in the literature on the application of expert system technology for the specific purpose of failure diagnosis, such as (a) Davis, R., "Diagnostic Reasoning Based on Structure and Behavior," *Artificial Intelligence,* 24 (1984), (b) Ayeb, B. E. and J. P. Finance, "On Cooperation Between Deep and Shallow Reasoning: SIDI—An Expert System for Troubleshooting Diagnosis in Large Industrial Plants," *Artificial Intelligence in Engineering: Diagnosis and Learning,* J. Gero, Ed., Elsevier, Amsterdam, 1988, (c) Hajek, J. and C. Haas, "Application of Artificial Intelligence in Highway Preventive Maintenance," *Artificial Intelligence in Engineering: Diagnosis and Learning,* J. Gero, Ed., Elsevier, Amsterdam, 1988. Most of these applications, however, involve the application of rule-based systems.

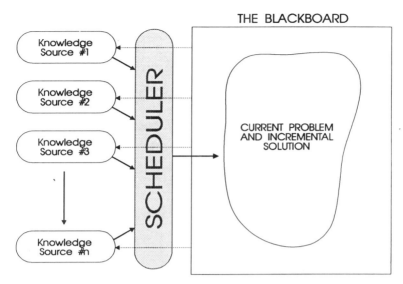

Figure 7.23. An illustrative blackboard architecture schematic.

1. *Reduced impact of employee turnover on effective maintenance operations.* This is because the knowledge of scarce expert maintenance technicians can be harnessed and replicated for use by the nonexperts through the development of expert systems. Further, expert systems may also reduce the training time expended on the new maintenance technicians. Experience and knowledge relative to specific maintenance tasks can be "carried" into the field by the inexperienced maintenance personnel through expert systems.

2. *Increased responsiveness in the performance of maintenance tasks since a greater number of maintenance technicians in the organization will be able to do the job with the aid of an expert system.* Responsiveness will also be positively impacted through a reduced dependence on voluminous maintenance guides, since information and data can be stored in a logical and structured manner within the knowledge base of an expert system and accessed more rapidly whenever needed.

3. *Expert maintenance knowledge sharing.* This is because knowledge about complex maintenance and diagnostic tasks, once encapsulated within an expert system, is portable and repeatable, and can be shared throughout an organization by nonexpert maintenance technicians and pooled between expert maintenance technicians. This results in significantly higher equipment effectiveness.

4. *Enhanced system availability and cost effectiveness.* This is the natural consequence of first pooling and then sharing experience and knowledge relative to system maintenance, troubleshooting, and diagnostics. The resulting consistent and reliable maintenance expertise will also reduce logistic support costs and maintenance delays.

Given the potential benefits described above, the utilization of expert systems as maintenance aids is likely to increase in the near future. However, before this technique is developed and implemented further, it is very important to define the specific requirements and expectations and assess the available "expert" knowledge. Further, over and above the technical compatibility issues, there are also the cultural and psychological dimensions to consider. The most significant "bottleneck" to the expeditious and effective development and implementation of expert systems is often the so-called expert and the process of acquiring his or her experience and knowledge in a logical and coherent manner.

7.3.4   Computer-Aided Design Applications

Computer-aided design (CAD), defined in a broad sense, refers to the application of computerized technology to the design process. With the availability of computer tools that can be utilized appropriately in the performance of certain design functions, the designer can accomplish more, at a faster pace, and earlier in the system/product life cycle. These tools, which include supporting software, incorporate graphics capabilities (vector and raster graphics, line and bar charts, x-y plotting, scatter diagrams, and three-dimensional displays), analytical capabilities (mathematical and statistical programs for analysis and evaluation), and data management capabilities (data processing, storage and retrieval, drafting, and reporting). These capabilities are usually combined into integrated packages for solving a specific design problem.[35]

The utilization of CAD capabilities in the design of systems/products, CAM in the design and implementation of manufacturing capabilities, and CALS in the development and acquisition of the elements of maintenance and support, is growing at a rapid pace as the technology becomes available. Figure 7.24 illustrates the application of these computerized methods in the context of the life cycle, and the interfaces between CAD, CAM, and CALS are many. While the maintainability engineer does not have to become an expert in these methods in terms of their construction, he or she needs to know about CAD (in particular), its capabilities and applications, and how he/she can interface and communicate with the designer and the design process.

On many projects, the various members of the design team (including the maintainability engineer) are located in different buildings or in different geographical areas. Figure 7.25 illustrates this situation. With the application of CAD technology, much of the design task can be accomplished rapidly and, if the maintainability engineer is to participate in the design process in a timely manner, he/she must be able to review and evaluate design data, accomplish design analysis activities, provide recommendations for design improvement (as applicable), and approve the

---

[35] Two good references covering some of the general aspects of CAD and CAM are Teicholz, E., Ed., *CAD/CAM Handbook*, McGraw-Hill Co., New York, 1985; and Krouse, J. K., *What Every Engineer Should Know About Computer-Aided Design and Computer-Aided Manufacturing (CAM)*, Marcel Dekker, New York, 1982.

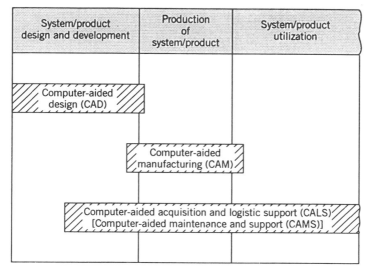

Figure 7.24. Application of computer methods in the life cycle.

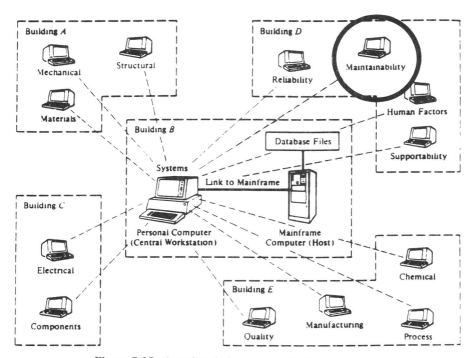

Figure 7.25. A project design communications network (example).

proposed design configuration on an expeditious basis. Otherwise, the responsible designer will proceed without an input from the maintainability engineer.

Currently, the maintainability engineer assigned to most large-scale projects is conversant with the utilization of computer models in the accomplishment of maintainability allocation, maintainability analysis, level of repair analysis, prediction, the FTA/FMECA, the MTA, life-cycle costing, and related trade-off analyses. However, most of these activities are accomplished locally within the organization and on an independent basis, and the results are not tied directly into the CAD system. Hence, the design input from these efforts is not always available in a timely manner. A challenge in the future is for the maintainability engineer to become an integral part of the CAD workstation communications network (if such does not already exist). Thus, he or she needs to become familiar with CAD capabilities, associated workstations, hardware, software, database structures, and so on. Additionally, the maintainability engineer needs to be familiar with both CAM and CALS. Continuous acquisition and life-cycle support is of particular significance since the output data from many of the maintainability program tasks is fed directly into the CALS database to fulfill logistic support objectives.

7.3.5 Condition Monitoring and Test Methods

Condition monitoring, or continuous inspection, is the ongoing surveillance of the operation of a product or process to ensure proper performance and to detect abnormalities indicative of an impending failure or of a failure that has already occurred. As an inherent capability within the system, this may lead directly to the accomplishment of corrective maintenance when an "out-of-specification" situation exists (i.e., a failure), or preventive maintenance when early system deterioration occurs and a scheduled component replacement, adjustment, or calibration is desired.

Condition monitoring, along with other frequent and/or infrequent intermittent inspection methods, is an important aspect of an applicable and effective system PM program developed through the application of analysis techniques like the FMECA, discussed in Section 7.2.2, and RCM, discussed in Section 7.2.4.

An effective and efficient PM program very often includes the capability of continuously monitoring certain critical aspects of a product or a process. Condition monitoring is ideal when it is not possible to accurately anticipate and predict the expected wear-out trends and characteristics of a product or process with age, and where the failure criticality justifies keeping a constant vigil on a particular product function or component, or process parameter. Different types of condition monitoring techniques and sensing apparatus exist and can be selected based upon the nature, characteristics, and functionality of the parameter being observed. Over and above the technical applicability and compatibility issues, certain condition monitoring techniques are fairly expensive and need to be amply justified through extensive system criticality, hazard, and economic trade-off analysis.

Numerous high-technology condition monitoring and nondestructive testing techniques have been developed and utilized in the past two decades. The purpose of this section is to review briefly some of the more modern testing technologies.

In the context of this review, a wide variety of testing technologies exist which differ in terms of applicability, resolution and accuracy, cost of application, and frequency of observations. The frequency of observations can span the entire spectrum between no observation and continuous observation, as depicted in Figure 7.26. In the event that a testing technique is capable of continuous observation or surveillance, it is classified as a condition monitoring technique. Whenever possible an attempt will be made to clearly define the concept underlying a particular condition testing technique. Further, the advantages and the disadvantages in each case will also be delineated.

Review of some of the selected testing techniques in the subsequent paragraphs may incorrectly imply that they are utilized in a mutually exclusive manner. Given the disadvantages and advantages of these techniques, more often than not, an optimal and cost-effective monitoring or inspection task may require a balanced and complementary implementation of multiple techniques. For example, the acoustic emission technique may first be used to locate internal discontinuities, and ultrasonic or radiographic techniques may thereafter be used to gauge the magnitude of these defects. This combination, more often than not, is likely to result in greater insight in a more economical manner than the use of any one of the two techniques in isolation.

1. *Visual inspection techniques.* Visual inspection is one of the oldest and most utilized testing and condition monitoring techniques. While the underlying concept remains unchanged, the techniques utilized to visually track product or process characteristics have evolved significantly over the years.

Traditionally, this mode of monitoring and inspection was limited by issues such as amount of available personnel access, noise levels, light conditions, inspection of environmental conditions (emissions, toxicity, etc.), and so on. Most of these limitations have now been neutralized through developments and advancements in fiber optics, minitelevisions, and video camera equipment. Increased image resolu-

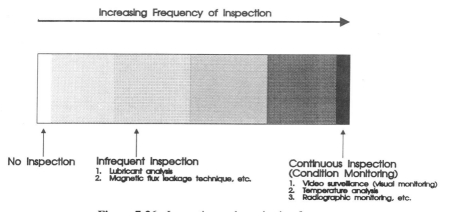

Figure 7.26. Inspection and monitoring frequency.

tion capabilities, the development of imaging equipment of consistently smaller size and lighter weight, and the ability of video cameras to produce good images under poor light conditions have made visual inspection an increasingly popular CM technique. Further, progress in optical imaging technology has been complemented by advances in the generation of visual images from sources such as heat and ultrasonics.

The development and evolution of supporting technologies notwithstanding, visual inspection has some inherent limitations which need to be clearly understood. Inspection and monitoring is limited to surface conditions and only surface defects and anomalies can be detected. Further, equipment or component surfaces often need to be prepared or cleaned in order to maximize the inspection effectiveness. In the case of monitoring a large surface area, small and highly localized defects may be missed. But the most significant limitation of the visual inspection technique is its inherent subjectivity. Two inspectors can arrive at two entirely different conclusions. Despite some of these limitations, this technique is often the most simple and effective to implement, and hence its popularity.

2. *Optical inspection techniques.* Optical holography is another testing technique utilized to detect equipment part and component surface defects and inconsistencies. Just as in the case of visual inspection, optical holography does not require direct contact with the system being inspected and thus there is little, if any, system functional interference. According to Bray and McBride,

> . . . optical holographic interferometry is a useful technique or tool for detecting flaws in material by making and recording measurements in unstressed and stressed conditions. Comparisons of holographic fringes created on a hologram reveal surface manifestations that signify flaws or discontinuities. The technique is capable of making stress and vibration analysis.[36]

Other optical testing approaches include the utilization of refractometers. In this case, light is passed through the material medium. Deviations, if any, in the refraction angle of the light waves are noted and measured. Careful analysis of these deviations is conducted to isolate material discontinuities and nonhomogeneities. High skill levels are needed for the effective utilization of this technique.

3. *Radiographic monitoring techniques.* These techniques can be classified into two groups, static radiography and dynamic radiography. Both involve generating a two-dimensional image which depicts the graded passage of X rays or gamma rays through the equipment being monitored or inspected. The image itself may be produced on photographic paper or a fluorescent screen. While an X-ray film or photographic paper may need to be developed before the image can be analyzed, a fluorescent screen allows instant observation (also called fluoroscopy).

As may be obvious, this technique involves access to two opposite sides of the equipment. Either X rays or gamma rays are targeted onto the equipment or compo-

---

[36] Bray, D. E. and D. McBride, *Nondestructive Testing Techniques,* Wiley Series in New Dimensions in Engineering, John Wiley and Sons, Inc., New York, 1992.

nent from one side, and the extent to which these rays penetrate and pass through the equipment is recorded on photosensitive film on the other side in the form of an image. While static radiography is the more traditional of the two approaches and is used to study stationary objects, dynamic radiography is used to analyze moving components within a piece of equipment. Experience and training is necessary to completely and competently study and evaluate images generated using this approach because of their low resolution. This technique is an effective way to detect internal cracks, bubbles, imperfections, nonhomogeneities, and material composition inconsistencies.

Radiography is fairly popular owing to its ability to detect a component's internal defects. Its disadvantages include access requirements to two opposite sides and the elaborate and expensive equipment requirements. Further, as mentioned earlier, advanced training and experience may be required to correctly assess a radiographic image.

Extensions to the basic approach and concepts described above have been developed for more specific applications; for example, X-ray tomography (where only a single plane of an object is analyzed without any interference from the adjacent planes on either side), microradiography (from the study of very small objects and components and minute defects and inconsistencies), neutron radiography (to study the characteristics of low-density materials flowing or encapsulated inside a high-density material), and stereoradiography (where two object images from slightly different angles are superimposed to generate a three-dimensional effect in the final image).

4. *Neutron analysis technique.* This is an approach to ascertain not only equipment part material or fluid constituents, but also the individual constituent quantities. This technique can be utilized to determine whether a process or product is within the required parameters with respect to material composition. The technique involves irradiating a fluid sample or piece part. Each constituting material will have a different resulting radioactivity, which is analyzed to first identify the constituents and then estimate their percentage constitution.

Although this technique is very accurate, the analysis process and the equipment necessary to conduct it is fairly expensive and must be justified. Moreover, there is also the hazard relative to personnel exposure to radiation, just as in the case of radiography.

5. *Ultrasonic monitoring technique.* This is a monitoring approach which makes possible the detection of both internal and external defects in a component or equipment part. Ultrasonic waves are sound waves of a frequency higher than the human audible range. This technique involves striking the surface of an equipment piece part with ultrasonic waves. These waves travel through the material of the component and are reflected back. Material conditions can be ascertained by analyzing the reflected ultrasonic waves.

Access to only one side of the equipment is needed in the case of ultrasonic monitoring, as compared with two opposite sides for radiographic approaches. Moreover, the monitoring equipment and the subsequent analysis requirements are significantly less expensive. On the other hand, the analysis gets considerably more tedious and difficult to accomplish in the case of irregularly shaped objects, and objects with a complex surface geometry. This technique is widely used in medicine, and should be applicable to hardware analysis.

6. *Acoustic emission technology.* Material transformations due to crack, tear and fissure propagation, corrosion, and so on generate energy in the form of sound waves. Acoustic emission involves "picking up" this tell-tale noise. The technique is able to isolate the number and location of these internal material transformations, which could be a result of applied stress, progressive wearing out, environmental degradation, or chemical exposure.

This technique is able to detect an "alive" material transformation; however, a preexisting condition or dormant crack may not be picked up. Further, this technique is unable to delineate the magnitude of the crack or any other material transformation and needs to be coupled with other ultrasonic and/or radiographic techniques for a more complete analysis.

7. *Vibration analysis techniques.* These techniques are based upon the concept that the vibration characteristics (e.g., sound frequency) of a component without a defect are different from the vibration characteristics of a similar component with a defect. This technique is very sensitive and is capable of "picking up" very small fissures or cracks. It can be utilized to detect component material transformations resulting from stressful conditions, environmental and chemical degradation, and age. Further, dormant inconsistencies internal to a component relative to homogeneities, for example, can also be delineated.

Once again, the skill level requirements to effectively use this technique are very high. Additionally, numerous tests and experiments need to be conducted to configure the apparatus and collect data on defect-free items before variations in defective items can be detected and correctly analyzed. The equipment necessary to generate the shock pulses and then to record the subsequent material behavior is also very expensive and its use must be justified.

8. *Lubricant analysis.* The objectives of equipment lubrication include decreased friction between moving parts, reduced heat generation, and inhibition of moving parts' wear-out characteristics. The lubricating oil in many applications "washes" away any loose or worn-out particles which may be generated. Analysis of this oil can provide insight into the inner workings of the overall equipment. While a certain amount of equipment wear-out is always expected, progressive wear-out trends may be estimated, potential failures preempted, contaminations detected, and any undue wearing out immediately detected and subsequently arrested by carefully analyzing this washed-out oil. Analysis of the lubricating oil can involve chemical spectroscopy, physical particle sizing and counting, and ferrography, which involves the measurement and projection of progressive machine wear.

The basic oil analysis technique can be enhanced further through the judicious use of advanced, small-pore filters. While oil analysis can be conducted in an intermittent manner, the presence of a filter, which can capture the "tell-tale" particles being washed out, can facilitate a constant vigil on the equipment or machine. Another extension of this technique is trace sensing. This is utilized to detect the presence of a particular contaminant or material in the oil.

9. *Magnetic flux leakage technique.* According to Bray and McBride,

Magnetic Flux leakage (MFL) techniques are ideally suited for the detection of surface or near-surface anomalies in ferromagnetic materials. The MFL techniques generally do not require mechanical contact with the piece being inspected. Further, the technique is very amenable to automatic signal recognition schemes. Both of these capabilities are of great benefit for automated, high-speed inspections.

This technique involves two stages. The surface of the equipment or component being studied is first magnetized or excited. Thereafter, deviations caused by surface discontinuities in the magnetic flux generated are detected, recorded and analyzed. However, this technique works best when the surface of the object or component being analyzed has a simple and smooth geometry and is devoid of too many sharp edges.

In the case of components and surfaces with more complex geometry, a more visual form of this technique can be applied, involving the use of fine ferromagnetic particles or dust. Under the influence of the magnetic flux, these fine magnetic particles tend to "gather" around the surface discontinuities. Hidden surface imperfections (due to paint, dirt, etc.), which may not otherwise be detectable, can be located with this technique. The technique is limited in that only ferromagnetic materials can be tested with it. Advances in the related technologies have resulted in the replacement of magnetic particles and dust with magnetic paints, magnetic rubber, and magnetic printing.

10. *Temperature analysis technique.* This approach is applicable when condition monitoring involves detecting any undue temperature deviations. For example, a failure mode that renders a cooling system valve inoperative may result in very high operational temperatures which, in turn, may cause a host of potentially undesirable effects such as thermal viscosity breakdown, leaking gaskets, and engine seizure. Various tools are available today to effectively implement this approach, including contact temperature sensors (thermocouples, thermometers, thermopiles, etc.), and infrared imaging.

The calibration of the temperature sensors is critical and could impact the ultimate effectiveness of this technique. Infrared imaging, on the other hand, cannot only be used to detect and isolate "hot spots," but can also be utilized to analyze circuits and detect breaks, shorts, and "cold spots." Another advantage of infrared imaging is that contact with the surface being monitored is not required, and distance sensing is possible. Further, infrared imaging and sensing, given the advances in infrared equipment development, is very sensitive and accurate. The downside of this monitoring technology is the high equipment cost involved.

11. *Eddy current testing technique.* Whenever a metallic material is introduced into an electromagnetic field and there is relative movement between the two, electric currents are induced in the metallic material. These currents are called eddy currents and their phase and magnitude are impacted by the presence of any material discontinuities such as bubbles, cracks, tears, or pores. The implementation of this technique is not only as simple as the underlying concept, but also fairly inexpen-

sive. Further, eddy current testing is highly sensitive and can be used to detect extremely minute defects. Once again, contact with the surface is not required. An obvious limitation of this technique is that only electrically conductive and ferromagnetic materials may be inspected and monitored.

12. *Leak detection techniques.* Depending upon the type of approach used, leak detection techniques can be classified as bubble testing, sonic or acoustic leak detection, spectrometry, and so on. Bubble testing involves coating the object surface with a solution that forms bubbles in the event of gases leaking out through the surface.

Sonic and acoustic leak detection works on the principle that sound waves are generated when fluids flow through a crack, a pore, or an orifice, and are accompanied by a certain amount of turbulence and/or cavitation. This detection technique involves "listening" for such sound waves with the help of highly sensitive sonic receivers. The characteristics of the sound generated are dependent not only upon the fluid, but also upon the size of the leakage. Spectrometry, on the other hand, involves introducing tracer gases into the fluid and then tracking these gases for potential tears or cracks in the equipment. The cost of these techniques is dependent upon the level of sensitivity desired.

7.3.6 Development of Mockups and Engineering Models

In the preliminary and detailed design and development phases, prior to the availability of preproduction prototype systems/products, there is often a requirement to demonstrate some facet of maintainability. For example, the maintainability engineer may wish to illustrate the incorporation of access doors to allow for the removal and replacement of components, the layout of components on a circuit board, a specific equipment modular packaging scheme, the layout of components on a maintenance test and diagnostics control panel, the application of fasteners to some structure, or some other feature of design. To accomplish this, he/she may wish to construct a mockup made of cardboard and/or wood (to some smaller scale) or have an engineering model made in the company's "model shop" (or equivalent facility). The objective is to develop some physical entity that will assist the maintainability engineer to demonstrate some desirable features in design which, for some reason, cannot adequately be illustrated through the evaluation of design data.

7.4 DESIGN DATA REVIEW, EVALUATION, AND RECOMMENDATIONS FOR CORRECTIVE ACTION

The design review and evaluation process includes two basic categories of activity. The first constitutes the informal day-to-day design participation effort, data/documentation review and evaluation, submission of recommendations for corrective action and/or product improvement, and so on. The second area of activity includes the formal design review process described in Chapter 10. Figure 7.27 presents these two activities in the context of the system life cycle. It is the informal

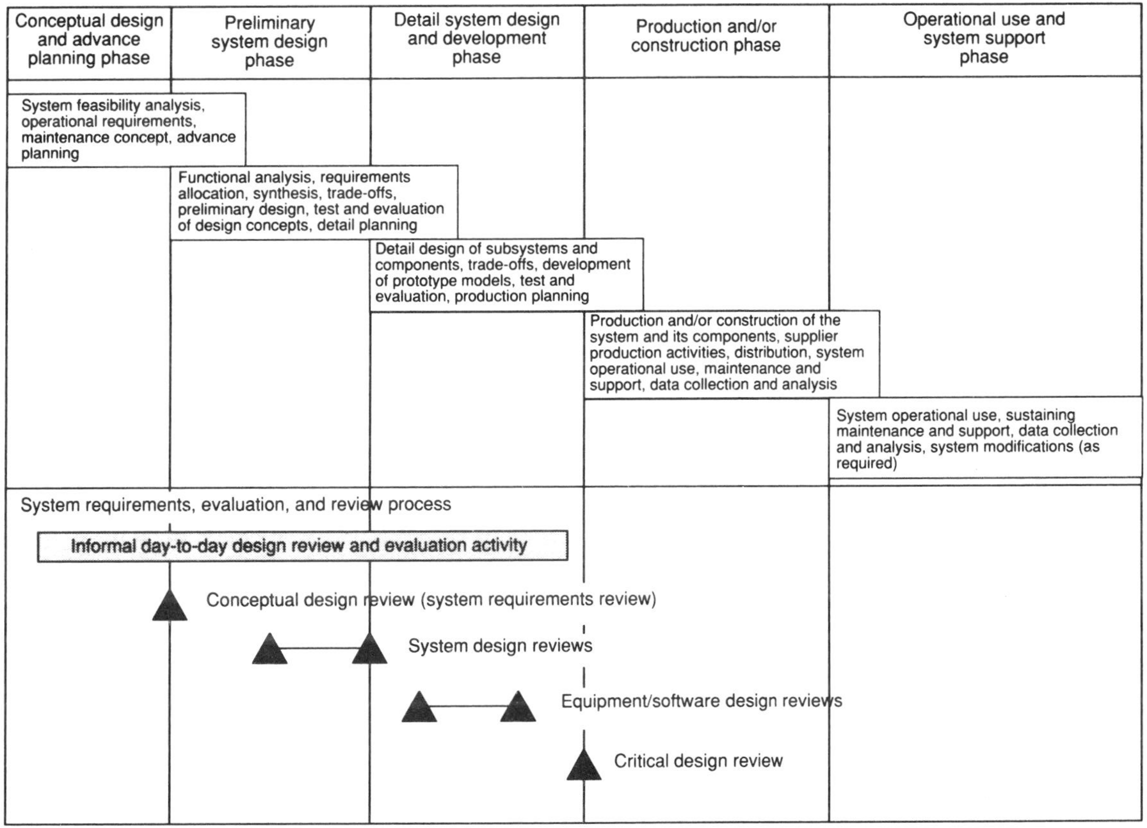

Figure 7.27. Design review and evaluation. (*Source:* Blanchard, B. S., *System Engineering Management*, John Wiley & Sons, N.Y., 1991, p. 168).

Conceptual design and advance planning phase	Preliminary system design phase	Detail system design and development phase	Production and/or construction phase	Operational use and system support phase
System feasibility analysis, operational requirements, maintenance concept, advance planning	Functional analysis, requirements allocation, synthesis, trade-offs, preliminary design, test and evaluation of design concepts, detail planning	Detail design of subsystems and components, trade-offs, development of prototype models, test and evaluation, production planning	Production and/or construction of the system and its components, supplier production activities, distribution, system operational use, maintenance and support, data collection and analysis	System operational use, sustaining maintenance and support, data collection and analysis, system modifications (as required)

System requirements, evaluation, and review process

Informal day-to-day design review and evaluation activity

Conceptual design review (system requirements review)

System design reviews

Equipment/software design reviews

Critical design review

review and evaluation activity, with the maintainability engineer participating as a member of the design team, that is discussed in this section.

One of the initial objectives of the maintainability engineer as a "team-member" participant in the design process is to develop the appropriate maintainability criteria, in response to the specified requirements, as an input to the design process. The responsible designer may solicit some guidance in this area, and the maintainability engineer may provide some general guidelines, as illustrated in Section 6.4. For a particular design, these criteria may be supplemented by addressing issues of a more specific nature—Figures 7.28 and 7.29 are examples.

Given the basic design requirements for the system, supported with some specific guidance in the form of criteria, various alternative design approaches are evaluated and the maintainability engineer will participate through the accomplishment of maintainability analysis tasks. This, of course, includes the activity described in Chapter 6, along with the utilization of the methods/techniques/tools described in the earlier sections of this chapter. The maintainability engineer should have access to the many design aids described.

As the design progresses, design data (i.e., drawings, layouts, part lists, lists of materials, electronic data, supplier data, trade-off study reports, prediction reports, etc.) are developed to describe the configuration envisioned at that point in time. The maintainability engineer will review the applicable data on a day-by-day basis, and evaluate the information available relative to the actual incorporation of the required maintainability characteristics in design. Does the design configuration described by the data meet the customer requirements as defined in the applicable specifications?

The maintainability engineer may wish to use a "checklist" to ensure that the proper questions are addressed! Figure 7.30 provides a sample checklist covering major features and areas of interest. When reviewing layouts, drawings, electronic data, and so on, the applicable items may be addressed and questioned, using more detailed criteria supporting the topics listed. Appendix A provides detailed questions backing up each item on the checklist. The list is so designed that the answer to each question (as applicable) should be YES. If additional information is desired, the designer should contact the maintainability engineer.

The process of design data review and evaluation is iterative, commencing during the conceptual design phase when system definition is covered by layout drawings and top-level block diagrams, and extending through the detail design and development phase when final production drawings, schematics, detailed parts lists, and so on are available. Not only is it appropriate to evaluate the information presented through the initial baseline documentation, but it is necessary to review and evaluate changes as well. Figure 7.31 illustrates the general process for data review and evaluation.

Referring to the figure, in the event that a problem is detected (i.e., the system does not comply with specification requirements or there is a need for a maintainability improvement—see block 3.0), the maintainability engineer may initiate a recommendation for a design change. Figure 7.32 shows an example of a "maintainability corrective action request," and Figure 7.33 presents two different exam-

1. Provide a signal that shows when test equipment is warmed up, or if this is not feasible, indicate warm-up time required next to the turn-on switch.

2. Provide quick, simple check to indicate when out of calibration or not functioning.

3. Provide for direct reading displays not requiring multiplication or other transformation of values.

4. If transformation of display value cannot be avoided, provide conversion table mounted on equipment.

5. If mounted conversion table is not feasible, provide transformation factor adjacent to each switch position or display requiring use of the factor.

6. When more than one scale must be in technician's view, differentiate by labeling and color coding to respective control positions.

7. Provide selector switches in lieu of a number of plug-in connections.

8. Provide circuit breakers or fuses to safeguard against damage if wrong switch or jack position is used.

9. Provide some device, such as warning light or automatic disconnect, which operates when test equipment lid is closed and written warning notice to ensure equipment turn-off when testing is completed.

10. Prominently display on outer surface a statement of purpose (use) for tester and any special precautions.

11. Label every item which technician must recognize, read, or manipulate.

12. Label outer case and all removable parts (loose items such as cords, plugs, probes, etc.) with their official nomenclature.

13. When adapters are required, be sure they are provided as a removable accessory.

14. Provide fasteners or holders in storage compartments for all accessories.

15. Recess handles on outer or carrying cases of portable test equipment for convenient and space-saving storage.

Figure 7.28. Statements of *M* criteria (example 1).

4.1 Maintainability Design Criteria

4.1.1 The design characteristics of the MPPG shall be such to permit fault isolation, repair, and test within a time frame of 30 minutes. Maintenance tasks shall not require the use of special equipment aside from what is normally supplied as part of test station 6811. A high impedance voltmeter and an accurate (1 part in 10^6) countertimer are not considered special equipment.

4.1.2 Design characteristics shall be consistent with operator skills and technician skills normally associated with Technician-3 (repair tasks) and Technician-5 (troubleshooting, test, and MPPG operation) classifications.

4.1.3 Self-test provisions shall be considered in the design of the MPPG.

4.1.4 Circuit functions should be logically grouped to permit rapid fault isolation. Physical packaging of specific circuit elements should be by functional grouping wherever feasible to enhance repair tasks.

4.1.5 Plug-in type modules and parts should be utilized where feasible in terms of operation requirements and cost. Guides and location keys should be provided on plug-in items.

4.1.6 Provisions shall be incorporated to ensure that parts are mounted (installed) properly during the repair tasks. Test points shall be provided wherever practical. Color coding shall be utilized where practical to aid in distinguishing parts.

4.1.7 Reference designators and part identification data shall be provided wherever practical. Test points shall be marked and coded.

4.1.9 Controls and indicators should be functionally grouped, sequentially positioned, and adequately labeled.

4.1.10 Front panel mounted test points shall be provided, as a minimum for the following functions:
a. Power supply output
b. Delay sync output
c. Pulse sync output
d. Test pulse output
e. Main pulse output
f. Start sync input
g. Stop sync input

4.1.11 Test points shall be provided on each major plug-in module to determine input and output functions.

4.1.12 A template containing adequate markings shall be provided as overlay to the wire-wrap connections.

4.1.13 Standard parts should be utilized where possible. The variety of parts should be minimized.

4.1.14 The MPPG shall have provisions to extend the item beyond the cabinet facing. Slides shall be considered. Access to all internal parts and circuits should be possible with the MPPG in the extended position, except where internal covers are included by necessity.

4.1.15 Covers shall be mounted with quick-disconnect fasteners.

4.1.16 Servicing requirements of the MPPG should be held to a minimum. Sealed "LIFE" bearings should be used where possible.

Figure 7.29. Statements of *M* criteria (example 2).

MAINTAINABILITY CHECKLIST

GENERAL

1. STANDARDIZATION MAXIMIZED
2. COMPONENTS FUNCTIONALLY GROUPED
3. CONSOLE LAYOUT OPTIMIZED
4. COMPLEXITY MINIMIZED
5. SELF-TEST INCORPORATED
6. MAX. TIME TO REPAIR MINIMIZED
7. TOOLS & TEST EQUIP. MINIMIZED
8. LABELING MAXIMIZED
9. WEIGHT MINIMIZED
10. CALIBRATION REQUIREMENTS KNOWN
11. REPAIR/REPLACE PHILOSOPHY KNOWN
12. MAINT. PROCEDURES KNOWN
13. PERSONNEL REQUIREMENTS MINIMIZED
14. TRADE-OFFS DOCUMENTED

HANDLING

1. EQUIPMENT LIFTING MEANS EMPLOYED
2. EQUIPMENT BASE REINFORCED (FORK-LIFT APP.)
3. DRAWER/PANEL HANDLES EMPLOYED
4. ASSEMBLY HANDLES EMPLOYED
5. CONSOLE CASTORS EMPLOYED (AS APPLICABLE)
6. DAMAGE SUSCEPTIBILITY MINIMIZED
7. WEIGHT LABEL ON CONSOLE

EQUIPMENT RACKS-GENERAL

1. DRAWERS ON ROLL-OUT SLIDES
2. PANELS HINGED
3. IN-POSITION MAINTENANCE POSSIBLE
4. CABLES CONNECTED WITH DRAWERS EXTENDED
5. PERMANENT CABLE INLETS ON FRONT AVOIDED
6. HEAVIEST ITEMS ON BOTTOM
7. OPERATOR PANELS OPTIMUM POSITION
8. AIR INTAKE/EXHAUST PROVISIONS ADEQUATE

PACKAGING

1. PLUG-IN COMPONENTS EMPLOYED
2. COMPONENT STACKING AVOIDED
3. ACCESSIBILITY BASED ON REPLACEMENT FREQ.
4. WRONG INSTALLATION OF UNIT PREVENTED
5. MODULES & MOUNTING PLATES LABELED
6. GUIDES USED FOR MODULE INSTALLATION
7. INTERCHANGEABILITY INCORPORATED

ACCESSIBILITY

1. ACCESS DOORS PROVIDED
2. ACCESS DOORS SELF-SUPPORTED
3. ACCESS DOORS LABELED
4. ACCESS OPENINGS ADEQUATE IN SIZE
5. ACCESS FASTENERS MINIMIZED
6. SPECIAL TOOLS MINIMIZED
7. COMPONENT ACCESSIBILITY ADEQUATE
8. GUIDES FOR DANGEROUS ACCESSES CONSIDERED

FASTENERS

1. QUICK RELEASE FASTENERS EMPLOYED
2. FASTENERS STANDARDIZED
3. QUANTITY OF FASTENERS MINIMIZED
4. HEXAGONAL SOCKET-HEAD FASTENERS USED
5. CAPTIVE NUTS/SCREWS EMPLOYED
6. MINIMUM NUMBER OF TURNS REQUIRED

CABLES

1. CABLES FABRICATED IN REMOVABLE SECTIONS
2. CABLES ROUTED TO AVOID SHARP BENDS
3. CABLES ROUTED TO AVOID PINCHING
4. PROTECTION FOR CABLES ROUTED THRU HOLES
5. CABLES IDENTIFIED
6. CABLE CLAMPING SUPPORT ADEQUATE
7. HANDHOLD/STEP PREVENTION CONSIDERED

CONNECTORS

1. QUICK DISCONNECT VARIETY
2. CONNECTOR SPACING ADEQUATE
3. LABELING ADEQUATE
4. CONNECTORS KEYED
5. CONNECTORS STANDARDIZED
6. SPARE PINS PROVIDED
7. MALE CONNECTORS CAPPED
8. RECEPTACLES "HOT" & PLUGS "COLD"
9. MOISTURE PREVENTION CONSIDERED

SERVICING/LUBRICATION

1. SERVICING REQUIREMENTS CONSIDERED
2. SERVICING POINTS ACCESSIBLE
3. SERVICING FREQUENCIES KNOWN

PANEL DISPLAYS/CONTROLS

1. CONTROLS STANDARDIZED
2. CONTROLS SEQUENTIALLY POSITIONED
3. CONTROLS PROPERLY SPACED
4. CONTROLS ADEQUATELY LABELED
5. CONTROLS ADJACENT TO APPLICABLE DISPLAY
6. RUGGEDIZED METERS EMPLOYED
7. METERS EXTERNALLY REMOVABLE
8. PANEL LIGHTING EMPLOYED
9. INDICATOR LIGHTS "PRESS-TO-TEST"
10. FUSE REQUIREMENTS SATISFIED
11. SPARE FUSES PROVIDED
12. WARNING LIGHTS EMPLOYED-CRITICAL FUNCTIONS
13. COLOR OF INDICATOR LIGHTS ADEQUATE
14. CONTROLS PLACED BY FREQUENCY OF USE

TEST POINTS

1. LOCATED ON FRONT PANEL
2. FUNCTIONALLY GROUPED
3. ADEQUATELY LABELED-NUMBER & SIGNAL VALUE
4. INTERNAL TEST POINTS ACCESSIBLE
5. DEGREE OF TEST INDICATED
6. ADEQUATELY PROTECTED
7. ADEQUATELY ILLUMINATED
8. LOCATED CLOSE TO APPLICABLE CONTROL

ADJUSTMENTS

1. ADJUSTMENT POINTS ACCESSIBLE
2. PERIODIC ADJUSTMENTS KNOWN
3. INTERACTION EFFECTS ELIMINATED
4. ADJUSTMENT LOCKING DEVICES PROVIDED
5. FACTORY ADJUSTMENTS SPECIFIED
6. ADJUSTMENT POINTS ADEQUATELY LABELED
7. FINE ADJUSTMENTS THROUGH LARGE MOVEMENTS
8. BUILT-IN JACKS FOR METER CALIBRATION
9. CLOCKWISE ADJUSTMENTS FOR INCREASING VALUES

PARTS/COMPONENTS

1. ARRANGED IN FAMILY GROUPS
2. ADEQUATELY LABELED
3. ADEQUATELY SPACED FOR TOOL ACCESS
4. INDIVIDUAL PARTS DIRECTLY ACCESSIBLE
5. DELICATE PARTS ADEQUATELY PROTECTED
6. NOT VULNERABLE TO EXCESSIVE SOLDER HEAT

ENVIRONMENT

1. TEMPERATURE/HUMIDITY RANGES CONSIDERED
2. ILLUMINATION ADEQUATE
3. TRANSPORTABILITY CONDITIONS CONSIDERED
4. MOBILITY CONDITIONS CONSIDERED
5. STORAGE CONDITIONS CONSIDERED

SAFETY

1. ELECTRICAL OUTLETS/JUNCTION BOXES LABELED
2. INTERLOCKS EMPLOYED
3. FUSE/CIRCUIT BREAKER PROTECTION ADEQUATE
4. WARNING DECALS ADEQUATE
5. GUARDS/SAFETY COVERS-HIGH POTENTIALS
6. PROTRUDING DEVICES ELIMINATED
7. EXTERNAL METAL PARTS ADEQUATELY GROUNDED
8. DRAWER/PANEL/STRUCTURE EDGES ROUNDED
9. TOOL USE CONSIDERED

RELIABILITY

1. ALLOCATED MTBF KNOWN
2. FAIL-SAFE PROVISIONS INCORPORATED
3. CRITICAL/SERVICE LIFE CONSIDERED
4. WEAR-IN/WEAR-OUT CYCLES CONSIDERED
5. FAILURES TRACEABLE BY TEST

Figure 7.30. Maintainability checklist (example).

| COMPANY XYZ | MAINTAINABILITY CORRECTIVE ACTION REQUEST (MCAR) | DATE ORIGINATED 9/6/93 | CONTROL NUMBER M105 |

TO John S. Stone   AREA 54   ROOM 18E   EXT. 5398   ITEM NAME IRU Switching Panel   ITEM NO.

ORIGINATOR Joe D. Smith   AREA 54   ROOM 18E   EXT. 5345   DWG/PART NO. A 11016-0011   REV/EFF. A

REQUEST AFFECTS: ☐ RELIABILITY ☐ MAINTAINABILITY ☒

☐ VALUE ENGINEERING. ☐ HUMAN FACTORS ☐ EVALUATION TEST ENG.

M ENDORSEMENTS:

SECTION SUPERVISOR Sam H. Brown   DATE 9/7/93

PROGRAM F-118 BNS Test Station Des & Dev   PROJECT M ADM. Robert A. Anderson   DATE 9/7/93

NO.	PROBLEM DESCRIPTION	CORRECTIVE ACTION REQUESTED	DISPOSITION SHOULD BE MADE AND THE MCAR COMPLETED AND RETURNED TO MAINTAINABILITY WITHIN FIVE DAYS OF RECEIPT.
1	The 36-Position, G-Wafer Switch on the IRU Switching Panel Requires an extensive amount of time for switch replacement and/or repair (54 wires to switch and 61 wires between wafers).	Mount Switch on a separate bracket which in turn should be mounted on the front panel. Route wires (54) from switch thru 60-PIN MS Connector. Corrective maintenance will involve removal of 4 screws and a connector break.	☒ ACCEPTED Problem No. 1 INCORPORATE IN: 1. Proposed switch design accepted 9/10/93 and will be incorporated in S/N 2 and on. S/N 1 will not be retrofitted.
2	Access to the numerous small switches, terminals, etc., mounted behind the IRU switching panel requires removal of 24 countersunk screws. An additional 20 countersunk screws are used to fasten a dust cover which must be removed for tool access.	Provide one-quarter (¼) turn fasteners in place of countersunk screws to facilitate access Reduce quantity of fasteners to 8 on the panel and 6 on the dust cover	☐ REJECTED REASON: ☒ PARTIAL ACCEPTANCE Problem No. 2 Use of ¼ turn fasteners accepted 9/12/93 S/N 2 and on. Reduction in Qty. of fasteners rejected due to R.F.I problems. SIGNATURE John S. Stone TITLE Sr. Design Eng. DATE 9/14/93

FORM 1116 E 9/66

Figure 7.31. Design data review, evaluation, and corrective-action process.

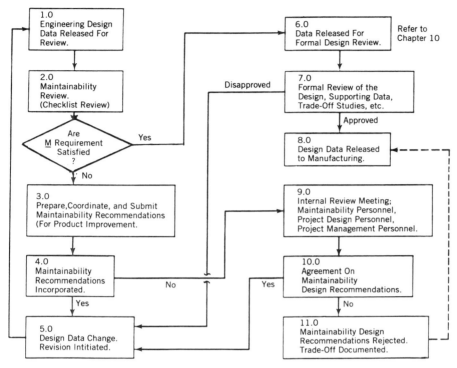

Figure 7.32. Maintainability corrective action request (example).

ples, with justification provided in terms of effectiveness gains and cost savings. For each recommendation, the maintainability engineer needs to be precise in defining the problem, to provide a recommended approach to problem resolution, and to justify the proposed course of action considering performance factors, effectiveness, and life-cycle cost. The proposed design recommendation is submitted to the responsible design engineer who, in turn, reviews the request and makes a decision whether to "accept" or "reject" the change.

If the design engineer is considering the selection of "off-the-shelf" equipment that is already produced and in a supplier's inventory, he/she may solicit the maintainability engineer's assistance in evaluating the feasibility of a given item from a maintainability perspective. Figure 7.34 shows an example of the results of an equipment evaluation. The specific categories and factors identified in the figure (e.g., category 4 covering "packaging" and "standardization" as a specific factor within) are supported through the development of checklist criteria. Figure 7.35 shows an example of evaluation criteria for Item 4.2. The design engineer will usually have access to a working model of the equipment item being considered, and will use this model to evaluate the adequacy of the performance characteristics

MAINTAINABILITY RECOMMENDATION
(Example *A*)

PROBLEM:

The 36 position, 6 wafer switch on the Test
Control Unit (A11471-00) requires an excessive amount of time for switch replacement and/or repair (54 wires to switch and
61 wires between wafers).

RECOMMENDATION:

Mount switch on a separate bracket which
in turn should be mounted on the front panel.
Route wires (54) from switch through 60 pin
MS connector. Corrective maintenance will
involve removal of 4 screws (required for
mounting switch assembly on panel) and a
connector break in lieu of individual wire
replacement with the switch installed on the
panel.

RESULTING EFFECTS:

Average Time Saving/Maintenance Action - -
1 hour, 22 minutes

Estimated Quantity of Maintenance Actions
Covering 30 Equipments for 10 Years - - 390

Estimated Cost Saving (Labor) - $2,949

Estimated Additional Material Expenditure - -
$1,920

Estimated Total Net Cost Saving - - $1,029

Increase in Equipment Availability (A_a) - -
2.4 percent

MAINTAINABILITY RECOMMENDATION
(Example *B*)

PROBLEM:

Troubleshooting the Serializer is difficult
due to lack of test points on front panel or
lack of tip jacks internally. In addition,
heavy use of the terminal points presently
provided will likely cause excessive maintenance-induced faults.

RECOMMENDATION:

Add the following test point coverage to the
Serializer front panel:

a.   Tape Levels 1 - 7
b.   Clear Pulse
c.   Enter Pulse
d.   Clock Pulse
e.   Three (3) Common Inputs

RESULTING EFFECTS:

Time Saving/Maintenance Action - - 7 min

Estimated Quantity of Maintenance Actions
Covering 30 Equipments for 10 Years - - 1770

Estimated Total Cost Saving - - $1,143

Figure 7.33. Maintainability design recommendations (example).

of the item. At the same time, the maintainability engineer will evaluate the *M*
characteristics by disassembling the equipment, accomplishing some simulated diagnostic tests, reviewing data supplied by the supplier (operating and maintenance
instructions, illustrated parts breakdown), and so on. The results of the maintainability evaluation are provided to the responsible design engineer in a timely
manner.

As the design progresses, recommendations for change may be initiated from a
number of different sources, with the maintainability organization being just one!
Such changes are prepared in the form of an Engineering Change Proposal (ECP).
Engineering change proposals and the procedures for the processing of changes are
addressed in Chapter 10 (Section 10.6).

PRELIMINARY MAINTAINABILITY EQUIPMENT EVALUATION

NO. M227

EQUIPMENT NAME Sweep Oscillator   MANUFACTURER Green Engineering

TYPE 38149-001   MODEL B   SERIAL NO. 321-09

POINTS 57   ☐ GOOD ☒ FAIR ☐ POOR   ENGINEER R. Krest   DATE 3-6-93

ADEQUACY, ACCESSIBILITY, ETC. DETERMINED ON FOLLOWING ITEMS AS APPLICABLE (0 TO 8 POINTS OBTAINABLE DEPENDING UPON THE DEGREE OF ACCEPTABILITY. 100 POINTS POSSIBLE).

1. GENERAL	PTS	4. PACKAGING	PTS	6. HUMAN FACTORS	PTS
1. WARMUP TIME	4	1. FAST VARIETY/TYPE	2	1. INST. POS. MAINT	2
2. LABELING	2	2. PARTS STANDARDIZED	3	2. DISPLAYS	2
3. MAINT. TOOLS/EQUIP.	1	3. CONNECTORS	4	3. CONTROLS	2
		4. WIRING/CABLING	4	4. WEIGHT/HANDLING	3
2. CIRCUITRY		5. PRINTED CIRCUITS	2	5. DAMAGE SUSCEPTIBLE	3
1. COMPLEXITY	3	6. PROBE/SOLD/IRON/OTHER		6. PERSONNEL TRAINING	2
2. FUNCTIONAL PKG'ING	3	TOOL SPACE	3		
		7. PLUG-IN-MODULES	0	7. SAFETY	
3. TEST POINTS				1. EQUIPMENT	4
1. AVAILABILITY	0	5. MAINT. AIDS		2. PERSONNEL	4
2. LABELING	0	1. MALF. EQUIP. KNOWN	2		
3. LOCATION/GROUPING	0	2. SELF CHK/CALIB.	1		

REMARKS:

1) Warmup time is approximately 2 minutes.

2) Labeling inadequate — not completely labeled with reference designations.

3) Packaging — P/C boards were used where possible, but some of them are almost impossible to remove. P/C boards are wired-in to chassis — not plug-in type.

4) Special tools are needed for P/C boards removal.

5) No test points are designed into the equipment. However, there are a few 3N connectors and terminal points that can be used as test points.

6) Equipment, when rackmounted, must be removed from the rack for fuse replacement (fuse is located on the rear surface of the chassis). No spare fuse is provided.

Figure 7.34. Off-the-shelf equipment evaluation (example).

Category/Item (Refer to Figure 7.34, Item 4.2.)	Numerical Rating (Points)
4.2 Parts Standardized	
a. All equipment parts are of the standard AN, NAS, or MIL-STD variety and are likely to be available in government supply.	4
b. Seventy-five percent of all equipment parts are of the standard AN, NAS, or MIL-STD variety and are likely to be available in government supply. Twenty-five percent are commercially available, but are not likely to be available in government supply.	3
c. Fifty percent of all equipment parts are of the standard AN, NAS, or MIL-STD variety and are likely to be available in government supply.	2
d. Seventy-five percent of all equipment parts are of the standard AN, NAS, or MIL-STD variety or are commercially available. Twenty-five percent are nonstandard or peculiar parts not likely to be readily available in either government supply or commercial supply.	1
e. Fifty percent of all equipment parts are of the standard AN, NAS, or MIL-STD variety or are commercially available. Fifty percent are nonstandard or peculiar parts not likely to be readily available in either government supply or commercial supply.	0

QUESTIONS AND PROBLEMS

1. What is meant by "design criteria?" Provide some examples. For what reason does the maintainability engineer develop "general" design criteria and "specific" design criteria?

2. What is the objective of reliability modeling? How are the results applied?

3. What is reliability stress–strength analysis? What information can one derive from such an analysis? How does this information impact maintainability (if at all)?

4. Briefly describe the FTA, the FMECA, the hazard analysis, and their respective applications. What are the significant differences? Identify some commonalities.

5. Select a system (or major subsystem) of your choice and perform a FMECA. Identify high-priority items from the results of the analysis.

6. What is the difference between a "process-oriented" FMECA and a "product-oriented" FMECA? Are they related? If so, how? Show an example.

7. Why is the FMECA important from a maintainability perspective? What information does it provide?

8. What is the purpose of the Ishikawa diagram?

9. What is the objective of the RCM?

10. Select a system (or major subsystem) of your choice and develop a PM program using the RCM as the basis for decision making.

11. What is the purpose of reliability prediction? When is it generally accomplished? How do the results impact maintainability?

12. Assume the following:

Component Part	Failure Rate/Part (%/1000 Hours)	Quantity of Parts
Part A	0.092	15
Part B	0.812	85
Part C	0.351	72
Part D	0.295	18
Part E	0.421	32
Part F	0.084	21
Part G	0.512	9
Part H	0.102	25
Part I	0.612	12

What is the predicted MTBF?

13. Describe what is meant by reliability growth modeling. How can it be applied in a typical program?

14. Describe what is meant by ESS. What is the purpose of ESS? How can the various "screens" be applied?

15. What type of maintainability information can be gained from a reliability qualification test?

16. What is the FRACAS? Why is it important to maintainability?

17. In the design for human factors, what characteristics in design are considered important? Describe some of the relationships between these characteristics and "maintainability in design."

18. In a typical human factors program, a number of different tasks should be accomplished to help ensure that the proper human–machine relationships are addressed. What specific tasks are considered important from a maintainability perspective? What information can be derived?

19. Select a system (or major component) of your choice and accomplish a detailed operator task analysis. Then accomplish a MTA on the same item. Describe some of the similarities.

20. Develop an OSD for a system (or subsystem) of your choice. Describe how the results can influence design. Provide some specific examples. What benefits can be derived from a maintainability perspective?

21. What is meant by "artificial intelligence?" How can it be applied? Provide some specific examples.

22. If you were assigned to develop a CM capability, what factors would you consider? How would you go about selecting the appropriate testing technologies to support CM objectives? How does the CM capability relate to the FMECA? How do the results impact the MTA?

23. What is meant by CAD? CAM? CALS? What "technologies" are included and how do they interrelate?

24. Identify a system (or major subsystem) of your choice and develop a design review checklist, incorporating the appropriate criteria, for use in evaluating the system in terms of "design for maintainability."

25. Select an "off-the-shelf" equipment item of your choice and accomplish a maintainability evaluation using the data format presented in Figure 7.32 (or develop and utilize your own form of an equivalent nature). Prepare at least three specific recommendations for improvement, following the examples in Figure 7.33 (i.e., justify your recommendations in terms of improved performance and effectiveness, reduced life-cycle cost, etc.).

26. Assume that you have just been assigned to a "design team" as the responsible maintainability engineer. What steps would you take to *ensure* that your objectives are ultimately fulfilled in an effective and efficient manner, that is, that the appropriate maintainability characteristics are incorporated into the final design configuration?

27. How are design changes initiated? What is an ECP? Why is "change control" so important? What is "configuration management?" What role does the maintainability engineer play in the context of the process illustrated in Figure 7.31?

8

MAINTAINABILITY PREDICTION

Maintainability prediction involves an assessment of the design of the system (and its components) from a maintainability perspective. The objective is to evaluate maintainability characteristics of the current system design configuration, and determine whether the initially specified/allocated system maintainability requirements are likely to be met. Any significant deviations are evaluated for possible redesign or compensation. Further, maintainability prediction can facilitate the establishment of confidence levels for successfully completing maintainability test and demonstration exercises at a later stage in the design and development process.

Maintainability prediction studies are initiated during the preliminary system design phase subsequent to the initial trade-offs and allocation of system level requirements (refer to Section 2.1 and Figure 2.1). This is the earliest point in time when sufficient engineering data are available to perform a meaningful quantitative evaluation of design characteristics in terms of performance and supportability. On the other hand, the system design process at this point has not advanced beyond the stage where the results of such predictions can influence system design in an effective and efficient manner.

The system maintainability prediction process evolves from the preliminary system design stage and is refined as increasingly accurate and applicable engineering data are made available to the analyst. Further, with enhanced knowledge of the system design, different and more accurate prediction methodologies are often utilized. The effectiveness of maintainability prediction as a system evaluation tool is a function of input data accuracy, extent of the data available, methodology used, and the technical knowledge and insight of the analyst. The potential benefits actually accrued from such studies are dependent on the expeditious feedback of the results to the design engineer (and program management), and the subsequent initiation of corrective action if the results of the prediction indicate a problem area. Additionally, the results of maintainability prediction constitute a key data input requirement in the accomplishment of design trade-off studies, MTA, life-cycle cost analyses, logistic support analysis, and related efforts. Above all, the prediction process should be undertaken with full objectivity on the part of the analyst.

Maintainability prediction, in general, involves the early and timely quantitative assessment of maintenance elapsed-time factors, maintenance labor-hour factors, maintenance frequency factors, and/or maintenance cost factors. This assessment process requires the analyst to study and review preliminary layouts, engineering drawings, component-part lists, and any other related data with the objective of anticipating necessary maintenance actions, identifying potential resource requirements (equipment, tools, skill levels, number of maintenance personnel), and estimating maintenance task completion times.

8.1   MAINTAINABILITY PREDICTION TECHNIQUES

Numerous maintainability prediction techniques and procedures are presently being used by industry and government agencies/organizations. These techniques and procedures vary depending upon the specific need for measurement, equipment type, extent and accuracy of information available, and applicable system design phase. Further, a certain analyst or organization may have a specific preference with respect to the applicable techniques and/or procedures. The purpose of this chapter is not to detail all the possible methodologies, but to briefly review some of the more commonly utilized prediction methods.[1]

8.1.1   Prediction Method 1

This maintainability prediction technique is utilized during the early phases of design, and is based on the principle of analyzing a certain number of replaceable system elements selected randomly. These elements are chosen from the complete set of replaceable elements constituting the system. Further, once the random sample size is ascertained, the replaceable elements are further classified into subgroups on the basis of component part types or classes. Typical component part classes include transistors, capacitors, and so on.

   The underlying philosophy of this technique is that system failures result from the failure of lower-level faulty replaceable items and, as such, the time to replace/repair these items reflects the likely overall system-level downtime. It is assumed that for repair by replacement, maintenance actions/procedures are likely to be fairly similar for component parts from any one class or type. For instance, the process of removing and replacing a specific resistor is likely to be similar to removing and replacing any other resistor. Assuming such uniformity in system design, analysis of an adequate and random selection of replaceable items from a particular type or class of components will be representative of that entire class.

   To compute completion times, each maintenance task is scored on the basis of

---

[1] Some of the techniques described herein are interpretations of the procedures included in MIL-HDBK-472, Military Handbook, *Maintainability Prediction,* Department of Defense, Washington DC, 1984. Three methods are presented to provide examples of how an analyst might approach the maintainability prediction process. However, the ultimate selection of a particular technique must be tailored to the specific system configuration being evaluated.

(1) the physical configuration of the system, (2) the facilities required to effect proper maintenance, and (3) the maintenance personnel skill level requirements. Each of these aspects is represented by a checklist, as depicted in Table 8.1.

Each checklist represents a set of criteria against which the current system design configuration is evaluated. For instance, checklist "A," which focuses attention on the physical characteristics of the design configuration, constitutes criteria such as internal and external accessibility, types of internal and external fasteners, configuration modularity, calibration and realignment requirements, and so on. Use of such checklists helps ensure a more comprehensive maintainability evaluation, and facilitates consistency through minimizing any scoring deviation between analysts. These checklists were developed from data covering the operation and support of systems in the field (particularly with regard to electronic systems in the defense sector). Maintenance times were correlated with the various maintainability features actually incorporated into the design. In instances where "accessibility" was poor, maintenance times are extended, and so on. The system configuration is rated on a scale of 0–4, for each of the criterion listed on the three checklists. If the design evaluated very well relative to a particular criterion, it is given a higher rating. For instance, a design configuration that provides excellent external accessibility from a visual and manual manipulation perspective for a particular maintenance action may be given a rating of "4" for the first criterion on checklist "A," and so on.

The steps involved in this prediction procedure are presented below:

1. *Determine the random sample size for replaceable system elements.* To rate and evaluate a design configuration from a maintainability perspective in accordance with the three checklists discussed earlier, a representative and random set of system elements and maintenance tasks is first selected. This step is based on the assumption that evaluating a representative sample of maintenance tasks and replaceable system elements is more efficient, and the results are likely to very closely approximate the analysis where all the replaceable elements and maintenance tasks were evaluated.

The sample size is selected on the basis of acceptable risk or desired confidence levels, the extent of population data dispersion, and finally, desired analysis accuracy. For instance, a large sample size will likely approximate actual results more closely than a small sample size. On the other hand, the prediction process may take significantly more time for a large sample size. The objective is to choose a sample size that will offer an efficient yet effective solution. This is accomplished through the use of the following relationship:

$$N = \left[C_x \frac{Z}{k} \right]^2 \qquad (8.1)$$

where N is the random sample size, C_x is the coefficient of variation (calculated as $C_x = \sigma / \overline{X}$, where $\overline{X}$ is the population mean), Z is the confidence level expressed in terms of a percentage, and k is the desired accuracy of the analysis expressed in

TABLE 8.1 Maintainability Design Checklists

Checklist A (Physical Design Characteristics)

	Criteria	Description
1.	Accessibility (External)	This refers to the external accessibility of the system with respect to visual inspection/manipulation.
2.	Fasteners (External)	This determines if fasteners on the system exterior require any special tool kits or excessive time.
3.	Fasteners (Internal)	This determines if fasteners on the system interior require any special tool kits or excessive time.
4.	Accessibility (Internal)	This refers to the internal accessibility and packaging of the system with respect to visual inspection/manipulation and ease of maintenance.
5.	Modularity & Packaging	This reflects the ease of access to component parts that need to be assembled or disassembled.
6.	Component Part Removeability	This determines the ease with which component parts can be removed and replaced.
7.	Sensors & Displays	This reflects the availability of vital information pertinent to system operation and failure through well placed displays.
8.	Built-In-Test Capability	This refers to the existence of ample indicators and alarms within the system to facilitate troubleshooting and fault isolation.
9.	Test Requirements	This determines if the maintenance task requires the utilization of test receptacles and points for completion.
10.	Testability	This reflects the clarity of test point/receptacle markings and ease of their use.
11.	Proper Marking (Labeling)	This reflects the proper identification and labeling of relevant parts, elements, and circuits to facilitate maintenance.
12.	Calibration/Adjustment Requirements	This reflects the extent to which a maintenance task requires system calibration, adjustment, and realignment.
13.	Testing Facility	This determines whether the potentially defective system element needs to be disassembled and removed before it can be tested.
14.	Damage Protection Capability	This determines if facilities exist within design to minimize damage to the overall system because of failures (e.g., through the use of fuses).
15.	Personnel Safety	This reflects the extent to which personnel safety is likely to be compromised while completing a maintenance task.

Checklist B (Facility Requirements)

	Criteria	Description
1	Test Equipment Requirements	This refers to whether external test equipment will be required to complete the maintenance task and, if so, then to what extent or how much?
2	Test Equipment Adapters/Connectors	This determines if the external test equipment requires any special adapters or connectors, and if special tools are needed for set-up, etc.
3	Maintenance Jigs & Fixtures	This refers to whether accessories such as jigs, fixtures, braces, etc., are necessary for maintenance task completion.
4	Visual Communication	This reflects whether visual communication is possible between maintenance personnel during maintenance.
5	Maintenance/Operation Personnel Interaction Requirements	This refers to the necessity of interacting with the operating personnel during the course of the maintenance task (e.g., to facilitate check-out).
6	Number Of Maintenance Personnel Needed	This reflects the number of maintenance personnel necessary for effective maintenance task completion.
7	Supervisory/Contractor Services Needed	This determines if the presence of supervisory or contractual personnel is necessary in order to effect the maintenance task.

Checklist C (Personnel Requirements)

	Criteria	Description
1.	Personnel Strength	This reflects the amount of arm, leg, and back strength required in order to complete the maintenance task.
2.	Endurance	This reflects the amount of endurance and energy required on the part of the maintenance personnel for maintenance task completion.
3.	Manual Dexterity, Reflex, And Eye-Hand Coordination	This determines the extent to which attributes such as manual dexterity, reflexes, and eye-hand coordination are required for the maintenance task in question.
4.	Visual Resolution/Accuracy	This reflects the extent to which visual alertness and accuracy is required for reading displays, isolating faulty component parts, etc.
5.	Memory	This reflects the extent to which familiarity and experience is required relative to procedures, tools, etc.
6.	Logic & Inference	This refers the need for inference and logical analyses on the part of the maintenance personnel.
7.	Organization	This reflects the amount of planning and organization needed for effective maintenance task completion.
8.	Alertness & Cautiousness	This reflects to the extent of awareness, readiness, and forethought needed on part of the maintenance personnel.
9.	Patience, Concentration, & Persistence	This refers to the amount of mental application and attention, and persistence needed on part of the maintenance personnel.
10.	Independence & Aptitude	This reflects the amount of initiative and independence the maintenance personnel is likely to need.

These checklists have been adapted from: MIL-HDBK-472, Military Handbook, *Maintainability Prediction*, Department of Defense, Washington D.C., 1984.

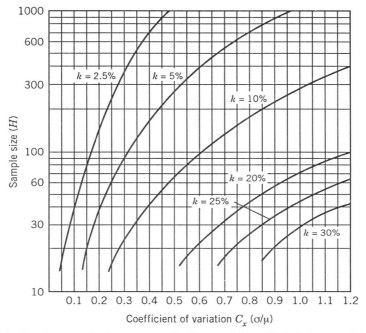

Figure 8.1. Random sample size nomograph. (*Source:* MIL-STD-472, Military Handbook, *Maintainability Prediction,* Department of Defense, Washington, D.C., 1984.)

terms of a percentage deviation. Further, $C_x = \sigma/\overline{M}ct$, where σ is the standard deviation of the population data (where standard deviation is a measure of the population data dispersion around the mean value), and $\overline{M}ct$ is mean corrective maintenance time. Equation 8.1 has been solved for numerous values of accuracy, k, and a constant confidence level of 95%. The solutions have been translated into a nomograph, as depicted in Figure 8.1. For instance, consider the data in Table 8.2. The sample size, N, is computed as

$$N = \left[0.262 \, \frac{1.645}{0.0975} \right]^2 = (4.42)^2 = 19.54 \text{ (assume 20)}$$

TABLE 8.2   Random Sample Size Computation

Z	95% or 1.645
k	9.75%
σ	17 minutes
$\overline{M}ct$	65 minutes

2. *Determine subsample size for component part categories.* The sample size derived in step 1 represents a random and representative mixture of replaceable items that constitute the current design configuration. This overall random sample, N, is next broken down into a number of subsamples of size n, where each subsample represents a certain class or category of component part types, for example, resistors, capacitors, motors, transistors, inductors, and so on. This is accomplished as follows:

1. Determine the quantity of parts in each component part category (refer to column 2 of Table 8.3). For instance, how many resistors are included in the design of the configuration? How many motors are included?

b. Determine average failure rate for each component part category in failures per million hours (refer to column 3 of Table 8.3). This is based on historical experience with a particular class of component types, and on other reliability prediction reports.

c. Determine the expected number of failures per million hours for each component part category. This is obtained by multiplying the quantity in column 2 by the failure rate in column 3 of Table 8.3 (refer to column 4).

d. Establish the percent contribution of each component part category type to the overall system failure rate (refer to column 5 of Table 8.3). This is the ratio of the expected number of failures per million hours for a particular component part category to the total number of failures per million hours of operation for the current design configuration.

e. Determine the quantity of failures per component part category based on sample size N. This is determined by multiplying the percent contribution of each category type by N (refer to column 6 of Table 8.3).

f. Determine the quantity of random samples per component part category. This is determined by rounding off the values in column 6 of Table 8.3 (refer to column 7).

3. *Select specific replaceable items.* Once the quantity of samples per component part category has been determined, as illustrated in Table 8.3, the next step is to randomly select derived quantities of specific replaceable system elements to be analyzed in more detail.

4. *Estimate maintenance completion times.* After selecting replaceable system elements in the correct quantities from the different component part categories, maintenance task completion times are next estimated in each case. System maintenance through faulty element replacement is assumed. This estimation is based on how well each maintenance task and replaceable item evaluates against the criteria listed in checklists "A"–"C" (refer to Table 8.1). However, before a maintenance task can be scored effectively, basic drawings and layouts need to be reviewed and evaluated in an attempt to delineate the necessary maintenance steps involved.

Each elemental step that constitutes the overall maintenance task is listed in proper sequence along with any pertinent comments that may have a bearing on the subsequent scoring. This process is illustrated in Table 8.4. Thereafter, each maintenance task is scored on a scale of "0"–"4" for each of the criterion in accor-

TABLE 8.3 Derivation of Component Part Category Subsample Size

(1) Component Part Category	(2) Quantity (Q)	(3) Part Failure Rate (λ)	(4) Expected Quantity of Failures ($Q\lambda$)	(5) Contribution to Total Expected Failures (%)	(6) Quantity of Failures per Category ($N=20$)	(7) Actual Subsample Size (n)
Capacitors	706	0.107	75.5	10.8	2.16	2
Diodes	458	0.105	48.0	6.8	1.36	1
Transistors	430	0.221	95.0	13.6	2.72	3
ICs	460	0.140	64.4	9.2	1.84	2
Connectors	89	0.140	12.4	1.7	0.34	0
Relays	3	0.600	1.8	0.5	0.10	0
Inductors	149	0.086	12.8	2.0	0.40	0
Transformers	65	0.333	21.6	3.0	0.60	1
Motors	5	25.000	125.0	17.9	3.58	4
Resistors	1090	0.179	195.1	28.0	5.60	6
Miscellaneous	252	0.181	45.6	6.5	1.30	1
Total	3112		697.2	100%		20

dance with the checklists shown in Table 8.1. If insufficient information is available for scoring the maintenance task relative to a particular criterion, the average score for that particular checklist is normally substituted. After the entire set of criteria has been addressed, the ratings are consolidated separately for the three checklists.

5. *Compute system maintenance downtime.* The last step in this prediction technique is to use the consolidated scores derived in step 4 to compute maintenance downtimes as a result of each maintenance task. This is accomplished by inserting the consolidated checklist scores in Table 8.4 into a regression equation. As with the checklists discussed earlier, regression equations used during the course of this prediction methodology have been developed based on prior experience and data collected from similar systems operating in the field in similar environments. A regression equation represents a correlation between maintenance action checklist scores and the expected system downtimes. This is based on the assumption that maintenance tasks remain largely similar across similar systems in terms of completion times, ease of completion, and so on, and as such, experience with previous similar systems can be utilized to evaluate the current system design configuration.

One such empirical regression relation is proposed by the military handbook, MIL-HDBK-472, and is expressed as:[2]

$$Mct_i = \text{antilog}(3.54651 - 0.02512A - 0.03055B - 0.01093C) \qquad (8.2)$$

where, A, B, and C represent consolidated scores from checklists "A," "B," and "C," and Mct_i represents the corrective maintenance completion time for the ith task. Substituting values from Table 8.4 in Equation 8.2 results in

$$
\begin{aligned}
Mct_i &= \text{antilog}[3.54651 - 0.02512(37) - 0.03055(19) - 0.01093(23)] \\
&= \text{antilog}(1.78543) = 61 \text{ minutes}
\end{aligned}
$$

Equation 8.2 is applied to all applicable maintenance tasks, as shown in Table 8.5. These maintenance completion times are next consolidated to reflect the mean corrective maintenance time for the overall system as

$$\overline{Mct} = \frac{\displaystyle\sum_{i=1}^{N} Mct_i}{N} = \frac{877}{20} = 43.8 \text{ minutes} \qquad (8.3)$$

The data in Table 8.5 can be further manipulated to compute the maximum maintenance time, $M_{\max}$, as

$$M_{\max} = \text{antilog}(\overline{\log Mct} + 1.645\sigma_{\log Mct}) \qquad (8.4)$$

[2] Military Handbook, MIL-HDBK-472, *Maintainability Prediction*, Department of Defense, Washington DC, 1984.

Task Number: CM-231   Description: Improper feedback in emitter circuit   Level: Intermediate

Date: 05/09/94   Assembly: VCO/Tripler   Unit/Part: Resistor R3   Equipment: Unit A

Failure Mode: Opened Resistor

Failure System: XYZ System indicates a NO-GO condition

Maintenance Steps	Pertinent Remarks
1. The equipment malfunction is initially indicated by a "no lock-on" condition in both the T/R mode and the A/A mode of operation. Initiating interruptive self-test switch on the Unit C front panel isolates the trouble to Unit A.	1. Isolation of fault to Unit A can be accomplished using only the built-in self-test features and a 50-ohm dummy load attached to the Antenna output of Unit A.
2. In the Field Electronics Shop, the faulty Unit A is connected to the Test Station for fault isolation. The Unit is then prepared for fault isolation within the Synthesizer-VCO-AFC Signal Loop.	2. Isolation of fault to Synthesizer-VCO-AFC Signal Loop is accomplished using the System Test Station control panel self-test readout lights. Digital word code on readout lights correlated with readings listed in Report No. 101A5 provide fault isolation.
3. VCO signal outputs at VCO Connector are observed and checked against correct values. No 1025 MHz sine wave output indicates trouble is in VCO.	3. Signal Generator SG-215 and Oscilloscope TKX-545 used to check VCO signals on subassembly Connector. Oscilloscope and Signal Generator set-up and adjustments are required. Proper reading listed in Report Number 101A5.
4. By removing the VCO output R. F. cables connected to the VCO Tripler assembly, the correct driver output from the VCO subassembly can be observed. Presence of the correct signal indicates trouble is probably in the Tripler Subassembly.	4. Location of VCO module requires the use of a module extender. Dense packaging within the VCO requires high manual dexterity in performing repair services on this subassembly. Some delay to be expected in repair action due to part location and necessity to use care in part removal to avoid head damage or solder contamination to adjacent parts.
5. Stage-by-stage probing of the L. O. Tripler board indicates an open resistor in Q1 emitter circuit.	

Criteria Number	1	2	3	4	5	6	7	8	9	10	11	12	13	14	15	Total
Checklist A	4	4	3	2	0	2	2	2	4	0	2	4	4	0	4	37
Checklist B	1	4	2	4	2	4	2	2	3							19
Checklist C	3	2	0	2	1	4	4	2	3	2						23

TABLE 8.5   Corrective Maintenance Task Times

Task Number	Mct_i	Log Mct_i	$(\text{Log } Mct_i)^2$
1	54	1.732	2.999
2	67	1.826	3.334
3	46	1.662	2.762
4	22	1.342	1.800
5	25	1.397	1.951
6	73	1.863	3.470
7	20	1.301	1.692
8	31	1.491	2.223
9	69	1.838	3.378
10	26	1.414	1.999
11	27	1.431	2.047
12	27	1.431	2.047
13	24	1.380	1.904
14	64	1.806	3.261
15	61	1.791	3.207
16	23	1.361	1.852
17	21	1.322	1.774
18	75	1.875	3.515
19	62	1.792	3.211
20	60	1.778	3.161
Total	877	31.733	51.587

where $\sigma_{\log Mct}$ is the standard deviation with respect to $\overline{M}ct$ and is given as

$$\sigma_{\log Mct} = \sqrt{\frac{\sum\limits_{i=1}^{N} (\log Mct_i)^2 - \left(\sum\limits_{i=1}^{N} \log Mct_i \right)^2 / N}{N-1}}$$

$$= \sqrt{\frac{51.587 - (31.733)^2/20}{19}} \tag{8.5}$$

$$= 0.3525$$

Further, $\overline{\log Mct}$ is given as

$$\overline{\log Mct} = \frac{\sum\limits_{i=1}^{N} \log Mct_i}{N} = \frac{31.733}{20} = 1.586 \tag{8.6}$$

Substituting values from Equations 8.5 and 8.6 into Equation 8.4 results in

$$M_{max} = \text{antilog}[1.586 + 1.645(0.3525)] = 146 \text{ minutes}$$

8.1.2 Prediction Method 2

This maintainability prediction technique focuses attention on the preventive maintenance aspect of system design. It involves the determination of preventive maintenance task requirements, task frequencies, and task times. The reliability-centered maintenance analysis discussed in Section 7.2.4 is dependent on this prediction technique and is utilized to delineate preventive maintenance task requirements and task frequencies.

A worksheet similar to the one illustrated in Table 8.6 may be used for recording the relevant information. All applicable preventive maintenance tasks are listed and described in column 1 and their frequencies in terms of number of tasks performed per million hours of system operation are listed in column 2. Next, the preventive maintenance task completion times are estimated and listed in column 3. Each preventive maintenance task may be broken down into the elemental activities that must be completed in sequence. This breakdown facilitates the estimation of completion times.

Given the data listed in Table 8.6, the mean system preventive maintenance time is computed as

$$\overline{Mpt} = \frac{\Sigma f(Mpt_i)}{\Sigma f} \tag{8.7}$$
$$= \frac{5,326.6}{32,764} = 0.1626 \text{ hours or } 9.75 \text{ minutes}$$

8.1.3 Prediction Method 3

This technique involves a detailed functional breakdown of the system or equipment. A functional breakdown facilitates the establishment of functional/physical levels at which maintenance tasks are performed based on equipment design features and item failure rates. This prediction method is based on the philosophy that the overall magnitude of a maintenance task completion time is equal to the sum of

TABLE 8.6 Preventive Maintenance Completion Times

(1) Preventive Maintenance Task Description	(2) Task Frequency (f)	(3) Task Completion Time (Mpt_i) (hr)	(4) (f) (Mpt_i)
1. Lubricate gear train	961	0.5	480.5
2. Calibrate board 1A6	1,940	0.2	388.0
3. Clean air filter	961	0.1	96.1
22. Align resolver	1,940	0.1	194.0
23. Ripple check of PS3	961	0.2	192.2
Total	32,764		5326.6

the individual maintenance task element completion times. In the context of this procedure, a maintenance task is comprised of elemental activities pertaining to localization, isolation, disassembly, interchange, reassembly, alignment, and checkout.

It is further assumed that the overall maintenance completion times are a function of the level within the system functional hierarchy where the actual repair or replacement is completed. For instance, time to effect repair at the subassembly level is likely to be greater than the time to effect repair at the assembly level. The steps involved in this technique are:

1. *Develop the functional/physical breakdown of the system.* This is accomplished during the system functional analysis and involves breaking down the system into its constituting functional/physical subelements. In the context of this prediction technique, the various elements of the system functional/physical hierarchy are defined as:

   a. *Part:* this is a single piece or multiple pieces joined together for a specific purpose. It is a primitive system element and is not disassembled further (e.g., resistors and transistors).

   b. *Stage:* this is a combination of two or more parts. For example, a transistor together with its directly associated parts (e.g., amplifier stage and detection stage).

   c. *Subassembly:* this is a combination of two or more stages (e.g., a circuit board together with the mounted parts).

   d. *Assembly:* this constitutes a combination of parts, stages, and subassemblies which as a whole perform a specific system function and are replaceable as a whole (e.g., frequency amplifier).

   e. *Unit:* this is a combination of parts, stages, subassemblies, and assemblies combined together in a manner capable of independent operation. It is accessible and removable without disassembly.

   f. *Group:* this is a collection of units, subassemblies, or assemblies, not capable of accomplishing a complete system function (e.g., antenna group, indicator group, or transmitter group).

   g. *Equipment or set:* this is a combination of one or more units, subassemblies, or assemblies which is capable of performing an operational function (e.g., radio receiving set or radar set).

   h. *Subsystem:* this is a combination of equipment, groups, and so on, which performs an operational function within the overall system. Such combinations constitute major subdivisions within the system.

   i. *System:* this is a combination of one or more subsystems or equipments, which may be physically separated or distributed when in operation (e.g., an automobile with all of its electrical, structural, transmission, and cooling subsystems).

Employing these definitions as a guide, a system functional/physical hierarchy is developed, similar to the one depicted in Figure 8.2. This breakdown should

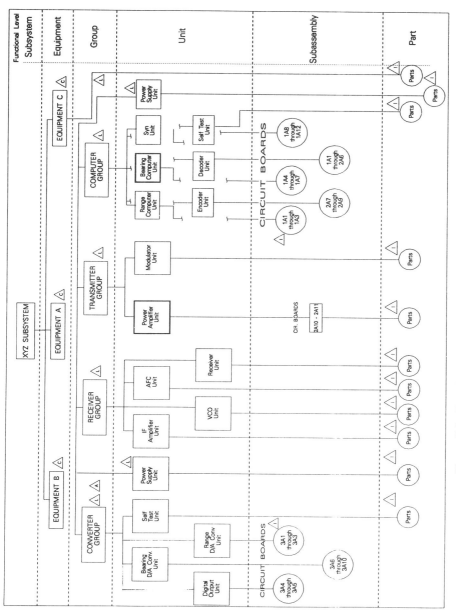

Figure 8.2. Functional/physical breakdown of the "XYZ" subsystem.

complement the system maintenance concept and the system functional analysis, as discussed in Chapter 6. A hypothetical subsystem is presented in Figure 8.2 showing the various constituting functional/physical elements. Each branch of the diagram terminates with a circle which represents the applicable replaceable item. For instance, if a malfunction exists in the bearing computer unit, the malfunction will be corrected by removing and replacing circuit board "1A4," "1A5," "1A6," or "1A7." Further, the connecting lines on this diagram indicate physical relationships only and do not necessarily depict all functional and data-flow interactions in a comprehensive manner.

2. *Determine functional/physical levels where localization, isolation, access, alignment, and checkout procedures occur.* These levels are assigned in accordance with the following descriptions:

1. *Level of localization:* this is the functional level to which a system failure can be traced without the need for external test and support equipment. Referring to Figure 8.2, this level is designated by $\triangle$.

b. *Level of isolation:* this is the functional level to which a system failure can be traced with the use of external test and support equipment at designated test points. Referring to Figure 8.2, this level is designated by $\triangle$.

c. *Level of access:* the access level for a replaceable item is the level to which system disassembly must be accomplished in order to gain access to the replaceable item for replacement, and from which reassembly must be accomplished after item replacement. For instance, in Figure 8.2, replacement of circuit board "1A4" in the bearing computer requires access to the unit level.

d. *Level of alignment:* this is the functional level at which alignment or calibration must be done subsequent to maintenance by replacement. This level is designated as $\triangle$.

e. *Level of checkout:* this is the functional level at which the system operation can be verified subsequent to maintenance completion. This level is designated as $\triangle$.

3. *Complete the maintainability prediction worksheet.* One maintainability prediction worksheet is prepared for every item delineated in the functional/physical breakdown of the system in question, except the piece parts. Two examples are depicted in Figure 8.3. Example "A" addresses the branch in Figure 8.2 containing the power amplifier unit made up of circuit boards "2A10" and "2A11." The circuit boards are assumed to be repairable, and piece parts are removed and replaced to effect repair. Example "B" addresses the branch in Figure 8.2 containing the bearing computer unit made up of circuit boards "1A4"–"1A7." In this example, faulty circuit boards are discarded and replaced with new boards. The maintainability prediction worksheet is completed as follows:

a. *Block 1—Item:* enter the item nomenclature.

b. *Block 2—Method of repair:* the type of item being replaced to correct the malfunction is indicated by a check mark in the appropriate block. Parts are replaced in Example "A."

1. Item Circuit Board No. 2A10
2. Method or Repair   Replace-Parts ■ Subassembly □ Assembly □   Unit □ SHEET 1 of 66

3 Part Category	4 N	5 $\lambda*$	6 Nλ	Maintainenance Task Times 8 LOC	9 ISO	10 ACC	11 ALIN	12 CHEC	13 INT	14 Rp	15 NλRp	17 LOG Rp	18 Nλ LOG Rp
Capacitors	2	2.48	4.96	0.106	0.772	0.427	0.000	0.108	0.081	1.494	7.410	0.1744	0.865
Electrolytic	2	0.46	0.92	0.106	0.772	0.427	0.000	0.108	0.081	1.494	1.374	0.1744	0.160
Mica	1	0.50	0.50	0.106	0.772	0.427	0.000	0.108	0.081	1.494	0.747	0.1744	0.087
Paper	3	0.28	0.84	0.106	0.772	0.427	0.000	0.108	0.081	1.494	1.255	0.1744	0.146
R. F. Coils	1	0.06	0.06	0.106	0.772	0.000	0.000	0.108	0.272	1.258	0.075	0.0997	0.006
Connectors	4	2.98	11.92	0.106	0.772	0.427	0.000	0.108	0.081	1.494	17.808	0.1744	2.079
Diodes	2	1.16	2.32	0.106	0.772	0.427	0.000	0.108	0.142	1.555	3.608	0.1917	0.445
Transformers													
Resistors	3	0.36	1.08	0.106	0.772	0.427	0.000	0.108	0.081	1.494	1.614	0.1744	0.188
Composition	2	0.39	0.78	0.106	0.772	0.427	0.000	0.108	0.081	1.494	1.165	0.1744	0.136
Carbon Film	1	1.40	1.40	0.106	0.772	0.427	0.000	0.108	0.081	1.494	2.072	0.1744	0.244
Wirewound	4	2.86	11.44	0.106	0.772	0.427	0.000	0.108	0.081	1.494	17.091	0.1744	1.995
Relays													
Total			(7) 36.22								(16) 54.293		(19) 6.351

*Failures/10^6 hrs.

Example A. Prediction of repairable subassembly.

1. Item Circuit Board No. 1A6
2. Method or Repair   Replace-Parts □ Subassembly ■ Assembly □   Unit □ SHEET 12 of 66

3 Part Category	4 N	5 $\lambda*$	6 Nλ	Maintainenance Task Times 8 LOC	9 ISO	10 ACC	11 ALIN	12 CHEC	13 INT	14 Rp	15 NλRp	17 LOG Rp	18 Nλ LOG Rp
Circuit Board	.	.	.	0.045	0.265	0.228	0	0.108	0.015	0.661	20.385	-0.1798	-5.545
Capacitors													
Electrolytic	2	2.48	4.96										
Mica	1	0.46	0.46										
Paper	2	0.50	1.00										
Connectors	1	0.06	0.06										
Diodes	6	2.98	17.88										
Resistors													
Composition	3	0.36	1.08										
Carbon Film	2	0.39	0.78										
Wirewound	1	1.40	1.40										
Transformer	1	1.16	1.16										
Transistor	2	1.03	2.06										
Total			(7) 30.84								(16) 20.385		(19) -5.545

*Failures/10^6 hrs.

Example B. Prediction of nonrepairable subassembly.

Figure 8.3. Maintainability prediction worksheet.

c. *Block 3—Part category:* list each part. Items can be combined into a part type (e.g., combine all composition resistors and include as one part type).

d. *Block 4—Quantity N:* enter the quantity of each part type.

e. *Block 5—Failure rate* λ: after each part type, indicate the applicable failure rate. If more than one part-numbered item is included in the part type, the listed failure rate will be an average value (combined). Failure rates are obtained from reliability prediction reports.

f. *Block 6—Part-type failure rate Nλ*: determine the reliability of each part type by multiplying the failure rate λ in block 5 by the quantity N of parts in block 4.

g. *Block 7—Item failure rate:* to obtain the overall reliability factor for the item covered by the worksheet, add the figures in block 6.

h. *Block 8—Maintenance time for localization (LOC):* maintenance times should be determined from the most reliable data source available to the maintainability engineer. Hopefully, the contractor has established a data bank containing experience data of systems already in operation. Data may stem from a combination of field reports, maintenance analysis reports, special studies involving utilization of design aids (mock-ups), and/or demonstrations. Reviews and comparisons of such data, applied to the system/ equipment being evaluated through this prediction, may result in realistic estimates of maintenance times. Such data, reflecting experience with similar systems in the past, is the preferred source of information required for completing the maintenance-task times (blocks 8–13 of the maintainability prediction worksheet). In the absence of such information, however, data may be obtained from charts such as those presented in Tables 8.7–8.10. Table 8.7 covers localization, isolation, disassembly, and reassembly, or access, alignment, and checkout. Tables 8.8–8.10 pertain to removal and replacement (or interchange) of parts.

To obtain localization time using Table 8.7, select the functional-level column headed by the type of replacement that will be accomplished. For example, replacement of a part would indicate column 1, replacement of a stage would indicate column 2, and so on. Relative to Example "A," power amplifier design is such that parts are replaced on circuit board "2A10"; therefore, column 1 is the correct column in this instance.

The appropriate corrective maintenance task time is found by determining the functional level at which maintenance features are effective. In this instance, Figure 8.2 indicates that LOC for the branch containing the power amplifier unit and circuit board "2A10" is effective to the group functional level, as designated by △. Referring to Table 8.7, the maintenance task time is determined by following column 1 on the chart down to the group level and extracting the appropriate quantitative value under localization. The value 0.106 hours is entered in block 8 of the maintainability prediction worksheet depicted in Figure 8.3 for Example "A." It is assumed that the localization value for each part on the circuit board "2A10" is the same; therefore, the value of 0.106 is entered after each part type.

i. *Block 9—Maintenance time for isolation (ISO):* determination of the maintenance time for fault isolation follows the same approach as employed for localization in block 8. In the absence of actual experience data, Table 8.7 may be used. Referring to the table, select the appropriate functional-level column for IOS for the branch containing the power-amplifier unit and cir-

TABLE 8.7 Average Maintenance Times (Functional Level-Time Matrix)

(Sheet 1)

Functional Levels									Corrective Maintenance Tasks					
									Diagnosis		Replacement		Test	

For DETERMINING LOCALIZATION AND ISOLATION TIME USE COLUMN BEGINNING WITH THE FUNCTIONAL LEVEL THROUGH WHICH FAILURE IS REMOVED

9	8	7	6	5	4	3	2	1	Localization	Isolation	Disassembly	Reassembly	Alignment	Checkout
System	Subsystem	Equipment	Group	Unit	Assembly	Subassembly	Stage	Part	0.021	0.772	1.281	1.223	0.156	0.175
None	System	Subsystem	Equipment	Group	Unit	Assembly	Subassembly	Stage	0.039	1.179	0.328	0.561	0.077	0.167
	None	System	Subsystem	Equipment	Group	Unit	Assembly	Subassembly	0.056	1.417	0.165	0.262	0.045	0.156
		None	System	Subsystem	Equipment	Group	Unit	Assembly	0.073	1.569	0.122	0.191	0.030	0.149
			None	System	Subsystem	Equipment	Group	Unit	0.089	1.700	0.094	0.134	0.021	0.138
				None	System	Subsystem	Equipment	Group	0.106	1.821	0.071	0.090	0.015	0.124
					None	System	Subsystem	Equipment	0.121	1.924	0.049	0.061	0.010	0.108
						None	System	Subsystem	0.136	2.022	0.032	0.037	0.007	0.091
							None	System	0.150	2.100	0.016	0.017	0.003	0.062
								None	0.165	2.172	0.000	0.000	0.000	0.000

(Sheet 2)

Applications, Notes, and Definitions

1. The average task times in the chart should be applied to fuses not on the front panel, transistors, and all electronic parts (resistors, capacitors, inductors, etc.).

2. The total replacement time should include disassembly time, interchange time, and reassembly time. However, since an average interchange time applicable to all situations is difficult to obtain, it has not been included in the tabulated average task time. Interchange times must be determined from a table such as Table 8.8.

3. The average time intervals in Sheet 1 do not include adminstrative time. Administrative time consists of time expended in part procurement (time spent by maintenance personnel in obtaining replacement items) and in non-technical procedures; for example, filing logs.

4. When equipment maintenance features enable localization to the functional level through which failure is being removed (the top row of FUNCTIONAL LEVELS columns), do not use the value shown in the ISOLATION column at this functional level, instead use 0.000 hours.

5. To determine the time for the disassembly, reassembly, alignment, and checkout tasks, the designer should only used column 1 of FUNCTIONAL LEVELS in the appropriate row at which the task is performed.

6. To properly approximate the checkout time, enter column 1 at the functional level at which the checkout is being made and multiply by the number of operational modes affected by the replaced functional level.

327

TABLE 8.8 Maintenance Interchange Task Times

Part Type	Average Time (Hours)
Plug-in fuses	0.010
Screw-in fuses	0.015
All fuses with screw cap	0.014
Parts with 2 wires or 2 tabs to be soldered	0.081
Parts with more than 2 wires or 2 tabs to be soldered with clamp	0.081 + 0.034/wire over 2 Add 0.027
Parts attached with screws, nuts, and washers	Add 0.022 for each screw, nut, and washer combination.

cuit board "2A10." Column 1, the part-level row, intersects with the "isolation" column, which contains a maintenance task-time value of 0.772 hours. This value is entered for each part type in block 9 of the worksheet presented in Figure 8.3 for Example "A."

j. *Block 10—Maintenance time for access (ACC):* maintenance time for access includes the disassembly time to gain access to the failed item and the reassembly time after item removal and replacement. Both the disassembly and

TABLE 8.9 Task Element Time Data

Element Descriptions	Element Time (Hours)
1. Fuse	
Insert into horizontal holder	0.0050
Insert into vertical holder	0.0075
2. Wiring and soldering	
Wire-wrapping and splicing	
Bare copper wire (a) End	0.0150
(b) Ends	0.0237
Jumper wire and cable leads (a) End	0.0134
(b) Ends	0.0265
Part with axial leads (includes part handling)	
(a) End	0.0178
(b) Ends	0.0289
Soldering	
Per joint	0.0058
3. Replacement with hardware	
Replace screw into tapped hole	0.0093
Replace screw through clearance hole	0.0023
Replace washer	0.0018
Replace nut	0.0071
Replace stop nut	0.0210
Replace set screw	0.0075
Apply glyptol screw	0.0018
4. Part Handling	
Pull up part and position in chassis for assembly	0.0025
5. Printed circuit wiring	
Replace (insert)	0.0033/End
Solder	0.0056/End

TABLE 8.10   Determination of Task Time with Use of Element Time Factors[a]

Detailed Interchange Steps	Element Description	Element Time (hours) (A)	Number of Times Performed (B)	Time for Detailed Interchange Steps (hours) (A × B)
Unsolder joint	Soldering (per joint)	0.0058	2	0.0116
Disconnect lead	Wire (jumper wire—1 end)	0.0134	2	0.0268
Remove nut	Remove nut	0.0071	2	0.0142
Remove washer	Remove washer	0.0018	2	0.0036
Remove screw	Remove screw through clearance hole	0.0023	2	0.0046
Remove failed resistor	Pull up part and position	0.0025	1	0.0025
Position new resistor	Pull up part and position	0.0025	1	0.0025
Replace screw	Replace screws through clearance hole	0.0023	2	0.0046
Replace washer	Replace washer	0.0018	2	0.0036
Replace nut	Replace nut	0.0071	2	0.0142
Connect lead	Wire (jumper wire—1 end)	0.0134	2	0.0268
Solder joint	Soldering (per joint)	0.0058	2	0.0116
Total	Hours of maintenance/part failure			0.1266

[a] This example is for a resistor with two terminals which are attached to the chassis by two screws, two nuts, and two washers.

reassembly maintenance times may be determined from Table 8.7. Referring to the table, select column 1 and proceed to the applicable functional level. Referring to Figure 8.2, access to the parts on circuit board "2A10" is to the subassembly level. The appropriate level is indicated by the first rectangular box in Figure 8.2 above the replacement item. Thus, the appropriate disassembly time is 0.165 hours and the reassembly time is 0.262 hours. Access time is the sum of the two values, or 0.427 hours. This value is entered against each part in block 10 of the worksheet depicted in Figure 8.3 for Example "A."

k. *Block 11—Maintenance time for alignment (ALIN):* the maintenance time for alignment or adjustment subsequent to item removal, replacement, and reassembly is determined by the same method as presented above. Select column 1 of Table 8.7 and proceed to the applicable functional level. Relative to Example "A," design is such that alignment or adjustment is not required; therefore, the value entered in block 11 of the worksheet is zero.

l. *Block 12—Maintenance time for checkout (CHEC):* the maintenance time involves checkout of the item after part replacement or repair has been accomplished (to verify proper function of the item). Referring to Table 8.7, select column 1 and proceed to the equipment level. The predicted value of 0.108 hours is entered in block 12 of the worksheet.

m. *Block 13—Maintenance time for interchange (INT):* the maintenance time for interchange involves the time to remove and replace the malfunctioning item. Table 8.8 contains interchange time information for some replaceable parts. For items having mounting methods which do not conform to the description given in Table 8.8, the interchange time can be determined from the data presented in Table 8.9. In determining the interchange time from Table 8.9, each detailed step involved in removing an item that has malfunctioned, and in positioning and attaching a new item, must be assigned a specific element of time. The interchange time for an item is equal to the sum of all time intervals. An example of determining interchange time for a typical part is illustrated in table 8.10.

n. *Block 14—Maintenance repair time (R_p):* to determine the repair time required for performing a corrective maintenance task, add the values recorded on the worksheet in columns 8–13 (Figure 8.3). The repair time, R_p, is equivalent to the active maintenance time per corrective maintenance task, Mct_i.

o. *Block 15—Weighted maintenance factor ($N\lambda R_p$):* determine the weighted factor by multiplying the value in block 6 by the repair time in block 14. The resultant value is entered in block 15.

p. *Block 16—Summation of weighted factors ($\Sigma\ N\lambda R_p$):* add the individual values in block 15.

q. *Block 17—Logarithm of repair time (Log):* enter the logarithm of each recorded value of R_p in block 17. When a value is less than 1, the logarithm should be expressed as a negative number. For instance, $\log 0.25 = 9.3979 - 10 = -0.6021$.

r. *Block 18—Product of log R_p and $N\lambda$:* the value entered in block 18 is obtained by multiplying the value in block 6 by that in block 17.

s. *Block 19—Total value of log R_p and $N\lambda$:* the value entered in block 19 is the summation of the individual values in block 18.

Referring to Figure 8.3, two examples are presented. Example "A," covering maintenance through the replacement of piece parts on circuit board "2A10," is illustrated by the data included in the steps listed above. Example "B," on the other hand, covers maintenance through the replacement of circuit board "1A6," a plug-in item in the bearing computer unit shown in Figure 8.2. The maintainability prediction worksheet for Example "B," also presented in Figure 8.3, is discussed briefly in the following steps:

a. *Blocks 1–7:* determination of data for these blocks is accomplished by following the same procedures described above. In this instance, however, the item checked in block 2 is subassembly, as the circuit board itself is replaced for corrective maintenance.

b. *Blocks 8–13:* since the circuit board is the item that is replaced in accomplishing corrective maintenance, and not individual parts mounted on the circuit board, only a single set of values is entered. These values are obtained using Table 8.7–8.10, as applicable.

c. *Blocks 14–19:* determination of data for these blocks is accomplished by following the same procedure as defined for Example "A." In this instance, however, the value indicated in block 15 is the product of the values in blocks 7 and 14.

3. *Complete the analysis process.* When individual prediction worksheets (Figure 8.3) have been prepared on each replaceable item (other than piece parts), the values presented in blocks 7, 16, and 19 of each worksheet are recorded on a form similar to that presented in Table 8.11. Based on the tabulated data, the MTTR ($\overline{M}ct$) of the XYZ subsystem can be computed as

$$\text{MTTR} = \frac{\Sigma N \lambda R_p}{\Sigma N \lambda} = \frac{\text{block 7 value}}{\text{block 6 value}} = \frac{3548.99}{1982.68} = 1.79 \text{ hours or 1 hour 47 minutes}$$

$$(8.8)$$

The geometric mean time to repair, MTTR_G, which is the median equipment repair time (ERT) when the repair times are log-normally distributed, is calculated as

$$\begin{aligned}
\text{MTTR}_G &= \text{antilog} \frac{\Sigma N \lambda \log R_p}{\Sigma N \lambda} = \text{antilog} \frac{\text{block 8 value}}{\text{block 6 value}} \\
&= \text{antilog} \frac{-11.892}{1982.68} = \text{antilog}(-0.00599) = \text{antilog}(9.99401) \quad (8.9) \\
&- 10 = 0.986 \text{ hours or 59 minutes}
\end{aligned}$$

The three maintainability prediction techniques presented in Sections 8.1.1–8.1.3 address the prediction of metrics such as $\overline{M}ct$, $\overline{M}pt$, M_{max}, and MTTR_G, and are by no means all-inclusive. The objective in discussing these techniques is to illustrate the process involved in addressing some of the basic characteristic maintainability metrics. As more information becomes available, models exist that can be used to address other metrics, such as MMH/OH and MMH/repair action, as discussed in Chapter 4.[3]

TABLE 8.11   Prediction Summary Worksheet

(1) Worksheet Number	(2) Item Designation	(3) WS Block 7	(4) WS Block 16	(5) WS Block 19
1	Ckt. Bd. 2A10	36.22	54.293	6.351
2				
12	Ckt. Bd. 1A6	30.84	20.385	−5.545
		(6)	(7)	(8)
Grand total		1982.68	3548.997	−11.892

---

[3] Some other prediction techniques, addressing these metrics, are included in MIL-HDBK-472, Military Handbook, *Maintainability Prediction*, Department of Defense, Washington DC, 1984.

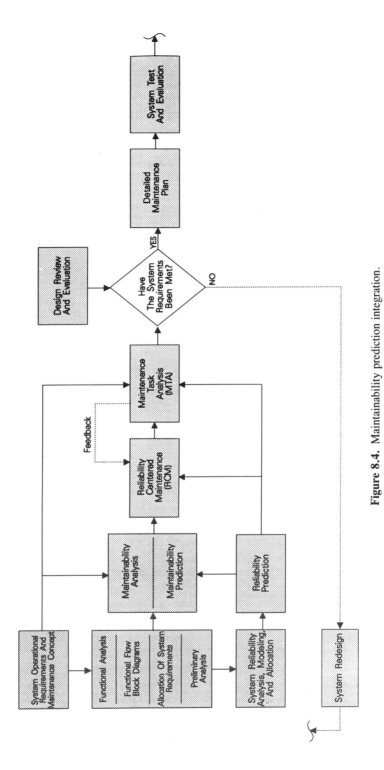

Figure 8.4. Maintainability prediction integration.

8.2   INTEGRATION OF MAINTAINABILITY PREDICTION RESULTS

The three prediction procedures discussed in Sections 8.1.1–8.1.3 address maintenance (corrective and preventive) task completion times in terms of mean values or most likely values. Obviously, experience with similar systems in the past is invaluable, and data and information collected from prediction studies conducted on earlier systems have an influence on the "most likely" values assumed during the course of working with the current design configuration.

It is important to keep in mind that while working with mean and most likely values simplifies the prediction process, parameters such as maintenance frequencies and maintenance completion times are stochastic in nature and their values may align with one of many probability distributions. Further, when dealing with maintenance frequencies and maintenance completion times in probabilistic terms, mathematical simulation tools such as Monte Carlo simulation can be very useful.[4]

System effectiveness is highly dependent upon system maintainability, and this results in a need to continuously assess and evaluate system design from a maintainability perspective. However, given specific requirements, the design phase, data and information requirements, the process and procedures utilized to effect this evaluation and assessment may differ. Three such procedures were outlined and discussed in this chapter. In order for such assessment to be more value-added and effective, it must not be performed in isolation, but must be integrated into the overall system analysis process, as depicted in Figure 8.4.

The prediction procedure utilized notwithstanding, prediction results need to be fed back to the system design process, and any significant deviations from the allocated requirements need to be addressed through a redesign effort or applicable compensatory provisions such as increasing built-in testing. Further, maintainability prediction is a direct input to the MTA and life-cycle cost analysis, and forms the basis for determining logistic support resource requirements for a given design configuration.

QUESTIONS AND PROBLEMS

1. What is the need for maintainability prediction? Define the benefits.

2. Discuss how the system maintenance concept influences maintainability prediction.

3. What purpose do system item/component failure rates serve in the maintainability prediction process?

4. When, during the system design and development, is maintainability prediction accomplished?

---

[4] An excellent reference that deals with maintainability analysis and modeling in probabilistic terms is Knezevic, J., *Reliability, Maintainability, and Supportability: A Probabilistic Approach*, McGraw-Hill Company, New York, 1993.

5. How does the maintenance concept influence the process of maintainability prediction?

6. Why is good system failure-rate information an important input to the maintainability prediction process?

7. Referring to the discussion pertaining to maintainability prediction method 1 (Section 8.1.1), what is the desired replaceable item sample size given:

   a. A confidence level of 95%
   b. A standard deviation of 10 minutes
   c. A mean time to repair of 40 minutes
   d. A desired accuracy of 10%

8. Referring to the discussion pertaining to maintainability prediction method 1 (Section 8.1.1) and Equation 8.2, calculate the system $\overline{M}ct$ given "A" = 40, "B" = 20, and "C" = 30.

9. Illustrate the usage of the three checklists pertaining to maintainability prediction method 1 (Section 8.1.1) by applying them to a maintenance task relative to a system of your choice.

10. Given that the overall system level failure rate is 1500 failures/million hours of system operation, and that the sum of all time-weighted repair times is 45,000 $\times$ 10^{-6}, calculate the weighted mean time to repair.

11. A system has six subsystems with failure rates (per million hours of operation equal to 0.0020, 0.0020, 0.0015, 0.0015, 0.0005, and 0.0025 respectively, and Mct_i values equal to 50, 50, 45, 30, 30, and 25 minutes respectively. Calculate the system level $\overline{M}ct$.

12. Given a system level $\overline{M}ct$ equal to 40 minutes, a confidence level of 95%, and a $\sigma_{\log Mct}$ of 2.4, calculate M_{max}.

13. Given preventive maintenance task frequencies of 10, 60, 120, 200, 320, and 1920 hours, and task times equal to 0.5, 0.5, 2.0, 3.0, 1.0, and 1.5 hours respectively, calculate the mean preventive maintenance task time $\overline{M}pt$.

14. Complete a maintainability prediction worksheet based on maintainability prediction method 3 (Section 8.1.3) for circuit board 3A7 within the bearing D/A convertor unit. Clearly state and justify any assumptions made during this prediction.

15. Discuss the various applications, tasks, methods, and so on that require the results of maintainability as an input.

9
MAINTENANCE TASK
ANALYSIS (MTA)

In addressing the various maintainability program activities, a certain amount of confusion may develop relative to the intent of the "maintainability analysis" described in Chapter 6 and the "maintenance task analysis (MTA)" described in this chapter. The maintainability analysis is defined as a "before-the-fact" process of translating system requirements into specific design criteria through the accomplishment of trade-off studies. It constitutes an ongoing iterative process applied during the early phases of system/product design and development. On the other hand, the MTA is the "after-the-fact" process of identifying detailed maintenance requirements through the analysis of a given design configuration, whether it is preliminary, and described through an electronic database and/or drawings during the early phases of design, or in the form of an operational system already in use. A MTA can be accomplished through an evaluation of design data and/or of an existing operational capability. As a result of this evaluation, specific maintenance resource requirements are defined for the configuration being addressed, that is, maintenance tasks, personnel and training, spares/repair parts and associated inventories, test and support equipment, maintenance facilities, transportation and handling requirements, maintenance data, and computer resources. Additionally, areas of poor design for maintainability can readily be detected, and recommendations for corrective action and/or improvement can be initiated to increase system effectiveness and reduce cost.

This chapter covers maintenance requirements (the levels and categories of maintenance), the identification of maintenance tasks for a given system, the steps in performing a MTA, and the process for submitting recommendations for corrective action and/or design improvement. The task analysis serves as an excellent tool for providing an early assessment of the extent to which maintainability characteristics are incorporated either in a proposed design configuration or an existing capability already in operational use.

9.1   MAINTENANCE REQUIREMENTS

The maintenance task analysis constitutes the process of evaluating a given system/ product design configuration to:

1. Identify the resources required for the sustaining support of that system/product throughout its planned life cycle. Such resource requirements may include the identification of anticipated functions/tasks for all categories of maintenance along with the necessary personnel and training, test and support equipment, spares/repair parts and associated inventories, transportation and handling requirements, facilities, technical data, and computer resources.

2. Provide an assessment of the given design configuration for the incorporation of maintainability characteristics, in the valuation of both the prime mission-oriented elements of the system and the proposed support capability. This may lead to the generation of recommendations for system/product improvement.

The MTA may be accomplished at a gross level during the conceptual design phase, when there is enough definition of specific elements of the system, or when an existing repairable item is being considered for incorporation into the design. As the design process evolves during the preliminary and detail design phases and the system becomes better defined, the MTA is accomplished through the review of design data, drawings, component part and material lists, and reports. The maintainability specialist, with knowledge of the maintenance concept and the results of the trade-off studies accomplished through the maintainability analysis, reviews the appropriate data and begins to define maintenance tasks, evolving from the maintenance functional analysis and down. Functions are broken down into sub-functions, groups of activity, tasks, subtasks, and so on. With the "WHATs" defined in terms of the identified tasks, the maintainability specialist evaluates the various alternative approaches for task accomplishment. Through the evaluation of alternatives at this level, one is able to begin with the identification of the resources needed for task completion (i.e., the "HOWs"). This process is similar to an operator task analysis (OTA), often a requirement in the implementation of a human factors program, except that the emphasis here is on the determination of *maintenance requirements*.

The MTA may also be accomplished on an *existing* system capability, that is, a system or product that is already in service and being utilized by the consumer. The purpose here is to evaluate the current configuration, identify the high-cost contributors through the projected system life cycle, determine the "cause-and-effect" relationships, identify areas where possible improvements can be economically incorporated, and make recommendations accordingly. This can be accomplished on an iterative basis, supporting the objective of "continuous process improvement."

The MTA evolves from the defined maintenance concept, the functional analysis and allocation, and the results of the trade-off studies accomplished through the maintainability analysis (Chapters 5 and 6). Further, the latest design data in the form of specifications, layouts, drawings, component part lists, material lists, reports, and supporting documentation are required as an input, along with the results from reliability analyses and predictions, human factors analyses, and so on. The depth of the MTA will, of course, depend on the degree of design definition. For an existing system capability, a good data collection and analysis capability should provide the necessary input.

The MTA output constitutes an engineering data package covering:

1. All significant repairable items (e.g., system, subsystem, equipment, unit, assembly, subassembly, and software package). Items identified through the level of repair analysis that are relatively complex and require an analysis to determine the type and extent of support needed should be addressed.
2. All unscheduled and scheduled maintenance requirements (e.g., troubleshooting, remove and replace, repair, servicing, alignment and adjustment, functional test and checkout, inspection, calibration, and overhaul).

The MTA must, of course, be "tailored" to meet a specific program requirement. The generation of too much data too early in a program may not be meaningful and certainly will be costly! On the other hand, the development of data too late in the design process will not allow for the timely incorporation of changes if required. It is essential that the maintainability specialist completely understand the concepts presented in this chapter and how they can be applied in terms of timeliness and depth of coverage. As a goal, the MTA should result in providing the information conveyed in Figure 9.1.

9.2   IDENTIFICATION OF MAINTENANCE TASKS

Referring to Figure 9.1, the development of maintenance requirements stems from the maintenance concept where one defines the levels of maintenance and the basic functions to be performed at each level. These requirements are then expanded through the functional analysis and the identification of maintenance functions, described in Section 6.1.1. Functional flow diagrams are developed as shown in Figures 6.5 and 6.6, covering both scheduled and unscheduled maintenance requirements. Gross-level functions are broken down into subfunctions, duties, tasks, subtasks, and so on to the depth necessary to provide for the identification of specific resource requirements. In accomplishing such, trade-off studies are initiated to ensure the best allocation of resources, given a number of alternative approaches. For maintenance tasks, one needs to consider whether a task should be accomplished using automation or by manual means. Maintenance tasks may be broken down to cover any one, or more, of the following requirements:

1. *Alignment/adjustment:* to restore a system or product to acceptable operation through the proper alignment of its components (e.g., the alignment of the front

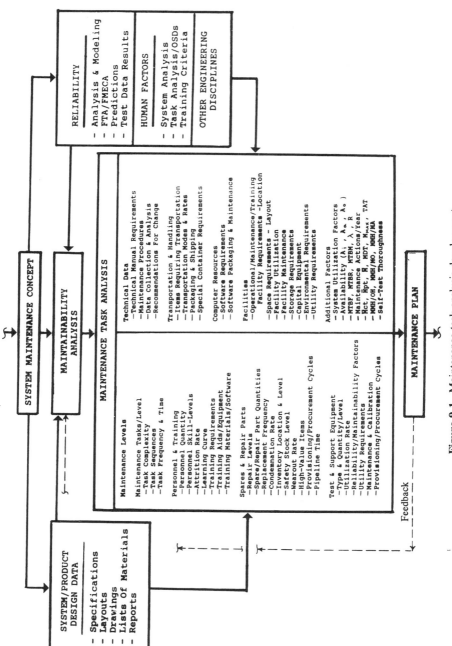

Figure 9.1. Maintenance task analysis development.

suspension of an automobile) and/or the adjustment of a control until a given performance parameter is brought into an acceptable range (e.g., the adjustment of a carburetor of an automobile or a voltage adjustment for an electrical system).

2. *Calibration:* the process of checking an item against a "working" standard, a "secondary" standard, or a "primary" standard. Calibration generally applies to precision measurement equipment (PME), and can be accomplished either on a scheduled basis or an unscheduled basis subsequent to the accomplishment of a repair action on a PME item. Calibration provides the necessary test accuracy, measurement accuracy, and traceability to the National Institute of Standards and Technology (i.e., an item is checked against a working standard that, in turn, is checked against a secondary standard, which is calibrated against a primary standard).

3. *Condition monitoring:* the ongoing surveillance of the operation of a system or product to ensure proper operation and to detect abnormalities indicative of an impending failure or of a failure that has already occurred (usually included as a preventive maintenance requirement).

4. *Functional test:* a system operational checkout, either as a condition verification subsequent to the accomplishment of item repair or as a periodic scheduled requirement.

5. *Inspection:* an action, or series of actions, taken to ensure that a requisite condition of quality exists. To inspect for a desired condition, it may be necessary to remove an item, to gain access by removing other items, or to partially disassemble an item. Inspection may include "condition assessment," or the evaluation of the physical condition of a component known to be subject to wear or other forms of deterioration.

6. *Overhaul:* an action, or series of actions, taken when an item is completely disassembled, refurbished, reworked, tested, and returned to a serviceable condition meeting all requirements set forth in applicable specifications. Overhaul may result from either a scheduled or unscheduled requirement, and is generally accomplished at the manufacturer/supplier/depot facility.

7. *Remove:* a maintenance requirement when the basic objective is to remove an item from the next higher entity in the system/product hierarchal structure (e.g., the removal of a unit from a system or the removal of an assembly from a unit).

8. *Remove and reinstall:* the removal of an item for maintenance and the reinstallation of that same item after repair has been accomplished.

9. *Remove and replace:* the removal of a designated item and the replacement with another like item (i.e., a spare item from the inventory). Such action can result from a component failure, or as a result of a preventive maintenance requirement (i.e., the periodic replacement of a critical-useful-life item).

10. *Repair:* constitutes a series of corrective maintenance tasks required to return an item to a serviceable condition. This may include the replacement of parts, the alteration of material, fixing, sealing, filling in, and so on.

11. *Servicing:* includes the maintenance tasks associated with the cleaning of system components and the application of lubricants, gas, oil, and so on. Servicing

may require removal, disassembly, reassembly, adjustment, and/or installation, and may be accomplished either on a scheduled basis or an unscheduled basis when a repair action is necessary.

12. *Troubleshooting:* involves the logical process (i.e., series of tasks) which leads to the positive identification of the cause of a system/product malfunction. It includes the initial "localization" of a problem utilizing built-in test means and the follow-on "fault isolation" using external test provisions. Troubleshooting, in this context, includes the entire spectrum of "diagnostics!"

These maintenance activities are typical, but should not be considered all-inclusive, and the specific nomenclature may vary from one organization to the next. However, this will serve as a frame of reference for further discussion.

In describing the MTA, an example may be helpful for the purposes of clarification.[1] Assume that the system in question is the manufacturing capability illustrated in Figure 9.2 (an extension of Figure 6.7), and that the maintainability specialist is interested in investigating the planned maintenance activities represented by block 13, an area of activity where the projected cost is high owing to anticipated corrective maintenance requirements.

In delineating the performance of corrective maintenance, the analyst must visualize what the maintenance technician will experience. At random points in time when the system is operating, failures are likely to occur and will be detected by the operator through visual, audio, and/or by physical means. The operator proceeds to notify the appropriate maintenance organization that a problem exists.

A maintenance technician is assigned to deal with the problem, and is dispatched to the production line to analyze the situation and to verify that the system is indeed faulty. In some instances, the fault will be obvious, particularly when dealing with mechanical or hydraulic systems when a structural failure has occurred or a fluid leak takes place. On other occasions, the technician must operate the system and attempt to repeat the condition leading to the failure occurrence. This is often the case for electronic equipment when the failure is not always obvious. In any event, corrective maintenance generally commences with the identification of a failure symptom such as the "system does not work," "the hydraulic system leaks," "there is no signal output," "the process is producing a faulty product," "the engine does not respond in terms of power output," "no voltage indication on the front panel," and so on. Based on a symptom of this nature, the maintenance technician proceeds to troubleshoot and accomplish the necessary repair actions.[2]

Troubleshooting may be extremely simple or quite complex. If a hydraulic leak

---

[1] The format illustrated herein has been utilized successfully in a number of different system applications. It is also described in Blanchard, B.S., *Logistics Engineering and Management,* 4th ed. Prentice-Hall, Inc., Englewood Cliffs, NJ, 1992.

[2] In the accomplishment of a functional analysis, performance parameters (i.e., input–output measures) are identified with each block in the functional block diagram. If these parameters are not apparent, through some indication or control, then there is a "NO-GO" action, which leads into the requirement for corrective maintenance. The FMECA, described in Section 7.2.3, is an excellent tool for identifying the various possible system/product functions and their failure modes, and should be used as an input to the preparation of the MTA.

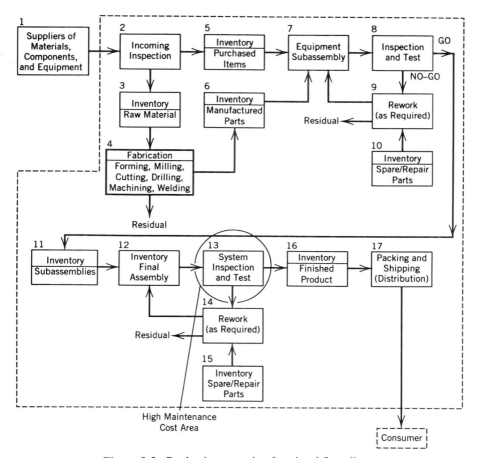

Figure 9.2. Production operation functional flow diagram.

is detected, the source of the leak is often quite easily traced. On the other hand, the failure of a small component in a radar or computer equipment may not readily be identified. In this instance, the technician must accomplish a series of steps in a logical manner which will lead him or her directly to the faulty item. At times, these steps are not adequately defined and the technician is forced into a "trial-and-error" approach to maintenance. A good example is when the technician starts replacing parts on a mass basis without analyzing cause-and-effect relationships, hoping that the problem will disappear in the process. This, of course, affects maintenance downtime and spare/repair part needs, as the technician may replace many parts when only one of them is actually faulty.

To preclude the possibility of wasting time and resources, the equipment design must incorporate the necessary characteristics to enable the maintenance technician to proceed in an accurate and timely manner in identifying the cause of failure.

Such characteristics may constitute a combination of go/no-go indicator lights, test points, meters, and other readout devices, providing the information necessary to allow the technician to go from step to step with a high degree of confidence that he or she is progressing in the right direction. This objective is one of the goals of maintainability. Given a design approach, the MTA is accomplished to verify that the design is supportable. This facet of the analysis (i.e., the diagnostic aspect) is best accomplished through the development of a logic troubleshooting flow diagram, a step-by-step, go/no-go series of tasks evolving from a maintenance functional flow. Figure 9.3 shows an abbreviated flow diagram covering the diagnostic tasks leading to the identification of a fault within the system inspection and test capability represented by block 13 in Figure 9.2.

The analyst should review the FMECA data to determine cause-and-effect relationships, and then proceed to list the major symptoms which the system is likely to experience. While it may be impossible to cover *all* symptoms of failure, the analyst should be able to identify the major ones involving the failure of a critical component or those occurring most frequently. For each symptom, various troubleshooting approaches are analyzed in terms of maintenance time and the anticipated logistic resources expended, and the best approach is selected. Referring to Figure 9.3, there may be more than one way to determine that Assembly A-7 is faulty, or that C.B. 1A5 has failed. The objective is to select the best approach, considering time, logistic resources, and overall cost. In doing so, not only is it possible to determine what is required in terms of built-in versus external test provisions, quantities and skill levels of personnel, and so on, but to assess the degree of maintainability incorporated within the design configuration being evaluated. This also serves as a useful technique to expand and/or update the maintenance concept.

In evaluating the diagnostics aspect of the system inspection and test capability (an activity accomplished at the organizational level of maintenance), the most feasible approach is assumed through the selected tasks identified in Figure 9.3 (i.e., Tasks 01, 02, . . . , 15). These tasks are presented in Figures 9.4 and 9.5 (the second being an extension of the first). Through the sample format shown, the analyst is able to evaluate each task further.

To clarify some of the considerations and methods used in accomplishing a MTA. The formats presented in Figures 9.4 and 9.5 will now be discussed. Referring to Figure 9.4, it is appropriate to first address some of the basic data entries in blocks 1–9.

1. *Block 1—System:* the basic system nomenclature should be entered here, for example, System "XYZ" for the overall production capability illustrated in Figure 9.2.

2. *Block 2—Item name/part number:* identify the name and part number (or identification number) of the system element, subsystem, unit, assembly, and so on being addressed by this particular task analysis. In this instance, the manufacturing test activity is the subject of the analysis (refer to Figure 9.2, block 13).

3. *Block 3—Next higher assembly:* include the name and part number (or identification number) of the next higher assembly. If the task analysis is being accom-

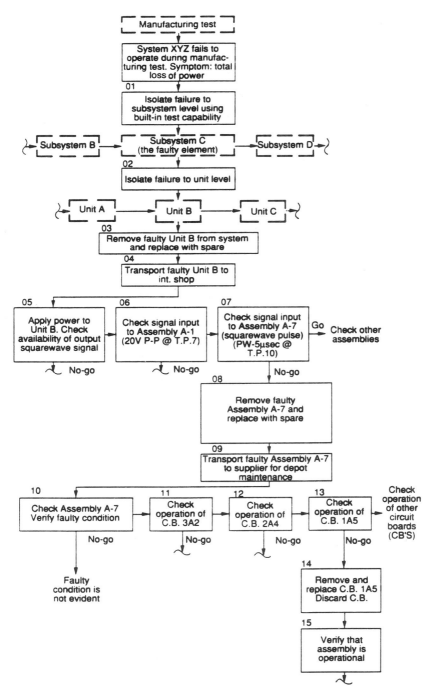

Figure 9.3. Abbreviated logic troubleshooting flow diagram.

Figure 9.4. Maintenance task analysis (Part 1).

| 1. System: XYZ | 2. Item name/part no.: Manufacturing Test/A4321 | 3. Next higher Assy.: Assembly and test | 4. Description of requirement: During manufacturing and test of Product 12345 (Serial No. 654), System XYZ failed to operate. The symptom of failure was "loss of total power output." Requirement: Troubleshoot and repair system |

| 5. Req. No.: 01 | 6. Requirement: Diag./Repair | 7. Req. Freq.: 0.00450 | 8. Maint. Level Org./Inter. | 9. Ma. Cont No.: A130000 |

10. Task Number	11. Task Description	12. Elapsed time - minutes	13. Total elap time	14. Task Freq	15. B	16. I	17. S	18. Total
01	Isolate failure to subsystem level (Subsystem C is faulty)		5	0.00450	5			5
02	Isolate failure to unit level (Unit B is faulty)	(2nd cycle)	25			25	25	25 / 25
03	Remove Unit B from system and replace with a spare Unit B		15		15			15
04	Transport faulty unit to int. shop		30		30			30
05	Apply power to faulty unit. Check for output squarewave signal	(3rd cycle)	20			20		20
06	Check signal input to Assembly A-1 (20v P-P @ T.P.7)	(4th cycle)	15			15		15
07	Check signal input to Assembly A-7 (squarewave. PW-5µsec @ T.P. 2)		20			20		20
08	Remove faulty A-7 & replace		10		10			10
09	Transport faulty Assembly A-7 to supplier for depot maintenance	14 calendar days in transit						
10	Check A-7 & verify faulty condition	(5th cycle)	25				25	25
11	Check operation of CB-3A2		15				15	15
12	Check operation of CB-2A4		10				10	10
13	Check operation of CB-1A5	(6th cycle)	20				20	20
14	Remove and replace faulty CB-1A5	(7th cycle)	40			40		40
15	Verify that assembly is operational and return to inventory		15				15	15
		Total	265		60	120	110	290

Elapsed time scale (12): 2 4 6 8 10 12 14 16 18 20 22 24 26 28 30 32 34 36 38

1. Item name/Part No.: Manufacturing Test/A4321		2. Req No.: 01		3. Requirement: Diagnostic Troubleshooting and repair		4. Req. Freq: 0.00450		5. Maint level: Organization, Intermediate, Depot	6. Ma. Cont. No.: A120000
7. Task No.	8. Qty per Assy	Replacement Parts		12 Qty	Test & support/Handling equipment		14. Use time (min)	16. Description of Facility Requirements	17. Special Technical Data Instructions
		9. Part nomenclature / 11. Part number	10. Rep Freq		13. Item part nomenclature / 15. Item part number				
01	-	-	-	1	Built-in test equip. A123456		5	-	Organizational Maintenance
02	-	-	-	1	Special system tester 0-2310B		25	-	-
03	1	Unit B B180265X	0.01866	1	Standard tool kit STK-100-B		15	-	-
04	-	-	-	1	Standard cart (M-10)		30	-	Intermediate maintenance
05	-	-	-	1	Special system tester I-8891011-A		20	-	
06	-	-	-	1	Special system tester I-8891011-A		15	-	
07	-	-	-	1	Special system tester I-8891011-A		20	-	
08	1	Assembly A-7 MO-2378A	0.00995	1	Special extractor tool EX20003-4		10	-	Refer to special removal instructions
09	-	-	-	1	Container, special handling T-300A		14 days	-	Normal trans. environment
10	-	-	-	1	Special system tester I-8891011-B		25	Clean room environment	Supplier (depot) maintenance
11	-	-	-	1	C B test set D-2252-A		15		
12	-	-	-	1	C B test set D-2252-A		10		
13	-	-	-	1	C B test set D-2252-A		20		
14	1	CB 1A5 GDA-221056C	0.00450	1	Special Extractor tool/EX45112 63 Standard tool kit STK-200		40		
15	-	-	-	1	Special system tester I-8891011-B		15		Return operating assy. to inventory

Figure 9.5. Maintenance task analysis (Part 2).

345

plished on a unit, the next higher assembly may be a subsystem, and so on. It is important to show "traceability," both upward and downward within the context of the overall analysis effort.

4. *Block 4—Description of requirement:* briefly describe the problem or the maintenance requirement justifying the MTA. The need for performing maintenance must be clearly established. Include references related to the requirement where appropriate. The description should support the requirement identified in block 6.

5. *Block 5—Requirement number:* a number is assigned for each requirement applicable to the item being analyzed. Requirements are identified in block 6, and may be numbered sequentially from 01 to 99 as necessary. In the example, the requirement is 01, troubleshooting and repair of the manufacturing test activity, which is part of System "XYZ."

6. *Block 6—Requirement:* the requirement nomenclature may be entered in this block. Maintenance requirements may include any one or combination of activities such as alignment/adjustment, calibration, condition monitoring, troubleshooting, or repair.

7. *Block 7, Requirement frequency;* the frequency at which a requirement is expected to occur is entered in this block. If the requirement is "repair" and the item being analyzed is the "power generating system," the analyst must predict or determine how often repair of the power-generating system is to be accomplished. This value is dependent on the task frequencies listed in block 14, and is generally expressed either in number of occurrences per hour or cycle of system/product operation (in event of corrective maintenance) or in terms of a given calendar time (for preventive maintenance).

A frequency factor for each requirement is necessary to provide the basis for determining associated logistics resource needs. The following questions are appropriate:

a. How often will the system be down for maintenance?
b. How often do we expect to need a spare/repair part?
c. How often do we need test and support equipment for maintenance?
d. How often will maintenance personnel be required?
e. How often will special maintenance facilities be required?
f. How often will servicing be required?
g. How often will overhaul be accomplished?

When determining the intervals or frequencies representing unscheduled maintenance actions, the inherent reliability characteristics of an item are certainly significant, but they are not necessarily always dominant! For some equipment, field data have indicated that less than one-half of all unscheduled maintenance actions are attributed to random catastrophic (i.e., primary) failures. Therefore, it is obvious that consideration must be given to other contributing factors as well. On the other hand, there are many instances where the inherent reliability failure rates dominate all considerations, and where a proportional amount of attention must be applied. Knowledge of the component characteristics and the physics of failure is important in the design of any system or product. In any event, the maintenance frequency factor must consider the following:

a. *Inherent reliability characteristics.* This category covers primary cata-strophic failures based on the physical makeup of the item and the stresses to which the item is subjected. Different modes of failure and out-of-tolerance conditions are included. Failure frequency data is initially derived from relia-bility allocation. As the design process evolves, results from reliability mod-eling and analysis, prediction, and so on, serve as the best source.

b. *Dependent failures (chain effect).* This category includes those secondary failures that occur as a result of a primary catastrophic failure. In other words, the failure of one item may cause other items to fail. The frequency of dependent failures is based on the extent of fail-safe characteristics inher-ent in the design. In electronic equipment, circuit protection provisions may be incorporated to reduce the probability of dependent failures. The reliabil-ity FMECA is the best data source for reflecting dependent failure character-istics.

c. *Manufacturing defects or burn-in characteristics.* Quite often when an equip-ment item first comes off the production line, there are a rash of failures until the system is operated for a short period of time and a constant failure rate is realized. This is particularly true for electronic equipment when certain cor-rective actions are necessary to attain system stabilization. The extent of dif-ficulty at this early point in the operational use phase is dependent on the amount of equipment operation and the type of testing accomplished in the production/construction phase. If enough hours of equipment operation are attained through testing, the quantity of failures after delivery will probably be less. In any event, when new equipment first enters the inventory, there may be more initial failures than indicated by the predicted reliability failure rate, and this factor must be considered in determining the requirements for initial system support.

d. *Wear-out characteristics.* After equipment has been in operational use for a period of time, various individual components begin to wear out. Some components (i.e., mechanical linkages, gears) wear out sooner than others. When this occurs, the resultant frequency of failure and corrective mainte-nance will increase.

e. *Operator-induced failures.* Overstress of equipment due to operator human error is a possibility and is very significant in determining frequency factors. Unplanned stresses can be placed on the equipment as a result of different operating methods (by different operators or by the same operator), and these stresses can accumulate to the extent that the equipment will fail more fre-quently. Hopefully, conditions such as this can be minimized or eliminated through the proper emphasis on human factors in the design process.

f. *Maintenance-induced failures.* Damage to equipment caused by human error during maintenance actions may result from not following the proper maintenance procedures, improper application and use of tools and test items, losing components or leaving parts out when reassembling items after repair, forcing replacement items to fit when physical interferences are present, causing physical damage to components lying adjacent to or near the item being repaired, and many others. Induced faults of this type may occur due to improper maintenance environment (inadequate illumination, uncomfortable temperatures, high noise level), personnel fatigue, inadequately trained personnel performing the maintenance tasks, equipment sensitivity or fragility, poor internal accessibility to components, and not having the right elements of logistic support available when needed. Maintenance-induced faults are possible both during the accomplishment of corrective maintenance and preventive maintenance. For instance, during a scheduled calibration when performing a fine adjustment, a screwdriver slip may cause damage in another area resulting in corrective maintenance.

g. *Equipment damage due to handling.* Another important aspect is the probability of equipment damage due to bumping, dropping, shoving, tossing, and so on. These factors are particularly relevant during transportation modes and often contribute to subsequent equipment failures. The extent to which consideration is given toward transportability in the equipment design will influence the effects of handling on the equipment failure rate. The analyst should evaluate anticipated transportation and handling modes, review the design in terms of effects, and assign an appropriate frequency factor.

System failures (of one type or another) may be detected through some form of condition monitory, built-in self-test "GO/NO-GO" indicators, performance evaluation devices, and/or through the actual visual indication of a faulty condition. Condition monitoring devices may allow for the measurement of certain designated system parameters on a periodic basis, thus permitting the observation of trends. The concept of condition monitoring is discussed in Section 7.3.4 (Chapter 7) and is illustrated in Figure 9.6, where measurement points are indicated in terms of system operating time. When certain "out-of-tolerance" conditions are noted, maintenance actions may be initiated even though an actual catastrophic failure has not occurred. In other words, the anticipation of future problems causes a maintenance action to be accomplished at an earlier time than what otherwise might have been expected. These instances must also be considered in determining maintenance frequency.

The foregoing considerations are reviewed on an individual basis and may be combined to provide an overall factor for a given corrective maintenance action, as illustrated in Figure 9.7. The frequency factor in block 7 (Figure 9.4) represents the overall number of instances per system operating hour (or calendar period) for the maintenance requirement defined in block 6. This factor is determined from the frequency of the individual tasks identified in blocks 10 and 11. The maintenance requirement may constitute a series of tasks, each one of which is accomplished

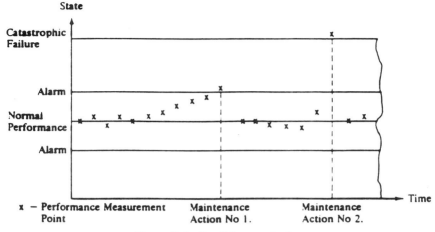

Figure 9.6. Condition monitoring.

whenever the requirement exists. In this instance, the frequency factor for each task (block 14) will be the same as the factor entered in block 7, Figure 9.4. On the other hand, it may be feasible to include a series of "either–or" tasks on a single MTA sheet. For example, when a troubleshooting requirement exists and there are different paths (with different frequencies) that the maintenance technician may wish to pursue, as shown in Figure 9.3, the tasks may be identified as follows:

Task No. (10)	Task Description (11)	Task Frequency (14)
0200	Isolate failure to subsystem	0.008600
X300	Repair Unit "A"	0.002900
X400	Repair Unit "B"	0.004500
X500	Repair Unit "C"	0.001200
0600	Check out subsystem	0.008600

CONSIDERATION	ASSUMED FACTOR (Instances/Operating Hour)
a. Inherent reliability failure rate	0.003620
b. Dependent failure rate	0.000690
c. Manufacturing defects	0.000020
d. Wearout failure	0.000010
f. Operator-induced failure rate	0.000020
g. Maintenance-induced failure rate	0.000110
h. Equipment/component damage rate	0.000030
TOTAL COMBINED FACTOR	0.004500

Figure 9.7. Maintenance frequency factor considerations.

The troubleshooting and repair requirement will apply to Unit "A" in 0.0029 instances per operating hour, Unit "B" in 0.0045 instances per operating hour, and so on. The sum of the various alternative paths ("X" tasks) will constitute the overall frequency, or the value in block 7 (0.0086). Tasks 0200 and 0600 are applicable for each occurrence of Tasks X300, X400, and X500. This method of data presentation provides a simplified approach for covering numerous possible alternatives often applicable to the diagnostics aspects of corrective maintenance, particularly for electronic equipment.

8. *Block 8—Maintenance level:* this refers to the level of maintenance at which the requirement is to be accomplished. Referring to Sections 1.3.7 and 5.3, this may refer to organizational, intermediate, manufacturer/depot, or supplier. The example in Figures 9.4 and 9.5 covers maintenance tasks accomplished at both the organizational and intermediate levels.

9. *Block 9—Maintenance analysis control number:* A maintenance analysis (MA) control number is entered for the item identified in block 2. The control numbering sequence is a numerical breakdown of the system by indenture level, and is designed to provide traceability from the highest maintenance-significant item to the lowest repairable unit. An example of MA control number assignment is shown in Figure 9.8. Maintenance tasks and logistic support resource requirements are identified against control number and may be readily tabulated from the lowest item to the system level. The MA control numbers are assigned by "function" and not necessarily by part number. If a modification of Assembly 6 is accomplished, or an alternative configuration of Unit "A" is employed, the manufacturer's part number will change; however, the MA control number remains the same as long as the basic function of the applicable item has not changed.

The maintenance requirements for a system are initially defined through the functional analysis described in Section 6.1.3. These functions are broken down into subfunctions and tasks, providing a basis for defining the specific resources required for system support. As the system definition process evolves through the development of design data, documentation, reports, component part and material lists, and so on, the maintainability analyst will evaluate the system configuration as it is described through this information and begin to define specific maintenance tasks that will be required. These tasks must be described within the context of the maintenance concept (refer to Chapter 5), the maintenance functions (refer to Chapter 6), and the maintenance requirements identified under block 6, Figure 9.4. The analyst, having had some prior field experience in the performance of maintenance and being knowledgeable about the user's environment, must be able to visualize (by looking at the design data/documentation) what will be required in the performance of maintenance tasks in the future. These tasks are recorded in block 11, Figure 9.4.

Maintenance tasks should be stated in a clear and concise manner, in the proper sequence, and defined to the extent necessary to enable the development of a com-

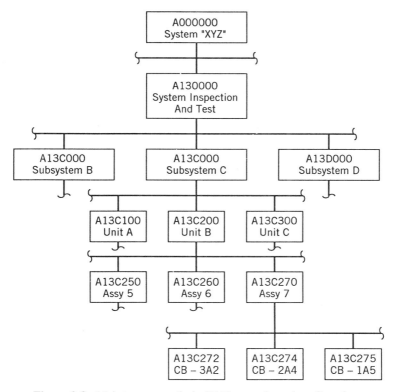

Figure 9.8. Maintenance analysis (MA) control number allocation.

plete maintenance checklist. Significant tasks include those whose accomplishment requires personnel with the appropriate skills, a spare part, a tool or item of test equipment, a special facility, an item of software, or some related element of logistic support. Including an extensive amount of detail is not necessary; however, enough detail should be provided to enable the identification of logistic support requirements. The MTA represents the process of transition from the functional analysis to the preparation of detailed maintenance procedures.

Referring to Figure 9.4, maintenance tasks are recorded in block 11, and task numbers are noted in block 10. Task numbers are included to provide the necessary traceability (both upward and downward), and to simplify task identification, particularly when maintenance analysis data are computerized. A numeric (or alphanumeric) designation may be assigned to each task, maintaining the proper ordering or sequencing of task accomplishment. If it is necessary to expand a designated task into a number of subtasks in order to gain better visibility through analysis, then the numeric designation should allow for the inclusion of an additional digit.

9.3   TASK TIME AND FREQUENCY REQUIREMENTS

For each task listed in blocks 10 and 11, Figure 9.4, the analyst should estimate the elapsed time for task completion and project this estimate in block 12. Individual tasks may be broken down into subtasks, task elements, and so on, and time standards may be applied to each segment of activity in much the same way as is accomplished in maintainability prediction. Such time standards may be developed from past experience, the evaluation of field data, and so on. In any event, the elapsed time for each individual is noted.

In presenting the estimated task times, task sequences are shown as well as tasks that may be accomplished in parallel. If two or more people are involved in a task, it may be feasible to construct a separate line for each individual (e.g., ■■① for worker 1, ■■② for worker 2, etc.). If the task extends beyond the 40-minute period indicated on the form, the timeline is continued on the next line commencing at zero and a new cycle is indicated. The 40-minute presentation is merely the result of a limitation of the form. If the task times are extensive, the time period may be extended and/or "hours" may be included in lieu of "minutes." The basic scale can readily be converted and tailored to the need.

The benefit of laying out task timelines is to allow for the evaluation of task sequences versus the accomplishment of tasks in parallel, and to assess personnel utilization requirements. If several system components require maintenance at the same time, the timeline analysis in block 12 will aid the analyst in determining labor-hour requirements. By pictorially presenting timeline data, it is possible to evaluate a number of sequences, arriving at an optimum arrangement. The objective is to minimize both elapsed time and labor time, and to make greater use of lower-skilled personnel where possible.

In block 13, Figure 9.4, the analyst should enter the numerical value for the elapsed timeline shown in block 12, and the total lapsed time for the overall maintenance requirement (i.e., 01) is presented at bottom of the column (i.e., 265 minutes). Individual task frequency factors are entered in block 14, and the frequency for the overall requirement is noted in block 7. These factors are derived from reliability information, and reflect the anticipated demand rates for the spare part requirements, test and support equipment utilization, facility utilization, and so on, reflected in Figure 9.5. In other words, the task requirements identified in Figure 9.4 lead directly to the resource requirements presented in Figure 9.5.

9.4   PERSONNEL QUANTITY AND SKILL-LEVEL REQUIREMENTS

The maintenance tasks, presented in block 12 of Figure 9.4, are broken down and evaluated in terms of both the quantity and the skill levels of the personnel required for task completion. In an approach similar to the accomplishment of the OTA as part of a human factors program (refer to Section 7.2.15), the maintainability analyst will evaluate all maintenance tasks to be performed by the human, assess each task on the basis of complexity level, and identify the number of personnel by skill-level category required. The analysis includes the consideration of human sensory and perceptual capacities, motor skills, mobility coordination, human physical dimensions (i.e., anthropometric factors), muscular strength, and so on. An appropriate skill-level requirement is assigned to each task, and the maintenance man-minutes by skill level are entered in blocks 15–18, with the total values for the overall requirement summarized at the bottom.

To determine personnel requirements, it is appropriate to establish a set of job standards and expectations for each applicable organizational entity. For instance, the skill-level categories of "Basic," "Intermediate," and "Supervisory" have been selected for the example presented in Figure 9.4. The expectations for each of these categories are noted as follows: [3]

1. *Basic skill level:* an individual 18–21 years in age; high school graduate; ninth grade reading level; and no regular work experience prior to training. After a limited amount of specialized training, this individual can perform routine checks, accomplish physical functions, use basic hand tools, and follow clearly presented instructions where interpretation and decision making are not necessary. Workers in the class usually assist more highly skilled personnel, and require constant supervision.

2. *Intermediate skill level:* an individual with a more formalized education consisting of approximately 2 years of college or equivalent course work in a technical institute. In addition, this individual will have had some specialized training, and 2–5 years of experience in the field related to the type of systems/products in question. Personnel in this class can perform relatively complex tasks using a variety of test instruments, and are able to make certain decisions pertaining to the maintenance and disposition of system components.

3. *Supervisory skill level:* an individual with 2–4 years of formal college or equivalent course work in a technical institute, and 10 years or more of related on-the-job experience. This individual is assigned to train and supervise basic and intermediate skill-level personnel, and is in a position to interpret procedures, accomplish complex tasks, and make major decisions affecting maintenance policy and the disposition of system components. He/she is knowledgeable about the operation and use of highly complex precision measurement equipment (e.g., PME used in calibration).

Once the personnel quantities and skills have been assigned for each task, they are combined with similar requirements for other tasks and presented in the form of

---

[3] There may be other classifications of personnel skills depending on the user's organization structure. The intent is to define specific classifications that will be applicable in system operational use, and to evaluate maintenance tasks in terms of these classifications. Some examples of personnel classification may be found in civil service job descriptions and in the following military documentation: (a) AFM 39-1, *U.S. Air Force Personnel Manual;* (b) NAVPERS 18068, *Manual of Navy Enlisted Manpower and Personnel Classifications and Occupation Standards;* and (c) AR 611-201, *Enlisted Military Occupational Specialties.*

a maintenance organizational objective. In other words, when considering all of the various maintenance requirements in total (for unscheduled and scheduled mainte-nance—block 6, Figure 9.4), the analyst can compile an initial projection of the personnel requirements for the future maintenance organization. The results must then be compared with the structure of the existing maintenance organization to establish the requirements for training. If the quantity and skill levels required for the new system are high, the organization must be upgraded to meet these new re-quirements.

Training requirements may involve a combination of informal "on-the-job train-ing (OJT)" and the conductance of a more formal training program to include partic-ipation in intensified short courses, seminars, workshops, and the like. This training must not only consider those who are already in the organization and are about to assume the responsibility for the maintenance and support of the new System "XYZ," but the training of replacement personnel as a result of attrition.

As an output from the MTA, a formal training program plan may be developed. This plan should not only include the type of training proposed but also the recom-mended program content, schedule, and training support. Training support consists of the necessary training manuals and documentation, training aids and devices, equipment and software, simulators, facility, and utility requirements.

With regard to maintainability objectives in design, the system being evaluated should incorporate the necessary characteristics that will result in the minimum consumption of personnel quantities and in the utilization of "basic" skills (in lieu of the higher "intermediate" and "supervisory" skill-level requirements). With this in mind, the maintainability analyst should review the design configuration reflected by the MTA data presented in Figures 9.4 and 9.5 and determine where the possible incorporation of changes might offer improvement by reducing the logistic support resources (and possibly life-cycle cost as well). Review of Figure 9.4 indicates that Tasks 2, 10–13, and 15 (in particular) reflect the requirement for high skill-level personnel and represent candidate areas for possible improvement. Is there any way that the design can be modified such that the tasks can be accomplished by a "basic" skill-level technician?

9.5   SPARE/REPAIR PART AND INVENTORY REQUIREMENTS

Spares and repair part requirements initially stem from the maintenance concept and the definition of levels of maintenance, the functions to be accomplished at each level, and so on. These requirements are expanded through the functional analysis, requirements allocation, and level of repair analysis and related trade-off studies. The results of these activities provide an input to the design.

As the design evolves, the MTA is accomplished to identify all anticipated re-placement parts and consumables associated with scheduled and unscheduled main-tenance. For each individual maintenance requirement, the "remove and replace" tasks generally result in the need for a spare or a repair part. Referring to Figure 9.4, Task 3 requires the removal of a faulty Unit "B" and replacement with a spare unit; Task 8 requires the removal of a faulty Assembly "A-7" and replacement with a like spare assembly; and Task 14 requires the removal of a faulty Circuit Board "1A5" and replacement with a spare circuit board. Thus, for this maintenance requirement (i.e., 01), there will be a need for at least one spare Unit "B," one Assembly "A-7," and one Circuit Board "1A5."

Referring to Figure 9.5, the spare/repair part resources necessary to accomplish each of the applicable tasks identified in Figure 9.4 are shown in blocks 8–11. Blocks 9 and 11 describe the replacement items by nomenclature and manufacturer's part number. Block 8 indicates the quantity of items used (actually replaced or consumed) in accomplishing the task. Block 10 specifies the predicted replacement rate based on a combination of primary and secondary failures, induced faults, wear-out characteristics, condemned items scrapped, and/or scheduled maintenance actions. If an item is replaced for any reason, it should be covered in this category.

In the event that an "either–or" situation exists (identified by by an "X" task— block 10 of Figure 9.4 and block 7 of Figure 9.5), the listing of replacement parts will be as shown in Figure 9.9. The individual replacement frequencies (i.e., replacements per hour of system operation) will indicate the demand for each component application. When these individual factors are compiled and integrated at the system level, it is then possible to determine spare/repair part requirements.

Task Number	Task Description	Replacement Parts		
		8. Qty. per Assy.	9 Part Nomenclature / 11 Part Number	10 Rep. Freq.
0010	Localize fault to Unit *A*	. . . .	. . . .	. . . .
0020	Remove Unit *A* access cover	. . . .	. . . .	. . . .
X030	Isolate to Assy. *A*	. . . .	. . . .	. . . .
X031	Remove and Replace Assy. *A*	1	Ampl.-Modular Assy. A160189-1	0.000133
X040	Isolate to Assy. *B*	. . . .	. . . .	. . . .
X041	Remove and Replace	1	Power Converter Assy. A180221-2	0.000212
X050	Isolate to Assy. *C*	. . . .	. . . .	. . . .
X051	Remove and Replace Assy. *C*	1	Power Supply Assy. A21234-10	0.000141
0060	Install Unit *A* access cover	. . . .	. . . .	. . . .
0070	Check out Unit *A*	. . . .	. . . .	. . . .

Figure 9.9. Maintenance task analysis (spares/repair parts).

The final determination of spares and repair part types and quantities must take into consideration a number of issues. The level of repair analysis, accomplished as part of the maintainability analysis discussed in Chapter 6, will lead to the determination of repairable versus nonrepairable items and the levels at which maintenance is to be accomplished. The FMECA will specify item criticality which, in turn may result in additional quantities of spares. The RCM analysis will lead to the identification of spares in support of preventive maintenance requirements. While many of these requirements are taken into consideration as part of the MTA, the ultimate determination of spare/repair parts must be accomplished within the context of the maintenance concept, supplemented by these various analysis efforts. Further, parameters such as maintenance process times, turnaround times, procurement lead times, transportation times, and so on, will have a great impact on the ultimate quantities of the spares and repair parts to be procured.

After consideration of the various factors that influence spare/repair part demand rates, the analyst needs to address the inventory requirements at different locations. Of significance are the factors described in Section 4.4.1. With these in mind, the logistic support organization can now proceed with preparation of the spares and repair parts provisioning (or procurement) package, that is, the initial source coding, the preparation of a drawing/data package, the evaluation and selection of suppliers, the procurement and receipt of inventory provisions, and so on. Although the ultimate procurement and handling of spares and repair parts are normally the function of the logistics or product support organization, the MTA does provide much of the input data in the initial determination of spare/repair part *requirements*. This type of information may also be developed through the LSA or equivalent.

9.6   TEST AND SUPPORT EQUIPMENT REQUIREMENTS

This category includes all tools, test equipment, handling equipment, fixtures, and other support equipment necessary to accomplish the maintenance tasks identified in block 11, Figure 9.4. The analyst, in evaluating design data, identifies the "HOWs" relative to task accomplishment, and lists the appropriate tools/equipment by item nomenclature and manufacturer's part number in blocks 13 and 15 (respectively) in Figure 9.5. The quantity of items needed is identified in block 12, and the anticipated tool/equipment utilization time is entered in Block 14. The objective is to identify the requirements for each task, combine and summarize these requirements at the top, and then determine what is actually needed for the new System "XYZ."

When determining the type of test and support equipment, the analyst should address the following issues in the order presented:

1. Determine that there is a definite need for the test and support/handling equipment. Can such a need be economically avoided through a design change? Given a need, determine the environmental requirements. Is the equipment to be used in a sheltered area or outside? How often will the equipment be deployed to other loca-

tions? If handling equipment is involved, what item(s) is to be transported, and where?

2. For test equipment, determine the parameters to be measured. What accuracies and tolerances are required? What are the requirements for traceability? Can another test equipment item at the same location be used for maintenance support or does the selected item have to be sent to a higher level of maintenance for calibration? It is desirable to keep test requirements as simple as possible and to a minimum. For instance, if 10 different testers are assigned at the intermediate level, it would be preferable to be able to check them out with one other tester (with higher accuracy) in the intermediate shop rather than send all 10 testers to the depot for calibration. The analyst should define the entire test cycle prior to arriving at a final recommendation.

3. Determine whether an existing off-the-shelf equipment item will do the job (through review of catalogs and data sheets) rather than contract for a new design, which is usually more costly. If an existing item is not available, solicit sources for design and development.

4. Given a decision on the item to be acquired, determine the proper quantity. Utilization, or use time, is particularly important when arriving at the number of test and support equipments assigned to a particular maintenance facility. When compiling the requirements at the system level, there may be 24 instances when a given test equipment item is required; however, the utilization of that item may be such that a quantity of two will fulfill all maintenance needs at a designated facility. Without looking at actual utilization and associated maintenance schedules, there is a danger of specifying more test and support equipment than what is really required. This, of course, results in unnecessary cost. The ultimate objective is to determine overall use time, allow enough additional time for support equipment maintenance, and procure just enough items to do the job.

The analyst, in determining the specific requirements in this area, needs to be familiar with the results of the level of repair analysis and any trade-off studies accomplished as part of the maintainability analysis (Chapter 6) and pertaining to built-in test versus external test, levels of self-test thoroughness, recommendations concerning condition monitoring, and so on. In this part of the task analysis effort (i.e., blocks 12–15, Figure 9.5), the analyst is identifying primarily *external* requirements. In doing such, there is a natural tendency to proceed in a "vacuum" and identify large quantities of newly designed "peculiar" items, as compared to common and standard equipment already available in the inventory.

With regard to the maintainability characteristics in design, the results of the MTA need to be evaluated with the objectives listed above in mind. Referring to Figure 9.5, for example, there are requirements for a special system tester in task 2, another special tester in tasks 5–7, another special tester in tasks 10 and 15, a special C.B. test set in tasks 11–13, a special handling container in task 9, and a special extractor tool in tasks 8 and 14. These requirements all lead to *new* design and development activity which, in turn, can be a very costly approach to system/product support. Can the basic design be modified in any way to reduce (if not eliminate) the need for all of the "special" items?

9.7   TRANSPORTATION AND HANDLING REQUIREMENTS

In the performance of system/product maintenance tasks, there is a need to return faulty items from the organizational level to the intermediate level, and from the intermediate level to the manufacturer/supplier/depot level (refer to the maintenance concept in Chapter 5). This flow not only includes those system-level components that are in the repair channel, but also items that are periodically replaced in accomplishing preventive maintenance, items that are condemned and being processed for disposal, items of test and support equipment that require a higher-level of maintenance and/or calibration, and so on. Further, there is a resupply requirement involving the replenishment of inventories with new spare/repair parts and the movement of personnel to support maintenance tasks at different locations.

The MTA, in support of these activities, must consider the various methods of packaging and handling of materials, the available modes of transportation (rail, air, highway, water, and pipeline), transportation rates and frequencies, allowable transportation times, and transportation costs. The analyst must address the requirements for transportation and handling, accomplish the necessary evaluation of alternatives, and specify a preferred approach. The results may be presented in a format similar to that shown for test and support equipment in blocks 12–15, Figure 9.5.

9.8   MAINTENANCE FACILITY REQUIREMENTS

When evaluating each task, the analyst must determine where the task is to be performed, and the facilities required for the accomplishment of such. This includes the determination of space requirements, capital equipment needs, equipment layout, storage space, power and light, telephone, computer requirements, water, gas, environmental controls, and so on. If a "clean room" is required for the accomplishment of precise maintenance and calibration, if a particular power capability is needed, if a special gas is necessary for testing purposes, and so on, it should be so specified. The analyst should solicit assistance from system design and human factors personnel and generate a facilities plan showing a complete layout of all essential items. The type of facility, a brief summary description, and reference to a facilities plan should be included in block 16, Figure 9.5.

In designing for maintainability, an objective is to develop a system/product that can be maintained using an existing facility, common and standard equipment, normal power sources, and so on. The requirement for a clean room for the accomplishment of task 10 (Figure 9.5) represents a special need, is likely to be costly, and should be avoided. Can Assembly "A-7" be designed such that it can be repaired without the use of a clean room, or can the assembly be designed such that it can be economically discarded at failure (versus being repaired)?

9.9   MAINTENANCE DATA AND COMPUTER RESOURCE REQUIREMENTS

In accomplishing a MTA, the analyst may wish to call attention to the need for special data coverage in critical situations. For example, the design configuration being analyzed may require that special safety-related procedures be included in the subsequent development of the regular maintenance procedures, normally prepared during the later stages of detail design; there may be some additional clarification required in the performance of a given maintenance task; there may be a requirement for the utilization of certain software in the troubleshooting of System "XYZ;" and so on. An indication of such a requirement, in the form of an abbreviated note, may be included in block 17, Figure 9.5.

9.10   MAINTENANCE TASK ANALYSIS SUMMARY

As previously mentioned, the MTA serves several purposes. It is accomplished, commencing during the preliminary design phase, to provide an early determination of the anticipated support requirements for a given system configuration. Individual task analyses are prepared to (1) cover all significant repairable components of the system/product, and (2) include consideration of all corrective and preventive maintenance requirements for the system being evaluated. Figures 9.4 and 9.5 represent a sample of the type of information developed for a small segment of the system. These (and other) elements of the overall. MTA are summarized, using a format similar to that presented in Figure 9.10, with the ultimate objective of providing the type of information conveyed in Figure 9.1. These requirements are combined upward until all of the maintenance elements of the system, shown in Figure 9.8, are covered. Through this definition of system/product support requirements, the analyst is able to provide an early assessment of the system in terms of anticipated lifecycle cost.

A second objective is to accomplish the MTA early enough in the system design and development process to allow for the timely feedback and possible incorporation of corrective changes (in the event that a specification requirement is not being met) or for overall product improvement. Referring to Figures 9.4 and 9.5, there are obvious areas where maintainability improvement can be realized. Some of these are noted:

1. With the extensive resources required for the repair of Assembly "A-7" (e.g., the variety of special test and support equipment, the necessity for a "clean-room" facility for maintenance, the extensive amount of time required for the removal and

1. System	2. Item Name/Part No.	3. Next Higher Assy.	4. Design Specification	5. Ma Control No.
CXXX Aircraft	Unit A Synchronizer / PN 1345	System XYZ	ZA 88446 (10/2/94)	B12000

6. Functional Description: System XYZ is an airborne navigation subsystem installed in the CXXX Aircraft. Unit A of the navigation subsystem provides finite range and bearing information for the overall aircraft. Unit A includes Assembly A, a power supply, and Circuit Boards 2A1 and 3A2.

7. Maintenance Concept Description: Unscheduled Maintenance - Upon detection of a fault at the system level, the applicable fault is isolated to Unit A, Unit B, or Unit C (as applicable) through built-in self-test. The unit is removed and replaced at the aircraft, and the faulty item is returned to the Intermediate Shop for corrective maintenance. Fault isolation is accomplished to CB2A1, CB3A2, Assy. A, or Pwr. Supply. Assy. A & Pwr. Supply are repairable while CBs are non repairable. Scheduled Maintenance-Periodic calibration.

8. Req. No.	9. Requirement	10. Maint. Level	11. Req. Freq.	12. Elap. Time	13. Per. Skills	14. Man-Min	15. Repl. Parts	16. Test and Support Equip.
O1	Troubleshooting	Intermediate	0.00184	27	Intermediate	27		Oscilloscope/HP1-34
					Basic	27		Voltmeter/BN-33
								Square-wave Gen/CPP33
O2	Repair	Intermediate	0.00184	18	Basic	18	Assy. A/A12345	Screwdrive, Soldering
							CB3A2/BN1456	Gun, Wire Clippers
							Power Supply/PP320	
							CB2A1/BN1576	
O3	Functional Test	Intermediate	0.00184	6	Intermediate	6		Oscilloscope/HP1-34
O4	Calibration	Depot	0.00139	480	Supervisory	590		Precision VTVM/ASI-13
								WS Bridge/SU-123
								Oscilloscope/Tek 462
								Common Hand Tools
Total			0.00321	5.31		668		

17. Notes: △ Refer to Addendum A for specific type and quantity of spare/repair parts. △ Refer to Addendum B for a detailed listing of test and support equipment.

Prepared by: Blanchard Date: 10/11/xx

Figure 9.10. Maintenance task analysis summary.

replacement of CB-1A5, etc.), it may be feasible to identify Assembly "A-7" as being nonrepairable! In other words, investigate the feasibility of whether the assemblies of Unit "B" should be classified as "repairable" or "discard at failure."

2. Referring to tasks 01 and 02, a "built-in test" capability exists at the organizational level for fault isolation to the "subsystem." However, fault isolation to the "Unit" requires a special system tester (0-2310B), and it takes 25 minutes of testing plus a highly skilled (supervisory skill) individual to accomplish the function. In essence, one should investigate the feasibility of extending the built-in test down to the unit level and eliminate the need for the special system tester and the high-skill-level individual.

3. The physical removal and replacement of Unit "B" from the system takes 15 minutes, which seems rather extensive. Although perhaps not a major item, it would be worthwhile investigating whether the removal/replacement time can be reduced (to less than 5 minutes for example).

4. Referring to tasks 10–15, a special clean-room facility is required for maintenance. Assuming that the various assemblies of Unit "B" are repaired (versus being classified as "discard at failure"), it would be worthwhile to investigate changing the design of these assemblies such that a clean-room environment is not required for maintenance. In other words, can the expensive maintenance facility requirement be eliminated?

5. There is an apparent requirement for a number of new "special" test equipment/tool items: special system tester (0-2310B), special system tester (I-8891011-A), special system tester (I-8891011-B), C.B. test set (D-2252-A), special extractor tool (EX20003-4), and special extractor tool (EX45112-6). Usually, these *special* items are limited as to general application for other systems, and are expensive to acquire and maintain. Initially, one should investigate whether these items can be eliminated; if test equipment/tools are required, can *standard* items be utilized (in lieu of special items)? Also, if the various special testers are required, can they be integrated into a "single" requirement? In other words, can a single item be designed to replace the three special testers and the C.B. test set? Reducing the overall requirements for special test and support equipment is a major objective.

6. Referring to task 09, there is a special handling container for the transportation of Assembly "A-7." This may impose a problem in terms of the availability of the container at the time and place of need. It would be preferable if normal packaging and handling methods could be utilized.

7. Referring to task 14, the removal and replacement of CB-1A5 takes 40 minutes and requires a highly skilled individual to accomplish the maintenance task. Assuming that Assembly "A-7" is repairable, it would be appropriate to simplify the circuit board removal/replacement procedure by incorporating plug-in components, or at least simplify the task to allow one with a basic skill level to accomplish.

While one objective is to accomplish the MTA during the design and development of a new system, it can also be applied appropriately in the evaluation of an *existing* system capability. Referring to Figure 9.2, assume that the manufacturing capability shown is already in operation and that one may wish to evaluate this existing capability in terms of overall productivity and effectiveness. The analyst may initially accomplish a life-cycle cost analysis, identify the various functions and the "high-cost contributors," determine the "cause-and-effect" relationships, and initiate recommendations for improvement. In this instance, the system inspection and test capability (i.e., block 13) is an excellent candidate for further evaluation. The MTA can be effectively applied to pinpoint some of the possible causes for the high cost.

9.11 LOGISTIC SUPPORT ANALYSIS (LSA)

Within the overall spectrum of "logistics," as it applies in the design and development of *systems* (versus consumable items), there is a requirement to accomplish a "logistic support analysis (LSA)." The LSA, which has been applied primarily in the defense sector, consists of the integration and application of various techniques and methods to ensure that supportability requirements are considered in the system design process. It is a process employed on an iterative basis throughout all phases of design and development, and it involves the utilization of different analytical methods to solve a wide variety of problems of varying magnitudes. More specifically, the LSA:[4]

1. Aids in the initial establishment of supportability requirements in conceptual design through the development of system operational requirements and the maintenance concept, and the conductance of feasibility studies involving different system support alternatives.

2. Aids in the evaluation of alternative system/equipment design configurations. Given different design approaches during the preliminary and detail design phases, the LSA is employed to analyze each of these approaches, evaluate the alternatives, and arrive at a preferred solution. Specific applications may include alternative repair policies, the consideration of reliability and maintainability considerations in design, the evaluation of two or more off-the-shelf equipment items being considered for a single support application, and so on.

3. Aids in the evaluation of a given design configuration ("fixed" or "assumed") relative to the determination of specific logistic support resources requirements.

---

[4] Prime references for the LSA are (a) MIL-STD-1388-1A, "Logistic Support Analysis," Department of Defense, Washington, DC; and (b) MIL-STD-1388-2B, "DOD Requirements for a Logistic Support Analysis Record," Department of the Defense, Washington, DC. The first reference includes all of the key logistics tasks applicable in the system design process (e.g., definition of system utilization requirements, the conductance of design trade-off studies, the functional analysis, and formal design reviews). The second reference deals primarily with the data/documentation evolving through the accomplishment of the design-related tasks in the MIL-STD-1388-1A, and includes much of the same type of information that is generated through the MTA. These two references, revised often, are being "tailored" and applied in the development of large-scale defense systems.

Once design data are available, it is possible to determine the type and quantity of test and support equipment, spare/repair parts and associated inventory requirements, personnel quantities and skills, training requirements, technical data, facilities, computer resources, and so on. This part of the LSA basically includes the same objectives as the MTA discussed in the previous sections of this chapter.

4. Aids in the measurement and evaluation of an operating system in terms of its effectiveness and supportability in the user's environment. Given a fully operational capability, can the system be effectively and economically supported through its planned life cycle? Field data are collected, LSA reports are updated, and the results are analyzed and compared against the initially defined system requirements. Problem areas are identified, and modifications for system improvement are initiated where appropriate.

The LSA basically constitutes a level of activity being implemented on numerous programs, and the accomplishment of many of its objectives requires the completion of tasks such as maintainability allocation, maintainability analysis, maintainability prediction, and the MTA. Thus, maintainability program requirements must be closely integrated with those of logistics.

9.12   CONTINUOUS ACQUISITION AND LIFE-CYCLE SUPPORT (CALS)

Continuous acquisition and life-cycle support (CALS) refers to the application of computerized technology to the entire spectrum of logistics. More specifically, CALS is a combined Department of Defense and industry strategy that enables a transition from paper-intensive acquisition and logistic processes to a highly automated and integrated mode of operation. It focuses on the development and generation, access, management, maintenance, distribution, and use of technical data.[5]

The objectives of the CALS program are to improve the timeliness of data preparation and processing, improve quality relative to data content and consistency, and reduce cost. The approach, in the short term, is to convert existing data from a paper product to a digital format using electronic databases. With the advent of computer-aided technologies, these digital data files can easily be transmitted both throughout a project organization and between the customer, contractor, and supplier operations. At this stage, the most appropriate applications involve the processing of system design data (or "product definition" data), the generation and processing of logistic support analysis records (LSARs), the preparation of spare/

---

[5] CALS represents a concept that is being promoted primarily by the Department of Defense to improve quality, timeliness, and responsiveness in the acquisition of future system support requirements. Major government references covering CALS include (a) MIL-HDBK-59, *Military Handbook,* "Computer-Aided Acquisition and Logistic Support (CALS) Program Implementation Guide," Department of Defense, Washington, DC; and (b) MIL-STD-1840A, Military Standard, "Automated Interchange of Technical Information," Department of Defense, Washington, DC.

repair parts provisioning data and associated procurement packages, the development of training materials, and the development of technical publications (e.g., operating and maintenance instructions, installation and test procedures, overhaul instructions, and calibration procedures). In the long term, the objective is to provide a fully integrated technical information system, appropriately tied in with CAD and CAM processes. With the continued implementation of the CALS approach, it is anticipated that the following benefits will be realized:

1. The preparation of technical data in a digital format, using an appropriate integrated electronic database configuration, can be accomplished in less time, should be more accurate and of higher quality, and should eliminate unnecessary redundancies and the extensive amount of paper required in the past.
2. The use of a shared database, with information distributed through local area networks and to suppliers located worldwide, should provide a common baseline definition of the system (and components thereof) to many different organizational groups *concurrently*. Not only should the communications improve, but the time(s) allotted for the transfer of information should be reduced significantly.
3. The processing of logistics provisioning data (and associated procurement information) can be accomplished in an expedited manner. This, in turn, will result in shorter lead times for the acquisition of system components. Additionally, the costs of provisioning should be reduced significantly.

The CALS initiative is designed to revolutionize the capabilities within the logistics field in terms of data development, access, processing, and utilization. With the implementation of computer-aided technologies, it should be possible not only to improve the accuracy and quality of the data generated, but to process such data in a timely manner. While the CALS activity is basically directed toward Department of Defense program applications, the concepts are applicable to commercial systems as well.

With regard to maintainability program requirements, the data generated from the accomplishment of maintainability allocation, the maintainability analysis, prediction, and the MTA are used in the development of LSARs. These, in turn, are being converted to the digital format on many programs. As the technology continues to be applied, it may be necessary to translate the results from these maintainability program tasks directly into a digital database such that the appropriate information can be fed directly to the logistics organization in an expeditious manner.

9.13   MAINTENANCE PLAN

The *maintenance plan* (as differentiated from the "maintenance concept") is a detailed plan, usually prepared during the latter stages of the detail design and development phase, specifying the methods, procedures, and major resources required for the distribution and support of the system throughout its programmed life cycle. The plan, which constitutes an evolutionary development from the maintenance concept, is based on the results of the MTA. Specifically, it includes a description of the proposed levels of maintenance, the prime responsibilities for maintenance (producer, consumer, and supplier functions), system-level criteria for the development of the various elements of support (spare/repair parts, test and support equipment, personnel quantities and skills), effectiveness factors pertaining to the support capability, the overall distribution and flow of materials, the maintenance environment, and so on. The plan serves as a basis for system life-cycle support, it covers the transition from producer to customer support, and it includes the basic procedures pertaining to maintenance operations in the field. In addition to its importance as a maintainability program output, it is also a key element of an ILS program.

The maintenance plan may include a number of subplans, each covering a major segment of the overall system support capability and "tailored" to the specific system requirement. Although it is not all-inclusive, the material conveyed below is presented to provide an indication of the type of information that might be covered:

1. *Distribution and customer service plan.* For large systems, in particular, the process of transitioning from the production and/or construction phase to a full-scale user capability is often a formidable one. A rapid buildup of equipment in the field occurs while, at the same time, there is a lack of adequately trained personnel and not enough spare/repair parts, test and support equipment, technical data, software, and so on, available for adequate system support. There is usually a designated period of time after the system becomes "operational" when an "interim contractor support capability" is required. The purpose of this plan is to (a) define the requirements for the initial distribution of system components to the appropriate user sites, (b) describe the extent to which and the resources required for the interim period of contractor support of the system until the consumer is able to fully operate and maintain the system as planned (time and level of support, quantity of personnel and spares, etc.), and (c) describe the level of customer service and the supporting resources required after the system has reached full operational status.

2. *Test and support equipment plan.* The basic requirements for test equipment, support equipment, ground handling equipment and PME for calibration, tools, and so on are dependent on the results of the MTA. This plan, in conjunction with LSA data, identifies these requirements and describes the methods and procedures for acquiring the necessary items, verifying their adequacy through test and evaluation, and distributing them to the consumer. Specifically, this may include:

   a. A summary listing of all recommended test and support equipment, ground handling and calibration equipment, and significant tools for each level of maintenance (i.e., organizational, intermediate, depot, and/or supplier).

   b. A plan for the procurement and acquisition of new items to be designed and developed, tested, produced, and delivered for consumer use. This should reference test requirements specifications (TRSs), the results of *make-or-buy* decisions, supplier contractual requirements, quality assurance provisions, warranty requirements, and maintenance requirements for each item being acquired (e.g., calibration requirements).

c. A plan for the procurement and acquisition of common and standard off-the-shelf test and support equipment and associated accessories. This should reference TRSs, supplier contractual requirements, quality assurance provisions, warranty requirements, and the maintenance requirements for each item (e.g., calibration requirements). Special provisions associated with government furnished equipment (GFE) for defense systems should be covered.

d. A plan for the acquisition of computer resources (software) as required to support the newly developed and common/standard test and support equipment identified under items 2 and 3.

e. A plan for integration, test, and evaluation to ensure compatibility between the prime elements of the system and the elements of support (equipment, software, personnel, and procedures).

f. A plan (including a schedule) for the delivery of test and support equipment, associated software, data, and so on to the consumer, and the follow-on installation and checkout at each geographical location.

Figure 9.11 presents a simplified flow of the test and support equipment acquisition process.

3. *Spares and repair parts plan.* The initial requirements for spares, repair parts, and associated inventories are dependent on the results of the MTA. This plan, in conjunction with LSA data, describes the methods/procedures for the provisioning (item identification, cataloging, source coding), procurement, and acquisition of spare/repair parts, consumable materials, and inventory for both the early interim contractor support period and the long-term support of the system throughout its planned life cycle. This plan, covering both situations as applicable, may include:

a. A summary listing of significant spare/repair parts and the consumable materials required for each level of maintenance (i.e., organizational, intermediate, depot, and/or supplier).

b. A plan for the procurement and acquisition of new (nonstocklisted) spares and consumable materials for supporting the prime mission-oriented equipment, test and support equipment, training equipment, facilities, and software. Special supplier requirements dealing with manufacturing and test, packaging and handling, transportation, and related issues should be addressed.

c. A plan for the procurement and acquisition of common and standard off-the-shelf spares and consumable materials for the prime mission-oriented equipment, test and support equipment, training equipment, facilities, and software. Special provisions associated with GFE for defense systems should be covered.

d. Warehousing and accountability functions associated with the ongoing maintenance and support of the system. This includes the initial cataloging and

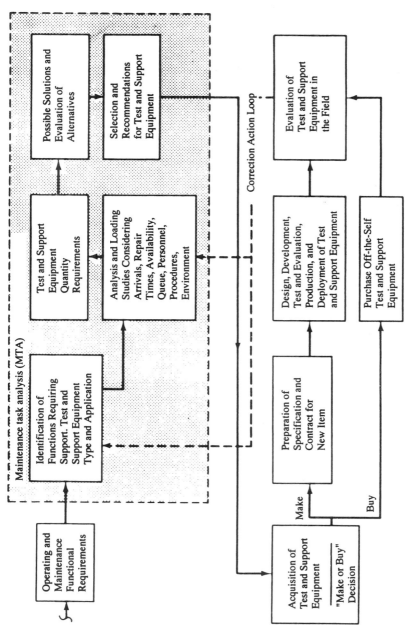

Figure 9.11. Test and support equipment development process.

stocking, inventory maintenance and control, procurement of replacement items, associated facility requirements, and disposition of condemned items and residual assets.

e. A plan for data collection, analysis, and the updating of spare/repair part (and consumable materials) demand factors necessary for improving the procurement cycles and reducing waste. This constitutes the necessary feedback process required for the true assessment of the system support capability and the updating of the LSA. Compatibility with the CALS requirements (as applicable) is essential.

f. A plan for the continuing acquisition of spare/repair parts beyond the production period for the expected life of the system. This is particularly critical when one is dependent on suppliers for spares deliveries.

While the spare/repair part and inventory requirements can be rather extensive for large-scale systems, a simplified flow of the overall process is presented in Figure 9.12.

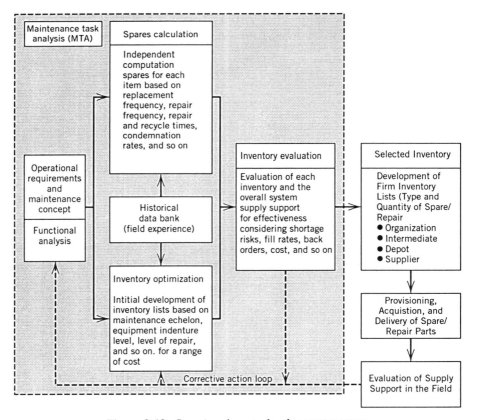

Figure 9.12. Spare/repair parts development process.

4. *Personnel and training plan.* The specific requirements for maintenance and support personnel in terms of quantities, skill levels, and job classifications by location are initially based on the results of the MTA. These requirements are compared with the quantities, skills, and job classifications currently within the user's organization, and the results lead to the development of a personnel training plan (i.e., the formal training necessary to bring the user personnel skills to the level specified for the system). While maintenance training is emphasized here, system operator training is sometimes included. The plan may cover:

a. The training of system operators: type of training, length, basic entry requirements, brief program/course outline, and output expectations. System operator requirements are often determined through system engineering and/or human factors program requirements.

b. The training of maintenance personnel for all levels: type of training, length, basic entry requirements, brief program/course outline, and output expectations.

c. Training equipment, devices, aids, simulators, computer resources, facilities, and data required to support operator and maintenance personnel training.

d. Proposed schedule for initial operator and maintenance personnel training, and for the accomplishment of replenishment training throughout the system life cycle (for replacement personnel).

5. *Packaging, handling, storage, and transportation plan.* This plan is developed to cover the basic distribution and transportation methods/procedures for the shipment of components from the producer to the consumer's operational site, for the shipment of the elements of support to the required consumer/customer location, and for the shipment of items requiring maintenance (i.e., from the operational site to the intermediate-level shop or to the manufacturer/depot). Specific transportation and handling requirements are initially derived from the maintenance concept, and supplemented through the development of the MTA and the LSA. This plan may include:

a. A summary listing of special categories of items requiring transportation.

b. Proposed mode(s) of transportation based on anticipated demand rates, available routing, weight and size of items, cost-effectiveness criteria, and so on. This shall consider both first destination requirements as dictated by material delivery schedules and recurring transportation requirements based on maintenance support needs.

c. Proposed methods of packaging items for shipment (types of container: reusable versus nonreusable, security measures, environmental protection provision).

d. Safety criteria, precautions, and provisions for handling, storage, and preservation of material.

In today's environment, logistics is an international business, and transportation requirements often involve the worldwide distribution and flow of materials. Systems and equipment are shipped across national boundaries, over land and sea, in the air, and so on. In planning for such, one needs to be conversant with international law, local transportation and customs requirements in different countries, and economic factors that relate to the transportation of goods and services across nationalistic boundaries.

6. *Facilities plan.* This plan is developed to identify all real property, plant, warehouse and/or special repair facilities, capital equipment, and utilities required to support system maintenance activities. Facility requirements initially stem from the maintenance concept, supplemented by the MTA and the LSA. This plan must contain sufficient qualitative and quantitative information to allow facility planners to:

   a. Identify requirements for facilities (location, environmental concerns, type of construction, space, and layout), and define utility requirements such as power and electricity, environmental controls, water, telephone, and so on.
   b. Identify capital equipment needs (material handling provisions, assembly and test processing equipment, warehousing shelves, etc.)
   c. Evaluate existing facilities and associated capital items, and assess adequacies/inadequacies in terms of meeting the need.
   d. Estimate the cost of facility acquisition (cost of constructing a new facility or modifying an existing facility), and the cost of capital equipment to meet the needs of the system.

The facilities plan should include appropriate criteria to ensure that facility design is completely compatible with the prime elements of the system, as well as with its support elements. The plan should include the necessary scheduling information to permit the proper and timely implementation of any required civil engineering activity.

7. *Technical data plan.* Technical data consist of system operating instructions, maintenance and servicing procedures, calibration procedures, overhaul instructions, change notices and data covering system modifications, and so on. The ultimate data format may vary from the traditional text material to a digital formatted database developed for compatibility with CALS requirements. This plan, which depends on the results of a MTA as an input, may include:

   a. A description of the technical data requirements for each level of activity (operator plus levels of maintenance) by system element. This may include the operating instructions for the system, maintenance procedures for Unit *B,* overhaul instructions for Assembly 1, and so on. It may be appropriate to present these data items in the form of a documentation tree, or to relate them in some form of a hierarchial structure, particularly if precedence exits.
   b. A schedule for the development of each significant data item.
   c. A plan for the verification and validation of operating and maintenance procedures.
   d. A plan for the preparation of change notices, and for the incorporation of changes/revisions to the technical manuals.

While both system operational and maintenance data are covered herein, the emphasis is on *maintenance* procedures and the data required to ensure that the system can be supported effectively and economically throughout its planned life cycle.

8. *Computer resources plan.* With the ever-increasing amount of software being utilized in modern systems, it is appropriate to include a computer resoures plan within the overall maintenance plan. This plan may cover:

a. Identification of all computer programs and software required for system support. This covers automated condition monitoring programs, maintenance diagnostic routines, information processing system involving logistics data, and so on.

b. Definition of computer language requirements, specifications, and compatibility requirements with existing programs (to include CALS).

c. Acquisition procedures for the development of and/or procurement of new software.

d. Software configuration management procedures and quality assurance provisions.

e. Software change procedures and change management.

f. Hardware (i.e., computers and associated accessories) required to interface with the software requirements.

The system maintenance plan constitutes a "vehicle" for the planning and implementation of a sustaining maintenance and support capability for the system throughout its planned life cycle. It includes a composite of the results from a number of tasks that may be accomplished for a given program, one of which is the MTA. While the information presented in this section may not completely cover all the elements of support peculiar to a specific system, it does indicate the nature of the material that may be included in such a plan.

QUESTIONS AND PROBLEMS

1. Describe in your own words the basic difference(s) between the maintainability analysis and the MTA. How does one relate to the other (if at all)?

2. When in the program life cycle is the MTA accomplished? Describe some of the "input" requirements for accomplishing a MTA. What type of information is provided by the MTA?

3. What is the scope of the MTA?

4. How does the maintenance concept influence the MTA? How does the MTA influence the maintenance concept? Provide a couple of examples for each answer.

5. Select a system of your choice and identify a particular component or element of that system. Assume that the component selected has failed, and develop a

MTA covering the troubleshooting, repair, and checkout (or condition verification) requirements.

6. Personnel requirements for the maintenance of a system are based on what factors? How are training requirements determined?

7. Spares and repair parts for system support are based on what factors? What considerations are necessary in the determination of inventory requirements?

8. Test and support equipment requirements are based on what factors? Identify some of the steps that should be followed in the selection and procurement of test equipment.

9. Why is the "calibration" of test equipment important? Select an item of your choice and show (in a flow process) "traceability" back to a primary standard.

10. What is the purpose of condition monitoring? How does it relate to the identification of maintenance resource requirements (address both corrective maintenance and preventive maintenance requirements)?

11. Transportation and handling requirements are based on what factors?

12. Maintenance facility requirements are based on what factors?

13. Maintenance task frequency requirements are based on what considerations?

14. What is the purpose of the "maintenance analysis control number?"

15. Why is the MTA important in terms of "maintainability in design?" Describe some of the benefits.

16. Assume that you have just completed a MTA on a system. What would you look for in assessing the extent to which maintainability characteristics have been incorporated in the design? Identify some criteria against which you would evaluate the results of the MTA.

17. What are the objectives of the LSA? How does the MTA relate to the LSA?

18. Describe what is meant by CALS. What are its objectives? How does the MTA relate to CALS?

19. What is the purpose of the maintenance plan? What is usually included?

20. Select a system of your choice (or an element thereof), and prepare a maintenance plan for that system.

21. What are some of the "cautions" that should be addressed when applying the MTA to a given system?

10

FORMAL DESIGN REVIEW

The system design evolves, in increasingly more and more detail, as progress is made from the original statement of need (or marketing description of consumer wants) to the development of design data, process and material specifications, selection of components, construction of a prototype model, and test and evaluation.[1] During this evolutionary process, it is necessary to pause at appropriate points to evaluate whether or not the design configuration at that time is responsive to the initially specified requirements. At these points, which represent natural evolutionary steps in design in terms of a "baseline" definition, formal design reviews are conducted. Figure 2.1 shows the scheduling of formal design reviews within the context of system life-cycle activities. These reviews are above and beyond the day-to-day informal design review and evaluation process described in Section 7.4.

Formal design reviews, scheduled at key points in the design process, usually include: (1) the conceptual design review at the end of the concept definition phase (this is also known as the system requirements review); (2) one or a series of *system design reviews* during preliminary design; (3) *equipment/software design reviews* during the detail design and development phase; and (4) the *critical design review* just prior to entering the production/construction phase. Figure 10.1 shows the relationships of these reviews with typical program activity.

Each of these reviews leads to the definition of a configuration "baseline" for the system design at a given phase in the life cycle. These baselines provide a single *common point of reference* for all members of the design team, program management, the suppliers, the customer, and so on. The first of these is called a "functional" baseline, which is described through the appropriate design data/documentation following the conceptual design review. It includes the results of feasibility studies, the definition of operational requirements and the maintenance concept, a top-level functional definition of the system, and applicable design criteria. These requirements are documented in the system Type "A" specification (refer to Figure

---

[1] Actually, the design process continues for the life of the system, through a process known as "sustaining engineering." This process is concerned with design problem solutions, changes, and improvements/ redesigns. Maintainability is a part of this continuing process.

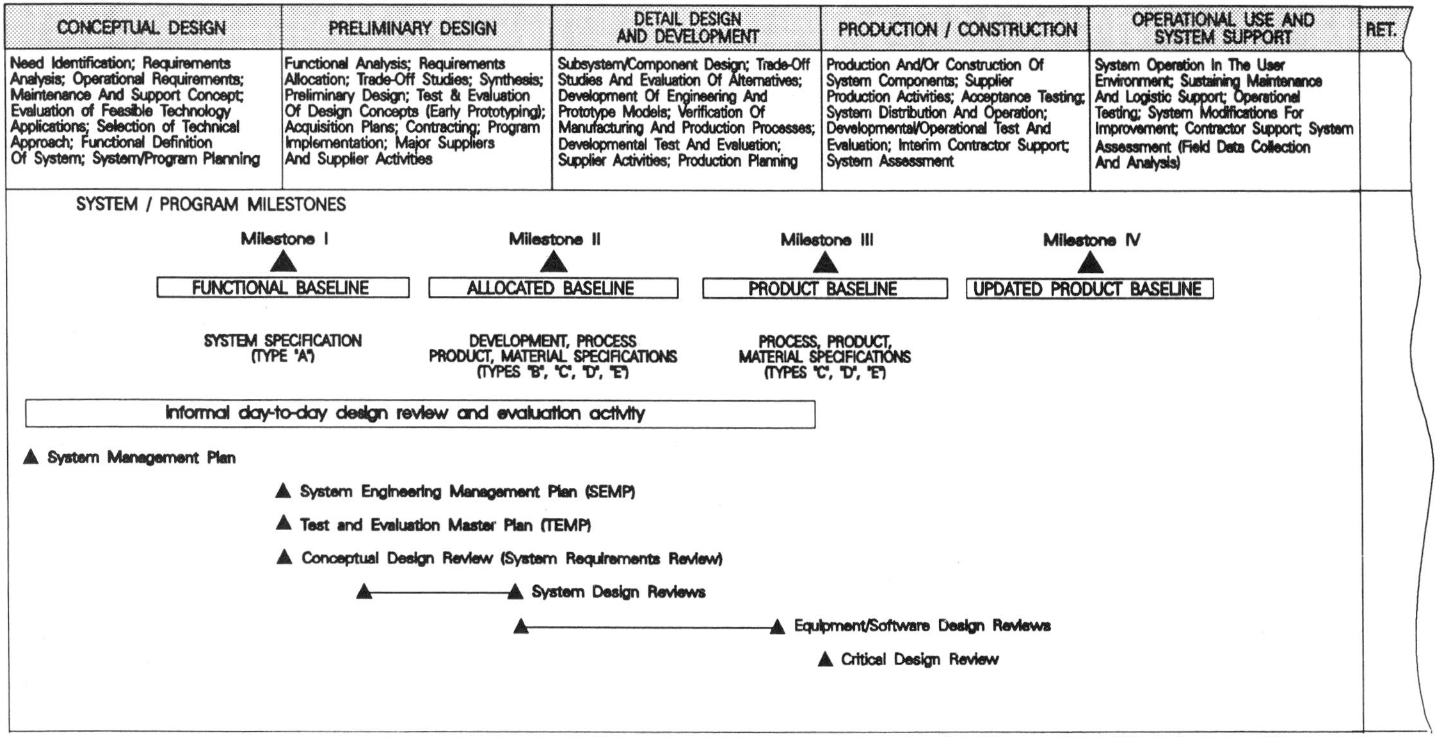

CONCEPTUAL DESIGN	PRELIMINARY DESIGN	DETAIL DESIGN AND DEVELOPMENT	PRODUCTION / CONSTRUCTION	OPERATIONAL USE AND SYSTEM SUPPORT	RET.
Need Identification; Requirements Analysis; Operational Requirements; Maintenance And Support Concept; Evaluation of Feasible Technology Applications; Selection of Technical Approach; Functional Definition Of System; System/Program Planning	Functional Analysis; Requirements Allocation; Trade-Off Studies; Synthesis; Preliminary Design; Test & Evaluation Of Design Concepts (Early Prototyping); Acquisition Plans; Contracting; Program Implementation; Major Suppliers And Supplier Activities	Subsystem/Component Design; Trade-Off Studies And Evaluation Of Alternatives; Development Of Engineering And Prototype Models; Verification Of Manufacturing And Production Processes; Developmental Test And Evaluation; Supplier Activities; Production Planning	Production And/Or Construction Of System Components; Supplier Production Activities; Acceptance Testing; System Distribution And Operation; Developmental/Operational Test And Evaluation; Interim Contractor Support; System Assessment	System Operation In The User Environment; Sustaining Maintenance And Logistic Support; Operational Testing; System Modifications For Improvement; Contractor Support; System Assessment (Field Data Collection And Analysis)	

SYSTEM / PROGRAM MILESTONES

Milestone I	Milestone II	Milestone III	Milestone IV

FUNCTIONAL BASELINE   ALLOCATED BASELINE   PRODUCT BASELINE   UPDATED PRODUCT BASELINE

SYSTEM SPECIFICATION (TYPE "A")

DEVELOPMENT, PROCESS PRODUCT, MATERIAL SPECIFICATIONS (TYPES "B", "C", "D", "E")

PROCESS, PRODUCT, MATERIAL SPECIFICATIONS (TYPES "C", "D", "E")

Informal day-to-day design review and evaluation activity

▲ System Engineering Management Plan (SEMP)

▲ Test and Evaluation Master Plan (TEMP)

▲ Conceptual Design Review (System Requirements Review)

▲ System Design Reviews

▲ Equipment/Software Design Reviews

▲ Critical Design Review

▲ System Management Plan

Figure 10.1. Design review and evaluation.

2.1). The second is called the "allocated" baseline, which primarily evolves from the results of the system design reviews. This includes the complete definition of the system in functional terms, the results from requirements allocation, a description of the major elements of the system, and so on. The third baseline is known as the "product" baseline, which evolves from the results of the equipment/software design reviews, and is updated as a result of the critical design review. This baseline constitutes a complete description of the production configuration, and serves as a frame of reference for all subsequent program activity.[2]

The design reviews serve as a "check and balance" on the design process to ensure that the developing design(s) remains responsive to the specified customer requirements. The main purpose of each design review is to ensure that the selected design meets these requirements in a cost-effective manner. Clearly, if the design review process is to be efficient, it must be well planned and managed, and the proper integration among the responsible design disciplines and supporting organizations must exist. Further, as design changes occur, they must first be "justified" and then incorporated expeditiously.

The purpose of this chapter is to describe the formal design review process, how formal design reviews are integrated with the developing design, the evaluation methods, and the procedure for implementing design changes.

10.1   DESIGN REVIEW REQUIREMENTS

The *formal design review* constitutes a coordinated activity (i.e., a structured meeting or a series of meetings) directed toward the final review and approval of a given design configuration, whether it be the overall system configuration, a subsystem, or an element of the system. While the informal day-to-day review process discussed in Section 7.4 covers specific aspects of the design, this coverage usually involves a series of independent fragmented efforts representing a variety of engineering disciplines. The purpose of the formal review is to provide a mechanism whereby ALL interested and responsible members of the design team can meet in a coordinated manner, communicate with each other, and agree on a recommended approach. The formal design review process usually includes the following steps:

1. A newly designed item, designated as being complete by the responsible design engineer(s), is selected for formal review and evaluation. The item may be the overall system configuration as an entity or a major element of the system, depending on the program phase and the category of review conducted.

2. A location, date, and time for the formal design review meeting are specified.

3. An agenda for the review is prepared, defining the scope and anticipated objectives of the review.

---

[2]"Functional," "allocated," and "product" baselines are identified in Figure 2.1 of this text, and are further delineated in the *Systems Engineering Management Guide,* Defense Systems Management College, Fort Belvoir, VA 22060.

4. A design review board (DRB) representing the organizational elements and the disciplines *affected* by the review is established. Representation from electrical engineering, mechanical engineering, structural engineering, reliability engineering, maintainability engineering, logistics engineering, manufacturing or production, component suppliers, management, and other appropriate organizations is included as applicable. This representation will, of course, vary from one review to the next. A well-qualified and unbiased chairman is selected to conduct the review.

5. The applicable specifications, drawings, parts lists, predictions and analysis results, trade-off study reports, and other data supporting the item being evaluated must be identified prior to the formal design review meeting, and made available during the meeting for reference purposes as required. Hopefully, each of the selected design review board members will familiarize himself/herself with the relevant data prior to the meeting.

6. Selected items of equipment (breadboards, service test models, prototypes), mock-ups, and/or software may be utilized to facilitate the review process. These items must, of course, be identified early.

7. Reporting requirements and the procedures for accomplishing the necessary follow-up actions(s) stemming from design review recommendations must be defined. Responsibilities and action-item time limitations must be established.

8. Funding sources for the necessary preparations, for conducting the formal design review meetings, and for the subsequent processing of outstanding recommendations must be identified.

The formal design review meeting generally includes a presentation (or a series of presentations) on the item being evaluated, by the responsible design engineer(s), to the selected DRB members. This presentation should cover the proposed design configuration, along with the results of trade-off studies and analyses that support the design approach. The objective is to summarize what had been established earlier through the informal day-to-day design activity. If the DRB members are adequately prepared, this process can be accomplished in an efficient manner.

The formal design review must be well organized and firmly controlled by the DRB chairman. Design review meetings should be brief and to the point, objective in terms of allowing for *positive* contributions, and must not be allowed to drift away from the topics on the agenda. Attendance should be limited to those who have a direct interest in and can contribute to the subject matter being presented. Design specialists who participate should be authorized to speak and make decisions concerning their area of speciality. Finally, the design review activity must make provisions for the identification, recording, scheduling, and monitoring of corrective actions. Specific responsibility for follow-up action must be designated by the DRB chairman.

With the conductance of formal design review meetings, a number of purposes are served:

1. The formal design review meeting provides a forum for communications across the board! The necessary coordination and integration are not adequately

accomplished through the informal day-to-day review process, even with the availability of computerized technology. The "person-to-person" contact is required.

2. It provides for the definition of a common configuration baseline for all project personnel; that is, everyone involved in the design process must work from the *same* baseline. The responsible design engineer(s) is given the opportunity to explain his/her design approach, and representatives from the various supporting disciplines are provided the opportunity to learn of the designer's problems. This, in turn, fosters better understanding between design and support personnel.

3. It provides a means for solving outstanding interface problems, and it promotes the assurance that all elements of the system are compatible. Those conflicts that have not been resolved through the informal day-to-day review process are addressed. Also, those disciplines not properly represented through earlier activity are provided the opportunity to be heard.

4. It provides a formalized check (i.e., audit) of the proposed system/product design configuration with respect to specification and contractual requirements. Areas of noncompliance are noted, and corrective action is initiated as appropriate.

5. It provides a formal report of major design decisions that have been made and the reasons for making them. Design documentation, analyses, predictions, and trade-off study reports that support these decisions are properly recorded.

The conductance of formal design review meetings tends to increase the probability of mature design, as well as the incorporation of the latest design techniques where appropriate. Group reviews may lead to the identification of new ideas, the application of simpler processes, and the realization of cost savings. A good "productive" formal design review activity can be very beneficial. Not only can it reduce the producer's risk relative to meeting specification and contractual requirements, but the results often lead to an improvement in the producer's method of operation.

As stated earlier, formal design review meetings are generally scheduled prior to each major evolutionary step in the design process; for example, after the definition of a functional baseline, but prior to the establishment of an allocated baseline. Although the quantity and type of design reviews scheduled may vary from program to program, four basic types are easily identifiable and common for many programs. They are the conceptual design review, the system design review, the equipment or software design review, and the critical design review. The relative time phasing of these reviews is illustrated in Figure 10.1.

10.2   CONCEPTUAL DESIGN REVIEW [3]

The *conceptual design review* (or the system *requirements review*) is usually scheduled toward the end of the conceptual design and prior to entering the preliminary system design phase of the program (preferable no longer than 1–2 months after

---

[3] This and subsequent sections have been adapted from Blanchard, B. S., *System Engineering Management*, John Wiley & Sons, Inc., 1991, Chapter 5.

program start). The objective is to review and evaluate the functional baseline for the system, and the material to be covered through this review should include the following:[4]

1. Feasibility analysis (the results of technology assessments and early trade-off studies justifying the system design approach being proposed)
2. System operational requirements (refer to Section 5.2)
3. System maintenance concept (refer to Chapter 5)
4. Functional analysis (top-level system definition)
5. Significant design criteria for the system (e.g., reliability factors, maintainability factors, and logistic factors)
6. Applicable effectiveness figures-of-merit (FOMs) and technical performance measures (TPMs)
7. System specification (to include maintainability design requirements at the system level)
8. System engineering management plan (early maintainability program planning information is often included herein)
9. Test and evaluation master plan (TEMP)
10. System design documentation (electronic data, layout drawings, sketches, parts lists, selected supplier components data)

The conceptual design review deals primarily with top *system-level requirements,* and the results constitute the basis for follow-on preliminary system design and development activity. Participation in this formal review should include selected representation from both the consumer and producer organizations. Consumer representation should involve not only those personnel who are responsible for the acquisition of the system (i.e., contracting and procurement), but also those who will ultimately be responsible for the operation and support of the system in the field. Individuals with experience in operations and maintenance should participate in this review. On the producer side of the spectrum, those lead engineers responsible for *system* design should participate, along with representation from various design disciplines and production (as necessary). A matrix, such as the one depicted in Figure 10.2, facilitates the delineation of specialty disciplines which are likely to have the greatest impact on the relevant system technical performance measures. Conducting such an exercise may provide additional insight into which design disciplines need to be involved when reviewing the system design configuration.

---

[4] It is recognized that some of these requirements may not adequately be defined during the conceptual design phase, and that the review of such may have to be accomplished later. However, in promoting the desired general approach described herein, maximum effort should be made to complete these requirements early, even though changes may be necessary as system design progresses. The object is to encourage (or "force") early system definition.

Technical Performance Measures (TPMs) / Engineering Design Functions	Aeronautical Engineering	Components Engineering	Cost Engineering	Electrical Engineering	Human Factors Engineering	Logistics Engineering	Maintainability Engineering	Manufacturing Engineering	Materials Engineering	Mechanical Engineering	Reliability Engineering	Structural Engineering	Systems Engineering
Availability (90%)	H	L	L	M	M	H	M	L	M	M	M	M	H
Diagnostics (95%)	L	M	L	H	L	M	H	M	M	H	M	L	M
Interchangeability (99%)	M	H	M	H	M	H	H	H	M	H	H	M	M
Life Cycle Cost ($350K / unit)	M	M	H	M	M	H	H	L	M	M	H	M	H
$\overline{\text{Mct}}$ (30 min.)	L	L	L	M	M	H	H	M	M	M	M	M	M
MDT (24 hrs.)	L	M	M	L	L	H	M	M	L	L	M	L	H
MMH/ OH (15)	L	L	M	L	M	M	H	L	L	L	M	L	H
MTBF (300 hrs.)	L	H	L	M	L	L	M	H	H	M	H	M	M
MTBM (250 hrs.)	L	L	L	L	L	M	H	L	L	L	M	L	H
Personnel Skill Levels	M	L	M	M	H	M	H	L	L	L	L	L	H
Size (150 ft. by 75 ft.)	H	H	M	M	M	M	M	H	H	H	M	H	M
Speed (450 mph.)	H	L	L	L	L	L	L	L	L	L	L	M	H
System Effectiveness (80%)	M	L	L	M	L	M	M	L	L	M	M	M	H
Weight (150K pounds)	H	H	M	M	M	M	M	H	H	H	L	H	M

H= high interest; M= medium interest; L= low interest

Figure 10.2. Relationship between technical performance measures and responsible design disciplines.

In summary, the conceptual design review is extremely important for all concerned, as it represents the first opportunity for formal communication relative to system requirements from the top down! It can provide an excellent baseline for all subsequent design effort. Unfortunately, for many projects in the past, the conductance of conceptual design reviews has not been readily evident. Further, if such a review were conducted, the results were not always made available to the responsible design personnel assigned to the project. This, in turn, has resulted in a series of "panic" efforts conducted later and in somewhat of a vacuum, and not well coordinated or integrated. Thus, with maintainability and other system objectives in mind, it is essential that a good functional baseline for the system be defined and properly evaluated through the conductance of an effective conceptual design review.

10.3   SYSTEM DESIGN REVIEWS

System design reviews are generally scheduled during the preliminary design phase when functional requirements and allocations are defined, preliminary design layouts and detailed specifications are prepared, system-level trade-off studies are conducted, and so on (refer to Figure 10.1). These reviews are oriented to the overall system configuration, in lieu of individual equipment items, software, and other components of the system. As the design evolves, it is important to ensure that the requirements described in the system specification are maintained. There may be one or more formal reviews scheduled depending on the size of the system and the design complexity. System design reviews cover a variety of topics, a few of which are noted:

1. Functional analysis and the allocation of requirements (beyond what is covered through the conceptual design review).
2. Development, process, product, and material specifications as applicable (Types "B"–"E").
3. Design data defining the overall system (electronic data, layouts, drawings, parts/material lists, supplier data).
4. Analyses, reports, predictions, trade-off studies, and related design documentation. This includes material that has been prepared in support of the proposed design configuration, and analyses/predictions that provide an assessment of what is being proposed. Reliability and maintainability predictions, logistic support analysis data, and so on, are included.
5. Assessment of the proposed system design configuration in terms of applicable TPMs.
6. Individual program/design plans (e.g., reliability and maintainability program plans, human factors program plan, and the integrated logistic support plan).

Participation in the system design reviews should include representation from both the consumer and producer organizations, as well as from major suppliers involved in the early phases of the system life cycle.

10.4   EQUIPMENT/SOFTWARE DESIGN REVIEWS

Formal design reviews covering equipment, software, and other components of the system are scheduled during the detail design and development phase of the life cycle. These reviews, usually oriented to a particular item, include coverage of the following:

1. Process, product, and material specifications (Types "C"–and "E"— beyond what is covered through the system design reviews).
2. Design data defining major subsystems, equipment, software, and other elements of the system as applicable (electronic data, assembly drawings, specification control drawings, construction drawings, installation drawings, logic diagrams, schematic diagrams, material and detailed parts lists, etc.).
3. Analyses, reports, predictions, trade-off studies, and other related design documentation as required in support of the proposed design configuration and/or for assessment purposes. Reliability and maintainability predictions, human factors task analysis, logistic support analysis data, and so on, are included.
4. Assessment of the proposed system design configuration in terms of the applicable technical performance measures (TPMs). An ongoing review and evaluation are required to ensure that these system-level requirements are maintained throughout the various stages of detail design and development.
5. Engineering breadboards, laboratory models, service test models, mock-ups, and prototype models used to support the specific design configuration being evaluated.
6. Supplier data covering specific components of the system as applicable (drawings, material and parts lists, analysis, prediction reports, etc.).

Participation in these formal reviews should include representation from the consumer (i.e., customer), producer (i.e., contractor), and supplier organizations.

10.5   CRITICAL DESIGN REVIEW

The *critical design review* is generally scheduled after the completion of detail design, but prior to the release of firm design data for production or construction. Design is essentially "frozen" at this point, and the proposed configuration is evaluated in terms of adequacy and producibility. The critical design review may address topics such as the following:

1. A complete set of final design documentation covering the system and its components (computer-aided design data, manufacturing drawings, material and parts lists, supplier component parts data, drawing change notices, etc.)
2. Analyses, predictions, trade-off study reports, test and evaluation results, and related design documentation (final reliability and maintainability predictions, maintenance task analysis, human factors and safety analyses, logistic support analysis records, maintainability demonstration reports, etc.)
3. Assessment of the final design configuration (i.e., the product baseline) in terms of applicable technical performance measures (TPMs)
4. A detailed production/construction plan (description of proposed manufacturing methods, fabrication processes, quality control provisions, supplier requirements, material flow and distribution requirements, schedules, etc.)
5. A final ILSP covering the proposed life-cycle maintenance and support of the system throughout the consumer utilization phase

The results of the critical design review describe the final system/product configuration baseline prior to entering into the production and/or construction phase. This review constitutes the last in a series of progressive evaluation efforts, reflecting design and development from a historical perspective and showing growth and maturity in design as the engineering project evolved. It is important to view the design review process in *total,* and to provide an overall evaluation of certain designated system attributes as the project progresses, particularly since close continuity is required between the various reviews. An example of designated system attributes that should be assessed on a continuing basis is presented in Figure 10.3.

10.6 SYSTEM DESIGN CHANGES AND MODIFICATIONS

The objective thus far has been to develop a system on a progressive basis and to establish a firm configuration baseline through the formal review and evaluation process. In essence, the results from the conceptual design review lead to the definition of system-level requirements, the results from the system design reviews constitute a more in-depth description of the system packaging concepts, and so on. As we progress through the series of design reviews described in Sections 10.2–10.5, the system definition becomes more refined, and the configuration baseline (updated from one review to the next) is established. This baseline, which constitutes a single point of reference for all individuals who are involved in the design process, is critical from the standpoint of meeting system objectives.

Once a configuration has been established, it is equally important that any variations, or changes, with respect to that baseline be tightly controlled. It is certainly not anticipated that a given baseline will remain as such forever, particularly during

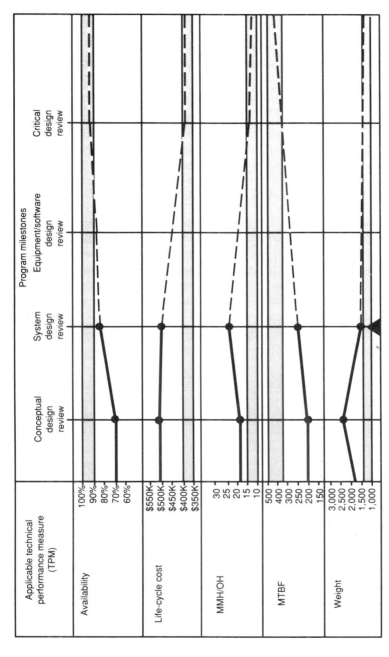

Figure 10.3. Typical parameter measurement and evaluation at a system design review (sample). (*Source:* Blanchard, B. S., *System Engineering Management*, John Wiley & Sons, N.Y., 1991, p. 182).

the early stages of system development. However, in evolving from one design configuration to the next, it is important that all changes be recorded and documented carefully in terms of their possible impact on the initially specified system requirements. The process of configuration identification, the control of changes, and maintaining the integrity and continuity of design are accomplished through *configuration management* (CM).[5]

In the defense sector, configuration management is often related to the concept of "baseline management." Referring to Figure 2.1, functional, allocated, and product baselines are established as the system development process evolves. These baselines are described through a family of specifications (Types "A"–"E"), electronic data, drawings and parts lists, reports, and related documentation. The formal design review process provides the necessary authentication of these baseline configurations, and the *configuration identification* (CI) function is accomplished. The CI relates to a particular baseline, while the *configuration status accounting* (CSA) function is a "management information system that provides traceability of configuration baselines and changes thereto, and facilitates the effective implementation of changes."[6] The CSA includes the documentation in evolving from one configuration baseline to the next.

Proposed design changes, or proposed changes to a given baseline (i.e., a CI design), may be initiated from any one of a number of sources during any phase in the overall system life cycle. Such changes, prepared in the form of an Engineering Change Proposal (ECP), may be classified as follows:[7]

1. *Class 1 changes*—design changes that will affect form, fit, and/or function (e.g., changes that will impact system performance, reliability, maintainability, safety, supportability, life-cycle costs, and/or any other system specification requirement)
2. *Class 2 changes*—design changes that are relatively minor in nature and which will not affect system specification requirements (e.g., changes covering material substitutions, documentation clarification, drawing nomenclature, and producer deficiencies)

Changes may be categorized as "emergency," "urgent," or "routine," depending on priority and on the criticality of the change.

A simplified version of the system control procedure is illustrated in Figure 10.4. Proposed changes to a given baseline may be initiated during any phase of system

---

[5]*Configuration management* (CM) constitutes the process that identifies the functional and physical characteristics of an item during its life cycle, controls changes to those characteristics, and records and reports change processing and implementation status. CM, as it is defined in the defense sector, is described in MIL-STD-483, Military Standard, "Configuration Management Practices for Systems, Equipments, Munitions, and Computer Programs," Department of Defense, Washington, DC.

[6]Defense Systems Management College (DSMC), *Systems Engineering Management Guide,* Fort Belvoir, VA 22060, Chapter 11.

[7]Engineering changes are covered in DOD-STD-480A, DOD Standard, "Configuration Control—Engineering Changes, Deviations, and Waivers," Department of Defense, Washington, DC.

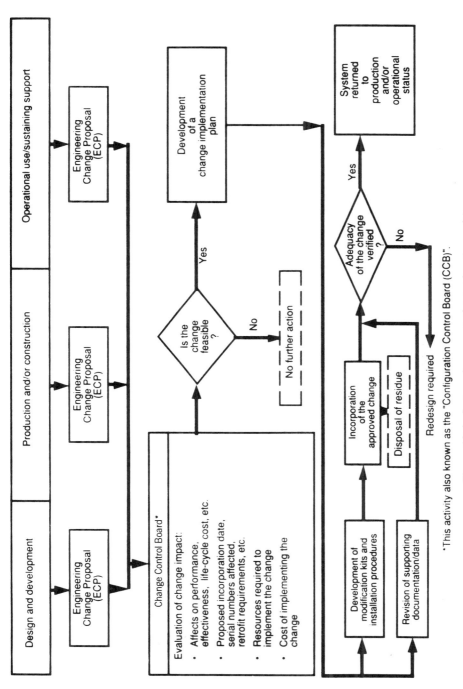

Figure 10.4. System change control procedure (example).

*This activity also known as the "Configuration Control Board (CCB)".

385

development, production, and/or operational use. Each proposal change is submitted for review, evaluation, and approval. In general, each ECP should cover the following issues: [8]

1. A statement of the problem and a description of the proposed change.
2. A brief description of alternatives that have been considered in responding to the need.
3. An analysis showing how the change will solve the problem.
4. An analysis showing how the change will impact system performance, effectiveness factors (reliability and maintainability requirements), packaging concepts, safety, elements of logistic support, life-cycle cost, and so on. What are the impacts (if any) on system specification requirements? What is the effect on life-cycle cost?
5. An analysis to assure that the proposed solution will not cause the introduction of new problems.
6. A preliminary plan for incorporating the change; that is, proposed date of incorporation, serial numbers affected, retrofit requirements, and verification test approach (as applicable).
7. A description of the resources required to implement the change.
8. An estimate of the costs associated with implementing the change.
9. A statement covering the impact on the system if the proposed change is *not* implemented; that is, an identification of the possible risks associated with a "do-nothing" decision.

Referring to Figure 10.4, ECPs are processed through the Change Control Board (sometimes known as the Configuration Control Board or the CCB) for review and evaluation. The CCB should function in a manner similar to the DRB discussed in Section 10.1. Board representation should cover those design disciplines impacted by the change, including customer and supplier representation as necessary. Not only is it necessary to review and evaluate the original design, but it is important to ensure that all proposed design changes are handled in a similar manner. On occasion, when project schedules are "tight," the designer will generate data just to have something available for the record, while the *real* design configuration will be reflected through the "change process." Although this is not a preferred practice, it does occur in a number of instances when the objective is to save time. In any event, the review of design changes must be treated with the same degree of importance as is specified for the formal design review.

Upon completion of the formal design change review by the CCB, approved ECPs will be supported with the development of a plan for incorporating the

---

[8] In many organizations, the procedures related to configuration management and change control are a little more complex than what is presented here. The procedure may involve engineering change requests (ECRs), design revision notices (DRNs), interface control documents (ICDs), and so on. The objective here is to present a *simplified* approach, providing a basic understanding of the importance of change control in the system design and development process.

change(s) in the system. This plan should include coverage of not only the modifications required for the prime equipment, but the modifications associated with test and support equipment, spares and repair parts, facilities, software, and technical documentation. All elements of the system must be addressed on an integrated basis.

The actual incorporation of changes to the system is accomplished using a variety of approaches depending on when the change is to be implemented. The time of implementation is a function of priority and/or criticality. Emergency or urgent changes may require immediate action, while routine changes may be grouped and incorporated at some convenient point in time later on. Approved changes initiated during system design and development, prior to the availability of any hardware or other physical components, may be incorporated through the preparation of design change notices (DCNs), or the equivalent, attached to the applicable drawings/documentation covering those areas of design affected by the change. As the project progresses, these "paper" (or database) changes will be reflected in the new design configuration.

In the event that changes are initiated during the production/construction phase, when multiple quantities of identical items are being produced, a designated serial-numbered item needs to be identified to indicate effectively that the change will be incorporated on the production line in Serial Number "X" and so on. This should ensure that all applicable items scheduled to be produced in the future will automatically reflect the updated configuration.

For those system components that are already in use, changes may be incorporated through the installation of a modification kit in the field at the consumer's operational site. Such kits are installed, and the system is tested, to verify the adequacy of the change. At the same time, the system support capability (e.g., test equipment, spares, and technical data) needs to be upgraded for compatibility with the prime mission-oriented segments of the system. Optimally, the installation process should take place at a time when the system is not in demand or being utilized in the performance of a function or mission.

This overall process is illustrated in Figure 10.4. With the incorporation of validated changes, the system configuration is updated and a new baseline is established. In situations where the adequacy of the change is not verified, some additional redesign may be required.

10.7 SUMMARY

This chapter primarily addresses the basic review, evaluation, and feedback process illustrated in Figure 10.1. This process, which is critical with regard to the objectives described throughout this text, must be tailored to the specific system development effort and must be properly controlled! An ongoing measurement and evaluation activity is essential, and must be initiated from the beginning. Performing a one-time "downstream" review and evaluation after the system has been produced and is in operational use may be costly in terms of possible modifications for corrective action. Also, the incorporation of design changes on a continuing basis without the proper controls may be costly from the standpoint of system maintainability and support. In essence, there must be a well-planned program approach, with the proper controls, in order to ensure a total integrated system configuration in the end.

QUESTIONS AND PROBLEMS

1. What is meant by "checks and balances?" Describe several examples as they are used in the design process.

2. What are the products of:

   a. The conceptual design review?
   b. The system design review?
   c. The equipment/software review?
   d. The critical design review?

3. How is design review and evaluation accomplished? Why is it important relative to meeting maintainability engineering objectives?

4. What is included in the establishment of a "functional" baseline? "Allocated" baseline? "Product" baseline? Why is baseline management important?

5. Identify some of the benefits derived through formal design review. Describe some of the concerns.

6. When developing an agenda in preparing for a formal design review, what considerations must be addressed in the selection of items to be covered in the review process? How are review and evaluation criteria identified? Describe the steps and resources required in preparing for the design review.

7. How are TPMs considered in the design review process?

8. In the event that a deficiency is identified during design review, what steps are required for corrective action?

9. How are design changes initiated? How are priorities established?

10. How are design changes implemented? Identify steps involved in system modification?

11. Describe the functions of the CCB.

12. What is Configuration Management (CM)? Define Configuration Identification (CI) and Configuration Status Accounting (CSA).

13. How does CM relate to maintainability engineering? Why is it important? What is likely to occur if CM practices are not followed?

11

MAINTAINABILITY TEST AND DEMONSTRATION

Up to this point, the text material has included the specification of maintainability requirements, the establishment of design criteria, the accomplishment of maintainability allocation and prediction, the accomplishment of maintainability analysis and trade-off studies, the completion of the MTA, and the design review. Through the accomplishment of these activities,the maintainability engineer has been able to initially specify design objectives and later evaluate a given design configuration relative to compliance with these objectives. The evaluation process thus far has been analytical in nature, providing some degree of confidence that both qualitative and quantitative maintainability requirements will be met. Although this analytical process fulfills a particular need, it is limited because it does not reflect practical "hands-on" experience with the applicable system and its components. A more realistic evaluation of whether or not the system complies with the initially specified customer requirements is necessary, and is accomplished through actual maintainability demonstrations involving the prime equipment elements of the system, along with supporting software, personnel, the elements of system support (spares, test equipment, technical data), and so on.

This chapter addresses the requirements for and the procedures associated with maintainability demonstration. As an introduction, it is first necessary to briefly discuss system test and evaluation requirements in general; then, it is appropriate to show how maintainability test requirements fit in. Test conditions, the test preparation phase, conductance of the formal demonstration test, and test reporting requirements are discussed.

11.1 SYSTEM/PRODUCT TEST REQUIREMENTS

A true test (that which is relevant from the standpoint of assessing maintainability characteristics in design) constitutes an evaluation of the system being utilized in an operating environment by the consumer, subjected to actual use conditions. For example, an equipment designed for aircraft use should be tested on board an aircraft flying in a typical operational profile. User personnel should accomplish operator and maintenance functions with the designated field test and support equipment, operating and maintenance procedures, and so on. In such a situation, actual operational and maintenance experience in a realistic environment can be recorded and subsequently evaluated to reflect a true representation of the system design for maintainability. A demonstration of this type can best be accomplished by the user during standard operations with the appropriate supporting resources (i.e., during the system utilization phase).

Idealistically it may be desirable to wait until the system is in operational use before accomplishing an evaluation of the effectiveness, maintainability, and other characteristics of that system. However, this is not practical from the standpoint of allowing for corrective action in an economical manner. In the event that the evaluation indicates a "noncompliant" situation (i.e., the system as presently designed will not meet the specified requirements), corrective action should be initiated as early as possible in the system life cycle so that changes can be incorporated on a more economical basis. Accomplishing corrective action after the system is in operational use can result in extensive modification programs, which can be quite costly.

Thus, in the development of a test and evaluation program, the objective is to accomplish a realistic assessment of the system and its components as early as practicable. One must review the requirements included in the system specification (Type "A"), identify the key technical performance measures, and then determine how the system is to be evaluated to ensure that these requirements will be met. Some characteristics may be verified early in a program through the appropriate use of analytical methods; the use of engineering or laboratory test models during preliminary design may provide additional information critical to system evaluation; and the utilization of preproduction prototype models during the detail design and development phase will, of course, result in a more accurate assessment of system characteristics. System test and evaluation *requirements* are defined during the conceptual design phase (as the technical performance measures are being identified for the system), and the subsequent test and evaluation process is accomplished on an evolutionary basis employing a combination of resources.

Figure 11.1 illustrates the test and evaluation approach. For the purposes of discussion, the levels of test are identified by categories. As design and development progresses, the system configuration becomes well defined, testing becomes more sophisticated and the effectiveness of evaluation increases, and the costs are higher. On the other hand, there are some limitations as to what can be accomplished earlier.

1. *Analytical evaluation.* With the advent of computer simulation, the design engineer can develop a three-dimensional solids model, a wire model, a graphical presentation of different segments of the system, and so on. One can view the system and its components in a hierarchal manner, can simulate the human being and show various human–equipment relationships, and can simulate operating and

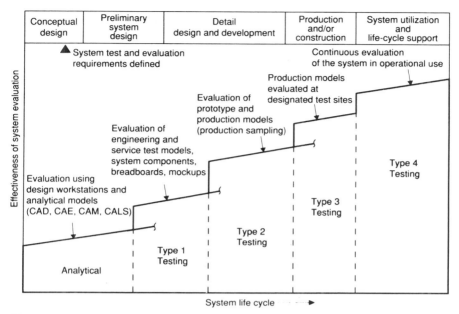

Figure 11.1. Stages of system evaluation during the life cycle. (*Source:* Blanchard, B. S., *System Engineering Management,* John Wiley & Sons, N.Y., 1991, p. 54).

maintenance functional activities (evaluate the accessibility to faulty components, simulate the removal and replacement of items, etc.). While the use of physical mock-ups and actual equipment were required in the past, a great deal can now be accomplished through the use of computer simulation. The evaluation of the system using these and comparable analytical methods provides an opportunity for an early assessment of maintainability characteristics in design. Not that one can completely satisfy demonstration requirements at this stage in the life cycle, but a great deal of insight as to the characteristics of system design can be gained.

2. *Type 1 testing.* During preliminary design, engineering models, service test models, bench test models, and so on are often built with the intent of verifying certain performance and physical characteristics of the system and/or its components. These models operate functionally (electrically or mechanically), but do not by any means represent production entities. Tests may involve equipment operational and maintenance actions which are directly comparable to tasks that will later be performed in the user's environment (e.g., measuring a performance parameter, accomplishing a remove–replace action, accomplishing a service requirement, etc.). Further, component life testing for the purposes of determining reliability may be initiated within this category. Although these tests are not formal demonstrations in a true operational environment, information pertinent to maintainability and supportability characteristics can be derived and used as an input to the accomplishment of maintainability prediction and/or the MTA. Such testing is generally

performed in the producer/supplier's facility by engineering laboratory technicians using "jury-rigged" test equipment and engineering notes for test procedures. Maintainability characteristics in design can be assessed to a limited extent, and it is during this initial phase of testing that changes in design can be incorporated on a minimum-cost basis.

3. *Type 2 testing.* Formal tests and demonstrations are accomplished during the latter part of the detail design and development phase when preproduction prototype hardware and software are available. Prototype items are similar to production items (i.e., that which will be distributed for operational use), but are not necessarily fully qualified at this point in time.[1] A test program may consist of a series of individual tests, tailored to the need, to include:

a. *Environmental qualification*—temperature cycling, shock and vibration, humidity, wind, salt spray, dust and sand, acoustic noise, explosion proofing, and electromagnetic interference. ESS can be included in this category. These factors are oriented to what the system will be subjected to during operation, maintenance, and transportation and handling functions.

b. *Reliability qualification*—tests accomplished on the system to determine the true MTBF and MTBM. The results are compared with the initially specified requirement, and are used to update the MTA.

c. *Maintainability demonstration*—tests accomplished on the system to determine the values of $\overline{M}$, $\overline{M}ct$, $\overline{M}pt$, $\overline{M}_{max}$, and MMH/OH. In addition, maintenance tasks, task times and sequences, diagnostic provisions, degree of testability, prime equipment–test equipment interfaces, maintenance personnel quantities and skills, maintenance procedures, and maintenance facilities are verified to varying degrees. The elements of logistic support are initially evaluated on an individual basis.

d. *Support equipment compatibility tests*—tests often accomplished to verify the compatibility between the prime equipment, test and support equipment, and transportation and handling equipment.

e. *Personnel test and evaluation*—tests often accomplished to verify the relationships between man and equipment, the personnel quantities and skill levels required, and training needs. Both operator and maintenance tasks are evaluated.

f. *Technical data verification*—the verification of operational and maintenance procedures.

g. *Software compatibility tests*—tests often accomplished to ensure that all system software meets performance requirements, is reliable, is compatible with other elements of the system, and meets all quality assurance provisions. Software for both operational and maintenance functions is evaluated.

---

[1] Qualified equipment refers to a production configuration which has been verified through the successful *completion* of environmental qualification tests (e.g., temperature cycling, shock, vibration), reliability qualification, maintainability demonstration, and compatiblity tests. Type 2 testing primarily refers to that activity associated with the qualification of a system.

The ideal situation is to plan and schedule these individual tests such that they can be accomplished on an integrated basis as "one" overall test. The intent is to provide the proper level of evaluation, consistent with the need, and eliminate redundancy and excessive cost. In some instances, it may be possible to acquire information from one test that can be used in fulfilling the requirements of another test. For example, when components fail during a reliability test, the subsequent repair actions can be evaluated from a maintainability standpoint. This, in turn, could lead to a reduction in the sample size required for maintainability demonstration. In any event, proper test planning is essential.

Another aspect of testing in this category includes "production sampling tests" when multiple quantities of an item are produced. The tests identified above basically "qualify" the system; that is, the configuration of the system components meets the requirements for production and for operational use. However, once an item is initially qualified, some assurance must be provided that all subsequent replicas of that item are equally qualified; thus, in a multiple-quantity situation, samples are selected from the production line and tested.[2]

Production sampling tests may cover certain critical performance characteristics, such as reliability, maintainability, or any other designated parameter which may significantly vary from one serial-numbered item to the next, or may vary as a result of the production process itself. Samples may be selected on the basis of a percentage of the total equipments produced, or may tie in with x number of items in a given calendar time period. This depends on the peculiarities of the system and the complexities of the production process. From production sampling tests, one can often measure system growth (or degradation) throughout the production/construction phase. An example of this pertains to the results of reliability testing where a growth in the measured MTBF may be realized as one proceeds through the overall production process. Of course, any changes in reliability will have an impact on maintainability.

Type 2 tests are generally performed in the producer/supplier's facility by personnel at that facility. Test and support equipment designated for operational use and preliminary operating and maintenance manuals should be utilized where possible. User personnel often observe and/or participate in the testing activities.

4. *Type 3 testing.* Formal tests and demonstrations, conducted after initial system qualification and prior to completion of the production process, may be accomplished at a designated test site by user personnel.[3] Test and support equipment, spares, procedures, software, and personnel designated for the consumer utilization phase are employed. Testing is generally continuous, accomplished over an extended period of time, and covers the evaluation of the various elements of the system as they are scheduled through a series of simulated operational exercises.

This is the first time that all elements of the system (i.e., prime equipment,

---

[2] These tests are in addition to the normal performance-oriented acceptance tests that are accomplished on every item after fabrication and assembly, but prior to delivery to the consumer.

[3] The test site may be a ship at sea, an aircraft or a space vehicle in flight, a facility located in the arctic or in the middle of the desert, a manufacturing plant, or a mobile land vehicle traveling over a designated route. The objective is to simulate a realistic operational environment to the extent practicable.

software, and the elements of logistic support) are operated and evaluated on an integrated basis. The compatibility between the prime equipment and software, the prime equipment and the elements of support, operator personnel compatibility, and so on are verified. Reliability and maintainability parameters, turnaround times, logistics supply times, inventory levels, test equipment availability, personnel effectiveness, and related factors can be measured.

5. *Type 4 testing.* During the system operational use phase, formal tests are sometimes conducted to gain further insight relative to a specific area of activity. It may be desirable to vary the mission profile or the system utilization rate to determine the impact on total system effectiveness, or it may be feasible to evaluate several alternative support policies to see whether system operational availability can be improved. Even though the system is designed and in the field, being utilized by the consumer, this is actually the first time that we really know its true capability. Hopefully, the system will accomplish its objective in an efficient manner; however, there is still the possibility that improvements can be realized by varying operational and maintenance support policies.

Type 4 testing is accomplished at one or more consumer operational sites, in a realistic environment, by user operator and maintenance personnel, and supported through a normal product support and logistics infrastructure. The elements of logistic support, as designated through predictions and analyses, earlier testing, and so on, are evaluated in the context of the total system.

Although there are variations in the test and evaluation requirements from one program to the next, the evolutionary concept illustrated in Figure 11.1 is still appropriate. One progresses from an analytical approach to a full demonstration of the system in the user environment. The objective is to develop a good plan where, hopefully, any potential problems can be detected early, with the appropriate corrective action being initiated at a time in the program when the incorporation of changes can be handled in an economical manner.

11.2   TEST PLANNING, ADMINISTRATION, AND CONTROL

Test planning actually commences early in the conceptual design phase when the basic system operational requirements and the maintenance concept are initially established (refer to Figure 11.1). If a requirement is to be specified for the system, there must be a way to evaluate the system later to ensure that the requirement has been met. This evaluation may be evolutionary, involving certain activities in each of the categories described earlier. In any event, testing considerations are intuitive from the beginning.

Throughout the various stages of system development, a number of individual tests may be specified. Often there is a tendency to design a test to measure one system characteristic, design another test to measure a different parameter, and so on. Before too long, the amount of testing specified may be overwhelming and quite costly. Test requirements must be considered on an integrated basis. Where possible, individual test requirements are reviewed in terms of resource needs and output results, and should be scheduled in such a manner as to gain the maximum benefits possible. For instance, maintainability data can be obtained through the use of simulation (i.e., analytical modeling), from reliability testing, and from certain performance tests. Test equipment compatibility data and personnel data can be obtained from reliability and maintainability testing. Thus, it might be feasible to schedule reliability testing first, maintainability demonstration second, test equipment compatibility third, and so on. In some instances, combining tests may be feasible as long as the proper characteristics are measured and the data output is compatible with the initially specified testing objectives.

For each program, an integrated test planning document (or a test and evaluation master plan) should be prepared during the conceptual design phase, but not later than 60 days prior to the scheduled start of any testing. Within this plan, there is a section addressing the requirements for maintainability demonstration, or the evaluation of the system/product to ensure that the initially specified maintainability requirements have been met. Test planning in this area should include:

1. The definition and scheduling of all test requirements (i.e., analytical, Types 1–4). Anticipated test output (in terms of what the test is to accomplish) is defined for each individual test and integrated where possible.
2. A description of organization, administration, and control responsibilities (organizational functions and interfaces, responsibilities for test conductance and the monitoring of test activities, data collection and analysis, cost control and reporting).
3. The definition of test conditions and resource requirements. This includes consideration of the test environment, facilities, test and support equipment, spares/repair parts and inventory requirements, maintenance software, test personnel, and test procedures. What conditions must exist in order to successfully accomplish a maintainability demonstration?
4. A description of the test preparation phase for each type of testing. This includes the selection of a specific test method, the training of personnel who will be participating in the test, and the acquisition of the necessary supporting resources (e.g., the specified test equipment, spares, and data).
5. A description of the formal demonstration and test process. This includes a description of the test procedures to be used and the data collection, reduction, and analysis methods.
6. A description of the conditions and provisions for a retest phase (if required). This includes the methods for conducting additional testing in the event of a "reject" situation.
7. The identification and scheduling of test documentation and reporting requirements.
8. A projection of anticipated test funding requirements (i.e., a cost projection).

The basic test plan describes the initial requirements for maintainability demonstration (stemming from a definition of the quantitative maintainability parameters for the system as included in the system Type "A" specification), the test methods to be implemented, the procedures for conducting the test, the resources required, and the expectations in terms of data output and reporting.

11.3   TEST CONDITIONS

Test conditions, as defined herein, encompass the necessary prerequisites to accomplishing a formal maintainability demonstration. These prerequisites include the initial definition of maintainability requirements for the system, the established maintenance concept, the accomplishment of a MTA or equivalent, the selection of a test model, and the acquisition of the appropriate elements of logistic support. While each of the categories described earlier will include some activities pertaining to maintainability evaluation, the emphasis here will be on maintainability demonstration as part of Type 2 testing.

11.3.1   Maintainability Requirements

As the maintainability demonstration process is basically a verification of compliance with a given set of qualitative and quantitative "design-to" requirements, the proper specification of these requirements is a necessary prerequisite. Referring to Chapter 5, the initial requirements for the system are developed through the definition of consumer need, the development of system operational requirements and the maintenance concept, the identification and prioritization of technical performance measures, and the development of the system Type "A" specification. Operational scenarios are described, functional sequences are defined, and quantitative effectiveness factors are applied as appropriate. This is where reliability factors (R, MTBF, λ), maintainability factors (MTBM, MDT, $\overline{M}ct$, $\overline{M}pt$, MMH/OH, $M_{\max}$), availability (A_o, A_a, A_i), and other comparable factors are initially specified. An objective of maintainability demonstration is to select one or a series of operational scenarios, simulate a failure, accomplish the necessary repair actions, and verify that the specified $\overline{M}ct$ (or $\overline{M}pt$, MDT, or $M_{\max}$, etc.) requirement has been met. The particular test method(s) selected for maintainability demonstration must be directly responsive to the measures described in the system Type "A" specification.

11.3.2   Maintenance Concept

Maintainability qualitative and quantitative requirements not only evolve directly from a system operational scenario, but they must be reflected throughout the maintenance concept. Referring to Figure 5.6, the illustration specifies three levels of maintenance, the proposed functions at each level, and some quantitative factors specified for each area of activity. The question is—will the demonstration apply to the organizational-level activities, to intermediate-level activities, and to manufacturer/depot activities; or just to the organizational level; or to what?

If the objective is limited to verifying *only* the maintainability requirements for the prime elements of the system, then it is likely that the maintainability demonstration plan selected will address only those maintenance functions/tasks accomplished at the organizational level of maintenance. On the other hand, if it is desired to evaluate not only the prime elements of the system but its maintenance and support capability as well, then the selected maintainability demonstration plan will likely include *all* of the specified levels of maintenance. Further, an essential (and initially specified) characteristic in design may include a built-in self-test requirement with fault isolation down to the assembly or module level. In accordance with the maintenance concept, a demonstration of this requirement will involve activities at both the organizational and intermediate levels of maintenance.

In the initial determination of maintainability demonstration requirements during the conceptual design phase (as system test requirements are being defined), it is important that demonstration objectives be completely defined within the context of the maintenance concept. Additionally, there must be complete agreement, about what is to be accomplished, between the customer and the contractor performing the test.

11.3.3 Demonstration-Model Configuration

The system configuration (equipment, software, etc.) used for maintainability demonstration must be representative of the operational system (i.e., that to be utilized by the consumer) to the maximum extent possible. For Type 1 tests, engineering and laboratory test models are used; however, these models are not usually directly comparable to a production configuration. For Type 2–4 tests, the models utilized should be representative of what the ultimate consumer will see. They contain the same parts, are manufactured and assembled employing the same processes as used for production items, are subjected to the same qualification tests, and so on.

A prerequisite for demonstration involves selecting the desired test model by serial number, defining the configuration by specifying the applicable drawings/documentation/data, including any previously unincorporated but approved engineering changes, and ensuring that the model is available at the location and time needed.

11.3.4 Test Environment

The goal is to accomplish maintainability demonstration in a realistic operational environment. If, for example, the system/product is to be installed and operated in a ground vehicle, then it is desirable to accomplish the demonstration using that vehicle while performing its mission. However, this is not always possible during the performance of Type 2 testing. These tests are usually conducted in a producer's facility, which does not offer the advantages of representing an operational situation. Nevertheless, the producer (or contractor) should attempt to simulate an "operational experience" to the maximum extent possible.

For systems/products that are designed to operate in an "outside" environment, a demonstration may be conducted in the contractor's back yard with the blowing wind, rain, cold temperatures, and so on. For smaller items, testing may be accomplished using environmental facilities and test chambers. In any event, the contractor must describe (in the demonstration test plan) the proposed environment in which the maintainability demonstration will be conducted.

11.3.5   Test Facilities and Resources

As part of the test planning process, there is a need to identify the requirements for special test facilities, environmental test chambers, capital equipment, special jigs and fixtures, special instrumentation, environmental controls, and associated resources (e.g., heat, water, air conditioning, gas, telephone, power, and lighting). In some instances, new design and construction is required which directly impacts the scheduling and duration of the test preparation phase. A detailed description of the test facility and the facility layout, the sources of materials, and the schedule must be included in test planning documentation and in the final test report.

11.3.6   Test Personnel and Training

Test personnel in this instance include those individuals who will actually operate and maintain the system/product during the formal test (and not the instrumentation technicians, data recorders, observers, or test management personnel). These individuals should possess backgrounds and skill levels similar to the consumer personnel who will normally be assigned to the system when it is in operational use. The various maintenance tasks to be demonstrated through the formal test activity must be accomplished in a realistic manner. A more highly trained individual from the producer's organization who has virtually "lived with the system" through design and development will obviously influence the results of the test in a different manner than an individual with lesser skills. Recommended personnel quantities and skill levels are derived from the MTA described in Chapter 9.

With regard to personnel selection for maintainability demonstration, three basic possibilities exist:

1. The selection of consumer or user personnel with the appropriate background and skill levels. Such personnel are assigned to the test site on a loan basis and are provided formalized training on the system being tested.

2. The selection of personnel from the producer's organization who have been assigned to programs not related to the system being tested. Such personnel are assigned to the test site on a loan basis, and are provided formalized training on the applicable system components.

Although these personnel have not had direct experience on the actual equipment being tested, they are familiar with the producer's design, test, and production methods. Through association with similar systems, their inherent ability may tend to influence test results by biasing operator and maintenance task proficiency and task times (i.e., higher proficiency and lower times for task accomplishment). This is particularly relevant when evaluating an equipment in terms of its design for supportability, and for Types 1 and 2 testing when the tests are conducted at the producer's facility.

On the other hand, these personnel are generally not familiar with the user's organization, operational methods, and environment. As a result, other factors (i.e., primarily those dealing with system interfaces and the elements of logistic support) may be biased in a different direction. This can be compensated for to a certain degree by providing formalized training covering user operations; however, full compensation requires actual experience or on-the-job training in the field.

3. The selection of personnel from the producer's organization who have been involved in the design and development of the system components being tested. In this instance, the personnel assigned will be thoroughly familiar with the equipment, and the test results may be highly biased (to a greater extent than with producer personnel previously assigned to other programs) in the performance of specific operator and maintenance tasks. Task proficiencies are likely to be high and task times lower than what would be expected if the equipment were in operational use in the field. In addition, nonfamiliarity with the user's operational methods and environment exists.

From the standpoint of approaching a realistic operational situation for test and demonstration, the preference in personnel selection involves items 1, 2, and 3 in that order. The realization of this preference is more likely as the system progresses through the various phases of testing. In addition, the desired approach is more probable if the test is accomplished at a user operational test site. In any event, the test plan must specify the approach to be used. Selected individuals being considered for participation in the test should be identified by name with the appropriate background and experience information noted. During the test preparation phase, these individuals are assigned to the test program and receive the necessary formalized training as prescribed. This training will vary somewhat depending on the background and skills of the personnel selected.

11.3.7 Test and Support Equipment

The successful completion of maintainability tasks within the specified requirements for the system (e.g., $\overline{M}ct$, $\overline{M}$, $\overline{M}pt$, and MMH/OH) is highly dependent on having the proper test and support equipment available. The requirements for test and support equipment evolve from the MTA, and may include any combination of manually operated, semiautomatic, and/or automatic test equipment. In accomplishing maintainability demonstration, it is important to have the right items for each task. If, for example, manually operated test equipment is utilized for a task where automatic test equipment is specified, then the demonstrated maintenance elapsed times will be extended. Utilizing a substitute item will significantly impact not only maintenance elapsed times, but MMH/OH as well. Also, there is likely to be a number of maintenance-induced errors during the demonstration of troubleshooting tasks in particular.

The appropriate test and support equipment items must be described in the test plan, acquired, adequately checked and calibrated, and scheduled at the proper time. In some instances, newly designed items may be required, and enough lead time should be provided to allow for the development and acquisition of these items in time for Type 2–4 testing.

11.3.8 Spare/Repair Parts and Inventory Requirements

The type and quantity of spares and repair parts, and the size and location of supporting inventory requirements, are determined from the maintenance concept, the results of the level of repair analysis, and the MTA. In conducting a maintainability demonstration on the prime elements of the system, it is desirable to have on hand those spares that are needed to complete maintenance "repair" and/or "remove and replace" tasks. One objective is to verify that the functional and physical interchangeability requirements have been met. If, on the other hand, an objective is to conduct a maintainability demonstration of the overall system support capability, then the appropriate spares/repair parts and inventory requirements must be established in accordance with the maintenance concept and the MTA.

More specifically, for smaller demonstrations, the primary objective is to measure only active elements of the maintenance cycle or factors applicable to that portion of the system being tested. In such cases, the requirements for spares may not be as critical. For example, maintenance tasks involving troubleshooting, disassembly, remove and replace, reassembly, and system checkout are evaluated. If the maintenance sequence requires the replacement of a major repairable assembly, the newly installed assembly may be the same one that was removed. Maintenance times other than for actual removal and replacement are discounted, and repair actions associated with the assembly are not applicable. For testing of this type (i.e., Types 1 and 2), the quantity of spares required may be negligible with careful planning. Recorded test data cover only to active maintenance task elements.

For large-scale tests (i.e., Types 3 and 4 testing), spare/repair parts will generally be required for all levels since these tests primarily involve an evaluation of the system as an entity and its total logistic support capability. The complete maintenance cycle, supply support provisions (the type and quantity of spares specified at each level), supply times, turnaround times, and related factors are evaluated. In certain instances, the producer's facility may provide depot level support. Thus, it is important to establish a realistic supply system (or a close approximation) for the item being tested.

The type and quantity of spare/repair parts and the procedures for spares inventory control throughout the test program should be specified in the test plan. Usage rates, reorder requirements, procurement lead times, and stock-out conditions during the test are recorded and included in the test report.

11.3.9 Test Procedures

Fulfillment of test objectives may involve the accomplishment of both operator and maintenance tasks. The completion of these tasks should follow the formally approved procedures which are generally available in the form of operating instructions, maintenance manuals, checklists, and so on, developed during the latter stages of the detail design and development phase. Following the approved procedures is necessary to ensure that the system is operated and maintained in a proper manner. Deviation from these approved procedures may result in the introduction of personnel-induced failures, and will cause a distortion in maintenance frequencies and task times as recorded in the test data. The identification of the procedures to be used in testing should be included in the maintainability demonstration plan.

11.3.10 System Software

In the conductance of maintainability demonstration, the system is usually operating in some mode when a failure occurs. This, in turn, will lead to the accomplishment of a maintenance action. All software used in system operation, plus the software used in the performance of maintenance, must be identified, acquired, and made available for testing purposes. The objective is to verify the compatibility between the equipment and the software, the maintenance technician and the software, and so on. The software requirements for the accomplishment of both corrective maintenance and preventive maintenance tasks must be described in the test plan and made available in a timely manner.

11.4 PREDEMONSTRATION PHASE

The predemonstration phase refers to the activities pertaining to the transition from the identification of test requirements to the actual conductance of the test. It includes the selection of demonstration methods, the selection of specific tasks to be demonstrated, the procurement and acquisition of the supporting resource requirements specified in Section 11.3, and the appropriate training of the test personnel.

11.4.1 Demonstration Method

As discussed earlier, the first step is to identify the proposed test site and the organization that will conduct the maintainability demonstration (i.e., customer/consumer at an operational site, at a contractor site with consumer personnel performing the demonstration, at a contractor site with contractor personnel performing the demonstration, or at an independent nonaffiliated test facility). The "demonstration method" defined here refers to the particular test procedure to be followed. The selection of a demonstration method is based on the specific requirements described in the system specification prepared during the conceptual design phase. For example, if there is a requirement for a 30-minute $\overline{M}ct$, then he test procedure selected must provide a $\overline{M}ct$ assessment as an output. A different approach may be appropriate for a $\overline{M}pt$, MDT, M_{max}, or MMH/OH requirement.[4]

---

[4]The material in this text is presented in the depth necessary to provide a basic understanding of test methods, input/output factors, procedures, and so on. For a more in-depth presentation, refer to Blanchard, B. S., *Logistics Engineering and Management,* 4th ed., Prentice-Hall, Inc., Englewood Cliffs, NJ, 1992. In the defense sector, a good reference is MIL-STD-471A, "Maintainability Verification, Demonstration, and Evaluation," Department of Defense, Washington, DC.

To illustrate the concept, two different test methods have been selected for discussion. The first follows a sequential test plan approach which is similar to the reliability qualification test described in Section 7.2.9. Two different sequential plans are implemented to demonstrate whether the system has met a $\overline{M}ct$ and a M_{max} requirement. An accept–reject decision for the system under test is reached when that decision can be made for both plans. The test plans assume that the underlying distribution of corrective maintenance task times is log-normal. The sequential test plan allows for an early decision when the maintainability of the system under test is either far better or far worse than the specified values of $\overline{M}ct$ and M_{max}.

The second method is applicable to the demonstration of $\overline{M}ct$, $\overline{M}pt$, and $\overline{M}$. The underlying distribution of maintenance task times is not restricted (no prior assumptions), and the sample size constitutes 50 corrective maintenance tasks for a verification of $\overline{M}ct$ and 50 preventive maintenance tasks for verification of $\overline{M}pt$. The value of $\overline{M}$ is determined analytically from the test results for $\overline{M}ct$ and $\overline{M}pt$; M_{max} can also be determined when the underlying distribution is log-normal. This method offers the advantage of a fixed sample size.

11.4.2 Demonstration Task Selection

The assurance that the results of the proposed maintainability demonstration will reflect the true design characteristics of the system being evaluated is facilitated through the task-selection process. A given demonstrated "task" includes a number of steps. A malfunction is introduced into the system; the operator will take the system through an operational scenario until a symptom of failure is noted; and the assigned maintenance technician(s) will proceed with the performance of corrective maintenance following the steps in Figure 4.8, that is, fault localization and isolation, disassembly, remove and replace (or repair in place), reassembly, alignment and adjustment, and condition verification. As the maintenance technician completes this cycle, test times are recorded and the results are compared with the initially specified $\overline{M}ct$ requirement. Additionally, data are collected in verifying tasks sequences and levels of difficulty, the compatibility of support resources, and so on. The completion of this cycle (evolving from a *single* induced failure) constitutes one demonstrated task, and the measure is Mct_i.

In demonstrating a corrective maintenance requirement, one needs to determine the sample size or the number of tasks to be demonstrated. The sample size should be based on the variance of the tasks to be represented, equipment complexity, the criticality of functions, and the probability of expected error (i.e., contractor/consumer risk factors). It is desirable to select a sample size large enough to be fully representative, and small enough to be compatible with program cost and schedule requirements. A large sample size will, of course, provide more definitive test results, but the time and costs of testing increase considerably as testing continues. A smaller number of samples involves increasingly larger risks and may produce inconclusive test results. Based on industrial experience, a quantity of 50 is sufficient for a fixed-sample-size test requirement (i.e., the second test method discussed in Section 11.5).

Corrective maintenance tasks are selected based on their expected contribution toward maintenance. High-failure-rate items will cause more maintenance actions and, thus, more tasks are demonstrated in this area. Table 11.1 shows the initial distribution of 50 tasks to the unit and assembly levels of a system, and Table 11.2 shows an extension of this distribution down to the subassembly or module level. Below this level, a component on a circuit board, a connector, or the equivalent is selected as the item where a failure will be introduced.

Referring to Table 11.1, one should list and categorize all significant assemblies, modules, or parts that make up the item being tested in column 1. The example covers the eight major assemblies of Units "A" and "B." In column 2, indicate the quantity (Q) of each item used, and in column 3 list the item failure rate (percent per 1000 hours). Column 4 represents the total failure rate including all applications. In column 5, determine the percent contribution of each item to the total, which is computed by dividing the figures in column 4 by the sum of all item failures per 1000 hours. In column 6, where applicable, items can be grouped when the percent contribution is low. Table 11.2 shows an example of some groupings. In column 7, apportion the number of corrective maintenance tasks to be demonstrated.

Table 11.1 apportions the 50 maintenance tasks among the eight major assemblies of Units "A" and "B." In order to ascertain specifically what maintenance tasks will be demonstrated below the assembly level, the same apportionment procedure is accomplished to the next level, as illustrated in Table 11.2. Each lower maintenance indenture level is analyzed until all 50 tasks scheduled to be demonstrated are completely identified. In instances where the lowest repairable item (made up of a number of parts) is covered by one or two sample demonstrations, the question arises as to what specific maintenance task sequences are to be selected, and what particular part(s) will be chosen for the purposes of inducing a failure. When a variety of possibilities exist, a table of random numbers can be employed to facilitate the selection process.

All corrective maintenance tasks to be demonstrated, and the specific items where failures are to be induced, are identified at a specific point in time and included in the test and demonstration plan (which should not be made available to the individuals serving as system operators during the test). Knowledge of the system, its elements, the various failure modes and effects, and so on, is definitely required in the selection of where failures are to be introduced. The objective is to simulate a real-world situation without either damaging the system or injuring the operator in the process.

In the selection of preventive maintenance tasks, there are two basic categories: (1) preventive maintenance tasks that can be performed while the system is operating (no maintenance downtime is involved); and (2) preventive maintenance tasks that require the system to be shut down for the duration of task accomplishment (maintenance downtime occurs). In the first instance, the demonstration of such tasks results primarily in an evaluation of tasks sequences and potential personnel problems due to possible difficulty in task accomplishment, and the compatibility of the logistic support resources required. There are no downtime values to be measured (i.e., $\overline{M}pt$); however, one can measure the consumed MMH/OH values.

TABLE 11.1 Corrective-Maintenance-Task Sample Selection (Example)

(1) Item	(2) Quantity of Each Item (Q)	(3) Failures/Item % 1000 hr (λ)	(4) Total Failures/ Item/1000 hr (Q) (λ)	(5) Item % Contribution to Total Maintenance Task	(6) Grouping of Block 5 Tasks (%)	(7) Allocation of Maintenance Tasks
			Unit A			
Power amplifier	1	0.55	0.55	25.2	25.2	13
Power supply	1	0.24	0.24	11.0	11.0	6
IF amplifier	1	0.05	0.05	2.4	2.4	1
Computer assembly	1	0.86	0.86	39.4	39.4	19
			Unit B			
Bearing converter	1	0.06	0.06	2.8	2.8	1
Range converter	1	0.11	0.11	5.0	5.0	3
Digital output	1	0.10	0.10	4.6	4.6	2
Power supply	1	0.21	0.21	9.6	9.6	5
Total			2.18	100		50

TABLE 11.2 Corrective-Maintenance-Task Sample Selection (Example)

(1) Item	(2) Quantity of Each Item (Q)	(3) Failures/Item % 1000 hr (λ)	(4) Total Failures/ Item/1000 hr (Q) (λ)	(5) Item % Contribution to Total Maintenance Task	(6) Grouping of Block 5 Tasks (%)	(7) Allocation of Maintenance Tasks
			Power Amplifier			
RF power amplifier	1	0.3800	0.380	69.1	69.1	9
Duplexer	2	0.0025	0.005	0.9	3.8	1
Coupler	2	0.0060	0.012	2.2		
Detector	4	0.0010	0.004	0.7		
Modulator control	1	0.0650	0.065	11.8	11.8	1
Modulator tuner	1	0.0510	0.051	9.3	9.3	1
Self-test oscillator	1	0.0300	0.015	2.7	6.0	1
Self-test amplifier	1	0.0030	0.018	3.3		
Total			0.550	100		13

With regard to preventive maintenance tasks that result in some downtime, if the quantity is small (i.e., less than 50), then all identified tasks should be demonstrated. If the quantity is large (more than 50), a representative sample should be selected, using a procedure similar to the selection of corrective maintenance tasks. Identify all applicable preventive maintenance tasks; determine the frequency of occurrence for each task listed; determine the number of preventive maintenance tasks with the same frequency of occurrence; determine the total frequency of occurrence and the percent contribution; group together those tasks where the contribution is low; and apportion the number of preventive maintenance tasks to be demonstrated.

The selection of both corrective and preventive maintenance tasks, as discussed thus far, has been based primarily on the frequency factors. In general, this may be a good approach! However, through completion of the FMECA, the RCM, and the MTA activities, there may be certain tasks that are highly "critical" from a mission-accomplishment perspective, but the anticipated frequency of occurrence may be relatively low. In such instances, because of the critical nature of the task, these tasks may be included in the demonstration.

11.4.3 Demonstration Test Preparation

The remaining effort in the predemonstration phase comprises final preparations at the test site. The system/product to be demonstrated, along with the specified elements of support, is deployed, installed, and operationally checked out at the test site. Any abnormalities (manufacturing defects, transportation-induced problems, etc.) must be eliminated prior to the start of demonstration. At the same time, the applicable elements of support are positioned in accordance with the recommendations specified in Section 11.3.

The maintenance technicians scheduled to participate in the demonstration (i.e., the individuals who will initially operate the system and then accomplish the required maintenance tasks) must be "trained" to the extent required to be equivalent to the quantities and skill levels recommended in the MTA. Such training, accomplished either at the test site or at the contractor's facility, may be limited to a short briefing and familiarization with the system to more in-depth instruction, depending on the current skills of the consumer personnel scheduled to operate and maintain the system, the complexity of the system design, and so on. In any event, the objective is to accomplish the demonstration using personnel with skills that are representative of what will be available to maintain and support the system throughout its planned utilization period.

A test team will be assigned to conduct the maintainability demonstration. The team will not only include the maintenance technicians who will actually accomplish the demonstration (i.e., those personnel described above), but the test team director, the technician who will induce the malfunctions into the demonstration model for each task, personnel who will be responsible for data collection and analysis, customer representatives monitoring the demonstration, and other observers. A formal briefing of test team personnel will be necessary to ensure complete communications and continuity in conducting the demonstration. The briefing should cover maintainability requirements for the system, the purpose and intent of the demonstration, the demonstration plan, specific organizational and personnel responsibilities, methods and the procedures to be employed during the demonstration, data recording methods, data analysis objectives, and report requirements. This briefing should be held at the test site just prior to the start of the formal demonstration.

11.5   FORMAL DEMONSTRATION PHASE

The formal demonstration includes the simulation (to the extent possible) of all system maintenance actions (i.e., corrective and preventive maintenance), at an early point in the life cycle, to verify that the initially specified maintainability requirements have been (or will be) met. Referring to Figure 11.1, the activities described herein are included as part of Type 2 testing, and two demonstration methods are covered—a sequential method and a fixed-sample method.

11.5.1   Demonstration Method 1 (Sequential Approach)

This method follows a sequential test plan similar in concept to the approach used in reliability qualification testing described in Section 7.2.9. Two different sequential test plans are used to demonstrate whether the system has met the $\overline{M}ct$ and M_{max} requirements (for corrective maintenance). An "accept" decision is reached when the data indicate that both of the requirements have been met. Figure 11.2 provides

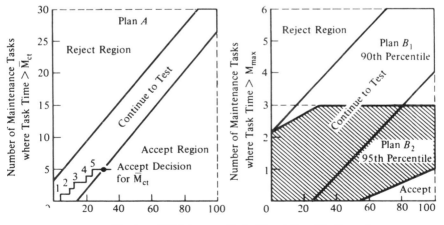

Number of Maintenance Tasks (Simulations) Performed

Figure 11.2. Graphical representation of $\overline{M}$ demonstration plans. (*Source:* Blanchard, B., *Logistics Engineering and Management*, 4th ed., Prentice-Hall, Englewood Cliffs, N.J., 1991, p. 284).

a graphical presentation of these test plans. Basically, one proceeds along the abscissa (to the right) as successful task demonstrations are completed, and along the ordinate (upward) as "noncompliant" tasks occur. The details of these plans will become more evident as one proceeds through the discussion. The basic steps in the test procedure are described below:

1. The maintainability test director will initially provide data covering the techniques governing failure simulation. A data sheet, such as the one illustrated in Figure 11.3, will be completed (through item A3) prior to the start of demonstration for each of the 100 potential tasks scheduled to be simulated for this testing method. Referring to Figure 11.2, if an "accept–reject" decision is made early, the actual number of demonstrated tasks may be less. However, initial planning must consider the possibility that testing will continue to the point of truncation, that is, the completion of 100 task demonstrations. The 100 simulation requirements stem from the sample selection process illustrated in Tables 11.1 and 11.2, and the method of failure simulation is determined during the predemonstration phase. Simulation can be accomplished through opening or shorting electrical connections (leads), the installation of known faulty parts, application of tape to connector pins, and so on. Simulations which could result in equipment damage or an unsafe condition should, of course, be avoided.

2. The maintenance technician(s) assigned to complete the tasks being demonstrated is requested to leave the immediate premises where the test is being conducted. During his/her absence, the test director and/or customer's representative may select at random any one of the 100 demonstration data sheets (i.e., Figure 11.3). The selected data sheet indicates the type of failure to be simulated and the method of simulation. A technician (assigned to assist the test director and maintainability test monitor) then proceeds to induce the failure into the system. The "bug" is checked to ensure that it provides the desired symptom of failure. Subsequently, the maintenance technician scheduled to complete the demonstrated task is recalled to the test site.

3. The maintenance technician then performs an operational check of the demonstration model using approved operating procedures. At some point during the performance of the operating function, a "symptom of failure" will occur (i.e., a "NO-GO" condition). At this time, the data recorder notes the symptom of failure, and the mode of system operation when the failure occurred, on the demonstration data sheet (Figure 11.3, items A4 and A5).

4. Once a symptom of failure has been detected, the maintenance technician proceeds to accomplish the necessary corrective maintenance tasks—fault localization and isolation, disassembly, repair, reassembly, checkout, and so on. The maintenance tasks are completed utilizing approved maintenance procedures. The data recorder monitors every detailed step in the maintenance process and records the appropriate information as shown in Figure 11.4. Task descriptions, start–stop times for active maintenance, administrative and logistics times, and related data are noted. While not highlighted in the figure, the data recorder should also note the logistic support resources needed to support the maintenance activities being accomplished. The technician must notify the test director when the cause of failure

DEMONSTRATION DATA SHEET

DEMO NO: ___29___ CM: ___X___ PM: ___ RECORDER: _J. SMITH_ DATE: _7/1/93_

EQUIPMENT	NOMENCLATURE	MFGR PART NO.
SYSTEM	System	A37700
UNIT	Unit A	A37701
ASSEMBLY	Computer	A58180
SUBASSEMBLY	C. B. No. 2A5	A58183

CORRECTIVE MAINTENANCE (CM):

    A.   FAULT SIMULATION:

        1.   Type of Failure Simulated: Open emitter lead on transistor

        2.   Method of Simulation: Replaced Q5 with known faulty transistor

        3.   Designation of Part Simulating Failure: F10

        4.   Symptom of Failure: Unit A front panel light indicates NO-GO condition. No bearing output data. Absence of signal at T. P. 1 of C. B. 2A4

        5.   Mode of operation at Detection: T/R Mode and A/A Mode

        6.   Additional Information:

    B.   TIME DATA:

MAINT. REQMT.	TIME	MAINT. REQMT.	TIME
LOCALIZATION	1.0	REASSEMBLY	17.0
ISOLATION	20.0	ALIGN/ADJUST	–
DISASSEMBLY	10.0	CHECKOUT	6.0
REMOVE/REPLACE	2.0	OTHER	–
REPAIR	5.0	TOTAL MAINT. (MIN)	61.0

PREVENTIVE MAINTENANCE (PM):

INSP:____ FUNCT. TEST:_____ CAL:____ SERV:_____ OH:_____ FORCE REM:_____

    A.   MAINT. PROCEDURE:

    B.   TOTAL MAINT. TIME (ACTIVE):

Figure 11.3. Demonstration data sheet (example).

has been identified. The test director must then ensure that the detected failure and the initially induced failure are one and the same before permitting the test to continue. In the event that the maintenance technician is unable to diagnose and find the induced failure (through some abnormality or equipment design deficiency), the point at which the test should be discontinued must be determined through mutual agreement of the test director and the customer's representative.

DEMONSTRATION WORKSHEET

DEMO NO: ___29___   CM: ___X___   PM: _____RECORDER: _J. SMITH_   DATE: 7/1/ 93

EQUIPMENT	NOMENCLATURE	MFGR. PART NO.
SYSTEM	System XYZ	A37700
UNIT	Unit A	A37701
ASSEMBLY	Computer	A58180
SUBASSEMBLY	C. B. No. 2A5	A58183

DETAILED MAINTENANCE TASKS

TASK NO.	TASK DESCRIPTION	QTY MEN	ACTIVE MAINT. TIME (HR:MIN) START	ACTIVE MAINT. TIME (HR:MIN) STOP	ADMIN/LOG TIME (HR:MIN) START	ADMIN/LOG TIME (HR:MIN) STOP
1	Verify System XYZ NO-GO Through Self-Test	1	9:15	9:16		
2	Remove Unit A From System XYZ	1	9:16	9:19	9:19	9:22
3	Remove Unit A Access Cover (six quick-disconnect fasteners)	1	9:22	9:26		
4	Connect Jumper Wires To Pins A & B	1	9:26	9:28		
5	Obtain Power Source	1	-	-	9:28	9:30
6	Apply Bias Voltage to Pin C	1	9:30	9:34		
7	Check for 15V p-p Signal at TP8 on C. B. 2A6	2	9:34	9:41		
8	Check For Logic Square-wave Pulse (4 μsec) at TP1, C. B. 2A4	2	9:41	9:48		
9	Remove Computer Assembly From Unit A	1	9:48	9:51		
10	Remove C. B. 2A5	1	9:51	9:52		
11	Repair C. B. 2A5	1	9:52	9:57		
12	Install Spare C. B. 2A5	1	9:57	9:58		
13	Break For Coffee	2	-	-	9:58	10:28
14	Replace Computer Assembly	1	10:28	10:34		
15	Accomplish Checkout of Unit A	2	10:34	10:39		
16	Install Unit A Access Cover	1	10:39	10:44		
17	Install Unit A In System XYZ	1	10:44	10:50		
18	Accomplish Final Check Utilizing Self-Test	1	10:50	10:51		

Figure 11.4. Demonstration worksheet (example).

5. While the maintenance technician is demonstrating the corrective maintenance tasks, the recorder is able to provide an assessment of the adequacies/inadequacies of the logistic support capability. Was the right type of support provided at the right time? Were there test delays due to inadequacies? Was there an overabundance of certain items and a shortage of others? Did each of the elements of logistic support accomplish its function in a satisfactory manner? Were the test procedures adequate? These and related questions should be in the mind of the data recorder during the observation of a demonstration.

This cycle is accomplished n times, where n is the selected sample size. For the sequential test, the number of demonstrations could possibly extend to 100, assuming that a "continue-to-test" decision prevails. In Table 11.3, the allocation of the 100 tasks to be demonstrated for a system composed of Units "A"–"C" is shown. The process is the same as that illustrated in Tables 11.1 and 11.2 where the sample size was 50.

With the tasks identified and listed in random order, the demonstration proceeds with the first task, then the second, the third, and so on. The criteria for "accept–reject" decisions are illustrated in Figure 11.2. Task times (Mct_i) are measured and compared with initially specified $\overline{Mct}$ and M_{max} values. When the demonstrated time exceeds the specified value, an event is noted upward along the ordinate of the graph, and the problem areas are described by the data recorder. Testing will continue until the event line either enters the "reject" region or the "accept" region.

An example of the results of demonstration is illustrated in Table 11.4. The accept–reject numbers support the decision lines in Figure 11.2 (refer to the $\overline{Mct}$ curve). In this instance, 29 tasks were completed before an "accept" decision was reached.

The sequential test requires that both the $\overline{Mct}$ and the M_{max} criteria be met before the system is fully acceptable. M_{max} may be based on either the ninetieth or ninety-fifth percentile depending on the specified system requirement and the test plan selected. If one test plan is completed with the event line crossing into the "accept" region, testing will continue until a decision is made in the other test plan.

TABLE 11.3 Corrective Maintenance Task Allocation

(1) Item	(2) Quantity of Items (Q_f)	(3) Failures/Item % 1000 hr (λ)	(4) Total Failures ($Q)(\lambda_f$)	(5) % Contribution	(6) Allocated Maintenance Tasks for Demonstration
Unit A	1	0.48	0.48	21	21 tasks
Unit B	1	1.71	1.71	76	76 tasks
Unit C	1	0.06	0.06	3	3 tasks
Total			2.25	100	100 tasks

TABLE 11.4   Demonstration Score Sheet

REQMT: $\overline{M}ct = 0.5$ hr = 30 Min.				Plan A
Maint. Task No.	Task Time Mct_i	Cum No. $Mct_i > \overline{M}ct$	Accept When Cum ≤ Than	Reject When Cum > Than
1	12 Min.	0		
2	6	0		
3	18	0		
4	32	1		
5	19	1		5
6	27	1		6
7	108	2		6
8	6	2		6
9	14	2		7
10	47	3		7
11	28	3		7
12	19	3	0	7
13	4	3	0	8
14	24	3	0	8
15	78	4	1	8
24	20	4	3	11
25	127	5	4	11
26	21	5	4	12
27	13	5	4	12
28	28	5	4	12
29	8	5	5	12
		Accept for $\overline{M}ct$		

Referring to Figure 11.2, the criteria for sequential testing specify that the minimum number of tasks possible for a quick decision for $\overline{M}ct$ is 12 (Test Plan A). For M_{max} at the ninetieth percentile, the least number of tasks possible is 26 (Test Plan B_1) while the figure is 57 for the 95th percentile (Test Plan B_2). Thus, if the maintainability of an item is exceptionally good, demonstrating complete sample of 100 tasks may not be necessary, thus saving time and cost. On the other hand, if the maintainability of an item is marginal and a *continue to test* decision prevails, the test program may require the demonstration of all 100 tasks. If truncation is reached, the equipment is acceptable for $\overline{M}ct$ if 29 or less tasks exceed the specified $\overline{M}ct$ value. Comparable factors for M_{max} are 5 or less for the ninetieth percentile and 2 or less for the ninety-fifth percentile.

TABLE 11.5   Maintenance Test Time Data (Partial)

Demonstration Task Number	Observed Time Mct_i	$Mct_i - \overline{Mct}$ $(Mct_i - 62)$	$(Mct_i - \overline{Mct})^2$
1	58	-4	16
2	72	$+10$	100
3	32	-30	900
50	48	-14	196
	3105		15,016

11.5.2   Demonstration Method 2 (Fixed-Sample Approach)

This method is applicable to the demonstration of $\overline{Mct}$, $\overline{Mpt}$, and $\overline{M}$. The underlying distribution of maintenance times is not restricted (no prior assumptions), and the sample size constitutes 50 corrective maintenance tasks for $\overline{Mct}$ and 50 preventive maintenance tasks for $\overline{Mpt}$. $\overline{M}$ is determined analytically from the test results for $\overline{Mct}$ and $\overline{Mpt}$. $\overline{M}_{max}$ can also be determined if the underlying distribution is assumed to be log-normal.

The method involves the selection and performance of maintenance tasks in a similar manner as described for Demonstration Method 1. Tasks are selected based on their anticipated contribution to the total maintenance picture, and each task is performed and evaluated in terms of maintenance times and required logistics resources. Illustration of this method is best accomplished through an example. It is assumed that a system is designed to meet the following requirements and must be demonstrated accordingly.

$\overline{M}$ $\quad = 75$ minutes
$\overline{Mct}$ $= 65$ minutes
$\overline{Mpt}$ $= 110$ minutes
$\overline{M}_{max} = 120$ minutes
Producer's risk $(\alpha) = 20\%$

The test is accomplished and the data collected are presented in Table 11.5.

The determination of $\overline{Mct}$ (upper confidence limit) is based on the expression

$$\text{upper limit} = \overline{Mct} + Z \frac{\sigma}{\sqrt{N_c}} \tag{11.1}$$

where

$$\overline{Mct} = \frac{\Sigma Mct_i}{N_c} = \frac{3105}{50} = 62.1 \text{ (assume 62)}$$

$$Z = 0.84 \text{ (refer to Chapter 4)}$$

$$\sigma = \sqrt{\frac{\sum_1^{N_c} (Mct_i - \overline{Mct})^2}{N_c - 1}} = \sqrt{\frac{15.016}{49}}$$

$$= 17.5$$

$$N_c = \text{corrective maintenance sample size} = 50$$

$$\text{upper limit} = 62 + \frac{(0.84)(17.5)}{\sqrt{50}} = 64.07 \text{ minutes}$$

The computed $\overline{Mct}$ statistic is compared to the corresponding accept–reject criterion, which is

accept if

$$\overline{Mct} + Z \frac{\sigma}{\sqrt{N_c}} \leq \overline{Mct} \text{ (specified)} \tag{11.2}$$

reject if

$$\overline{Mct} + Z \frac{\sigma}{\sqrt{N_c}} > \overline{Mct} \text{ (specified)} \tag{11.3}$$

Applying demonstration test data, it can be seen that 64.07 minutes (the upper value of $\overline{Mct}$ derived by test) is less than the specified value of 65 minutes. Therefore, the system passes the $\overline{Mct}$ test and is *accepted*.

For preventive maintenance the same approach is used. Fifty preventive maintenance tasks are demonstrated and task times (Mpt_i) are recorded. The sample mean preventive downtime is

$$\overline{Mct} = \frac{\Sigma Mpt_i}{N_p} \tag{11.4}$$

The "accept–reject" criterion is the same as stated in Equations (11.2) and (11.3) except that preventive maintenance factors are used. That is,

accept if

$$\overline{Mpt} + Z \frac{\sigma}{\sqrt{N_p}} \leq 110 \text{ minutes} \tag{11.5}$$

reject if

$$\overline{M}pt + Z\,\frac{\sigma}{\sqrt{N_\mathrm{p}}} > 110 \text{ minutes} \tag{11.6}$$

Given the test values for $\overline{M}ct$ and $\overline{M}pt$, the calculated mean maintenance time is

$$\overline{M} = \frac{(\lambda)(\overline{M}ct) + (fpt)(\overline{M}pt)}{\lambda + fpt} \tag{11.7}$$

where

λ = corrective maintenance rate or the expected number of corrective maintenance tasks occurring in a designated period of time

fpt = preventive maintenance rate or the expected number of preventive maintenance tasks occurring in the same time period

Using test data, the resultant value of $\overline{M}$ should be equal to or less than 75 minutes. Finally, $M_{\max}$ is determined from

$$M_{\max} = \text{antilog } [\log \overline{M}ct + Z\sigma_{\log Mct_i}] \tag{11.8}$$

The calculated value should be equal to or less than 120 minutes for acceptance. An example of calculation for $M_{\max}$ is presented in Chapter 4.

If all the demonstrated values are better than the specified values, following the criteria defined above, then the system is *accepted*. If not, some retest and/or redesign may be required depending on the seriousness of the problem.

11.5.3   Retest Phase

The "retest phase" is designed to allow for the repeat of certain facets of the formal maintainability demonstration in order to investigate any deficiencies detected during the initial demonstration. In general, a retest may be required as a result of the following conditions:

1. The system/product fails to meet the requirements of the formal demonstration, and the margin of failure is relatively small. In this instance, the customer may request that an additional quantity of tasks be demonstrated to increase confidence in the test data and the results. If, after additional testing, the data indicate compliance with the initially specified requirements, the customer may choose to accept the system and testing is concluded. If, on the other hand, the data still indicate noncompliance, the contractor may be required to accomplish some redesign, or to accept a financial penalty.

2. The system/product may fail to meet the formal demonstration requirements

by a considerable margin. In this instance, the contractor may be forced to accomplish some redesign (at his expense) and/or be penalized financially.

In the event that system redesign is required, possible areas of concentration are easily identified by carefully reviewing and analyzing the information recorded on the demonstration data sheet (Figure 11.3) and the demonstration worksheet (Figure 11.4). Extensive maintenance times will, of course, indicate problem areas. The results are reviewed and evaluated by the design engineer and the maintainability engineer for corrective action.

Subsequent to the approval (by the customer) and incorporation of the required design modifications, the system may be subjected to additional demonstration testing. Possibilities may include either a complete retest (i.e., 50 new sample simulations), or a partial retest covering only specific areas where problems were identified. In either case, additional testing will be accomplished following the same procedures as described for demonstration methods 1 and 2.

An additional option may be to defer any retesting requirements associated with Type 2 testing, and evaluate the maintainability characteristics of the system/product as part of Types 3 and/or 4 testing. While it may be more advantageous to evaluate the system in a more realistic consumer environment, the potential cost of any required modifications at this late stage in the life cycle will be high.

11.6 SYSTEM TEST AND EVALUATION

The reliability testing, maintainability demonstration, and related areas of testing have been accomplished primarily on an independent basis up to this point. It has not been possible to demonstrate the entire system and its overall support capability in a realistic consumer environment.

System test and evaluation, accomplished as Types 3 and 4 testing (refer to Figure 11.1), often provides the first real opportunity to view the system as a "whole." Prime equipment, software, people, and all of the elements of logistic support are delivered to the operational test site, integrated, and scheduled through a series of mission scenarios, during which operational, performance, reliability, maintainability, and supportability characteristics are measured and evaluated. Through the accomplishment of successive mission scenarios over an extended period of time, the system can be measured in terms of effectiveness, availability, dependability, and so on. Reliability (MTBF) and maintainability (MTBM, $\overline{M}ct$, MMH/OH) can be measured on a more realistic basis than is possible during Type 2 testing. Additionally, the elements of logistic support can be assessed in terms of capability and adequacy. This includes an evaluation of the maintainability characteristics in the design of test and support equipment, the interchangeability of replacement parts (spares), the functional packaging and modularization of software, the verification of maintenance procedures, and so on. So often, one finds that a poor design of any one of the elements of logistic support will cause problems of a "feedback" nature for the prime equipment. For example, if an item of test equipment is not "maintainable," it probably will not be calibrated properly, and thus it may induce failures into the item of prime equipment it was designed to check. In essence, these interaction effects are more likely to be identified during the system test and evaluation, as compared to what can be accomplished as part of Type 2 testing.

As an example of system test and evaluation, it is assumed that an aircraft system has been designed to accomplish three basic mission profiles. The three profiles and the performance requirements of the various individual components of the aircraft during each profile were defined as part of the operational requirements when the system was first conceived. In addition, the following factors were specified as requirements:

Operational availability of the system	0.95
Dependability of the system during the mission	0.85
Maintenance man-hours per system operating hour	1.5
Turnaround times:	
1. Organizational maintenance	12 minutes
2. Intermediate maintenance	2 hours
3. Depot maintenance	80 hours

A test program may be designed to evaluate the system and measure these factors. A representative number of flights are scheduled over a designated period of time with specific flights assigned to each of the three missions. There must be enough flights scheduled to ensure that valid data are obtained. By virtue of the mission definition, the numerous functions and operational modes of the aircraft will be exercised to varying degrees. A realistic operational situation should be simulated to the greatest extent possible. One or more aircraft may be assigned to the test program depending on the number of test flights scheduled and the time period allowed for testing. Figure 11.5 illustrates the basic test cycle.

As the test program progresses, the availability of the aircraft to commence with its assigned mission is measured. Assuming that the aircraft is available and takes off on schedule, the probability that the aircraft will complete its mission successfully becomes significant. When failures occur, it is then necessary to perform corrective maintenance at the organizational level within the prescribed time to get the aircraft back into an *operationally ready* state and meet the availability requirement. Organizational maintenance is supported by labor and material at the intermediate level which, in turn, is supported by the depot level. The supply system, test and support equipment, personnel quantities and skill levels, facilities, software, and data at each level are evaluated to ensure responsiveness to the turnaround time, maintenance manhours, and operational availability requirements.

Data collection throughout the test program is somewhat extensive, and is based on the type and depth of information required from the test. Both system operational data and maintenance data are collected, and good and bad points are noted. Dis-

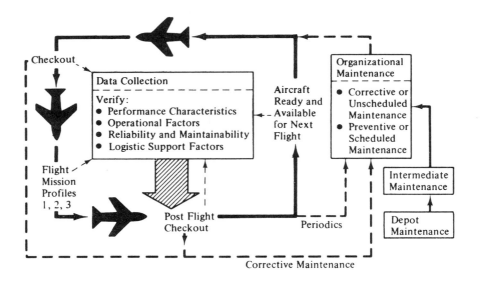

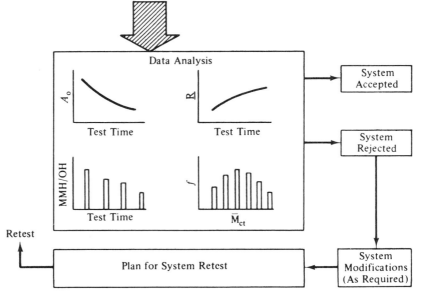

Figure 11.5. System test and evaluation. (*Source:* Blanchard, B., *Logistics Engineering and Management,* 4th ed., Prentice-Hall, Englewood Cliffs, N.J., 1991, p. 289).

crepancies are recorded and events are analyzed to determine cause-and effect relationships. Failure trends are identified and should result in corrective action leading to a change in an operating procedure or a modification to the equipment.

When the required number of test flights has been completed, the recorded data are analyzed and the results are compared with the specified system requirements. These requirements include not only operational availability, dependability, maintenance man-hours per operating hour, but also such lower-level requirements as reliability MTBF and maintainability $\overline{M}ct$. The type of requirements specified will dictate the type of data recorded, which in turn is included in the basic test planning document.

Given an assemblage of test data, the analyst can determine operational availability from the expression

$$A_o = \frac{\text{MTBM}}{\text{MTBM} + \text{MDT}} \qquad (11.9)$$

Assuming that the proper type of data has been collected, the following factors can be derived (refer to Chapter 4):

$$\lambda = \frac{\text{number of failures}}{\text{total system operating time}} \qquad (11.10)$$

$$\text{MTBF} = \frac{1}{\lambda} = \text{MTBM}_u \qquad (11.11)$$

$$fpt = \frac{\text{number of preventive maintenance actions}}{\text{total system operating time}} \qquad (11.12)$$

$$\text{MTBM}_s = \frac{1}{fpt} \qquad (11.13)$$

$$\text{MTBM} = \frac{1}{1/\text{MTBM}_u + 1/\text{MTBM}_s} \qquad (11.14)$$

$$\text{MDT} = \overline{M} + \overline{\text{administrative time}} + \overline{\text{logistics time}} \qquad (11.15)$$

$$\overline{M} = \frac{(\lambda)(\overline{M}ct) + (fpt)(\overline{M}pt)}{\lambda + fpt} \qquad (11.16)$$

$$\overline{M}pt = \frac{\Sigma(fpt_i)(Mpt_i)}{\Sigma \, fpt_i} \qquad (11.17)$$

$$\overline{M}ct = \frac{\Sigma(\lambda_i)(Mct_i)}{\Sigma \lambda_i} \qquad (11.18)$$

$$\text{MMH/OH} = \frac{\overline{\text{maintenance manhours}}}{\text{total system operating time}} \qquad (11.19)$$

Other factors may be determined as deemed feasible. For instance, when evaluating the logistic support capability, the analyst may wish to know:

1. Turnaround times at the organizational, intermediate, and depot levels of maintenance.

2. Spare part supply times between the levels of maintenance. This includes the evaluation of a *nonoperationally ready* system due to a supply shortage or delay.

3. Spare/repair part stock levels and the demand rates for individual items.

4. Personnel effectiveness factors (organizational efficiency, assignment of the proper skill levels by function, adequacy of the prescribed formal training program, human error rates).

5. Adequacy of the operating and maintenance procedures (comprehensiveness and clarity).

6. Compatibility and utilization of test and support equipment (will it do the job, and is it being effectively utilized?).

7. Adequacy of operational and maintenance facilities (proper environment, effective use of facilities, personnel–facility interfaces, adequate storage for spares and test equipment).

8. Total cost-effectiveness of logistic support capability.

Discrepant areas are recorded and reviewed in terms of possible corrective action. Corrective action may constitute a modification to the equipment, software, procedures, and/or elements of logistic support. Depending on the seriousness of the problem, the system may be subjected to retest or the applicable modification may be incorporated into the operational system without requiring a retest.

11.7 DATA ANALYSIS AND CORRECTIVE ACTION

Needless to say, the data analysis and corrective-action loop is a significant aspect of the program. Without it, all of the testing in the world would not be of much value. If a system is to meet the operational requirements (which is the purpose of the system to begin with), it must be evaluated and areas of deficiency must be corrected; thus, the continuous feedback of data and a means by which corrective action is accomplished should be identified. In addition, the corrective-action loop must have teeth!

It is realized that corrective action may at times be rather costly. Unforeseen events often arise. On the other hand, the lack of adequate consideration of maintainability (or other comparable factors) in the design phase is predictable and will definitely become evident through testing. In this instance, neglecting to do what should have been done earlier in the program may cause no end of problems later. Costly modifications may be required (as a result of testing) which could have been avoided if the necessary steps had been taken.

The corrective-action loop is illustrated in Figure 11.6. Test data are analyzed and the results are compared with the system requirements. Areas of noncompliance are corrected and the system may be retested (or not) depending on the seriousness of the problem. The retest phase is designed to repeat certain tests because either the system, when initially tested. failed to meet all requirements by a significant margin, or the system met all requirements by a narrow margin. In the first instance, the following may occur:

1. Redesign of the prime equipment, an element of software, or a change in supplier
2. A major modification to a manufacturing or quality control process; and/or
3. A change in the basic logistic support policy or a modification to a particular element of logistic support

The system may be retested to verify that the incorporated change corrected the identified problem(s).

In the second case (i.e., requirements are met, but by a narrow margin), it may be desirable to gain more experience through formal testing in order to provide added confidence that system requirements will be met in the operational use phase. When retesting is specified, a plan for the retest is required and should contain the same basic information included in the initial test planning document.

The final effort in the test program consists of the preparation of a test report. The final test report should reference the initial test planning document and should

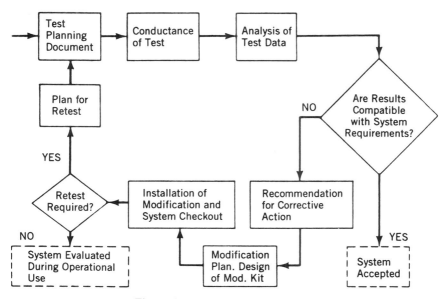

Figure 11.6. Corrective-action loop.

describe test conditions, test data, and the results of data analysis. These results may include appropriate recommendations for future system changes.

QUESTIONS AND PROBLEMS

1. What is the purpose of maintainability demonstration? When is it accomplished?

2. Briefly describe the differences between Types 1, 2, 3, and 4 testing. How does maintainability demonstration fit into these categories?

3. When in the system life cycle are maintainability demonstration requirements first established?

4. In determining specific maintainability demonstration requirements, what information must be provided?

5. How does reliability qualification testing relate to the verification of maintainability requirements?

6. Describe some of the test conditions (i.e., the test "environment") that must exist to conduct a successful maintainability demonstration.

7. Describe the criteria that you would apply in selecting personnel to perform the maintenance tasks to be demonstrated. What criteria would you apply in specifying the test and support equipment for the demonstration? In the event that "substitutes" are used (e.g., the use of higher skills, or manual test equipment instead of automatic), how would this possibly impact the test results?

8. Describe how you would select the specific maintenance tasks to be demonstrated.

9. In preparing for a maintainability demonstration, what would be your primary areas of concern if you had been appointed as the test director?

10. Refer to the two demonstration methods described in this chapter. Compare the advantages and disadvantages between the sequential test approach and the fixed-sample approach.

11. In addition to maintenance elapsed-time factors, what other information can be derived from maintainability demonstration?

12. Under what conditions can a retest requirement be specified?

13. Describe the corrective-action process in the event that the test results indicate a noncompliant situation.

14. What considerations need to be addressed in selecting the "model" for maintainability demonstration?

15. When selecting tasks for maintainability demonstration, how does the selection process result in a representative sample of what is to be expected in an operational situation?

16. Refer to maintainability demonstration method 2 and determine whether the system will meet the $\overline{M}ct$ requirement if the assumed producer's risk is 5%.

17. How does the confidence factor in maintainability demonstration affect the accept–reject criteria?

18. Select a system of your choice and specify the appropriate operational, performance, reliability, maintainability, and logistic support requirements. Assume that you have been appointed as test director and are responsible for demonstrating that all requirements have been met. Develop a test plan that you intend to implement.

19. The $\overline{M}ct$ requirement for an equipment item is 65 minutes and the established risk factor is 10%. A maintainability demonstration is accomplished and yields the following results for the 50 tasks demonstrated.

39	57	70	51	74	63	66	42	85	75
42	43	54	65	47	40	53	32	50	73
64	82	36	63	68	70	52	48	86	36
74	67	71	96	45	58	82	32	56	58
92	91	75	74	67	73	49	62	64	62

(Task times are in minutes.)

Did the equipment item pass the maintainability demonstration?

20. The $\overline{M}pt$ requirement for the equipment in Problem 19 is 120 minutes and the risk is the same. A maintainability demonstration is accomplished and yields the following results for the 50 tasks demonstrated.

150	120	133	92	89	115	122	69	172	161
144	133	121	101	114	112	181	78	112	91
82	131	122	159	135	108	95	67	118	103
78	93	144	152	136	86	113	102	65	115
113	101	94	129	148	118	102	106	117	115

(Task times are in minutes.)

Did the equipment item pass the maintainability demonstration?

21. What is the calculated mean maintenance time for the equipment in Problems 19 and 20 if the equipment operation time is 1000 hours?

22. Discuss the "pros" and "cons" of accomplishing maintainability demonstration as part of Type 2–4 testing.

23. How can the results of maintainability demonstration be effectively used?

12

MAINTAINABILITY APPLICATIONS FOR OPERATING SYSTEMS

The earlier chapters in this text primarily address the application of maintainability concepts, principles, methods, and techniques to the design, development, and acquisition of new systems/products. Maintainability requirements are initially established, specific design criteria are developed, various maintainability analysis techniques are applied in the accomplishment of design trade-off studies, maintainability predictions and a maintenance task analysis are accomplished for the purposes of assessment and in the determination of system support requirements, and maintainability demonstration is accomplished to verify that the initially specified customer requirements have been met. Maintainability is a design characteristic, and maintainability engineers are assigned as part of the "design team" to help ensure that newly developed systems do incorporate the appropriate maintainability features in the ultimate design configuration.

While the emphasis in maintainability has been directed to the design of new systems, there is also a need to apply maintainability principles both in the evaluation of currently operating systems already in the inventory and in the improvement of those systems for greater effectiveness and reduced cost. The approach is to first evaluate an existing system/product in terms of life-cycle cost, identify the high-cost contributors from a functional perspective, determine the major cause-and-effect relationships, identify specific areas where the design of a given product or process can be improved, and initiate the necessary recommendations for system/product improvement. This is usually a continuous process, and the results are highly beneficial in terms of increased customer satisfaction while costs and the expenditure of unnecessary resources are reduced. This, in turn, can lead to increased profits for commercial enterprises in which the costs of operating and maintaining existing systems has been high.

The objective of this chapter is to discuss the implementation of maintainability program requirements for the purposes of system/product improvement. Initially, a good data collection and analysis capability must be established to properly assess the actual effectiveness and efficiency of the system being utilized by the consumer. Given an initial baseline, it is then appropriate to apply selected maintainability analysis tools in the evaluation of possible alternatives for accomplishing systems objectives. Finally, modifications for system improvement must be addressed. This iterative, continuous process for product improvement is discussed herein.

12.1   THE CURRENT STATUS OF OPERATING SYSTEMS

Referring to Figure 1.2 (Chapter 1), many systems in use today are not living up to the expectations of the consumer. The overall effectiveness is low, the costs from a life-cycle perspective are high, and an imbalance exists, as illustrated. On one side of the spectrum are the technical characteristics of the system, which include the applicable performance factors (e.g., range, accuracy, speed, and capacity), availability, dependability, reliability and maintainability, and so on. On the other side is the aspect of life-cycle cost. An objective is to attain the appropriate balance and, while there are many factors to consider, maintainability is a key parameter that must be addressed in meeting this objective. Maintainability, as a characteristic in design, significantly impacts both sides of the balance in Figure 1.2.

With regard to the cost side of the balance, experience has indicated that for many large systems the costs associated with system operation and support constitute from 60 to 75% of the total life-cycle cost for that system. Further, these costs are not always visible when those early design and management decisions that will ultimately have the greatest impact on these costs are made. The cost "visibility" problem illustrated by the "iceberg" in Figure 1.3 is *real,* although the specific cost elements (and their percent contributions) will vary from one system configuration to another.

In the commercial sector, numerous studies have been conducted dealing with maintenance and support problems in manufacturing, that is, the commercial factory, which is a "system" in itself! According to a study by R. K. Mobley, from 15 to 40% (with an average of 28%) of the total cost of finished goods can be attributed to maintenance activities in the factory.[1] Given the ongoing addition of new technologies, the introduction of more automation, robotics, and so on, maintenance costs in the factory are likely to be even *higher* in the future with the continuation of existing practices. T. Wireman reports from a study conducted in 1989 that the estimated costs of maintenance for a selected group of companies increased from $200 billion in 1979 to $600 billion in 1989.[2] In S. Nakajima's book *Total Productive Maintenance (TPM),* he indicates that it is not uncommon for many companies to experience less than a 50% overall equipment effectiveness rate, that is, a very low productivity rate with a high level of required maintenance.[3]

---

[1] Mobley, R. K., *An Introduction to Predictive Maintenance,* Van Nostrand Reinhold, Inc., New York, 1990.

[2] Wireman, T., *World Class Maintenance Management,* Industrial Press, Inc., New York, 1990.

[3] Nakajima, S., *Introduction to Total Productive Maintenance (TPM),* Productivity Press, Inc., Cambridge, MA, 1988 (translated into English from the original text published by the Japan Institute for Plant Maintenance, Tokyo, Japan, 1984).

With regard to the issue of costs due to maintenance activities, these costs may be broken down into (1) *direct costs* required to keep the factory equipment operating (both labor and material in accomplishing corrective and preventive maintenance), (2) *secondary costs* associated with the maintenance of standby equipment, and (3) *loss-of-production costs* incurred when both the primary and secondary equipment are unavailable for the manufacture of a product (i.e., production losses or the loss of potential revenue). The last category of "production losses" often ranges from 2 to 15 times the direct costs for the maintenance and repair of the factory equipment. In many instances, factory equipment maintenance is accomplished on a "reactive" or "emergency" basis, and there is a great deal of waste involved! According to T. Wireman, in companies where reactive or emergency types of maintenance make up 50% or more of the maintenance organization's workload, technicians average only 2–3 hours of hands-on activities in an 8-hour day. The rest of the time is spent in nonproductive activities such as looking for parts, drawings, instructions, and/or some form of authorization.

In essence, whether we are dealing with large defense systems, commercial communications or transportation systems, manufacturing systems, or whatever, the costs associated with system maintenance and support continue to be significant! In evaluating "cause-and-effect" relationships, one can look at both the inherent characteristics of the original design of the system and the configuration of the current maintenance and support structure. In other words, there are two basic approaches: (1) to evaluate and improve the design of the prime elements of the system and (2) to improve the organizational and procedural approaches associated with the sustaining maintenance and support of the existing configuration (i.e., increasing the effectiveness and productivity of the existing maintenance organization).

It is the first of these two approaches that is being emphasized herein. Through the proper assessment and evaluation of currently operating systems, the analyst can identify areas in design where improvements can be made to help reduce some of the downstream maintenance burden and associated costs. The discipline of maintainability, and the incorporation of the proper maintainability characteristics in design, can offer great benefits toward improving the overall cost-effectiveness of many of the systems in use today.

12.2 THE IMPLEMENTATION OF MAINTAINABILITY PROGRAM REQUIREMENTS FOR SYSTEM/PRODUCT IMPROVEMENT

Maintainability program requirements, presented in the context of the system life cycle, are illustrated in Chapter 1 (Figure 1.5). While the earlier chapters of this text concentrate on the activities in blocks 1–3, the emphasis here is on the activities in block 5. Figure 12.1 covers some of these activities.

One of the first steps in the process illustrated in the figure is to accurately assess how well the system is performing for the customer. This may seem to be rather "basic;" however, when pursuing the point, one is often surprised by the number of users, engineers, and/or managers who really do not know just how well their sys-

Reference: Figure 1.5 (block 5)

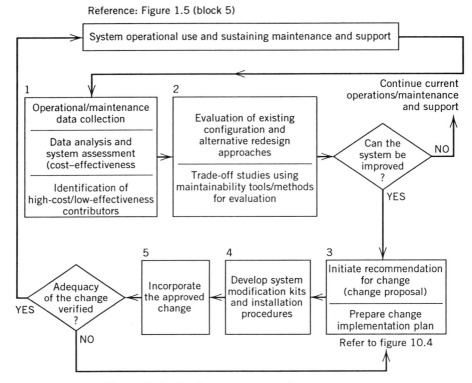

Figure 12.1. Continuous system/product improvement.

tems are performing. In most instances, they do know that equipment is working, that the factory is producing a product, or that the customer is willing to "live" with the system as it is now performing. However, they often do not have any idea about whether the equipment is operating at the proper "designed speed," whether the productivity in the factory is optimum, whether the costs of current operations and support are minimized, and so on. Further, while the system user does have a general "feel" as to the major problems being experienced in the field, he/she usually does not know the specifics in terms of system "weaknesses." Often when system failures occur, the maintenance organization is tasked with accomplishing the necessary repair actions in order to get the system back into operation as soon as possible. Some information is recorded to indicate that maintenance tasks have been performed. However, the data collected are usually not very complete, and it is impossible to respond to such questions as: how was the system being utilized when the failure occured (i.e., stresses on the system)? What component of the system actually failed? How did the component fail (i.e., failure mode)? Did any other components fail as a result of the first component failure? What was the cause of failure? Are there any specific failure trends (i.e., a series of similar failures in the past)? What specific tasks were performed in the accomplishment of system repair (i.e., maintenance task sequences)? What was the Mct_i value for the maintenance cycle completed? What specific logistic support resources were consumed in the performance of maintenance (i.e., spare/repair parts, test and support equipment, or quantity of personnel)? Did the technicians performing the maintenance tasks make any mistakes, or were there any induced problems? The responses to these questions, and others of a similar nature, are necessary to understand how well the system is performing and the major problem areas that currently exist. Thus, a good maintenance data system is an essential first step for the initiation of an effective continuous process improvement program.

12.2.1 Maintenance Data Requirements

The purpose of a good maintenance data information feedback capability is twofold:

1. It provides ongoing data that are analyzed to evaluate and assess the performance, effectiveness, operations, maintenance, logistic support capability, and so on, for the system in the field. The consumer needs to know exactly how the system is doing and needs the answer in a timely manner. Thus, certain types of information must be provided at designated times.

2. It provides historical information that can be utilized effectively not only in the modification of existing systems for the purposes of improvement, but in the design and development of new systems/products in the future. While there are many good analytical models, tools, and so on available and being utilized in the design process, there is often a lack of good historical input data. Thus, the results of the many predictions and analyses described in the earlier chapters are highly "suspect!" Our engineering growth and potential in the future certainly depends on our ability to capture past experiences and subsequently apply the results in terms of "what to do" and "what not to do" in new system design.

The design of a maintenance data capability must consider the availability of the right type of data at the right time. It is necessary to determine the specific elements of data required, the frequency of need, and the associated format for data reporting. These factors are combined to identify total volume requirements and the type, quantity, and frequency of data reports.

A format for maintenance data collection must be developed and should include both "success" and "maintenance" data. Success data consists of information covering system utilization on a day-to-day basis, and the information should be comparable to the factors listed in Table 12.1. It is important to describe how the system is being utilized (the mission scenario, utilization profile, etc.) at the time when a failure occurs. Maintenance data, on the other hand, cover each event involving scheduled and unscheduled maintenance. The events are recorded and referenced in system operational information reports, and the factors recorded in each instance are illustrated in Table 12.2.

The format for data collection may vary considerably from one application to

TABLE 12.1 System Success Data

System Operational Information Report
1. Report number, report date, and individual preparing report
2. System nomenclature, part number, manufacturer, serial number
3. Description of system operation by data (mission type, profiles and duration)
4. Equipment utilization by date (operating time, cycles of operation, etc.)
5. Description of personnel, transportation and handling equipment, and facilities required for system operation
6. Recording of maintenance events by date and time (reference maintenance event reports)

another, as there is no set method for the accomplishment of such, and the information desired may be different for each system. However, most of the factors in Tables 12.1 and 12.2 are common for all systems and must be addressed in the design of a new capability. In any event, the following considerations should apply:

1. The data collection forms should be simple to understand and complete, preferably on single sheets, as the task of recording the data may be accomplished under adverse environmental conditions by a variety of personnel skill levels. If the forms are difficult to understand, they will not be completed properly (if at all), and the required data will not be available.
2. The factors (i.e., entry requirements) on each form must be clear and concise, and not require a lot of interpretation and manipulation to obtain. The right type of data must be collected.
3. The factors specified must have a meaning in terms of direct application. The usefulness of each factor must be verified.

These considerations are extremely important and cannot be overemphasized. All of the analytical methods, prediction techniques, models, and so on discussed earlier have little meaning without the proper input data. Our ability to evaluate alternatives and predict future results depends on the availability of good historical data, and the source of such stems from the type of information feedback capability developed. This capability must not only incorporate the forms for recording the right type of data, but must consider the personnel factors involved in the data recording process. The individual(s) who is assigned to complete the appropriate form(s) must understand the system and the purposes for which the data are being collected. If the person is not properly motivated to do a good thorough job in recording events, the resulting data will of course be highly suspect.

Once the appropriate data forms are distributed and completed by the responsible line organizations, a means must be provided for the retrieval, formatting, sorting, and processing of data for reporting purposes. Field data are collected and sent to a designated facility for analysis and processing. The results are disseminated for evaluation and entered into a data bank for retention and possible future use.

TABLE 12.2   System Maintenance Data

Maintenance Event Report

1. *Administrative data*
   a. Event report number, report date, and individual preparing report
   b. Work order number
   c. Work area and time of work (month, day, hour)
   d. Activity (organization) identification
2. *System factors*
   a. Equipment part number and manufacturer
   b. Equipment serial number
   c. System operating time when event occurred (when discovered)
   d. Segment of mission when event occurred
   e. Description of event (describe symptom of failure for unscheduled actions)
3. *Maintenance factors*
   a. Maintenance requirement (repair, calibration, servicing, etc.)
   b. Description of maintenance tasks
   c. Maintenance downtime (MDT)
   d. Active maintenance times (Mct_i, and Mpt_i)
   e. Maintenance delays (time awaiting spare part, delay for test equipment, work stoppage, awaiting personnel assistance, delay for weather, etc.)
4. *Logistics factors*
   a. Start and stop times for each maintenance technician by skill level
   b. Technical manual or maintenance procedure used (procedure number, paragraph, data, comments on procedure adequacy)
   c. Test and support equipment used (item nomenclature, part number, manufacturer, serial number, time of item usage, operating time on test equipment when used)
   d. Description of facilities used
   e. Description of replacement parts (type and quantity)
      i. Nomenclature, part number, manufacturer, serial number, and operating time on replaced item; describe disposition
      ii. Nomenclature, part number, manufacturer, serial number, and operating time on installed item
5. *Other information*
   Reference reliability analysis report covering item failure (failure mode, cause of failure, effects of failure)

12.2.2   System/Product Evaluation

As a first step in the ongoing system evaluation effort, the information acquired through the maintenance data collection capability should be analyzed with the objective of not only providing some level of assessment at the time of evaluation, but also providing an indication of trends. While the actual measures of effectiveness should evolve directly from the technical performance measures established through the definition of operational requirements and the maintenance concept (refer to Chapter 5), an example of a few evaluation factors is presented in Figure 12.2. In

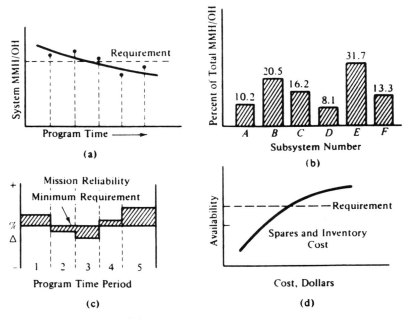

Figure 12.2. Typical evaluation factors (example).

the figure, parts "a" and "c" reflect specific measures at designated points in the life cycle, while "b" and "d" reflect the results of a one-time investigation. Proceeding further, Subsystem "E" (Figure 12.2b) is a likely candidate for investigation since the recorded MMH/OH for that item constitutes 31.7% of the total system value. Referring to Figure 12.2c, one may wish to investigate Program Time Period 3 to determine why the mission reliability was so poor at that time. In any event, the analyst can make appropriate use of the data collected as a good starting point in the overall system evaluation process.

With the results of the field data analysis providing an *initial* indication of potential areas where system/product improvements can be realized, the analyst may wish to pursue the evaluation further! For example, what are the "high-cost contributors" in terms of elements of the system projected over its planned life cycle? In response, it may be appropriate to perform a life-cycle cost analysis, following the procedure described in Appendix B. Assuming that the results of such an analysis are as indicated for Example 5, Section 6.3.5, Figure 6.32, the high-cost contributors in the area of operations and maintenance (C_o) include the categories of "maintenance personnel and support (C_{omm})" representing 23.4% of the total cost for Configuration "A" and "spare/repair parts (C_{OmX})" representing 11.5% of the total.

When evaluating the cause-and-effect relationships associated with maintenance personnel and support costs, the analyst may identify various possible contributors. First, such costs may be generated by high-failure-rate items where the frequency of corrective maintenance is high, the quantity of personnel and the MMH/OH are high, the skill levels required in task accomplishment are high, and/or a combination thereof! Second, the analyst may wish to know what specific elements of the system are causing this high consumption of maintenance resources? One can develop a functional flow diagram of the system, showing the steps in product development and the associated processes; identify the input–output factors related to each block in the diagram; and identify the resources required for the completion of each function (refer to Section 6.1, Chapter 6). A high-cost contributor may be Unit "B" of System "XYZ," Assemble 2 of Unit "C," or a given process in accomplishing a specific function. This same approach can be followed when evaluating the "causes" associated with spare/repair parts costs, or the costs related to any other item identified in the CBS of a life-cycle cost analysis.

Given the identification of a high-cost element of the system, the analyst may wish to delve further into the "cause-and-effect" relationships, that is, beyond that which was initially evident from results of a preliminary life-cycle cost analysis. In the event that more information is desired, covering anticipated system failures and their possible impacts on the overall system and the accomplishment of its mission, the analyst may wish to apply a FMECA to selected elements of the system. There is no better way of learning more about a particular item of equipment or a given process than through the accomplishment of a FMECA (refer to Section 7.2.2).

If, on the other hand, a high-cost contributor refers to the PM program for a given system, or a specific PM requirement, it may be appropriate to review this area of activity through the accomplishment of a RCM analysis, which depends on the FMECA as an input, and can be applied with the intent of further "optimizing" the ongoing PM program requirements, while reducing the life-cycle costs for a system (refer to Section 7.2.4).

To assist in the determination of specific resource requirements in the accomplishment of system functions, the analyst may wish to accomplish a MTA in accordance with the procedures in Chapter 9. The MTA is directed toward the evaluation of specific functions (i.e., maintenance requirements) for the purposes of identifying task times and sequences, maintenance personnel quantities and skill levels, test and support equipment, spares/repair parts and associated inventories, facility requirements, technical data, and so on. If, through the task analysis, there is a requirement for a nonstandard piece of test equipment, personnel skills at the supervisory level, and/or a special facility such as a "clean room," this would indicate a high maintenance cost area of activity. By identifying the specific "cause," one can analyze the situation and initiate a recommendation for improvement. The MTA is an excellent tool for making *visible* the actual maintenance requirements for one or a series of functions. This maintainability tool can be used very effectively, as a supplement to the life-cycle cost analysis, in the implementation of an ongoing program for system/product improvement (refer to Figures 9.3–9.5, Chapter 9).

As the analyst works through the system evaluation process, there may be any number, or combination, of tools used to fulfill a specific objective. A reliability model may be used to assist in developing the necessary maintenance frequency factors in the event that the field data collected are not adequate; growth models may be developed to help describe specific trends; a human factors task analysis may be accomplished to help evaluate a given maintenance procedure; a special maintainability demonstration may be accomplished to verify a poor design feature, and so on. The many maintainability tools and techniques described throughout the earlier chapters may be used very effectively, not only in the design and developent of a new system, but in the evaluation and improvement of existing systems already in operational use.

12.2.3   Recommendations for System/Product Improvement

When addressing specific goals for system/product improvement, the analyst should review and reevaluate the operational requirements and the maintenance concept initially specified to ensure that the objectives for the system (in terms of mission accomplishment) are still the same. Very often, the threat may change when considering the requirements for defense systems, or the business objectives may change yearly when dealing with commercial products. In any event, system goals must be current and firmly established.

Given that the existing system configuration is not optimum in terms of fulfilling the desired objective(s), it is necessary to implement a formal program activity whereby the system can be modified for improvement. From a maintainability perspective, improvements can be realized through a reduction in (1) the number of maintenance actions required (i.e., improving reliability), (2) maintenance task times and MMH/OH, (3) the logistic support resources needed in the accomplishment of maintenance, (4) the cost of performing maintenance, and so on. These measures can be impacted in a number of ways, including increasing the standardization of components, improving testability and system diagnostics, improving the accessibility to critical components, incorporating improved component mounting provisions for interchangeability, eliminating unstable alignment/adjustment requirements, and improving component labeling. Many of the design characteristics of interest to the maintainability are included in the checklist in Appendix A.

As a start in the implementation of a design improvement effort, the analyst should establish some type of "benchmark," or a set of specific goals, as a "target" for the future. The objective may include improving the system configuration to meet a desired level of maintainability while reducing cost at the same time, and the analyst should develop a plan for accomplishing this objective, similar to the one illustrated in Figure 12.3. The intent is to incorporate system improvements over time, realizing a growth in maintainability and a gradual reduction in lifecycle cost. The approach here is similar to the reliability growth model described in Section 7.2.7.

With a growth plan developed, the analyst should identify, categorize, and list in order of priority the major problems that are highlighted as a result of the system/product evaluation process described in Section 12.2.2. A Pareto analysis may be accomplished, similar to the one described in Section 7.2.2, identifying the problems that should be addressed first! An example of the results of such an analysis is shown in Figure 12.4.

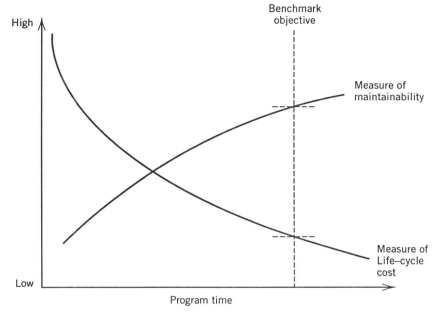

Figure 12.3. Effectiveness relationships.

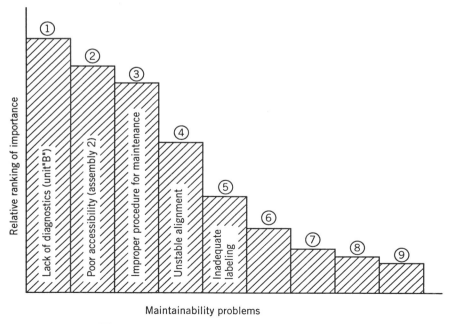

Figure 12.4. Pareto ranking of major problem areas.

In addressing each of the identified problem areas (in order of priority), a number of different design approaches should be considered for possible problem resolution. Trade-off studies may be initiated, and each of the proposed candidate solutions needs to be evaluated in terms of its impact on performance, reliability and maintainability, life-cycle cost, and so on. The evaluation factors included in Section 10.6 for proposed design changes are appropriate here.

Specific recommendations for improvement may be documented in a format similar to the maintainability design recommendations shown in Figures 7.30 and 7.31. The justification for each recommendation for change must be included, along with the proposed plan for system modification. Additionally, the impact of the change on the growth plan illustrated in Figure 12.3 must be well defined.

12.2.4 System/Product Modifications

Referring to Figure 12.1, given the system consumer's approval for the installation of the change, the next step is the development and incorporation of the necessary modification kits. Subsequently, the system may be tested to verify that the incorporated change does indeed reflect the maintainability improvement that has been proposed.

12.3 SUMMARY

While it is recognized that the preferred approach for incorporating the appropriate maintainability characteristics in a system configuration is to accomplish such as part of the initial design process, this does not always happen! In fact, many systems in operational use today are not very effective in terms of accomplishing their intended functions, are not maintainable in terms of ease and accuracy in the performance of maintenance, and are costly in terms of system logistics and support. In such instances, maintainability (as a design characteristic) was not considered in the initial design process.

On the other hand, given such a system configuration, there are many opportunities for system/product improvement. The consumer (or the user of such a system) can realize numerous benefits through the implementation of a maintainability program effort involving the initial evaluation of the system and the identification of potential high-cost contributors in the maintenance area, the identification of cause-and-effect relationships, and the incorporation of changes for system/product improvement. The approach to a "continuous system/product improvement" activity, as defined herein, can lead to system growth in terms of both increased effectiveness and reduced life-cycle cost, that is, the proper balance of the factors in Figure 1.2 (Chapter 1).

QUESTIONS AND PROBLEMS

1. Assume that you were just assigned the task of implementing a maintainability program for the "continuous improvement" of a given operating system. Identify the steps that you take in fulfilling this objective.

2. In developing a field data collection capability, what type of data would you like to acquire? Describe the considerations that must be addressed in determining the requirements in this area.

3. In the evaluation of an existing operating system, the analyst may wish to use a variety of different analytical tools/methods to aid in the evaluation process. Identify some of these tools and describe the benefits of each (i.e., to include input–output requirements).

4. Describe the benefits of using the FMECA as an aid in the evaluation process. What is the difference between a "product" FMECA and a "process" FMECA?

5. Refer to Figure 12.4. What considerations should be addressed in determining the "ranking" of maintainability problems?

6. Select a system of your choice and perform an evaluation of that system from a cost-effectiveness standpoint. Identify, categorize, and rank in order of priority the maintainability-related problems detected (construct a Pareto diagram).

7. In developing recommendations for system/product improvement, what factors need to be considered in justifying the proposed change?

8. What benefits can be derived from accomplishing a life-cycle cost analysis on an existing operating system?

9. Describe the benefits that can be derived from implementing an an overall "continuous improvement program" for an existing operating system/product.

APPENDIX A
MAINTAINABILITY DETAILED CHECKLIST[1]

In accomplishing an informal maintainability review, specific questions which reflect good maintainability principles in design serve as a basis in determining whether the item being reviewed incorporates the appropriate maintainability characteristics. These questions are listed below. Not all questions are applicable in all reviews; however, the answer to those questions that are applicable should be "YES" to reflect good maintainability. This detailed list of questions supports the abbreviated checklist shown in Figure 7.30 (Chapter 7).

A. General
    1.0   Standardization
          1.1   Are standard commercial equipment items incorporated to the maximum degree possible (except for such items not compatible with system time constraints)?
          1.2   Are the same commercial equipment items used in similar applications?
          1.3   Are corresponding rack/console positions used throughout all workstations for the same equipment items?
          1.4   Are corresponding rack/console positions used throughout all workstations for peculiar control panels?
          1.5   Are preferred standard parts, AN, NAS, or MIL-STD parts utilized to the maximum extent possible?
          1.6   Are control-panel layouts (from panel to panel) the same or similar?
          1.7   Is panel nomenclature/color coding the same for all panels?
          1.8   Are like modular units employed throughout rack/console drawers?

---

[1] Much of this information was taken from Blanchard, B. S. and E. E. Lowery, *Maintainability Principles and Practices,* McGraw-Hill Book Co., New York, 1969.

    1.9   Are maintainability checklists, test procedures, and technical manual formats standardized?

    1.10 Are electrical/electronic circuit types kept to a minimum?

2.0 Components functionally grouped

Packaging of hardware should be by direct association with primary signal flow. The following are in order of preference:

    2.1   Are electrical circuit functions packaged in individual plug-in modules?

    2.2   Are electrical circuit functions, not packaged in individual plug-in modules, packaged in a *minimum* number of modules or printed circuit boards?

    2.3   Are electrical circuit components, not packaged in plug-in modules or printed circuit boards, functionally grouped?

3.0 Console Layout

Related rack/console features are covered under Equipment Racks– General

    3.1   Is rack/console layout optimized from the human-factors standpoint (consult human-factors personnel)?

    3.2   Are rack/console drawers optimally placed according to functional relationships, frequency of schedule maintenance, etc?

4.0 Complexity

    4.1   To the degree ascertainable, are the number of parts minimized?

    4.2   Does each part have an absolute function or need?

    4.3   Are difficult maintenance procedures, adjustments, etc., minimized? If not, are they justified?

5.0 Self-test

    5.1   Have self-test provisions been incorporated?

    5.2   Is the degree or depth of self-testing in accordance with specification requirements?

    5.3   Are self-test provisions automatic?

    5.4   Have direct fault indications been provided? (this means that either a fault light or an audio signal is provided, or there is a direct indication of a malfunction through a meter reading or similar means.)

6.0 Maximum-time-to-repair

    6.1   Has equipment turnaround time been minimized through:

        a. Capability for rapid and positive fault localization?

        b. Maximum accessibility to enable rapid removal and replacement of assemblies or subassemblies?

        c. Selection of assemblies or subassemblies requiring minimum time to self-test (warm-up time included)

        d. Rapid equipment-performance-verification capability?

    6.2   Does equipment maximum-time-to-repair meet the required time constraint?

7.0   Auxiliary tools and test equipment
    7.1   Are the number of required auxiliary tools and test-equipment items minimized?
    7.2   Are required tools and test-equipment items standard?
    7.3   Are required tools and test-equipment items simple to operate? Can applicable maintenance tasks be accomplished in a minimum span of time?
    7.4   Have peculiar and nonstandard auxiliary tools and test-equipment items been eliminated?

8.0   Labeling
    8.1   Is labeling maximized?
    8.2   Is labeling adequate in the following areas?
        a. Racks/consoles
        b. Panel controls and meters
        c. Test and adjustment points
        d. Access openings
        e. Safety areas
        f. Servicing/lubrication areas
        g. Cables and connectors

Labeling relative to specific areas is covered in other sections of this checklist. Size and type of labeling should be discussed with human-factors personnel.

9.0   Weight

Equipment should be of the lightest weight possible consistent with sturdiness, safety, and reliability.
    9.1   To the degree ascertainable, is equipment weight minimized through the elimination of unnecessary structural members, cases, heavy parts, etc.?
    9.2   Are assemblies and subassemblies which are removed for maintenance minimized relative to weight (preferably less than 10 lb)?

10.0   Calibration requirements
    10.1   Are calibration requirements known?
    10.2   Do standards exist for calibrating test equipment?
    10.3   Are calibration tolerances known?
    10.4   Are calibration frequencies known?
    10.5   Are calibration procedures prepared?
    10.6   Are calibration requirements possible within the existing capabilities of a standard calibration laboratory?

11.0   Repair/replace philosophy
    11.1   Is the basic maintenance concept known?
    11.2   Has the level of maintenance for each repairable item been defined (whether the item will be repaired at intermediate level or at depot level)?
    11.3   Has the maintenance frequency (MTBM) been defined or predicted?

11.4   Has the maintenance-downtime requirement $(\overline{M}ct)$ been specified?

11.5   Have the maintenance-facility and resource requirements been specified?

11.6   Have the maintenance personnel-quantity and skill-level requirements been specified?

11.7   Are test-equipment requirements compatible with the maintenance concept?

12.0   Maintenance procedures

12.1   Have maintenance procedures and task sequences been identified (maintenance task analysis)?

12.2   Have maintenance checklists been prepared?

12.3   Have maintenance technical publications been prepared?

12.4   Are the maintenance procedures compatible with the maintenance concept?

13.0   Personnel requirements

13.1   Are anticipated maintenance tasks simple enough to enable accomplishment by an average technician?

Average Technician

a. Age	21 years
b. Experience	No regular work experience prior to training
c. Time on the job	On first 3- or 4-year tour
d. Educational level	High school graduate
e. Schooling	A total of 19 weeks of for-formal training
f. General level of interest	Varies, but usually low
g. General level of morale	Varies, but usually low
h. General reading level	About 9th grade

13.2   Are anticipated maintenance tasks such that no more than two technicians are required for accomplishment?

14.0   Trade-off studies

Design decisions are made based on weighing factors such as performance, reliability of equipment, weight, schedule, maintainability considerations, and cost. When design decisions are made which compromise maintainability, the trade-off considerations as applicable should be documented.

14.1   Have maintainability trade-off studies been documented?

14.2   Have trade-offs been justified/validated?

B. Handling

1.0   Equipment lifting

1.1   Are lifting eyes installed on each rack console which is handled separately? Is the inside diameter of each lifting eye at least 1 in?

1.2   Is the base of each console reinforced to enable lifting with a forklift truck?

    1.3   Are hoisting fittings identified relative to hoisting capacity and forklift application points?

2.0   Rack/console drawer or panel handles

    2.1   Are rack/console drawers provided with handles?

    2.2   Are two handles provided on each drawer front panel?

    2.3   Are drawer handles mounted vertically?

    2.4   Are drawer handles adequate in size?

    2.5   Is one handle provided on each rack/console panel:

    2.6   Are handles optimally located (provide at least 2-in. clearance between handles and obstructions)?

3.0   Assembly handles

    3.1   Do assemblies weighing over 10 lb within equipment items incorporate handles?

    3.2   Are handles optimally located from the weight-distribution aspect (handles should be located over the center of gravity)?

4.0   Console castors

    For heavy racks or consoles which are likely to be moved periodically within the maintenance shop, castors are sometimes desirable. However, when castors are employed, jacks should be installed adjacent to each castor to enable leveling and to preclude undesirable movement.

    4.1   Are castors provided on racks or consoles which are likely to be moved periodically?

5.0   Damage susceptibility

    5.1   Are covers or cases provided to protect equipment vulnerable areas from damage?

    5.2   Are connectors and other parts which normally extend beyond outside covers or cases recessed or adequately protected by some other means?

6.0   Weight labeling

    6.1   Are weight labels provided on each rack and console?

    6.2   Are weight labels provided on each equipment weighing over 50 lb?

    6.3   Are weight labels provided on three external surfaces? Three surfaces should be front, top, and rear.

C. Equipment racks–General

    1.0   Drawer mounting

        1.1   Are rack/console drawers mounted on roll-out slides?

        1.2   Are drawer slides easily and quickly detached for removal from the rack or console without use of hand tools?

        1.3   Do slides incorporate provisions for locking in the open position?

    2.0   Panel mounting

        2.1   Are panels hinge mounted?

        2.2   Are separable hinges used in panel mounting?

        2.3   Are provisions provided for bracing the panel when in the open position?

3.0 In-position maintenance
    3.1 Is in-position maintenance possible with the rack/console drawer or panel in the open position?
    3.2 Can rack/console drawers or panels be extended in the open position with power cables attached?

4.0 Cable inlets
    4.1 Are test station power or interconnecting cables routed such as to avoid connection on the front of the rack or console?

5.0 Equipment mounting (weight)
    5.1 Are the heavier items on the bottom of each rack or console? Unit weight should decrease with the increase in installation height above the floor.

6.0 Operator panels optimally positioned
Optimum panel positioning is based on a human-factors system analysis. However, as a guide, the following questions are applicable:
    6.1 For maintenance personnel in the standing position, are panels located between 40 and 70 in. above the floor? Are key, critical, or precise controls and displays located between 48 and 64 in. above the floor?
    6.2 For maintenance personnel in the sitting position, are panels located 30 in. above the floor?

7.0 Air intake/exhaust provisions
    7.1 Are air intake/exhaust provisions adequate? Are hot spots covered?
    7.2 Are air intake ports insulated from exhaust ports?
    7.3 Are blower motors of the self-lubricating type?
    7.4 Are air filters used?
    7.5 Are air filters readily accessible?

D. Packaging
  1.0 Plug-in modules and components
    1.1 Are drawer modules and components of the plug-in type?
    1.2 Are plug-in modules and components removable without the use of tools? If tools are required, are they of the standard variety?
    1.3 Are accesses between plug-in modules and components adequate to allow for hand grasping?

  2.0 Module/component stacking
    2.1 Are modules and components (not of the plug-in type) mounted with no more than four fasteners? Are the fasteners standardized?
    2.2 Are modules and components mounted such that the removal of any single unit for maintenance should not require the removal of other units?

  3.0 Accessibility based on replacement frequency
    3.1 Are modules and components mounted such that access priority has been assigned according to predicted removal failure rates?

(The component with the greatest anticipated removal frequency should be readily accessible.)

4.0   Wrong installation prevention

    4.1   Are modules and components designed to preclude installation in the wrong position?

5.0   Module/component and mounting-plate labeling

    5.1   Are modules and components labeled?

    5.2   Are module/component labels located on top of each unit or in plain sight?

    5.3   Is label information adequate (full identification nomenclature)?

    5.4   Are labels attached to the module/component mounting plate?

    5.5   Are mounting-plate labels adjacent to the applicable module or component?

    5.6   Are labels etched or embossed into the applicable surface rather than merely painted or stamped on the surface?

6.0   Guides for module/component installation

    6.1   Are guide pins provided on modules and components requiring alignment during installation?

    6.2   Are guides provides on the chassis to facilitate printed-circuit-board installation and removal?

7.0   Interchangeability

    7.1   Are modules and components having like functions both electrically and mechanically interchangeable?

    7.2   Are module/component mounting fasteners interchangeable (i.e., fastener type, size, and length)?

    7.3   Are cable harness sections interchangeable?

E. Accessibility

  1.0   Access doors provided

    1.1   Are access openings provided in areas (not readily accessible through an open drawer or panel in front of the rack or console) where periodic maintenance is anticipated?

    1.2   Are access openings optimally located for the access required?

    1.3   Are hinged doors used where physical access is required (instead of a cover plate held in place by screws or other fasteners)?

    1.4   Is a transparent window or quick-opening cover used for visual inspection accesses?

    1.5   Are access doors shaped such as to permit passage of components and implements which must pass through?

  2.0   Access-door support

    2.1   On hinged console panels and access doors, is there some means of holding the panel or door in the open-position?

    2.2   Are small access doors which are not hinge mounted (held with fasteners) attached with a chain or by some other means to preclude loss?

3.0 Access-door labeling
    3.1 Is each access door or opening labeled to indicate the items which are accessible?
    3.2 Does each access label indicate what auxiliary equipment is to be used at the access?
    3.3 Is each access labeled uniquely so each one can be clearly identified in maintenance instructions?
    3.4 Does each access label indicate the recommended time period for performing maintenance operations?

4.0 Access size
    4.1 Are access openings adequate in size?
    4.2 Do hinged panels open wide enough to allow access to components?

5.0 Access fasteners minimized
    5.1 Are access panels and door held in the closed position with a minimum quantity of fasteners? No more than four fasteners should be used.
    5.2 Are access panel door fasteners of the quick-release variety? (Hand-operated fasteners are preferred.)

6.0 Special tools minimized
    6.1 Are the number of tools required to gain access held to a minimum?
    6.2 Are required tools of the standard variety? Have special tools been eliminated?

7.0 Component accessibility
    7.1 Are accesses between plug-in modules and components adequate to allow for hand grasping?
    7.2 Are accesses between fastener-mounted modules and components adequate to allow for proper tool use?

8.0 Guides for dangerous accesses
    8.1 Are screwdriver guides provided on adjustment points near high voltages?
    8.2 When blind accesses exist, are protective features provided to preclude personnel injury?

F. Fasteners

1.0 Quick-release fasteners
    1.1 Are quick-release fasteners employed?

2.0 Fasteners standardized
    2.1 Are the number of different types of fasteners minimized?
    2.2 Is the same type and size of fastener used for a given application (i.e., all mounting bolts for a given type of item)? Are bolt/screw lengths the same? Are bolt screw thread sizes the same?
    2.3 If fasteners require torque application, are the number of differing torque requirements minimized?

2.4 Have fasteners been chosen based on the requirement for standard tools, in lieu of requiring special tools?

2.5 Are all external fasteners which are manipulated during normal maintenance of a contrasting color to the surface on which they appear? Are all other fasteners and assembly screws of the same color as the surface on which they appear?

3.0 Fasteners minimized

    3.1 Is the quantity of fasteners minimized? No more than four fasteners per application is preferable.

    3.2 Are a few large fasteners used in lieu of many small fasteners?

4.0 Hexagonal socket-head fasteners

    4.1 Are hexagonal socket-head screws used?

5.0 Captive fasteners

    5.1 Are captive fasteners used wherever lost screws, bolts, or nuts might cause a malfunction or excessive maintenance time?

    5.2 Can captive fasteners be operated by hand or using a standard tool?

    5.3 Can captive fasteners be easily replaced in case of damage?

6.0 Fasteners turns

    6.1 Are bolt/screw lengths no more than required for the application?

    6.2 Is the number of turns required to tighten a bolt screw less than 10?

    6.3 When tightened, does the bolt or screw extend two threads beyond the nut?

G. Cables

  1.0 Cables fabricated in removable sections

    1.1 Are cable harnesses designed so they can be fabricated in removable sections and installed as a unit?

    1.2 Are cables routed so they are accessible and not behind panels that are difficult to remove?

    1.3 Can cables routed through walls, cabinet holes, and so on, be easily removed?

  2.0 Cables routed to avoid sharp bends

    2.1 Are cables routed to avoid sharp bends?

    2.2 Are cables routed so they need not be bent sharply when connected or disconnected?

  3.0 Cables routed to avoid pinching

    3.1 Are cables routed so they cannot be pinched by doors, rack/console drawers, panels, and so on.

  4.0 Protection of cables routed through holes

    4.1 Are cables routed through holes, bulkheads and so on adequately protected with grommets, pads, or in some other way?

  5.0 Cable labeling

    5.1 Is cable labeling provided?

    5.2   Are individual wire numbers (wires within the cable bundle) labeled on the outside of the cable bundle?

    5.3   Are cable labels repeated (at 2- or 3-ft intervals) such that searching for cable identification is minimized?

    5.4   Are different cable sections labeled differently (i.e., two sections of the same cable are labeled 101A and 101B respectively)?

6.0   Cable clamping

    6.1   Are cables adequately supported by cable clamps (not more than 24 in. apart)?

    6.2   Are the clamps used of the type that will distribute pressure over a wide area of the cable?

    6.3   Are the clamps used of the correct size so that the cable will not be squeezed too tightly and cause possible damage?

    6.4   Are the clamps used of the quick-disconnect variety?

7.0   Handhold/step-prevention considered

    7.1   Are cables routed so they will not be walked on or used for handholds?

H. Connectors

1.0   Quick-disconnect variety

    1.1   Are quick-disconnect connectors used (no more than one full turn for connection/disconnection)? Quick-disconnect provisions should be incorporated when auxiliary test equipment is connected, when units require frequent disconnection or replacement, and when units require replacement within critical readiness times.

    1.2   Are connectors removable by hand?

2.0   Are connectors far enough apart from other obstructions so they can be grasped firmly for connecting and disconnecting? In general, minimum separations are 0.75 in. for bare fingers of 1.25 in. for the bare hand or gloved fingers.

3.0   Connector labeling

    3.1   Are connectors and receptacles labeled?

    3.2   Are plugs and their corresponding receptacles coded alike?

    3.3   Do plugs and receptacles have painted arrows or other indications to show proper position of keys for aligning pins?

    3.4   Are the pins in each plug clearly identified?

4.0   Connectors keyed

    4.1   Are plugs keyed or designed so that it is impossible to insert a plug in the wrong receptacle?

    4.2   Do aligning pins or keys extend beyond electrical pins?

5.0   Connectors standardized

    5.1   Is the same style or type of connector used? Standardization should not be carried to the extent where a plug can be inserted into the wrong receptacle.

6.0 Spare pins provided

    6.1 Are spare pins provided in each plug? The total quantity of spare contact should be two for up to 35 contacts in the connector, four for 26–100 contacts in the connector, and six for 101 or more contacts in the connector.

7.0 Male connectors capped

    7.1 Are male connectors capped to preclude damage and unsafe conditions?

    7.2 Are connector caps chained to the equipment item to preclude possible loss?

8.0 Receptacles "hot" and plugs "cold"

    8.1 Is the equipment designed such that receptacles are "hot" and plugs are "cold"?

9.0 Moisture prevention

    9.1 Where applicable, have cable "drip-loops" been provided to preclude moisture from entering the connector?

    9.2 Where possible, have moisture-proof connector provisions been considered?

I. Servicing and lubrication

  1.0 Servicing requirements

    1.1 Have servicing requirements been minimized?

    1.2 When servicing is needed, are specific requirements identified?

    1.3 Are servicing resources identified?

  2.0 Servicing points accessible

    2.1 Are servicing points accessible?

  3.0 Servicing frequencies

    3.1 Are servicing frequencies known?

J. Panel displays and controls

Panel displays and controls are defined as a result of a human-factors system analysis. Many factors influence panel arrangements, and the human-factors group should be consulted relative to criteria. The following questions cover some of the areas of maintainability interest (this checklist, covering panel displays and controls, is not intended to represent the entire picture):

  1.0 Controls standardized

    1.1 Are preferred commercial and standard parts utilized to the maximum extent possible?

    1.2 Do knobs performing the same function have the same shape?

    1.3 Are pointer-type knobs used to indicate a marking or to indicate relative position from a fixed point?

    1.4 Are bar-type knobs used where rotation is more than 360°?

    1.5 Are control knobs secured by two setscrews?

    1.6 Are goggle switches employed when control functions require no more than two discrete positions?

    1.7 Are toggle switches mounted vertically (i.e., the switch should move upward for ON, START, or INCREASE, and downward for OFF, STOP, or DECREASE)?

1.8 Is there a guard for those toggle switches susceptible to accidental activation which is considered serious?

1.9 Do values increase with clockwise control rotation?

1.10 Are display scales standardized?

1.11 Are circular display scales used in lieu of straight scales?

2.0 Controls sequentially positioned

2.1 Are controls and displays that are used sequentially aligned horizontally from left to right or vertically from top to bottom, or in rows from top to bottom row and from left to right within each row?

2.2 Are controls and displays that are used sequentially arranged in an order analogous to the actual order of events?

3.0 Control spacing

3.1 Are controls properly spaced?

4.0 Control labeling

4.1 Are controls properly labeled?

4.2 Are controls and displays identified as to their function?

4.3 Are labels located either on or immediately adjacent to (preferably above) the controls and displays to be identified?

4.4 Is the location of control and display labels consistent on all equipment ?

4.5 Is the labeling brief?

5.0 Control/display relationship

5.1 Are related controls and displays located on the same panel?

5.2 Are controls adjacent to the applicable display?

5.3 Is the control and display relationship unambiguous and unmistakable?

6.0 Meters

6.1 Are ruggedized meters incorporated?

6.2 Are meters externally removable from the outside of the panel?

6.3 Can meters be calibrated from the front panel?

7.0 Panel lighting

7.1 Is panel lighting employed?

7.2 Is panel lighting adequate?

7.3 Are panel indicator lights of the "press-to-test" variety?

8.0 Fuse requirements

8.1 Are all line fuses mounted on the applicable front panel?

8.2 Are fuses replaceable without use of tools and without moving other parts?

8.3 Are fuse holders of the self-indicating type (light indication for blown fuse)?

8.4 Are fuse-holder caps of the quick-disconnect type rather than screw-in type?

8.5 Are spare fuses provided?

8.6 Are spare fuses located adjacent to the applicable line fuse?

8.7 Are fuses labeled both as to function and value? If space is limited, provide fuse value rather than fuse function.

9.0 Warning lights employed

9.1 Are warning lights employed when a malfunction has occurred or when equipment marginal operation exists?

10.0 Color of indicating lights

10.1 Are the following colors used for the applicable indications as stated?

White—indicates those conditions that are not identified to provide a right or wrong implication (i.e., power-on light).

Green—indicates that a unit is in tolerance or that a condition is satisfactory and all right to proceed (i.e., go-ahead, in tolerance, ready, acceptance, normal, etc.).

Yellow—operation in progress transitory condition, or a condition that exists which is marginal (i.e., adjustment required or tuning in process).

Red—equipment inoperable or dangerous condition exists.

11.0 Controls placed according to frequency of use

11.1 Are emergency controls and displays placed in readily accessible positions with critical emergency controls and displays located in the optimum position?

11.2 Are primary controls aid displays placed in optimal manual and visual areas?

K. Test points (testability beyond built-in test)

1.0 Front panel location

1.1 Are test-point locations (major, intermediate, minor) compatible with the maintenance concept?

1.2 Are major (primary) test points located on the front panel?

2.0 Functionally grouped

2.1 Are test points grouped conveniently to allow for sequential testing (signal flow), testing of similar functions (power supply voltages, i–f amplifier stages, etc.), or frequency of use (when access is limited)?

3.0 Test-point labeling

3.1 Is each test point labeled with the proper signal (with tolerance limits) that should be measured at the test point?

3.2 Is each test point identified with a unique number (i.e., TP-24)?

3.3 Are test points which are located in groups sequentially numbered from left to right, or from top to bottom.

3.4 Are test-point numbers located above each test point, and are signal values located on the right side of each test point?

3.5 Are test points color-coded so that they are easily distinguishable?

4.0   Internal test points accessible
4.1   Are secondary test points readily accessible within each assembly? Are they brought out to a single point?
    4.2   Are secondary test points which are not brought out to a single point adjacent to the item being tested?
5.0   Degree of test
    5.1   Are test points provided for direct test of all replaceable parts?
    5.2   Are test points provided at the input and output for each major unit or assembly?
6.0   Adequately protected
    6.1   Are test points recessed or otherwise protected from damage by personnel, dust, or moisture?
7.0   Adequately illuminated
    7.1   Is adequate illumination provided to allow the technician to see the test-point number and labeled signal value?
    7.2   Are luminescent markings used so test points can be read in very low illumination?
8.0   Adjacent to applicable control or display
    8.1   Are test points located adjacent to the controls and displays that are used with these test points?
L.   Adjustments
  1.0   Adjustment points accessible
    1.1   Are adjustment requirements minimized (if not eliminated)?
    1.2   Are adjustment-point locations keyed to the maintenance level at which the adjustment will be made?
    1.3   Are adjustment points required for overall equipment adjustment located on the front panel?
    1.4   Are routine alignment/adjustment points located on the front panel so that individual equipment assemblies do not require removal?
    1.5   For hermetically sealed units requiring periodic adjustment, are adjustment points externally located to permit adjustment without breaking the hermetic seal?
  2.0   Periodic adjustment known
    2.1   Are periodic adjustment requirements known? Is the extent of adjustment known?
    2.2   Are periodic adjustment frequencies known?
    2.3   Have equipment *internal* adjustments been eliminated?
    2.4   Have equipment calibration adjustments during the duty cycle been eliminated?
    2.5   Can adjustments compensate for tolerance change?
  3.0   Interaction effects eliminated
    3.1   When maintenance adjustments are required, are single adjustments for a given function possible?
    3.2   Have interaction effects been eliminated?

     3.3   If interaction adjustments are required, they should be bench-type?

4.0   Adjustment locking devices

     4.1   Are adjustment locking devices provided to preclude inadvertent changes to adjustment-point settings?

5.0   Factory adjustments

     5.1   Are factory adjustments specified?

     5.2   Are critical adjustments which, if improperly made, could result in serious degradation to system performance or reliability, identified as factory adjustments?

6.0   Adjustment-point labeling

     6.1   Is each adjustment point labeled with the adjustment-point number and he adjustment tolerances?

7.0   Fine adjustments through large movements

     7.1   Are fine adjustments caused by large adjustment increments?

     7.2   Are stopping devices employed to prevent overriding the limit of adjustment?

8.0   Built-in jacks for meter calibration

     8.1   Are built-in jacks provided to allow for meter calibration with the meter in the installed position?

9.0   Clockwise adjustments for increasing values

     9.1   Does clockwise rotation of an adjustment point result in an increased adjustment value?

M. Parts and components

1.0   Grouping

     1.1   Are parts and components functionally grouped?

     1.2   Are parts such as resistors, capacitors, tube sockets, and so on mounted on subassemblies rather than on the unit chassis?

     1.3   Are parts which are arranged in family groups outlined by painted borders or outlined in some other manner?

2.0   Labeling

     2.1   Are parts and components adequately labeled (i.e., resistor number and resistor value)?

     2.2   Are potted parts labeled with current, voltage, impedance, terminal information, and so on?

     2.3   Are part/component labels adjacent to the applicable item, and visible after assembly?

3.0   Adequate space for tool access

     3.1   Is direct tool access provided to allow removal of a single part or component? This includes space required to solder a joint with a soldering iron and to remove or install a fastener with a screwdriver or socket wrench.

4.0   Individual parts directly accessible

     4.1   Are individual parts directly accessible (do not require the removal of other parts for access)?

        4.2   If individual parts are not directly accessible, is the number of parts requiring removal to gain access minimized?

    5.0   Delicate parts

        5.1   Are delicate parts located where they will not be damaged during maintenance operations? If not, are they adequately protected?

    6.0   Part vulnerability

        6.1   Are parts far enough apart so that adjacent parts will not be damaged by a hot soldering iron or by the use of other tools?

        6.2   Are leads and terminals far enough apart so that adjacent items will not be damaged by a hot soldering iron or by the use of other tools?

        6.3   Is the clearance between the soldered connection and the part between $1/4$ and $1/2$ in?

N. Environment

    1.0   Temperature and humidity ranges

        1.1   Have temperature and humidity requirements been considered?

    2.0   Illumination

        2.1   Is illumination adequate to enable maintenance operations? The average electronic shop maintains 20 footcandles, and small detailed-type maintenance work usually requires 50 footcandles. Good illumination should include suitable brightness, uniform lighting, suitable contrast between task and background, freedom from glare, and suitable quality and color for illuminants and surfaces.

        2.2   Has built-in rack/console lighting been provided?

    3.0   Transportability

        3.1   Have transportation considerations been incorporated?

        3.2   Is equipment designed for air transportation? Are attachment/tie-down provisions incorporated on the equipment rack or console outside packing cases? Are hoisting fittings provided?

        3.3   Is labeling adequate? Are weight and hoist fitting labels provided?

        3.4   Will equipment in shipping cases pass through intermediate shop doorways?

    4.0   Mobility

        4.1   Have mobility provisions been considered?

    5.0   Storage conditions

        5.1   Can the equipment be stored for extended periods of time without degrading affects?

        5.2   Have special servicing requirements for stored equipment been minimized?

        5.3   Have servicing requirements for stored equipment been defined?

        5.4   Have servicing resources been identified?

        5.5   Have storage environments been identified?

6.0   External impacts
    6.1   Have all detrimental impacts on the environment been eliminated?

O. Safety
    1.0   Electrical outlets/junction boxes
        1.1   Are all electrical outlets or junction boxes labeled as to the voltages contained therein?
        1.2   Are electrical outlets or junction boxes designed or covered to preclude inadvertent hazardous contact?
    2.0   Interlocks
        2.1   Are interlocks provided where potentials exceed 40 volts?
        2.2   Where interlocks are provided, is there a means of bypassing for servicing with a proper warning indicator?
    3.0   Fuse/circuit-breaker protection
        3.1   Are circuit breakers provided in lieu of fuses?
        3.2   Is fuse/circuit-breaker protection adequate (protection on both sides of a line)?
        3.3   Are spare fuses provided?
        3.4   Are fuses or circuit breakers provided to safeguard against damage which might result from the wrong switch or jack position?
    4.0   Warning decals
        4.1   Are adequate warning labels provided in high-voltage areas and in other areas where hazards exist?
        4.2   Are labels placed adjacent to the hazardous area and in plain sight?
    5.0   Guards/safety covers
        5.1   Are guards, safety covers, and warning plates provided for high potentials (in excess of 350 volts) on contacts, terminals, and like devices?
        5.2   On terminals that are used as test points, are adequate barriers between adjacent terminals provided to alleviate the possibility of a probe's slipping to the adjacent terminal?
    6.0   Protruding devices
        6.1   Are unnecessary protruding devices eliminated?
        6.2   Are delicate protruding parts suitably protected?
        6.3   When performing bench maintenance, are legs or standoffs provided to protect parts?
    7.0   External metal parts
        7.1   Are all external metal parts adequately grounded?
    8.0   Drawer/panel/structure edges
        8.1   Are drawer/panel/structure edges rounded?
        8.2   Are accesses provided with fillets, rubber, fiber, or plastic protection to preclude personal injury?
        8.3   Are covers on outside cases adequately grounded?

9.0 Tools

    9.1 Are tools which must be used near high-voltage areas adequately insulated at the handle or at other parts of the tool which the technician is likely to touch?

    9.2 Are screwdriver guides used for adjustments which are near high-voltage areas?

P. Reliability

    1.0 Has the system/equipment wear-out period been defined?

    2.0 Have failure modes and effects been identified?

    3.0 Are item failure rates known?

    4.0 Have parts with excessive failure rates been identified?

    5.0 Has mean life been determined?

    6.0 Have adequate derating factors been established and adhered to where appropriate?

    7.0 Has equipment design complexity been minimized?

    8.0 Is protection against secondary failures (resulting from primary failures) incorporated where possible?

    9.0 Has the use of adjustable components been minimized?

   10.0 Has the use of friction or pressure contacts in mechanical equipment been avoided?

   11.0 Have all critical-useful-life items been eliminated from the equipment design?

   12.0 Have cooling provisions been incorporated in design "hot spot" areas?

   13.0 Is cooling directed toward the most critical items?

   14.0 Have all reliability program requirements been met?

Q. Software

    1.0 Have all system software requirements for maintenance activities been met?

    2.0 Is the maintenance software complete in terms of scope and depth of coverage?

    3.0 Is the software compatible relative to the equipment with which it interfaces? Is the maintenance software compatible with the operating software? Other elements of the system?

    4.0 Is the maintenance software language compatible with system language requirements in general?

    5.0 Has the software been adequately tested and verified for accuracy and reliability?

    6.0 Have software reliability and maintainability requirements been incorporated?

    7.0 Has the software been packaged in easily removable modules?

    8.0 Are the software packages easy to maintain?

APPENDIX B
LIFE-CYCLE COST ANALYSIS

System maintainability is one of the primary criteria used in the evaluation of design alternatives. For maximum benefit and completeness, system design evaluation in general, and system maintainability in particular, must be considered and addressed within the context of the overall system life cycle. In other words, the criteria selected to affect the evaluation of alternative configurations must reflect the relevant life-cycle considerations. Since cost is a major parameter in the evaluation process, it must also be considered in a similar manner (i.e., life-cycle cost). The material in this appendix is included to supplement the discussions in Section 4.7 on economic factors and in Section 6.3.5 on life-cycle cost analysis.

Life-cycle cost analysis consists of the process of evaluating alternative system design configurations from an economic perspective. Such an evaluation activity commences as part of the early decision-making process in conceptual design and extends throughout the subsequent design and development effort leading to an operational system configuration. This process basically follows the analysis approach described briefly in Section 6.3.5 (refer to Figure 6.29), and is often applied as an inherent part of the LSA discussed in Section 9.11.

To illustrate the life-cycle cost analysis process in more detail, some of the major activities involved are described in the following paragraphs.

B.1 DEFINITION OF SYSTEM REQUIREMENTS

In evaluating alternatives on the basis of life-cycle cost, the analyst needs to project each alternative in terms of life-cycle activities or events. Such activities (or events) generally evolve from a combination of stated requirements, concepts, plans, and so on, which have a significant impact on follow-on design, production, operation, and support requirements. In other words, the life-cycle cost analysis needs to be based on a definition of system operational requirements, a definition of the maintenance concept (Chapter 5), and a program plan and profile illustrating major life-cycle activities and the projected operational and sustaining maintenance and support horizon for the system.

These requirements need to be addressed to the extent deemed necessary to define the problem at hand in specific terms. For a large complex system, this definition may be rather extensive, involving a significant level of detail. On the other hand, for small-scale systems, or components of a system, one still needs to address requirements in terms of the functions to be performed, quantities required and geographical distribution, reliability factors, maintainability factors, the maintenance and support concept, operational horizon period, and so on. However, the level of detail may be very limited. In other words, the extent to which system requirements need to be defined is a function of the system being evaluated, its nature, prior organizational experience with similar systems, and the time in the life cycle when the analysis is being performed.

In any event, regardless of the type of problem at hand, the configuration(s) being evaluated must be projected in terms of system-level requirements. Further, these requirements may be modified and changed as the program evolves and as more information becomes available. However, an initial baseline must be established. Changes to this initial baseline may be evaluated systematically and in a controlled manner, and integrated to form a new baseline in accordance with the configuration management procedures discussed in Chapter 10.

B.2   COST BREAKDOWN STRUCTURE (COST CATEGORIES)

When accomplishing a life-cycle cost analysis, the analyst must develop a cost breakdown structure (i.e., a cost tree) showing the numerous categories that are combined to provide the total cost. There is no set method for breaking down costs as long as the method used can be tailored to the specific application. However, the cost breakdown structure should exhibit the following basic characteristics:

1. All system cost elements must be considered.

2. Cost categories are generally identified with a significant level of activity or some major item of hardware. Cost categories must be well defined. The analyst, manager, customer, supplier, and so on must all have the same understanding of what is included in a given category and what is not! Cost doubling (i.e., counting the same cost in two or more categories) and omissions must both be eliminated. Lack of adequate definition causes inconsistencies in the evaluation process, and could lead to a wrong decision.

3. The cost structure and categories should be coded in such a manner as to allow for the analysis of certain specific areas of interest (e.g., system operation, energy consumption, equipment design, spares, maintenance personnel and support, and maintenance equipment and facilities) while virtually ignoring other areas. In some instances, the analyst may wish to pursue a designated area in depth while covering other areas with gross top-level estimates. This will certainly occur from time to time as a system evolves through the different phases of its life cycle. The areas of concern (for decision-making purposes) will vary.

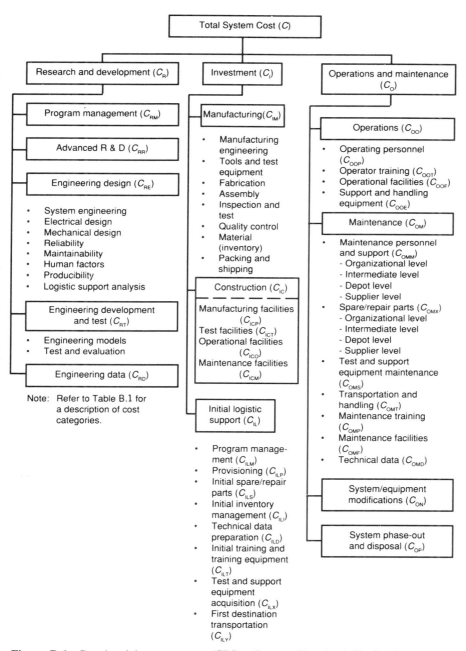

Figure B.1. Cost breakdown structure (CBS). (*Source:* Blanchard, B. S., *System Engineering Management,* John Wiley & Sons, N.Y., 1991, p. 322).

TABLE B.1 Description of Cost Categories (Partial)

Cost Category (Figure B.1)	Method of Determination (Quantitative Expression)	Cost Category Description and Justification
Total system cost (C)	$C = (C_R + C_I + C_O)$ $C_R = $ R and D cost $C_I = $ Investment cost $C_O = $ Operations and maintenance cost	Includes all future costs associated with the acquisition, utilization, and subsequent disposal of system/equipment.
Research and development (C_R)	$CR = (C_{RM} + C_{RR} + C_{RE} + C_{RT} + C_{RD})$ $C_{RM} = $ Program management cost $C_{RR} = $ Advanced R&D cost $C_{RE} = $ Engineering design cost $C_{RT} = $ Equipment development/test cost $C_{RD} = $ Engineering data cost	Includes all costs associated with conceptual/feasibility studies, basic research, advanced research and development, engineering design, fabrication and test of engineering prototype models (hardware), and associated documentation. Also covers all related program management functions. These costs are basically nonrecurring.
Investment (C_I)	$C_I = (C_{IM} + C_{IC} + C_{IL})$ $C_{IM} = $ System/equipment manufacturing cost $C_{IC} = $ System construction cost $C_{IL} = $ Cost of initial logistic support	Includes all costs associated with the acquisition of systems/equipment (once design and development has been completed). Specifically, this covers manufacturing (recurring and nonrecurring), manufacturing management, system construction, and initial logistic support.
Operations and maintenance (C_O)	$C_O = (C_{OO} + C_{OM} + C_{ON} + C_{OP})$ $C_{OO} = $ Cost of system/equipment life-cycle operations $C_{OM} = $ Cost of system/equipment life-cycle maintenance $C_{ON} = $ Cost of system/equipment life-cycle modifications $C_{OP} = $ Cost of system/equipment phase-out and disposal	Includes all costs associated with the operation and maintenance support of the system throughout its life cycle subsequent to equipment delivery in the field. Specific categories cover the cost of system operation, maintenance, sustaining logistic support, equipment modifications, and system/equipment phase-out and disposal. Costs are generally determined for each year throughout the life cycle.

Equipment development and test (C_{RT})

$$C_{RT} = \left(C_{RDL} + C_{RDM} + \sum_{i=1}^{N} C_{RDT_i} \right)$$

C_{RDL} = Cost of prototype fabrication and assembly labor

C_{RDM} = Cost of prototype material

C_{RDT_i} = Cost of test operations and support associated with specific test i

N = Number of identifiable test

The fabrication, assembly, test and evaluation of engineering prototype models (in support of engineering design activity—C_{RE}) is included herein. Specifically, this constitutes fabrication and assembly, instrumentation, quality control and inspection, material procurement and handling, logistic support (personnel, training, spares, facilities, support equipment, etc.), data collection, and evaluation of prototypes. Initial logistic support for operational system/equipment is covered in C_{IL}.

Initial logistic support (C_{IL})

$$C_{IL} = (C_{ILM} + C_{ILP} + C_{ILS} + C_{ILI} + C_{ILD} \\ + C_{ILT} + C_{ILX} + C_{ILY})$$

C_{ILM} = Logistic program management cost

C_{ILP} = Cost of provisioning

C_{ILS} = Initial spare/repair part material cost

C_{ILI} = Initial inventory management cost

C_{ILD} = Cost of technical data preparation

C_{ILT} = Cost of initial training and training equipment

C_{ILX} = Acquisition cost of operational test and support equipment

C_{ILY} = Initial transportation and handling cost

Includes all integrated logistic support planning and control functions associated with the development of system support requirements, and the transition of such requirements from supplier(s) to the applicable operational site. Elements cover:

a. Logistic program management cost—management, control, reporting, corrective action system, budgeting, planning, and so on.

b. Provisioning cost—preparation of data which is needed for the procurement of spare/repair parts and test/support equipment.

c. Initial spare/repair parts cost—spares material stocked at the various inventory points to support the maintenance needs of prime equipment, test and support equipment, and training equipment. Replenishment spares are covered in C_{OMX}.

d. Initial inventory management cost—cataloging, lighting, coding, and so on of spares entering the inventory.

461

TABLE B.1 (*Cont.*)

Cost Category (Figure B.1)	Method of Determination (Quantitative Expression)	Cost Category Description and Justification
		e. Technical data preparation cost—development of operating and maintenance instructions, test procedures, maintenance cards, tapes, and so on. Also includes reliability and maintainability data, test data, and so on covering production and test operations.
		f. Initial training and training equipment cost—design and development of training equipment, training aids/data, and the training of personnel initially assigned to operate and maintain the prime equipment, test and support equipment, and training equipment. Personnel training costs include instructor time, supervision, student pay and allowances, training facilities, and student transportation. Training accomplished on a sustaining basis throughout the system life cycle (due to personnel attrition) is covered in C_{OOT} and C_{OMP}.
		g. Test and support equipment acquisition cost—design, development, and acquisition of test/support equipment plus handling equipment needed to operate and maintain prime equipment in the field. The maintenance of test and support equipment throughout the system life cycle is covered in C_{OOE} and C_{OMS}.
		h. Initial transportation and handling cost (first destination transportation of logistic support elements from supplier to the applicable operational site).

parts, test and support equipment, transportation and handling, replenishment training, support data, and facilities necessary to meet the maintenance needs of the prime equipment throughout its life cycle. Such needs include both corrective and preventive maintenance requirements at all echelons—organizational, intermediate, depot, and factory.

Initial spare/repair part costs are covered in C_{ILS}. This category includes all replenishment spare/repair parts and consumable materials (oil, lubricants, fuel, etc.) that are required to support maintenance activities associated with prime equipment, operational support and handling equipment (C_{OOE}), test and support equipment (C_{OMS}), and training equipment at each level (organizational, intermediate, depot, supplier). This category covers the cost of purchasing; the actual cost of the material itself; and the cost of holding or maintaining items in inventory. Costs are assigned to the applicable level of maintenance. Specific quantitative requirements for spares (Q_M) are derived from the MTA and LSA. The optimum quantity of purchase orders (Q_A) is based on the EOQ criteria. Support equipment spares are based on the same criteria used in determining spare part requirements for prime equipment.

$C_{OMP} + C_{OMF} + C_{OMD}$

C_{OMM} = Maintenance personnel and support cost

C_{OMX} = Cost of spare/repair parts

C_{OMS} = Test and support equipment maintenance cost

C_{OMT} = Transportation and handling cost

C_{OMF} = Cost of maintenance facilities

C_{OMD} = Cost of technical data

Spare/repair parts costs (C_{OMX})

$C_{OMX} = (C_{SO} + C_{SI} + C_{SD} + C_{SS} + C_{SC})$

C_{SO} = Cost of organizational spare/repair parts

C_{SI} = Cost of intermediate spare/repair parts

C_{SD} = Cost of depot spare/repair parts

C_{SS} = Cost of supplier spare/repair parts

C_{SC} = Cost of consumables

$$C_{SO} = \sum_{N_{MS}} [(C_A)(Q_A) + \sum_{i=1} (Q_{Mi})(C_{Mi}) + \sum_{i=1} (C_{Hi})(Q_{Hi})]$$

C_A = Average cost of material purchase order (\$/order)

Q_A = Quantity of purchase orders

C_M = Cost of spare item i

Q_M = Quantity of i items required

C_H = Cost of maintaining spare item i in the inventory (\$/\$ value of the inventory)

Q_H = Quantity of i items in the inventory

N_{MS} = Number of maintenance sites

C_{SI}, C_{SD}, and C_{SS} are determined in a similar manner.

TABLE B.2 Summary of Terms

C	Total system life-cycle cost
C_A	Average cost of material purchase order (\$/order)
C_{DC}	Cost of maintenance documentation/data for each corrective maintenance action (\$/MA)
C_{DIS}	Cost of system/equipment disposal
C_{DP}	Cost of maintenance documentation/data for each preventive maintenance action (\$/MA)
C_H	Cost of maintaining spare item i in the inventory or inventory holding cost (\$/dollar value of the inventory)
C_I	Total investment cost
C_{IC}	Construction cost
C_{ICA}	Construction fabrication labor cost
C_{ICB}	Construction material cost
C_{ICC}	Capital equipment cost
C_{ICM}	Maintenance facilities acquisition cost
C_{ICO}	Operational facilities acquisition cost
C_{ICP}	Manufacturing facilities cost (acquisition and sustaining)
C_{ICT}	Test facilities cost (acquisition and sustaining)
C_{ICU}	Cost of utilities
C_{IL}	Initial logistic support cost
C_{ILD}	Cost of technical data preparation
C_{ILI}	Initial inventory management cost
C_{ILM}	Logistics program management cost
C_{ILP}	Cost of provisioning (preparation of procurement data covering spares, test and support equipment, etc.)
C_{ILS}	Initial spare/repair part material cost
C_{ILT}	Cost of initial training and training equipment
C_{ILX}	Acquisition cost of operational test and support equipment
C_{ILY}	Initial transportation and handling cost
C_{IM}	Manufacturing cost
C_{IN}	Nonrecurring manufacturing/production cost
C_{INA}	Quality assurance cost
C_{INM}	Manufacturing engineering cost
C_{INP}	Manufacturing management cost
C_{INQ}	Cost of qualification test (first article)
C_{INS}	Cost of production sampling test
C_{INT}	Tools and factory equipment cost (excluding capital equipment)
C_{IR}	Recurring manufacturing/production cost
C_{IRE}	Recurring manufacturing engineering support cost
C_{IRI}	Inspection and test cost
C_{IRL}	Production fabrication and assembly labor cost
C_{IRM}	Production material and inventory cost
C_{IRT}	Packing and initial transportation cost
C_M	Cost of spares item i
C_{MHC}	Cost of material handling for each corrective maintenance action (\$/MA)
C_{MHP}	Cost of material handling for each preventive maintenance action (\$/MA)
C_O	Operations and maintenance cost
C_{OCP}	Corrective maintenance labor cost (\$/M_{MHC})

TABLE B.2   (*Cont.*)

C_{OM}	Cost of system/equipment life-cycle maintenance
C_{OMD}	Cost of technical data
C_{OMF}	Cost of maintenance facilities
C_{OMM}	Maintenance personnel cost
C_{OMP}	Cost of replenishment maintenance training
C_{OMS}	Test and support equipment maintenance cost
C_{OMT}	Transportation and handling cost
C_{OMX}	Spare/repair parts cost (replenishment spares)
C_{ON}	Cost of system/equipment modifications
C_{OO}	Cost of system/equipment life-cycle operations
C_{OOE}	Cost of support and handling equipment
C_{OOF}	Cost of operational facilities
C_{OOO}	Cost of operation for support and handling equipment
C_{OOP}	Operating personnel cost
C_{OOS}	Cost of equipment preventive (scheduled) maintenance
C_{OOT}	Cost of replenishment training
C_{OOU}	Cost of equipment corrective (unscheduled) maintenance
C_{OP}	Cost of system/equipment phase-out and disposal
C_{OPP}	Preventive maintenance labor cost ($\$/M_{MHP}$)
C_P	Cost of packing
C_{PO}	Cost of operators labor (\$/hour)
C_{PPE}	Cost of operational facility support (\$/operational site)
C_{PPF}	Cost of operational facility space (\$/square foot/site)
C_{PPM}	Cost of maintenance facility support (\$/maintenance site)
C_{PPO}	Cost of maintenance facility space (\$/square foot/site)
C_R	Total research and development cost
C_{RD}	Engineering data cost
C_{RDL}	Prototype fabrication and assembly labor cost
C_{RDM}	Prototype material cost
C_{RDT}	Prototype test and evaluation cost
C_{RE}	Engineering design cost
C_{REC}	Reclamation value
C_{RM}	Program management cost
C_{RR}	Advanced research and development cost
C_{RT}	Equipment development and test cost
C_{SC}	Cost of consumables
C_{SD}	Cost of depot spare/repair parts
C_{SED}	Cost of depot test and support equipment
C_{SEI}	Cost of intermediate test and support equipment
C_{SEO}	Cost of organizational test and support equipment
C_{SI}	Cost of intermediate spare/repair parts
C_{SO}	Cost of organizational spare/repair parts
C_{SS}	Cost of supplier spare/repair parts
C_T	Cost of transportation
C_{TOM}	Cost of maintenance training (\$/student-week)
C_{TOT}	Cost operator training (\$/student-week)
C_{TP}	Packing cost (\$/pound)
C_{TS}	Shipping cost (\$/pound)

TABLE B.2   *(Cont.)*

C_U	Cost of utilities (\$/operational site)
fc	Condemnation factor (attrition)
fpt	Frequency of preventive maintenance (actions/hour of equipment operation)
M_{MHC}	Corrective maintenance man-hours/maintenance action
M_{MHP}	Preventive maintenance man-hours/maintenance action
N_{MS}	Number of maintenance sites
N_{OS}	Number of operational sites
N_{PO}	Number of operating systems
Q_A	Quantity of purchase orders
Q_{CA}	Quantity of corrective maintenance actions
Q_H	Quantity of i items in the inventory
Q_M	Quantity of i items required or demanded
Q_{PA}	Quantity of preventive maintenance actions
Q_{PO}	Quantity of operators/system
Q_{SM}	Quantity of maintenance students
Q_{SO}	Quantity of student operators
Q_T	Quantity of one-way shipments
S_O	Facility space requirements (square feet)
T_O	Hours of system operation
T_T	Duration of training program (weeks)
W	Weight of item (pounds)

4. When related to a specific program, the cost structure should be compatible (through cross-indexing, coding, etc.) with the contract WBS and with the management accounting procedures used in collecting costs (refer to Section 2.3.1). Certain costs are derived from accounting records and should be a direct input to the life-cycle cost analysis.

5. For programs, where subcontracting is prevalent, it is often desirable and necessary to separate supplier costs (i.e., initial bid price and follow-on program costs) from the other costs. The cost structure should allow for the identification of specific work packages that require close monitoring and control.

An example of a cost breakdown structure is presented in Figure B.1. Some of the particular categories in the structure are supported by the descriptive material and quantitative expressions in Table B.1 for illustration purposes. A summary listing of relevant terms is represented in Table B.2

Referring to Figure B.1, costs may be accumulated at different levels depending on the areas of interest and the depth of detail required. In some instances, the analyst may wish to thoroughly investigate maintenance cost (C_{OM}) while roughly estimating operations cost (C_{OO}). On the other hand, engineering design cost (C_{RE} may be amplified while looking at operations and maintenance cost (C_O) in terms of an estimated total value. The cost breakdown structure should incorporate the flexibility to allow for cost collection both horizontally and vertically, and in functional terms. This can readily be accomplished through the use of computer methods and the proper coding.

Referring to Table B.1, the quantitative expressions which represent the costs in each of the categories define the requirements for cost estimating. In some instances, specific known cost factors are used, while in other cases the establishment of cost estimating relationships may be required.

B.3   COST ESTIMATING

A cost estimate is an opinion or judgment concerning the expected cost to be accrued in the acquisition and/or utilization of an item. Cost estimates are derived from:

1. Know factors or rates
2. Estimating relationships (analogous and/or parametric)
3. Expert opinion

The first category is fairly straightforward. The analyst (or cost estimator) uses the current cost of an incremental unit of some commodity or service and multiplies it by the quantity of units involved. Some examples are presented in Table B.3.

The second category (estimating relationships) seeks to establish some higher level of estimation. Included in this category is the use of analytic tools that relate various cost categories to cost-generating or explanatory variables. For instance, it may be feasible to relate life-cycle cost in terms of unit weight, cost per mile of range, cost per maintenance action, cost per equipment module, cost per quantity of maintenance support personnel or in terms of quantity of operating personnel, cost per unit of volume, cost per unit of reliability, and so on.

Cost estimating relationships may assume numerous forms, varying from informal rules of thumb or simple analogies to more formal mathematical relationships derived through a statistical analysis of empirical data. Generally, cost and related data are collected on existing systems in the inventory, analyzed, converted to the form of some relationship, and applied to a new system (which is similar in form and function) as a predicting tool. Given an identifiable database, the analyst assumes some theoretical relationship and then proceeds to test that relationship for validity (hypothesis testing). Testing may range from the use of simple graphics to a complex statistical test using well-defined data samples.

The estimating relationship may be a simple linear function identified by the equation

$$\text{cost} = (\text{some constant})\ (\text{variable } X) \tag{B.1}$$

From a series or mass of data points, a relationship can be established through curve-fitting techniques such as "least squares" or through a conventional "linear regression" analysis.

TABLE B.3   Typical Cost Factors[a]

1. Personnel labor	
Operators (C_{PO})	$27.50/hr
Preventive maintenance (C_{OPP})	
Organizational	$21.98/hr
Intermediate	$24.81/hr
Depot	$25.80/hr
Corrective maintenance (C_{OCP})	
Organizational	$21.98/hr
Intermediate	$23.81/hr
Depot	$25.80/hr
2. Transportation	
Shipping (C_{TS})	$63/lb-mi
Packing (C_{TP})	$65/lb-mi
3. Training	
Operational (C_{TOT})	$675/student-wk
Maintenance (C_{TOM})	$900/student-wk
4. Facilities	
Operational (C_{PPF})	$45/ft^2
Maintenance (C_{PPO})	$60/ft^2
5. Inventory	
New item entry	
Repairable	$435/item
Nonrepairable	$372/item
Maintenance	20% of inventory value ($/$/year)
6. Data	
Technical manuals	$1000/page
Maintenance documentation	$500/MA
7. Consumables	
Oil	$1.55/qt
Gasoline	$1.12/gal

[a] Referring to the information in Table B.3, it is obvious that the analyst must be continually aware of changing prices or rates.

Nonlinear cost relationships may be stated in the form of some distribution such as normal, log-normal, exponential, and so on. Sometimes the relationship assumes one form between two discrete values of a variable and another form between other values of the same variable. A simple example is the step function, where cost is constant over a certain range, then suddenly jumps to a higher level before becoming constant once again over another range of values. Determining production costs for a varied quantity of equipment items is a good example of the step-function concept.

Still another form of the estimating relationship is the multivariate function. In all situations, it may not be possible to express cost in terms of a single variable. Cost may be expressed as a function of system range and weight, utilization and the number of system elements, spare parts and inventory level, and so on. The cost function may take the form

$$\text{cost} = 100 + (K_1) \,(\text{variable } X_1) + (K_2) \,(\text{variable } X_2) \tag{B.2}$$

where K_1 and K_2 are some fixed constants.

The cost-estimating relationship must be reasonable and it must have predictive value. It is a highly significant factor in the analysis and, when improperly applied, can lead to the introduction of a great deal of error in the results. The type of technique used is dependent on the problem being solved and the type of data available to the analyst. If unsure of this area, the student is advised to review additional material before proceeding.[1]

The last category of cost estimating is the one dictated by expert opinion. Although argumentative, it is often the only method available to the analyst since backup data are sometimes scarce, if it exists at all. In such instances, when expert opinion is used in the analysis, the analyst should be sure to include assumptions and the rationale that support his or her position. Further, when estimating costs, the following issues need to be addressed:

1. *Discounting.* Discounting relates to the time value of money. Time is valuable, and the analyst cannot treat benefits and costs as being equal without considering the time element. When comparing two or more alternatives, a common base or equivalency is necessary to ensure a fair evaluation. Since the common base is usually the present point in time, all future costs must be adjusted to the present value. Discounting refers to the application of a selected rate of interest to a cost stream such that each future cost is adjusted to the present time, the point when the decision is made.

The procedure for discounting is simple; however, selecting the proper discount rate is often difficult. The proper choice of a discount rate used in comparing future dollars with today's dollars depends on the options available for exchanging one for the other. In other words, all the investment opportunities should be evaluated. Through using the present-value concept, these opportunities are evaluated on a comparable basis.

Thus, when accomplishing a life-cycle cost analysis, costs are determined by category (see Table B.1) for each year in the life cycle. A discount rate may be applied to adjust those costs back to the decision point. An example of the calculation for discounting is presented in Chapter 4.

2. *Inflation.* Inflation is another manner in which the time element can impact the value of money and, therefore, present and future costs. Consideration of inflation is necessary in the performance of a life-cycle analysis. Costs are estimated for each year, inflated to cover similar activities in future years of the life cycle, and related in terms of a cost profile for the projected life cycle. Note that different

---

[1] Fisher, G. H., *Cost Considerations in Systems Analysis,* American Elsevier Publishing Company, Inc., New York, 1971, Chapter 6.

inflation rates may be applied to different categories of activity; (personnel, materials, etc.).

3. *Learning curves.* When accomplishing a process on a repetitive basis, learning takes place and the experience gained often results in reduced cost. This is particularly true in the performance of the same maintenance task over time or production of a large quantity of a given item. The cost of the first unit is generally higher than the cost of the fiftieth unit, which may be higher than the cost of the hundreth unit, and so on. This is primarily owing to job familiarization by workers in the production facility, development of more efficient methods of assembly, use of more efficient tools, improvement in overall management, and so on. This continues until a leveling off takes place.

In such instances, it may be appropriate to apply a learning curve in order to prepare a more realistic cost profile. For example, if the cost of producing the tenth unit is 80% of the fifth unit cost, the cost of producing the twentieth unit is 80% of the tenth unit cost, and the cost of producing the fortieth unit is 80% of the twentieth unit cost, then the production process is said to follow an 80% *unit* learning curve. This cost learning curve can be applied to the manufacturing cost category under investment. An illustration of several learning curves is presented in Figure B.2. If it turns out that the average cost of producing the first 20 units is 80% of the average cost of producing the first 10 units, then the process follows an 80% *cumulative average* learning curve.

The application of learning curves must not only consider labor costs, but should also cover material costs which may vary because of shifts in procurement methods (large lot quantity purchases) and inventory policies. There are many factors involved and different learning curves may be applied depending on the situation. To gain more insight on the subject, the review of additional material is recommended.[2]

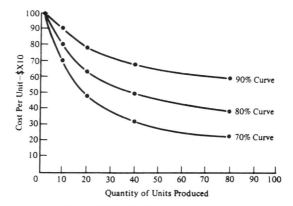

Figure B.2. Unit learning curves.

[2]Fabrycky, W. J., P. M. Ghare, and P. E. Torgersen, *Applied Operations Research and Management Science,* Prentice-Hall, Inc., Englewood Cliffs, NJ, 1984.

To further facilitate the process of cost estimating, it is recommended that an attempt be made to first identify the functions and activities that need to be accomplished. This can be achieved through a functional analysis and through the utilization of graphical modeling languages such as functional flow diagrams and IDEF models (discussed in Chapter 6). Thereafter, delineate the type and quantity of resources required to complete each function and activity. These resources could be in terms of facilities, equipment, raw material, personnel, skill levels, procedures and methodologies, fuel and energy, and so on. Linking resources (types and quantities) to functional activity areas in some detail provides an opportunity to more accurately estimate costs, and further, allows for a more effective cause-and-effect analysis when investigating the high-cost contributors to effect improvements or reductions in overall system life-cycle cost. This methodology is also called *activity based costing* (ABC).

In summary, cost estimating involves a variety of techniques, and depends on the availability of various categories of data. The sources for many of these data are spread throughout a project organization, and Figure B.3 illustrates some of the interfaces that exist. The data are not always available when required, and cost estimating methods are utilized. However, as the program progresses and system definition occurs, more and more of these sources become available. The interfaces are many, and the interrelationships that prevail must be thoroughly understood.

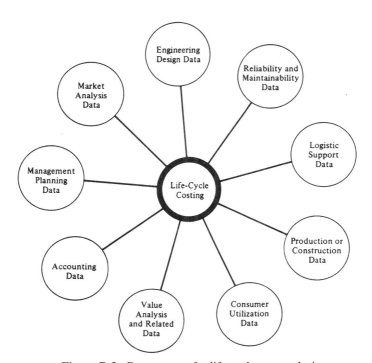

Figure B.3. Data sources for life-cycle cost analysis.

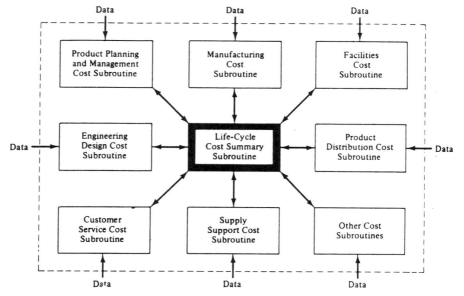

Note: Although not shown here, many of the subroutines actually interact
with each other in terms of data input and output.

Figure B.4. Life-cycle cost model configuration (sample).

B.4   COST MODEL

A model is a simplified representation of a real-world situation, and is an analytical
tool employed in the decision-making process. Cost models can be classified in a
number of ways—according to the function they serve, according to the anticipated
repetitive type usage in terms of the subject area they are intended to represent, and
so on.

With reference to the material covered in this text, the model is the analytical
tool which combines the effectiveness elements and cost categories to generate life-
cycle cost data in a timely manner for evaluation purposes. Figure B.4 illustrates a
typical model makeup in terms of subroutine interfaces, and some model applications are discussed in the summary information presented in Appendix C.

B.5   COST PROFILE

Once costs are determined for each year in the projected life cycle and inflated to
reflect real budgetary estimates, the results can be presented in the form of a cost
profile as illustrated in Figure B.5. This illustration is similar to the approach presented in Chapter 6.

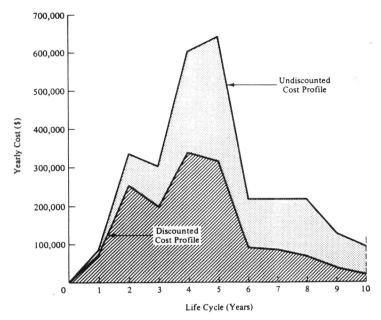

Figure B.5. Life-cycle cost profiles.

Referring to Figure B.5, two profiles are shown. First, the inflated undiscounted profile is shown to reflect future *budgetary* requirements. Second, for *comparative* purposes, a discounted cost profile is presented (assuming a 15% discount rate). Economically speaking, the comparison of alternatives requires the evaluation of various discounted profiles.

B.6   EVALUATION OF ALTERNATIVES

In the evaluation of alternatives, a discounted cost profile is developed for each candidate. The appropriate profiles are compared, a breakeven analysis is accomplished, and an approach is selected (refer to Example 5 described in Section 6.3.5).[3]

---

[3] The comparative evaluation of alternative system design configurations from an economic perspective and life-cycle costing in general is covered in further detail in Fabrycky, W. J., and B. S. Blanchard, *Life-Cycle Costing and Economic Analysis,* Prentice-Hall, Inc., Englewood Cliffs, NJ, 1991.

APPENDIX C
MAINTENANCE AND SUPPORT MODELS

Experience through the years has justified the development and utilization of a wide variety of models in support of system maintainability and maintenance-related analyses. The accomplishment of maintainability predictions, level of repair analyses, spare/repair parts and inventory policy analyses, FMECAs, RCM analysis, maintenance shop repair and material flow analyses, life-cycle cost analyses, and so on, has in the past and will continue to be a requirement for most system developmental efforts. Further, given that most system design activities today are being performed in a computer-aided environment, the maintainability engineer has to respond in order to better integrate his/her activities with the mainstream design process. Proper use of selected computer-based methods and tools facilitates the integration of maintainability principles into the total design effort and contributes to the achievement of the overall objectives. These principles and objectives guide the development of relevant and applicable methodologies and practices, which in turn support a customized system engineering process and the underlying activities as shown in Figure C.1.

Various contractors and vendors have developed computer-based methods and tools to facilitate the conductance of maintainability prediction, assessment, and other system maintainability-related analyses. When evaluating these tools for a particular program, various issues need to be kept in mind. Some of these issues pertain to data requirements or the resolution of the applicable maintainability analysis/evaluation activity (this is a function of the system design and development phase), integration with other analysis/evaluation activities, applicable design and development paradigm (object-oriented development versus functional), and so on, as depicted in Figure C.2. An illustrative set of criteria to evaluate computer-based methods and tools is presented in Table C.1. While a majority of the criteria listed in Table C.1 are self explanatory, the terms scaleability and flexibility, or adaptability, need further clarification.

In the context of this discussion, scaleability refers to the tools capability, which allows its application across multiple design and development phases. In other

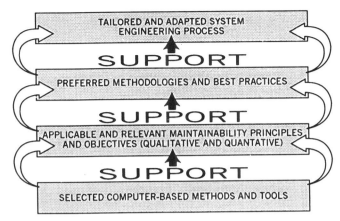

Figure C.1. Integration of maintainability principles and objectives in a computer-aided system design and development environment.

words, the ability of the tools to "deal with" increasing resolution, more detailed information, more detailed design parameters and variables, and so on. For example, while a system engineer may have to contend with a system-level metric such as availability during the early stages, this metric may be broken down into maintainability, reliability, and supportability during the more detailed phases. Each of

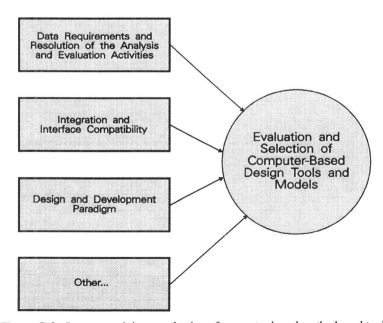

Figure C.2. Issues pertaining to selection of computer-based methods and tools.

TABLE C.1   List of Criteria for Evaluating Computer-Based Methods and Tools

Criteria for Evaluating Computer-Based Models and Tools	Relative Criterion Priorities[a]
Conformance to organizational/program constraints pertaining to:	
1. Design and development approaches	1
2. Integration with existing design workstations	1
3. Hardware platform	1
4. Size of systems	1
5. Standards compliance	1
Scaleability	2
Flexibility and adaptability	4
Data and database requirements	5
User friendliness and user interface	8
Skill-level requirements	9
Cost and upgrade policy	10
Technical maintenance and support	6
Training requirements and support	7
Documentation	3
Network support	11

[a] A lower criterion priority number signifies greater importance, or a higher priority.

these parameters (maintainability, reliability, and supportability) may be broken down further into more detailed variables and parameters with the progression of the system design and development process. A highly scaleable tool should be able to address this continuously decreasing level of abstraction. Very often commercially available tools need much too much information to be of any significant use during the nascent phases of the system design process.

The terms flexibility and adaptability are intended to convey the extent to which a tool allows an analyst to better address changing requirements, different types and sizes of system configurations, and so on.

Subsequent to the development of a consistent and clearly defined set of evaluation criteria, their relative importances must be addressed as depicted in Table C.1. These priorities could very well change during the course of a program, or from one program to the next. However, it is important to define these in order to objectively evaluate the available computer-based methods and tools.

C.1   SELECTED COMPUTER-BASED MODELS FOR MAINTAINABILITY AND RELATED ANALYSIS

In recent years numerous commercial, easy to use, personal computer-based models and tools have been developed. These tools are of a generic nature, or conform to popular standards, which allows for their portability and common use. On the other hand, many organizations involved in system design have also developed "proprietary" computer-based models for their own specific system applications and/or in response to a particular contractual requirement.

We have identified a few models that are representative of the tools that are available and may be utilized to facilitate maintainability analyses. Note that the following list of computer-based models and tools is by no means all-inclusive, and serves only to illustrate the availability of useful tools. A brief description of each, along with reference to the source, is included:

1. *Maintainability prediction program (MPP).* The MPP can be used to predict the maintainability of electrical, electronic, electromechanical, and mechanical systems. Predictions of MTTR $(\overline{M}ct)$, M_{max} (sixtieth to ninetieth percentile), $\overline{MMH/OH}$, and $\overline{MMH/MA}$ factors associated with maintenance activities at the organizational, intermediate, and depot/supplier levels can be accomplished. Repair time factors include localization, fault isolation, disassembly, interchange, reassembly, alignment, and checkout. The quantity of replaceable items, which may be a mixture of assemblies, subassemblies, and component parts, is limited only by disk space. MPP, using a DOS-based personal computer, can be utilized to accomplish a maintainability prediction in accordance with Military Handbook, MIL-HDBK-472 (Notice 1, Procedure V, Method B), *Maintainability Prediction,* Department of Defense, Washington DC. An abridged university edition has also been developed which can serve as an excellent instructional tool. (Reference: Powertronic Systems, Inc., 13700 Chef Menteur Highway, P. O. Box 29109, New Orleans, LA 70189.)

2. *Maintainability effectiveness analysis program (MEAP).* This model is used to compute maintenance times for electronic and electromechanical components, and to accomplish maintainability predictions for systems in accordance with Military Handbook, MIL-HDBK-472, *Maintainability Prediction,* Department of Defense, Washington DC. (Reference: Systems Effectiveness Associates, Inc., 20 Vernon Street, Norwood, MA 02062.)

3. *Failure mode, effects, and criticality analysis (FME).* The FME can be used in the analysis and development of a FMECA for a system, and can be applied to a user-defined hierarchy including up to 25 levels of assemblies, subassemblies, components, and so on. The FME allows for the determination of failure modes, percent contributions, local effects and effects on the system, frequencies of occurrence, severity classifications, failure detection methods, and recommended compensating maintenance provisions. The data structure is presented in a "tree" form, and operations may be accomplished by highlighting individual tree elements. The FME can be implemented using a DOS-based personal computer. The program includes all features necessary for accomplishing a FMECA in accordance with Military Standard, MIL-STD-1629A, *Procedures for Performing a Failure Mode, Effects, and Criticality Analysis,* Department of Defense, Washington DC. All outputs from the FME program can be applied directly to an LSAR database for future retrieval and subsequent use. A university edition of FME is also available for

instructional use. (Reference: Powertronic Systems, Inc., 13700 Chef Menteur Highway, P. O. Box 29109, New Orleans, LA 70189.)

4. *Failure Mode and Effects Analysis (FMEAplus)*. The FMEAplus implements the Ford Motor Company's FMEA process standard. It implements both the "design" and "process" orientations of the failure modes and effects analysis. Further, glossaries are provided to facilitate the standardization of the terminology used within an organization. (Reference: Adistra Corporation, 100 Union Street, Plymouth, MI 48170.)

5. *Failure mode, effects, and criticality analysis (Relex-FMECA)*. The Relex-FMECA facilitates the accomplishment of system FMECA in accordance with MIL-STD-1629A, *Procedures for Performing a Failure Mode, Effects, and Criticality Analysis,* Department of Defense, Washington DC. The program allows a structured input of information pertaining to failure modes, local effects and effects on the system, frequencies of occurrence, severity classifications, failure detection methods, and recommended compensating maintenance provisions. This information is stored in a "tree" form, and operations may be accomplished by selecting individual tree elements. Reference: Innovative Software Design, Inc., One Kimball Ridge Court, Baltimore, MD 21228.

6. *Failure mode, effects, and criticality analysis (FailMode)*. FailMode is a computer-based implementation of a failure mode, effects, and criticality analysis process in accordance with MIL-STD-1629A, *Procedures for Performing a Failure Mode, Effects, and Criticality Analysis,* Department of Defense, Washington DC. It has two separate worksheets for failure mode and effect analysis and criticality analysis. (Reference: Item Software Limited, Fareham, Hampshire, United Kingdom, PO 15 5Su; distributed in the United States by: Mitchell and Gauthier Associates (MGA), Inc., 200 Baker Avenue, Concord, MA 01742–2100.)

7. *Equipment designer's cost analysis system (EDCAS)*. The EDCAS is a design tool that can be used in the accomplishment of a level-of-repair analysis (LORA). It includes a capability for the evaluation of repair versus discard-at-failure decisions, and it can handle up to 3500 unique items concurrently (i.e., 1500 line replaceable units and 2000 shop replaceable units). Repair-level analyses can be accomplished at two indenture levels of the system, and the results can be used in determining optimum repair/spare part requirements and in the accomplishment of life-cycle cost analysis (LCCA). EDCAS is also available in the form of an abridged university edition, EDCAS-UE, and is compatible with the requirements of MIL-STD-1390B, *Level of Repair,* Department of Defense, Washington, DC. (Reference: Systems Exchange, Inc., 170 17th Street, Pacific Grove, CA 93950.)

8. *Optimum repair level analysis (ORLA) model*. A LORA model used to examine the economic feasibility of maintenance and support alternatives. Up to four alternatives can be evaluated, with life-cycle cost broken down into 13 distinct logistics areas. Reference: U. S. Army–MICOM, Code AMSMI-LC-TA-L, Readstone Arsenal, AL 35898.

9. *Level of repair (Navy)*. This model is used in the performance of a LORA during the design and development of new systems and equipment. Two objectives

are to establish a least-cost maintenance policy, and to influence design in order to minimize logistic support costs. (Reference: MIL-STD-1390B, *Level Of Repair,* Department of Defense, Washington, DC.)

10. *Network repair level analysis (NRLA).* The NRLA model is used in establishing equipment and component repair levels, and for making repair versus discard-at-failure decisions in both the design of new equipment and in provisioning. (Reference: US Air Force, Air Force Acquisition Logistics Center, AFALC/LSS, Wright-Patterson AFB, OH 45433.)

11. *Optimum supply and maintenance model (OSAMM).* A LORA model used to determine the optimum economic maintenance policy for each item that fails, to identify items in terms of repair versus discard, to identify economic screening criteria, and to evaluate support equipment/repairmen options. Four levels of support and four indenture levels of equipment can be evaluated. (Reference: US ARMY-CECOM, Code AMSEL-PL-SA, Fort Monmouth, NJ 07703.)

12. *Computed optimization of replenishment and initial spares based on demand and availability (CORIDA).* A model that computes initial and replenishment spare/repair part requirements, for multiple levels of maintenance, in terms of costs and distribution. It addresses organizational and budgeting constraints, and is utilized in support of provisioning activities. (Reference: Thomson–CSF Systems Canada, 350 Sparks Street, Suite 406, Ottawa, Ontario K1R 7S8, Canada.)

13. *OPUS-9 model.* A versatile model used primarily for spare/repair part level, allocation, and location optimization in terms of overall investment and system availability. It considers different operational scenarios and system utilization profiles in determining demand patterns for spares, and it aids in the evaluation of various design packaging schemes. This tool has a facility to conduct numerous cost-effectiveness studies and trade-off analyses between various system design factors, repair policies, alternative logistic support structures, and stocking policies and constraints. (Reference: Systecon AB, Box 1381, S-17127, Solna, Sweden.)

14. *VMetric model.* VMetric is a spares model that can be used to optimize system availability by determining the appropriate individual availabilities for system components and the stockage requirements for three indenture levels of equipment (e.g., line replaceable units, shop replaceable units, and subassemblies) at all echelons of maintenance. Output includes optimum stock levels at each echelon of maintenance, EOQ quantities, and optimal reorder intervals. (Reference: Systems Exchange, Inc., 5504 Garth Avenue, Los Angeles, CA 90056.)

15. *Repairable equipment population system (REPS).* The REPS model evaluates a homogeneous population of repairable equipment deployed to meet a projected demand. As equipment units fail, they join a queue, are repaired in a multiple channel maintenance facility, and subsequently returned to service. As equipment ages, the older units are removed from the system and replaced with new ones. The objective of REPS is to determine the number of equipment units to deploy in response to a certain estimated demand, the number of parallel and equivalent maintenance facilities to set up, and the replacement age of the equipment units so that the sum of all costs associated with the system will be minimized. (Reference:

Systems Engineering Design Laboratory, 146 Whittemore Hall, College of Engineering, Virginia Tech, Blacksburg, VA 24061.)

16. *Life-cycle cost calculator (LCCC).* The LCCC model aids in the accomplishment of a life-cycle cost analysis, using a CBS methodology. Objectives and activities are linked to resources, and constitute a logical breakdown of cost by functional activity area, major element of a system, and/or one or more discrete classes of common or like items. It provides a mechanism for the initial allocation of cost, cost categorization, and finally, cost summation. It has the capability of identifying cost contributions, in both real and discounted dollars, at any level in the CBS. (Reference: Systems Engineering Design Laboratory, 146 Whittemore Hall,College of Engineering, Virginia Tech, Blacksburg, VA 24061.)

17. *Cost analysis strategy assessment (CASA).* The CASA model is utilized to develop life-cycle cost estimates for a wide variety of systems and equipment. It incorporates various analysis tools into one functioning unit and allows the analyst to generate data files, perform life-cycle costing, sensitivity analysis, risk analysis, cost summaries, and the evaluation of alternatives. (Reference: Defense Systems Management College, DSS Directorate (DRI-S), Fort Belvoir, VA 22060.)

18. *Life-cycle cost model for defense material systems (Marine Corps).* This model provides the structure for the calculation of LCC for a system or equipment being developed. (Reference: MIL-HDBK-276-1 (MC), *Life-Cycle Cost Model for Defense Material Systems Data Collection Workbook,* Department of Defense, Washington DC.)

19. *AECMA material and order support system (AMOSS 2000).* The MOSS 2000 program is the implementation of the European defense community CALS specification 2000M to formalize the method and format for information exchange between the contractor and customer, to automate parts identification, planning, ordering, progress inquiry, and invoicing requirements of a contract. Aecma stands for Association Europeenne des Constructeurs de Materiel Aerospatial. (Reference: OBS Organisationsberatung und Softwareentwicklung GmbH, Geschwister-Scholl-Ring 1, D-8034, Germany, or Logistic Engineering Associates, 2700 Navajo Road, Suite 1, El Cajon, CA 92020–2123.)

20. *Systems and logistics integration capability (SLIC).* This program is an integrated logistic support analysis data management system designed to respond to CALS objectives in producing mini LSAR in accordance with MIL-STD-1388-2B, *Requirements for a Logistic Support Analysis Record.* (Reference: Integrated Support Systems, Inc., P. O. Box 1842, Clemson, SC 29679.)

21. *Logistics support analysis records (Omega2B).* Omega2B is a distributed database processor designed to produce a LSAR. It incorporates reliability, maintainability, and logistics data, and creates LSAR reports in accordance with MIL-STD-1388 2B, *Requirements for a Logistics Support Analysis Record.* (Reference: Omega Logistics International, 2700 Navajo Road, El Cajon, CA 92020.)

22. *Logistic support analysis (Max2000).* Max2000 is a tool kit to facilitate the implementation logistic support analysis throughout the system design and development process. It creates LSAR reports in accordance with MIL-STD-1388 2B, *Re-*

quirements for a Logistic Support Analysis Record. It also includes a module to assist the cost estimation of acquisition programs. (Reference: Management Action Corporation, 9275 Corporate Circle, Manassas, VA 22110.)

23. *System design utility (SDU).* The SDU model can be applied during the early stages of the system design in the accomplishment of requirements allocation (i.e., the top-down apportionment of system-level requirements to lower-level elements in the system), and in the later stages of development in the accomplishment of design trade-off studies. It incorporates the flexibility to allow for the evaluation of different system architectures, as well as providing database management capability for configuration control purposes. The output from the SDU model can be utilized in a variety of situations, to include providing an input for the EDCAS model. (Reference: Systems Exchange, Inc., 5504 Garth Avenue, Los Angeles, CA 90056.)

24. *Requirements driven development (RDD-100) system designer/requirements editor (RE).* RDD is a modeling approach employed to evaluate various design alternatives and to assess system behavior. This approach utilizes functional block diagrams to describe sequential and concurrent activities, input–output requirements, data flow, and physical item flow, and to assist in system decomposition. RDD is a system engineering method, supported by an executable graphic modeling language and computer-based tools developed to assist the engineer in designing a complex system through functional analysis and allocation. The requirements editor module features an improved method for information extraction which allows the edition of master requirements database for subsequent modeling. After modification, the edited requirements can be uploaded into the master database with a resolution of preexisting requirement relationships. System platforms include selected DECstation®, HP Apollo 9000 Series 700®, IBM RISC System/6000™, Macintosh®, PCs with MS-Windows®, and SPARCstation® workstations. (Reference: Ascent Logic Corporation, 180 Rose Orchard Way, Suite 200, San Jose, CA 95134.)

25. *Computer-aided system engineering (CORE).* CORE is a modeling approach and automated tool used to evaluate design alternatives and assess system behavior via functional behavior diagrams and an executable modeling language. CORE assists the system engineer(s) with accomplishing system functional analysis and allocation. It is implemented on PC/Windows and Macintosh. (Reference: Vitech Corporation, 2422 Rocky Branch Road, Vienna, VA 22181.)

26. *Reliability prediction program (RPP).* The RPP constitutes a series of computer software routines designed to aid in the accomplishment of reliability predictions based upon component part stress factors. A system may contain any number of assemblies, subassemblies, and component parts (limited only by available disk space); component failure rate information can be introduced through a series of menu prompts; part application data and stress factors can be addressed (temperature, power, etc.); and the prediction of MTBF or λ values can be determined for the system in question. Implementation of RPP can be accomplished using a DOS-based personal computer and is available for all current versions of MIL-HDBK-

217, *Reliability Prediction of Electronic Equipment,* Department of Defense, Washington DC. (Reference: Powertronic Systems, Inc., 13700 Chef Menteur Highway, P. O. Box 29109, New Orleans, LA 70189.)

27. *PC availability.* A model that utilizes Markov analysis to study the influence of failure rates, repair rates, and logistic support on system availability. The objective is to provide assistance in the development of optimum system configuration design and repair policies. (Reference: Management Sciences, Inc., 6022 Constitution Avenue, NE, Albuquerque, NM 87110.)

28. *PC predictor.* A reliability model that automatically applies the part stress analysis or parts count methods of MIL-HDBK-217, *Reliability Prediction of Electronic Equipment,* to produce equipment failure rate estimates and to accomplish reliability predictions. (Reference: Management Sciences, Inc., 6022 Constitution Avenue, NE, Albuquerque, NM 87110.)

29. *Mechanical reliability prediction program (MRP).* The MRP can be used in accomplishing a reliability prediction of mechanical equipment. The program calculates the reliability of components such as static and dynamic seals, springs, solenoids, valve assemblies, bearings, gears and splines, pump assemblies, filters, brakes and clutches, and actuators. (Reference: Powertronic Systems, Inc., 13700 Chef Menteur Highway, P. O. Box 29109, New Orleans, LA 70189.)

30. *Reliability prediction and modeling (R1).* The R1 program performs both parts stress and parts count reliability predictions in accordance with MIL-HDBK-217, *Reliability Prediction of Electronic Equipment.* Further, it is capable of modeling a number of redundant component part and assembly configurations. (Reference: Systems Exchange, Inc., 5504 Garth Avenue, Los Angeles, CA 90056.)

To further illustrate the application and capabilities of these models, a few student exercises have been included in Section C.3 of this appendix. The objective is to acquire the appropriate model from the reference source and to utilize it in solving the problems as stated.

C.2   UTILIZATION AND INTEGRATION OF COMPUTER-BASED MODELS

The models described thus far are primarily being utilized on a "stand-alone" basis. They were developed by independent suppliers for specific applications and, in many instances, these suppliers are in competition with others. As such, the models are independently operated and are not integrated into any higher-order information flow network. At best, computer-based models developed by the same vendor may "talk" to each other.

At some point in the future, it will be necessary to develop an overall integrated approach. Referring to Figure 6.19, one can establish a structure and an information flow, tying the models together in terms of input–output requirements and characteristics. From the example illustrated in Figure 6.19, the analyst may wish to ac-

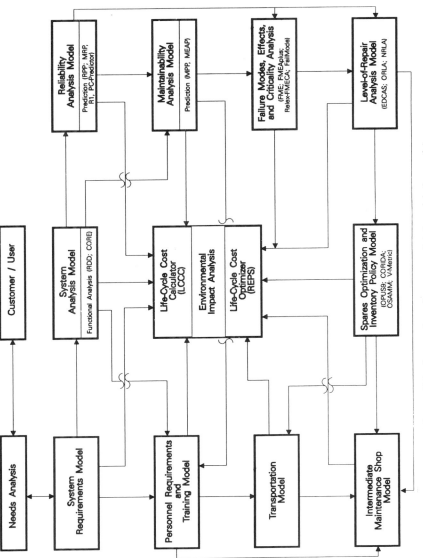

Figure C.3. Integration of support-related models (example).

483

complish a life-cycle cost analysis for the system (the output from block 5); perform a level of repair analysis independently (block 11); identify maintenance shop requirements based on the results of the level of repair analysis (blocks 9 and 11); and project the results from these lower levels of analysis into the system LCC analysis (block 5). It will be necessary to view the system as an entity; identify the high cost areas of activity and optimize these for the purposes of improvement (i.e., reduction in LCC); and project the results back in the context of the whole! The interrelationships among the system components are numerous, and there are many interactions between the activities represented by the blocks in Figure 6.19.

As an ultimate objective, it is essential not only that the models "talk to each other," but that the collective group of models be appropriately integrated into the design workstation configuration concept. If the maintainability engineering and maintenance planning and organization activities are to be successful in influencing the system/product design process, the appropriate lines of communication must be established. An essential segment of this communication is through the timely back and forth transfer of relevant information as noted.

With this objective in mind, an initial attempt to conceptually integrate some of the models identified in Section C.1 has been made. Referring to Figure C.3, there have been some efforts to improve communications between RPP, MPP, FME, and EDCAS. Further, some commercial software developers are working toward the "common or shared database" concept, where all the model and tools tap into and feed-off from a single database. This facilitates a more active interface between the interdependent analyses and avoids redundant information storage. The CALS initiative has contributed to the enhanced sense of integrating the analyses that facilitate design robustness from a "design for supportability" perspective. These efforts are continuing, with expansion, as this text goes to print. It is essential that the desired level of information flow be available in a timely and effective manner throughout the network.[1]

C.3 SUGGESTED PROBLEM EXERCISES UTILIZING SELECTED MODELS

To illustrate the application of selected computer-based models in a simplistic manner, special problem exercises have been developed and are presented here. These problems may be completed using the personal computer and the model identified. It is highly recommended that you acquire access to the necessary tools and complete these exercises, as the results should provide a more thorough understanding of some of the material presented in this text. The special exercises included are:

1. Exercise 1: a reliability allocation problem using RPP.
2. Exercise 2: a reliability prediction problem using RPP.

---

[1] Figure C.3 illustrates an objective in terms of computer-based models and tools "talking to each other. While some progress has been made in this direction, much still remains to be done!

3. Exercise 3: a maintainability prediction problem using MPP.

4. Exercise 4: a LORA problem using EDCAS.

5. Exercise 5: a LCCA problem utilizing LCCC.

6. Exercise 6: a design FMECA problem utilizing FME.

7. Exercise 7: a process FMECA problem using PFM.

1. *Problem Exercise 1: reliability allocation (RPP).* In view of the increased commercial activity in a certain sector, a company conducted a survey and recognized the need for a quick, reasonably priced mail delivery service. Thereafter, a fleet of six airplanes was acquired in an effort to capitalize on and meet the demand. The management is presently considering the installation of a new communications device in the planes to boost the present range of communications. One of the activities being undertaken toward this objective is the reliability analysis of the configurations under consideration. The requirement is for a system level MTBF of 250 hours.

Figure C.4 represents the system structure of one of the configurations under study, along with an incomplete reliability allocation.

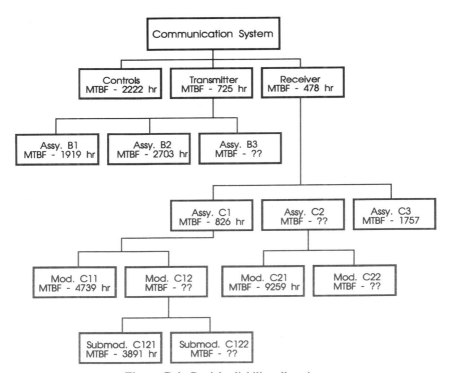

Figure C.4. Partial reliability allocation.

TABLE C.2

System Element	Mean Time Between Failure (Hours)
Submodule C121	3012
Submodule C122	1202
Module C11	4545
Module C21	9434
Module C22	4762
Assembly C3	1912
Assembly B1	1739
Assembly B2	2899
Assembly B3	1953
Unit A—Control	1550

a. Complete the reliability allocation using RPP.
b. Does the calculated MTBF at the system level meet the requirement?

2. *Problem Exercise 2: reliability prediction (RPP).* Keeping the requirement allocation from problem exercise 1 in mind, detail design activity of the communication device has commenced. To verify how well the designing activity is "tracking" the requirements specified through the reliability allocation, a prediction study is accomplished. Table C.2 reflects the results of the design activity to date.

Propagate the results of the detail design, as presented in the table, to the system level. Is the detail design proceeding satisfactorily from the reliability requirements and allocation point of view?

3. *Problem Exercise 3: maintainability prediction (MPP).* Referring to the communication device structure in Figure C.5 and the results from the reliability prediction program, accomplish the following using MPP as an aid:

a. Create a "scenario" of Fault Isolation and Detection modes (at least one for each repairable element in the system structure).
b. Follow the "scenario" above with a systematic creation of a set of repair tasks corresponding to each fault mode. Each repair task will address, wherever applicable, all significant maintenance phases like preparation, isolation, disassembly, interchange, reassembly, alignment, and checkout.

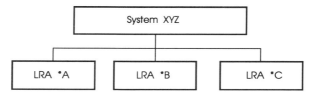

Figure C.5. System XYZ structure.

c. Complete the question above, first for a single level of maintenance, on board the aircraft itself, where all faulty elements are replaced with readily available spares. Secondly, attempt the problem from the perspective of two maintenance levels. What is the predicted $\overline{M}ct$ for the first case and the second case? What is the deviation in M_{max} values between the two scenarios at the ninetieth percentile?

4. *Problem exercise 4: level-of-repair analysis (EDCAS-UE).* System "XYZ" is currently in conceptual design. The system is comprised of the following three lowest removable assemblies (LRAs): *A, *B, and *C. During negotiations you truly impressed the prime contractor with your discussion on the necessity to conduct a LORA. Therefore, you are awarded the task of determining the optimal level of repair for each LRA (see Figure C.5). After reviewing the system specifications the following information was ascertained:

a. There will be 528 System "XYZs" distributed throughout two major geographical areas—eastern and western United States. The three LRAs comprising System "XYZ" will be functioning the entire time that System "XYZ" operates (i.e., each LRA has a duty cycle equal to unity). There are 8 sites within the eastern US region and 7 in the western region. The 8 eastern site and the 7 western sites contain 2 and 4 organizations each respectively. In both the eastern and western regions there are 12 equipments per organization. The "XYZs" located at the eastern sites will operate an average of 15 hours per week, while peaking at 24 hours per week. Similarly, the systems located in the western sites will average 10 hours per week, but will have a peak utilization of 36 hours. The eastern US sites will operate an average of 26 weeks per year, while the western US sites will be utilized an average of 48 weeks per year.

b. System "XYZ" is required to maintain a 95% operational availability over its 20-year life cycle. The MTBF for each LRA is estimated as follows:

LRA	MTBF (hrs)
*A	2222
*B	4500
*C	2100

c. Upon failure, System "XYZ" will be restored to operation at the organizational level by removing its failed LRA and replacing it with a fully functional spare. The expected length of time required to restore System "XYZ" to operation once it has failed is 0.75 hours. The 0.75 hours include only the time required to restore "XYZ" to operation by fault isolation and on-equipment repair. At this time the failed LRA will either be repaired or discarded. The LRA can either be repaired at the local level or at the depot maintenance shop. The failed LRA can be repaired at one of four repair depots and the functional LRAs are stored as spares at one of five storage depots. MTTR for the LRAs is:

LRA	MTTR(hr)
*A	2.0
*B	1.0
*C	1.5

The MTTR includes only the time required to restore the LRA to a "ready-for-issue" condition. This time includes fault isolation, removal and replacement of failed components, subsequent test to verify the repair, data recording, and administrative delay.

d. The average cost of all piece parts and materials consumed in the course of repair is $33.5, $535, and $75 for the *A, *B, and *C LRAs, respectively.

e. The proportion of failures of repairable assemblies which are ultimately not repaired (i.e., condemnation rate) is as follows:

LRA	Condemnation Rate
*A	0.0000
*B	0.0015
*C	0.0150

f. The average distance from the operating sites to the depot is 1300 miles. The cost to ensure successful transportation of the LRAs is $0.025 per pound per mile. The following is a listing of each LRA by weight, and the cost of a reusable shipping container suitable for transporting the LRA between the intermediate and depot repair facilities:

LRA	Weight (lb)	Container ($/unit)
*A	5.0	0
*B	4.6	325
*C	25.0	75

The weight of each LRA does not include packing materials. A factor of 14% is added internally (i.e., within EDCAS) to account for packing materials in the transportation algorithms. The packing department has approved the 14% added as an appropriate cost-estimating relationship.

g. There are three hours per week of scheduled maintenance activities for System "XYZ." That is, three maintenance personnel man-hours are required per "XYZ" per operating week to perform all "XYZ"-level scheduled maintenance tasks. *B is the only LRA requiring scheduled maintenance. One tenth of an hour per week is the necessary scheduled maintenance time for *B.

h. In the course of this analysis, the optimal level of repair will be determined. The delay times associated with the various repair policies are an internal component in determining the optimal level of repair. Five critical delay times have been estimated:

(1) *Local repair response time.* The local response time is estimated at 20 days. The local response time is the average elapsed time from failure of the assembly to its return to the unit, ready for use. This includes all administrative waiting time, batching and delays for repair resources at the intermediate facility. This is the waiting time over and above the MTTR for System "XYZ" in the event that a ready-for-issue spare is unavailable in local stock.

(2) *Order and shipping time.* It is assumed that the average elapsed time from request to receipt of a replacement assembly from a supply facility (assuming no stockout at the supply depot) is 14 days.

(3) *Depot repair response time.* The elapsed time from arrival of a failed assembly at a depot to its availability in stock at the depot is 183 days. The depot response time is the waiting over and above that of "XYZ" in the event that a ready-to-issue spare is unavailable at the depot.

(4) *Consumables replenishment lead time.* The average elapsed time from request to receipt of a replenishment or consumption item is 60 days.

(5) *Repairables procurement lead time.* The average elapsed time required to procure a replacement assembly from the manufacturer when the assembly is normally expected to be repairable is 450 days.

i. The annual labor cost for the depot, local, and organizational maintenance technician is $50,000, $40,000, and $28,000, respectively. The productive work hours for the local repair technician are 34 hours per week and 30 hours for the technicians at the depot level. Past history has shown a 33% turnover among the local maintenance shop workers, while the depot maintenance shop has experienced a 25% job turnover.

j. There is a daily fee of $135 to train the workers to perform their intended maintenance tasks. Six hours of training is necessary for an already trained maintenance technician to perform scheduled and unscheduled maintenance on System "XYZ." For the LRAs, the following are the marginal classroom times for scheduled and unscheduled maintenance training. Note that only the *B LRA requires scheduled maintenance.

LRA	Marginal Classroom Repair (hr)	Marginal Classroom Scheduled (hr)
*A	2.0	0
*B	35.4	1
*C	15.0	0

k. The LRA unit costs and corresponding lot sizes are listed below:

LRA	Unit Cost ($)	Lot Size
*A	550	1
*B	7850	300
*C	2950	500

After discussions with the production personnel and management you have determined that System "XYZ" will experience approximately a 95% learning curve reduction rate.

1. The cost of technical publication development is $850 per page and the cost of technical drawings is $3000 per sheet. The number of pages of technical documentation for assembly repair and maintenance training are as follows:

LRA	Pages
*A	5
*B	15
*C	20

Sixty-seven pages of additional technical documentation is required to describe System "XYZ" beyond the assembly level repair. Since all assembly-level documentation is maintenance dependent, the additional 67 pages include all documentation that is not maintenance dependent. Twenty sheets of technical drawings are required for System "XYZ." The annual maintenance of publications development and technical drawings is estimated at 5% of their development cost.

m. The fully burdened recurring cost per unit of System "XYZ" integration, assembly, and acceptance testing is estimated at $30,000 for a lot size of 500.

n. There are no external support equipment requirements for System "XYZ;" however, LRA external support equipment is needed. The purchase price of all external support equipment used only for repair and scheduled maintenance of the assembly is as follows:

LRA	External Support Equipment ($/Suite)
*A	0
*B	16,300
*C	25,000

o. There will be an estimated $155,000 of common module test unit costs per suite and a one-time software development cost of $250,000. The $155,000 of common module test unit costs is the hardware unit purchase price of a test unit capable of performing all automatic test functions for the three LRAs that comprise System "XYZ." The $250,000 is the cost to develop overhead software to drive the module test unit. The LRA specific software development is accounted for in the LRA data set variable, diagnostic software development cost. The fully burdened cost to develop diagnostic software to assist in the repair of a specific LRA is as follows:

LRA	Diagnostic Software Development Cost ($)
*A	85,000
*B	0
*C	100,000

Be aware that the diagnostic software costs will only be added to life-cycle cost in the event that the assembly is replaced. The total software development cost (both from the diagnostic software development and the common module test unit software) is used as the basis for software maintenance cost. The annual software maintenance is estimated at 8% of the total software development costs.

p. The allocated fuel cost is $0.015 /lb/operating hour.

q. Finally, a 5% inflation rate is forecasted.

Answer the following questions:

a. What is the operational availability of System XYZ?: Does the operational availability meet the intended target?

b. How many maintenance personnel are required at the organizational, intermediate, and depot locations over the 20-year lifetime of the system?

c. What is the optimal repair policy for each LRA (i.e. *A, *B, and *C)?

d. How many spares of each LRA are required over the 20-year life cycle at the intermediate maintenance facility and the depot maintenance facility? Why do the total number of spares required differ for each LRA?

e. How many *C LRAs will be produced over the 20-year life cycle?

f. Briefly discuss the impact of MTTR on LCC analysis.

g. Does the scheduled maintenance requirement affect the sparing results? If so, how?

h. What is the major cost of training?

5. *Problem Exercise 5: life-cycle cost analysis problem (LCCC).* Two alternate configurations for a communications device to be installed on board a fleet of airplanes are being evaluated. The basic configuration is presented in Figure C.6. A decision has to be made as to which configuration is more economical to pursue. Relative to the information provided below, the following tasks have to be performed:

a. Develop a CBS for the scenario at hand.

b. Compute the discounted life-cycle cost for each of the two configurations (assume an interest rate of 11%).

c. Select a preferred approach.

d. Plot the undiscounted and discounted cost streams for the selected approach.

e. What is the difference in the life-cycle cost of the selected approach assuming an inflation rate of 6%?

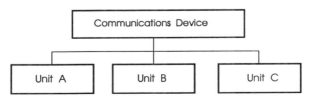

Figure C.6. System structure.

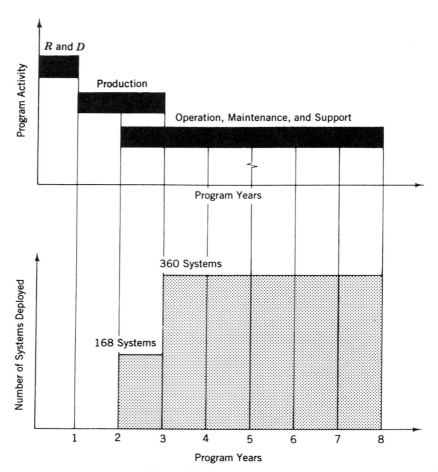

Figure C.7. Program schedule.

The system is to be installed on board a fleet of airplanes throughout two major geographical areas—eastern and western United States. There are 8 eastern US sites and 7 western US sites. The 8 eastern sites and the 7 western sites include 2 organizations each. There are 12 systems installed in each organization. Figure C.7 outlines the life-cycle profile of the systems deployed in order to meet the expected demand.

Assume that all of the installed systems are operated an average of 2.5 hours each day, 365 days a year. It will take a single crew member 3% of his time to operate the system. The system needs to have a MTBF of 240 hours and a MTTR of 2.25 hours. The system will not undergo any preventive maintenance. For the sake of simplicity, the analysis should be limited to the first indenture level (see Figure C.6). Relative to the support and maintenance of the system, a plug-in diagnostic tool enables system checkout and fault isolation to the unit level. The faulty unit is removed and replaced with a spare. The damaged unit is sent to the intermediate level repair shop, where isolation is accomplished to the submodule level and repair is accomplished by replacement. There is one intermediate repair shop in the eastern region and one in the western.

The following data are required for calculating the system life-cycle cost:

a. The crew member assigned to operate each system is paid at the rate of $32 per hour.

b. Design and development costs for the system are:

Configuration A—$73,000

Configuration B—$89,500

c. Design and development costs for the repair equipment needed at each intermediate shop are:

Configuration A—$11,000

Configuration B—$8,750

Parameter	Configuration A (hr)	Configuration B (hr)
	System Level	
MTBF	255	295
MTTR	2.1836	2.245
	Unit Level	
Unit A		
MTBF	2222	1667
MTTR	2.0	2.0
Unit B		
MTBF	725	1010
MTTR	1.0	1.0
Unit C		
MTBF	478	555
MTTR	1.5	1.5

d. The production costs for each system produced are:

Configuration A—$18,000

Configuration B—$16,600

Assume that all the systems in the western region along with the spares and backups are installed by the end of year 2, and all in the eastern region along with the spares and backups are installed by the end of year 3.

e. Special support equipment is needed at each intermediate shop at the beginning of year 3. The one-time cost of this equipment is:

Configuration A—$5,000

Configuration B—$4,750

f. Spares are required on board the airplanes for replacement of the units that need repair. Assume that one Unit A, one Unit B, and one Unit C constitute a set of spares located in each airplane. The cost of each set of spares is equal to the cost of one system. Moreover, an additional two sets of spares are located at each intermediate shop.

The turnaround time on spares is ignored for the sake of simplicity. Assume that it costs $97 for each maintenance action (corrective) for both the configurations. This cost includes the material costs, inventory maintenance costs, and maintenance data costs.

g. For maintenance on board an airplane, the crew member shall need a kit that also includes the diagnostic tool. The cost of this kit is:

Configuration A—$900

Configuration B—$850

h. Maintenance at the system level is performed by the operator. In the intermediate shop, the maintenance is performed by a low-skilled technician at $20 per man-hour.

All assumptions must be stated and justified in a clear, concise manner.

6. *Problem Exercise 6—FMECA system.* In the system illustrated in Figure C.3, identify some hypothetical failure modes at the lowest level and trace the failure mode and effect pairs up to the system level. Examples of failure modes include intermittent operation, loss of output, displacement of the calibration, and so on. The student is expected to follow this exercise with a criticality analysis of the system under study. Also, discuss how FMECA results can contribute to a maintainability analysis.

7. *Problem Exercise 7—process FMECA.* Define a process by completing its functional flow diagram to a reasonable level of detail. Delineate the necessary process requirements, allocate these requirements to the elemental activities that constitute the overall process, and subsequently conduct a process-oriented FMECA. The student is expected to complete a causal pareto analysis in conclusion to identify the most critical failure mode causes that need to be addressed within the process being analyzed. Given the generic applicability of this tool, the process selected could relate to a production line, purchasing, distribution, or any other process with which the student is familiar.

APPENDIX D
NUMERICAL TABLES 1

D.1   Interest Tables

[1] Tables D.1–D.12 are from Thuesen, G. J. and W. J. Fabrycky, *Engineering Economy,* 8th ed., Prentice-Hall, Inc., Englewood Cliffs, NJ, 1993.

TABLE D.1 6% Interest Factors for Discrete Annual Compounding

	Single Payment		Equal Payment Series				Uniform Gradient-Series Factor
	Compound-Amount Factor	Present-Worth Factor	Compound-Amount Factor	Sinking-Fund Factor	Present-Worth Factor	Capital-Recovery Factor	Uniform Gradient-Series Factor
	To Find F Given P	To Find P Given F	To Find F Given A	To Find A Given F	To Find P Given A	To Find A Given P	To Find A Given G
n	$F/P, i, n$	$P/F, i, n$	$F/A, i, n$	$A/F, i, n$	$P/A, i, n$	$A/P, i, n$	$A/G, i, n$
1	1.060	0.9434	1.000	1.0000	0.9434	1.0600	0.0000
2	1.124	0.8900	2.060	0.4854	1.8334	0.5454	0.4854
3	1.191	0.8396	3.184	0.3141	2.6730	0.3741	0.9612
4	1.262	0.7921	4.375	0.2286	3.4651	0.2886	1.4272
5	1.338	0.7473	5.637	0.1774	4.2124	0.2374	1.8836
6	1.419	0.7050	6.975	0.1434	4.9173	0.2034	2.3304
7	1.504	0.6651	8.394	0.1191	5.5824	0.1791	2.7676
8	1.594	0.6274	9.897	0.1010	6.2098	0.1610	3.1952
9	1.689	0.5919	11.491	0.0870	6.8017	0.1470	3.6133
10	1.791	0.5584	13.181	0.0759	7.3601	0.1359	4.0220
11	1.898	0.5268	14.972	0.0668	7.8869	0.1268	4.4213
12	2.012	0.4970	16.870	0.0593	8.3839	0.1193	4.8113
13	2.133	0.4688	18.882	0.0530	8.8527	0.1130	5.1920
14	2.261	0.4423	21.015	0.0476	9.2950	0.1076	5.5635
15	2.397	0.4173	23.276	0.0430	9.7123	0.1030	5.9260
16	2.540	0.3937	25.673	0.0390	10.1059	0.0990	6.2794
17	2.693	0.3714	28.213	0.0355	10.4773	0.0955	6.6240
18	2.854	0.3504	30.906	0.0324	10.8276	0.0924	6.9597
19	3.026	0.3305	33.760	0.0296	11.1581	0.0896	7.2867
20	3.207	0.3118	36.786	0.0272	11.4699	0.0872	7.6052

21	3.400	0.2942	39.993	0.0250	11.7641	0.0850	7.9151
22	3.604	0.2775	43.392	0.0231	12.0416	0.0831	8.2166
23	3.820	0.2618	46.996	0.0213	12.3034	0.0813	8.5099
24	4.049	0.2470	50.816	0.0197	12.5504	0.0797	8.7951
25	4.292	0.2330	54.865	0.0182	12.7834	0.0782	9.0722
26	4.549	0.2198	59.156	0.0169	13.0032	0.0769	9.3415
27	4.822	0.2074	63.706	0.0157	13.2105	0.0757	9.6030
28	5.112	0.1956	68.528	0.0146	13.4062	0.0746	9.8568
29	5.418	0.1846	73.640	0.0136	13.5907	0.0736	10.1032
30	5.744	0.1741	79.058	0.0127	13.7648	0.0727	10.3422
31	6.088	0.1643	84.802	0.0118	13.9291	0.0718	10.5740
32	6.453	0.1550	90.890	0.0110	14.0841	0.0710	10.7988
33	6.841	0.1462	97.343	0.0103	14.2302	0.0703	11.0166
34	7.251	0.1379	104.184	0.0096	14.3682	0.0696	11.2276
35	7.686	0.1301	111.435	0.0090	14.4983	0.0690	11.4319
40	10.286	0.0972	154.762	0.0065	15.0463	0.0665	12.3590
45	13.765	0.0727	212.744	0.0047	15.4558	0.0647	13.1413
50	18.420	0.0543	290.336	0.0035	15.7619	0.0635	13.7964
55	24.650	0.0406	394.172	0.0025	15.9906	0.0625	14.3411
60	32.988	0.0303	533.128	0.0019	16.1614	0.0619	14.7910
65	44.145	0.0227	719.083	0.0014	16.2891	0.0614	15.1601
70	59.076	0.0169	967.932	0.0010	16.3846	0.0610	15.4614
75	79.057	0.0127	1300.949	0.0008	16.4559	0.0608	15.7058
80	105.796	0.0095	1746.600	0.0006	16.5091	0.0606	15.9033
85	141.579	0.0071	2342.982	0.0004	16.5490	0.0604	16.0620
90	189.465	0.0053	3141.075	0.0003	16.5787	0.0603	16.1891
95	253.546	0.0040	4209.104	0.0002	16.6009	0.0602	16.2905
100	339.302	0.0030	5638.368	0.0002	16.6176	0.0602	16.3711

TABLE D.2 7% Interest Factors for Discrete Annual Compounding

	Single Payment		Equal Payment Series				Uniform Gradient-Series Factor
	Compound-Amount Factor	Present-Worth Factor	Compound-Amount Factor	Sinking-Fund Factor	Present-Worth Factor	Capital-Recovery Factor	
n	To Find F Given P $F/P, i, n$	To Find P Given F $P/F, i, n$	To Find F Given A $F/A, i, n$	To Find A Given F $A/F, i, n$	To Find P Given A $P/A, i, n$	To Find A Given P $A/P, i, n$	To Find A Given G $A/G, i, n$
1	1.070	0.9346	1.000	1.0000	0.9346	1.0700	0.0000
2	1.145	0.8734	2.070	0.4831	1.8080	0.5531	0.4831
3	1.225	0.8163	3.215	0.3111	2.6243	0.3811	0.9549
4	1.311	0.7629	4.440	0.2252	3.3872	0.2952	1.4155
5	1.403	0.7130	5.751	0.1739	4.1002	0.2439	1.8650
6	1.501	0.6664	7.163	0.1398	4.7665	0.2098	2.3032
7	1.606	0.6228	8.654	0.1156	5.3893	0.1856	2.7304
8	1.718	0.5820	10.260	0.0975	5.9713	0.1675	3.1466
9	1.838	0.5439	11.978	0.0835	6.5152	0.1535	3.5517
10	1.967	0.5084	13.816	0.0724	7.0236	0.1424	3.9461
11	2.105	0.4751	15.784	0.0634	7.4987	0.1334	4.3296
12	2.252	0.4440	17.888	0.0559	7.9427	0.1259	4.7025
13	2.410	0.4150	20.141	0.0497	8.3577	0.1197	5.0649
14	2.579	0.3878	22.550	0.0444	8.7455	0.1144	5.4167
15	2.759	0.3625	25.129	0.0398	9.1079	0.1098	5.7583
16	2.952	0.3387	27.888	0.0359	9.4467	0.1059	6.0897
17	3.159	0.3166	30.840	0.0324	9.7632	0.1024	6.4110
18	3.380	0.2959	33.999	0.0294	10.0591	0.0994	6.7225
19	3.617	0.2765	37.379	0.0268	10.3356	0.0968	7.0242
20	3.870	0.2584	40.996	0.0244	10.5940	0.0944	7.3163

21	4.141	0.2415	44.865	0.0223	10.8355	0.0923	7.5990
22	4.430	0.2257	49.006	0.0204	11.0613	0.0904	7.8725
23	4.741	0.2110	53.436	0.0187	11.2722	0.0887	8.1369
24	5.072	0.1972	58.177	0.0172	11.4693	0.0872	8.3923
25	5.427	0.1843	63.249	0.0158	11.6536	0.0858	8.6391
26	5.807	0.1722	68.676	0.0146	11.8258	0.0846	8.8773
27	6.214	0.1609	74.484	0.0134	11.9867	0.0834	9.1072
28	6.649	0.1504	80.698	0.0124	12.1371	0.0824	9.3290
29	7.114	0.1406	87.347	0.0115	12.2777	0.0815	9.5427
30	7.612	0.1314	94.461	0.0106	12.4091	0.0806	9.7487
31	8.145	0.1228	102.073	0.0098	12.5318	0.0798	9.9471
32	8.715	0.1148	110.218	0.0091	12.6466	0.0791	10.1381
33	9.325	0.1072	118.933	0.0084	12.7538	0.0784	10.3219
34	9.978	0.1002	128.259	0.0078	12.8540	0.0778	10.4987
35	10.677	0.0937	138.237	0.0072	12.9477	0.0772	10.6687
40	14.974	0.0668	199.635	0.0050	13.3317	0.0750	11.4234
45	21.002	0.0476	285.749	0.0035	13.6055	0.0735	12.0360
50	29.457	0.0340	406.529	0.0025	13.8008	0.0725	12.5287
55	41.315	0.0242	575.929	0.0017	13.9399	0.0717	12.9215
60	57.946	0.0173	813.520	0.0012	14.0392	0.0712	13.2321
65	81.273	0.0123	1146.755	0.0009	14.1099	0.0709	13.4760
70	113.989	0.0088	1614.134	0.0006	14.1604	0.0706	13.6662
75	159.876	0.0063	2269.657	0.0005	14.1964	0.0705	13.8137
80	224.234	0.0045	3189.063	0.0003	14.2220	0.0703	13.9274
85	314.500	0.0032	4478.576	0.0002	14.2403	0.0702	14.0146
90	441.103	0.0023	6287.185	0.0002	14.2533	0.0702	14.0812
95	618.670	0.0016	8823.854	0.0001	14.2626	0.0701	14.1319
100	867.716	0.0012	12381.662	0.0001	14.2693	0.0701	14.1703

TABLE D.3 **8% Interest Factors for Discrete Annual Compounding**

	Single Payment		Equal Payment Series				Uniform Gradient-Series Factor
	Compound-Amount Factor	Present-Worth Factor	Compound-Amount Factor	Present-Worth Factor	Sinking-Fund Factor	Capital-Recovery Factor	
	To Find F Given P F/P, i, n	To Find P Given F P/F, i, n	To Find F Given A F/A, i, n	To Find P Given A P/A, i, n	To Find A Given F A/F, i, n	To Find A Given P A/P, i, n	To Find A Given G A/G, i, n
n							
1	1.080	0.9259	1.000	0.9259	1.0000	1.0800	0.0000
2	1.166	0.8573	2.080	1.7833	0.4808	0.5608	0.4808
3	1.260	0.7938	3.246	2.5771	0.3080	0.3880	0.9488
4	1.360	0.7350	4.506	3.3121	0.2219	0.3019	1.4040
5	1.469	0.6806	5.867	3.9927	0.1705	0.2505	1.8465
6	1.587	0.6302	7.336	4.6229	0.1363	0.2163	2.2764
7	1.714	0.5835	8.923	5.2064	0.1121	0.1921	2.6937
8	1.851	0.5403	10.637	5.7466	0.0940	0.1740	3.0985
9	1.999	0.5003	12.488	6.2469	0.0801	0.1601	3.4910
10	2.159	0.4632	14.487	6.7101	0.0690	0.1490	3.8713
11	2.332	0.4289	16.645	7.1390	0.0601	0.1401	4.2395
12	2.518	0.3971	18.977	7.5361	0.0527	0.1327	4.5958
13	2.720	0.3677	21.495	7.9038	0.0465	0.1265	4.9402
14	2.937	0.3405	24.215	8.2442	0.0413	0.1213	5.2731
15	3.172	0.3153	27.152	8.5595	0.0368	0.1168	5.5945
16	3.426	0.2919	30.324	8.8514	0.0330	0.1130	5.9046
17	3.700	0.2703	33.750	9.1216	0.0296	0.1096	6.2038
18	3.996	0.2503	37.450	9.3719	0.0267	0.1067	6.4920
19	4.316	0.2317	41.446	9.6036	0.0241	0.1041	6.7697
20	4.661	0.2146	45.762	9.8182	0.0219	0.1019	7.0370

n							
21	5.034	0.1987	50.423	0.0198	10.0168	0.0998	7.2940
22	5.437	0.1840	55.457	0.0180	10.2008	0.0980	7.5412
23	5.871	0.1703	60.893	0.0164	10.3711	0.0964	7.7786
24	6.341	0.1577	66.765	0.0150	10.5288	0.0950	8.0066
25	6.848	0.1460	73.106	0.0137	10.6748	0.0937	8.2254
26	7.396	0.1352	79.954	0.0125	10.8100	0.0925	8.4352
27	7.988	0.1252	87.351	0.0115	10.9352	0.0915	8.6363
28	8.627	0.1159	95.339	0.0105	11.0511	0.0905	8.8289
29	9.317	0.1073	103.966	0.0096	11.1584	0.0896	9.0133
30	10.063	0.0994	113.283	0.0088	11.2578	0.0888	9.1897
31	10.868	0.0920	123.346	0.0081	11.3498	0.0881	9.3584
32	11.737	0.0852	134.214	0.0075	11.4350	0.0875	9.5197
33	12.676	0.0789	145.951	0.0069	11.5139	0.0869	9.6737
34	13.690	0.0731	158.627	0.0063	11.5869	0.0863	9.8208
35	14.785	0.0676	172.317	0.0058	11.6546	0.0858	9.9611
40	21.725	0.0460	259.057	0.0039	11.9246	0.0839	10.5699
45	31.920	0.0313	386.506	0.0026	12.1084	0.0826	11.0447
50	46.902	0.0213	573.770	0.0018	12.2335	0.0818	11.4107
55	68.914	0.0145	848.923	0.0012	12.3186	0.0812	11.6902
60	101.257	0.0099	1253.213	0.0008	12.3766	0.0808	11.9015
65	148.780	0.0067	1847.248	0.0006	12.4160	0.0806	12.0602
70	218.606	0.0046	2720.080	0.0004	12.4428	0.0804	12.1783
75	321.205	0.0031	4002.557	0.0003	12.4611	0.0803	12.2658
80	471.955	0.0021	5886.935	0.0002	12.4735	0.0802	12.3301
85	693.456	0.0015	8655.706	0.0001	12.4820	0.0801	12.3773
90	1018.915	0.0010	12723.939	0.0001	12.4877	0.0801	12.4116
95	1497.121	0.0007	18701.507	0.0001	12.4917	0.0801	12.4365
100	2199.761	0.0005	27484.516	0.0001	12.4943	0.0800	12.4545

TABLE D.4 9% Interest Factors for Discrete Annual Compounding

	Single Payment		Equal Payment Series				Uniform Gradient-Series Factor
	Compound-Amount Factor	Present-Worth Factor	Compound-Amount Factor	Sinking-Fund Factor	Present-Worth Factor	Capital-Recovery Factor	Uniform Gradient-Series Factor
	To Find F Given P	To Find P Given F	To Find F Given A	To Find A Given F	To Find P Given A	To Find A Given P	To Find A Given G
n	F/P, i, n	P/F, i, n	F/A, i, n	A/F, i, n	P/A, i, n	A/P, i, n	A/G, i, n
1	1.090	0.9174	1.000	1.0000	0.9174	1.0900	0.0000
2	1.188	0.8417	2.090	0.4785	1.7591	0.5685	0.4785
3	1.295	0.7722	3.278	0.3051	2.5313	0.3951	0.9426
4	1.412	0.7084	4.573	0.2187	3.2397	0.3087	1.3925
5	1.539	0.6499	5.985	0.1671	3.8897	0.2571	1.8282
6	1.677	0.5963	7.523	0.1329	4.4859	0.2229	2.2498
7	1.828	0.5470	9.200	0.1087	5.0330	0.1987	2.6574
8	1.993	0.5019	11.028	0.0907	5.5348	0.1807	3.0512
9	2.172	0.4604	13.021	0.0768	5.9953	0.1668	3.4312
10	2.367	0.4224	15.193	0.0658	6.4177	0.1558	3.7978
11	2.580	0.3875	17.560	0.0570	6.8052	0.1470	4.1510
12	2.813	0.3555	20.141	0.0497	7.1607	0.1397	4.4910
13	3.066	0.3262	22.953	0.0436	7.4869	0.1336	4.8182
14	3.342	0.2993	26.019	0.0384	7.7862	0.1284	5.1326
15	3.642	0.2745	29.361	0.0341	8.0607	0.1241	5.4346
16	3.970	0.2519	33.003	0.0303	8.3126	0.1203	5.7245
17	4.328	0.2311	36.974	0.0271	8.5436	0.1171	6.0024
18	4.717	0.2120	41.301	0.0242	8.7556	0.1142	6.2687
19	5.142	0.1945	46.018	0.0217	8.9501	0.1117	6.5236
20	5.604	0.1784	51.160	0.0196	9.1286	0.1096	6.7675

n						
21	6.109	0.1637	56.765	0.0176	9.2923	0.1076	7.0006
22	6.659	0.1502	62.873	0.0159	9.4424	0.1059	7.2232
23	7.258	0.1378	69.532	0.0144	9.5802	0.1044	7.4358
24	7.911	0.1264	76.790	0.0130	9.7066	0.1030	7.6384
25	8.623	0.1160	84.701	0.0118	9.8226	0.1018	7.8316
26	9.399	0.1064	93.324	0.0107	9.9290	0.1007	8.0156
27	10.245	0.0976	102.723	0.0097	10.0266	0.0997	8.1906
28	11.167	0.0896	112.968	0.0089	10.1161	0.0989	8.3572
29	12.172	0.0822	124.135	0.0081	10.1983	0.0981	8.5154
30	13.268	0.0754	136.308	0.0073	10.2737	0.0973	8.6657
31	14.462	0.0692	149.575	0.0067	10.3428	0.0967	8.8083
32	15.763	0.0634	164.037	0.0061	10.4063	0.0961	8.9436
33	17.182	0.0582	179.800	0.0056	10.4645	0.0956	9.0718
34	18.728	0.0534	196.982	0.0051	10.5178	0.0951	9.1933
35	20.414	0.0490	215.711	0.0046	10.5668	0.0946	9.3083
40	31.409	0.0318	337.882	0.0030	10.7574	0.0930	9.7957
45	48.327	0.0207	525.859	0.0019	10.8812	0.0919	10.1603
50	74.358	0.0135	815.084	0.0012	10.9617	0.0912	10.4295
55	114.408	0.0088	1260.092	0.0008	11.0140	0.0908	10.6261
60	176.031	0.0057	1944.792	0.0005	11.0480	0.0905	10.7683
65	270.846	0.0037	2998.288	0.0003	11.0701	0.0903	10.8702
70	416.730	0.0024	4619.223	0.0002	11.0845	0.0902	10.9427
75	641.191	0.0016	7113.232	0.0002	11.0938	0.0902	10.9940
80	986.552	0.0010	10950.574	0.0001	11.0999	0.0901	11.0299
85	1517.932	0.0007	16854.800	0.0001	11.1038	0.0901	11.0551
90	2335.527	0.0004	25939.184	0.0001	11.1064	0.0900	11.0726
95	3593.497	0.0003	39916.635	0.0000	11.1080	0.0900	11.0847
100	5529.041	0.0002	61422.675	0.0000	11.1091	0.0900	11.0930

TABLE D.5 10% Interest Factors for Discrete Annual Compounding

| | Single Payment | | Equal Payment Series | | | | Uniform Gradient-Series Factor |
| | Compound-Amount Factor | Present-Worth Factor | Compound-Amount Factor | Sinking-Fund Factor | Present-Worth Factor | Capital-Recovery Factor | |
n	To Find F Given P $F/P, i, n$	To Find P Given F $P/F, i, n$	To Find F Given A $F/A, i, n$	To Find A Given F $A/F, i, n$	To Find P Given A $P/A, i, n$	To Find A Given P $A/P, i, n$	To Find A Given G $A/G, i, n$
1	1.100	0.9091	1.000	1.0000	0.9091	1.1000	0.0000
2	1.210	0.8265	2.100	0.4762	1.7355	0.5762	0.4762
3	1.331	0.7513	3.310	0.3021	2.4869	0.4021	0.9366
4	1.464	0.6830	4.641	0.2155	3.1699	0.3155	1.3812
5	1.611	0.6209	6.105	0.1638	3.7908	0.2638	1.8101
6	1.772	0.5645	7.716	0.1296	4.3553	0.2296	2.2236
7	1.949	0.5132	9.487	0.1054	4.8684	0.2054	2.6216
8	2.144	0.4665	11.436	0.0875	5.3349	0.1875	3.0045
9	2.358	0.4241	13.579	0.0737	5.7590	0.1737	3.3724
10	2.594	0.3856	15.937	0.0628	6.1446	0.1628	3.7255
11	2.853	0.3505	18.531	0.0540	6.4951	0.1540	4.0641
12	3.138	0.3186	21.384	0.0468	6.8137	0.1468	4.3884
13	3.452	0.2897	24.523	0.0408	7.1034	0.1408	4.6988
14	3.798	0.2633	27.975	0.0358	7.3667	0.1358	4.9955
15	4.177	0.2394	31.772	0.0315	7.6061	0.1315	5.2789
16	4.595	0.2176	35.950	0.0278	7.8237	0.1278	5.5493
17	5.054	0.1979	40.545	0.0247	8.0216	0.1247	5.8071
18	5.560	0.1799	45.599	0.0219	8.2014	0.1219	6.0526
19	6.116	0.1635	51.159	0.0196	8.3649	0.1196	6.2861
20	6.728	0.1487	57.275	0.0175	8.5136	0.1175	6.5081

21	7.400	0.1351	64.003	0.0156	8.6487	0.1156	6.7189
22	8.140	0.1229	71.403	0.0140	8.7716	0.1140	6.9189
23	8.954	0.1117	79.543	0.0126	8.8832	0.1126	7.1085
24	9.850	0.1015	88.497	0.0113	8.9848	0.1113	7.2881
25	10.835	0.0923	98.347	0.0102	9.0771	0.1102	7.4580
26	11.918	0.0839	109.182	0.0092	9.1610	0.1092	7.6187
27	13.110	0.0763	121.100	0.0083	9.2372	0.1083	7.7704
28	14.421	0.0694	134.210	0.0075	9.3066	0.1075	7.9137
29	15.863	0.0630	148.631	0.0067	9.3696	0.1067	8.0489
30	17.449	0.0573	164.494	0.0061	9.4269	0.1061	8.1762
31	19.194	0.0521	181.943	0.0055	9.4790	0.1055	8.2962
32	21.114	0.0474	201.138	0.0050	9.5264	0.1050	8.4091
33	23.225	0.0431	222.252	0.0045	9.5694	0.1045	8.5152
34	25.548	0.0392	245.477	0.0041	9.6086	0.1041	8.6149
35	28.102	0.0356	271.024	0.0037	9.6442	0.1037	8.7086
40	45.259	0.0221	442.593	0.0023	9.7791	0.1023	9.0962
45	72.890	0.0137	718.905	0.0014	9.8628	0.1014	9.3741
50	117.391	0.0085	1163.909	0.0009	9.9148	0.1009	9.5704
55	189.059	0.0053	1880.591	0.0005	9.9471	0.1005	9.7075
60	304.482	0.0033	3034.816	0.0003	9.9672	0.1003	9.8023
65	490.371	0.0020	4893.707	0.0002	9.9796	0.1002	9.8672
70	789.747	0.0013	7887.470	0.0001	9.9873	0.1001	9.9113
75	1271.895	0.0008	12708.954	0.0001	9.9921	0.1001	9.9410
80	2048.400	0.0005	20474.002	0.0001	9.9951	0.1001	9.9609
85	3298.969	0.0003	32979.690	0.0000	9.9970	0.1000	9.9742
90	5313.023	0.0002	53120.226	0.0000	9.9981	0.1000	9.9831
95	8556.676	0.0001	85556.760	0.0000	9.9988	0.1000	9.9889
100	13780.612	0.0001	137796.123	0.0000	9.9993	0.1000	9.9928

TABLE D.6 11% Interest Factors for Discrete Annual Compounding

	Single Payment		Equal Payment Series				Uniform Gradient-Series Factor
	Compound-Amount Factor	Present-Worth Factor	Compound-Amount Factor	Sinking-Fund Factor	Present-Worth Factor	Capital-Recovery Factor	
	To find F Given P	To find P Given F	To find F Given A	To find A Given F	To find P Given A	To find A Given P	To find A Given G
n	$F/P, i, n$	$P/F, i, n$	$F/A, i, n$	$A/F, i, n$	$P/A, i, n$	$A/P, i, n$	$A/G, i, n$
1	1.110	0.9009	1.000	1.0000	0.9009	1.1100	0.0000
2	1.232	0.8116	2.110	0.4739	1.7125	0.5839	0.4739
3	1.368	0.7312	3.342	0.2992	2.4437	0.4092	0.9306
4	1.518	0.6587	4.710	0.2123	3.1024	0.3223	1.3700
5	1.685	0.5935	6.228	0.1606	3.6959	0.2706	1.7923
6	1.870	0.5346	7.913	0.1264	4.2305	0.2364	2.1976
7	2.076	0.4817	9.783	0.1022	4.7122	0.2122	2.5863
8	2.305	0.4339	11.859	0.0843	5.1461	0.1943	2.9585
9	2.558	0.3909	14.164	0.0706	5.5370	0.1806	3.3144
10	2.839	0.3522	16.722	0.0598	5.8892	0.1698	3.6544
11	3.152	0.3173	19.561	0.0511	6.2065	0.1611	3.9788
12	3.498	0.2858	22.713	0.0440	6.4924	0.1540	4.2879
13	3.883	0.2575	26.212	0.0382	6.7499	0.1482	4.5822
14	4.310	0.2320	30.095	0.0332	6.9819	0.1432	4.8619
15	4.785	0.2090	34.405	0.0291	7.1909	0.1391	5.1275
16	5.311	0.1883	39.190	0.0255	7.3972	0.1355	5.3794
17	5.895	0.1696	44.501	0.0225	7.5488	0.1325	5.6180
18	6.544	0.1528	50.396	0.0198	7.7016	0.1298	5.8439
19	7.263	0.1377	56.939	0.0176	7.8393	0.1276	6.0574
20	8.062	0.1240	64.203	0.0156	7.9633	0.1256	6.2590
21	8.949	0.1117	72.265	0.0138	8.0751	0.1238	6.4491
22	9.934	0.1007	81.214	0.0123	8.1757	0.1223	6.6283
23	11.026	0.0907	91.148	0.0110	8.2664	0.1210	6.7969
24	12.239	0.0817	102.174	0.0098	8.3481	0.1198	6.9555
25	13.585	0.0736	114.413	0.0087	8.4217	0.1187	7.1045
26	15.080	0.0663	127.999	0.0078	8.4881	0.1178	7.2443
27	16.739	0.0597	143.079	0.0070	8.5478	0.1170	7.3754
28	18.580	0.0538	159.817	0.0063	8.6016	0.1163	7.4982
29	20.624	0.0485	178.397	0.0056	8.6501	0.1156	7.6131
30	22.892	0.0437	199.021	0.0050	8.6938	0.1150	7.7206
31	25.410	0.0394	221.913	0.0045	8.7331	0.1145	7.8210
32	28.206	0.0355	247.324	0.0040	8.7686	0.1140	7.9147
33	31.308	0.0319	275.529	0.0036	8.8005	0.1136	8.0021
34	34.752	0.0288	306.837	0.0033	8.8293	0.1133	8.0836
35	38.575	0.0259	341.590	0.0029	8.8552	0.1129	8.1594
40	65.001	0.0154	581.826	0.0017	8.9511	0.1117	8.4659
45	109.530	0.0091	986.639	0.0010	9.0079	0.1110	8.6763
50	184.565	0.0054	1668.771	0.0006	9.0417	0.1106	8.8185

TABLE D.7 12% Interest Factors for Discrete Annual Compounding

	Single Payment		Equal Payment Series				Uniform Gradient-Series Factor
	Compound-Amount Factor	Present-Worth Factor	Compound-Amount Factor	Sinking-Fund Factor	Present-Worth Factor	Capital-Recovery Factor	
	To find F Given P	To find P Given F	To find F Given A	To find A Given F	To find P Given A	To find A Given P	To find A Given G
n	$F/P, i, n$	$P/F, i, n$	$F/A, i, n$	$A/F, i, n$	$P/A, i, n$	$A/P, i, n$	$A/G, i, n$
1	1.120	0.8929	1.000	1.0000	0.8929	1.1200	0.0000
2	1.254	0.7972	2.120	0.4717	1.6901	0.5917	0.4717
3	1.405	0.7118	3.374	0.2964	2.4018	0.4164	0.9246
4	1.574	0.6355	4.779	0.2092	3.0374	0.3292	1.3589
5	1.762	0.5674	6.353	0.1574	3.6048	0.2774	1.7746
6	1.974	0.5066	8.115	0.1232	4.1114	0.2432	2.1721
7	2.211	0.4524	10.089	0.0991	4.5638	0.2191	2.5515
8	2.476	0.4039	12.300	0.0813	4.9676	0.2013	2.9132
9	2.773	0.3606	14.776	0.0677	5.3283	0.1877	3.2574
10	3.106	0.3220	17.549	0.0570	5.6502	0.1770	3.5847
11	3.479	0.2875	20.655	0.0484	5.9377	0.1684	3.8953
12	3.896	0.2567	24.133	0.0414	6.1944	0.1614	4.1897
13	4.364	0.2292	28.029	0.0357	6.4236	0.1557	4.4683
14	4.887	0.2046	32.393	0.0309	6.6282	0.1509	4.7317
15	5.474	0.1827	37.280	0.0268	6.8109	0.1468	4.9803
16	6.130	0.1631	42.753	0.0234	6.9740	0.1434	5.2147
17	6.866	0.1457	48.884	0.0205	7.1196	0.1405	5.4353
18	7.690	0.1300	55.750	0.0179	7.2497	0.1379	5.6427
19	8.613	0.1161	63.440	0.0158	7.3658	0.1358	5.8375
20	9.646	0.1037	72.052	0.0139	7.4695	0.1339	6.0202
21	10.804	0.0926	81.699	0.0123	7.5620	0.1323	6.1913
22	12.100	0.0827	92.503	0.0108	7.6447	0.1308	6.3514
23	13.552	0.0738	104.603	0.0096	7.7184	0.1296	6.5010
24	15.179	0.0659	118.155	0.0085	7.7843	0.1285	6.6407
25	17.000	0.0588	133.334	0.0075	7.8431	0.1275	6.7708
26	19.040	0.0525	150.334	0.0067	7.8957	0.1267	6.8921
27	21.325	0.0469	169.374	0.0059	7.9426	0.1259	7.0049
28	23.884	0.0419	190.699	0.0053	7.9844	0.1253	7.1098
29	26.750	0.0374	214.583	0.0047	8.0218	0.1247	7.2071
30	29.960	0.0334	241.333	0.0042	8.0552	0.1242	7.2974
31	33.555	0.0298	271.293	0.0037	8.0850	0.1237	7.3811
32	37.582	0.0266	304.848	0.0033	8.1116	0.1233	7.4586
33	42.092	0.0238	342.429	0.0029	8.1354	0.1229	7.5303
34	47.143	0.0212	384.521	0.0026	8.1566	0.1226	7.5965
35	52.800	0.0189	431.664	0.0023	8.1755	0.1223	7.6577
40	93.051	0.0108	767.091	0.0013	8.2438	0.1213	7.8988
45	163.988	0.0061	1358.230	0.0007	8.2825	0.1207	8.0572
50	289.002	0.0035	2400.018	0.0004	8.3045	0.1204	8.1597

TABLE D.8 13% Interest Factors for Discrete Annual Compounding

	Single Payment		Equal Payment Series				Uniform Gradient-Series Factor
	Compound-Amount Factor	Present-Worth Factor	Compound-Amount Factor	Sinking-Fund Factor	Present-Worth Factor	Capital-Recovery Factor	
	To find F Given P	To find P Given F	To find F Given A	To find A Given F	To find P Given A	To find A Given P	To find A Given G
n	$F/P, i, n$	$P/F, i, n$	$F/A, i, n$	$A/F, i, n$	$P/A, i, n$	$A/P, i, n$	$A/G, i, n$
1	1.130	0.8850	1.000	1.0000	0.8850	1.1300	0.0000
2	1.277	0.7831	2.130	0.4695	1.6681	0.5995	0.4695
3	1.443	0.6931	3.407	0.2935	2.3612	0.4235	0.9187
4	1.630	0.6133	4.850	0.2062	2.9745	0.3362	1.3479
5	1.842	0.5428	6.480	0.1543	3.5172	0.2843	1.7571
6	2.082	0.4803	8.323	0.1202	3.9975	0.2502	2.1468
7	2.353	0.4251	10.405	0.0961	4.4226	0.2261	2.5171
8	2.658	0.3762	12.757	0.0784	4.7988	0.2084	2.8685
9	3.004	0.3329	15.416	0.0649	5.1317	0.1949	3.2014
10	3.395	0.2946	18.420	0.0543	5.4262	0.1843	3.5162
11	3.836	0.2607	21.814	0.0458	5.6869	0.1758	3.8134
12	4.335	0.2307	25.650	0.0390	5.9176	0.1690	4.0936
13	4.898	0.2042	29.985	0.0334	6.1218	0.1634	4.3573
14	5.535	0.1807	34.883	0.0287	6.3025	0.1587	4.6050
15	6.254	0.1599	40.417	0.0247	6.4624	0.1547	4.8375
16	7.067	0.1415	46.672	0.0214	6.6039	0.1514	5.0552
17	7.986	0.1252	53.739	0.0186	6.7291	0.1486	5.2589
18	9.024	0.1108	61.725	0.0162	6.8399	0.1462	5.4491
19	10.197	0.0981	70.749	0.0141	6.9380	0.1441	5.6265
20	11.523	0.0868	80.947	0.0124	7.0248	0.1424	5.7917
21	13.021	0.0768	92.470	0.0108	7.1016	0.1408	5.9454
22	14.714	0.0680	105.491	0.0095	7.1695	0.1395	6.0881
23	16.627	0.0601	120.205	0.0083	7.2297	0.1383	6.2205
24	18.788	0.0532	136.831	0.0073	7.2829	0.1373	6.3431
25	21.231	0.0471	155.620	0.0064	7.3300	0.1364	6.4566
26	23.991	0.0417	176.850	0.0057	7.3717	0.1357	6.5614
27	27.109	0.0369	200.841	0.0050	7.4086	0.1350	6.6582
28	30.633	0.0326	227.950	0.0044	7.4412	0.1344	6.7474
29	34.616	0.0289	258.583	0.0039	7.4701	0.1339	6.8296
30	39.116	0.0256	293.199	0.0034	7.4957	0.1334	6.9052
31	44.201	0.0226	332.315	0.0030	7.5183	0.1330	6.9747
32	49.947	0.0200	376.516	0.0027	7.5383	0.1327	7.0385
33	56.440	0.0177	426.463	0.0023	7.5560	0.1323	7.0971
34	63.777	0.0157	482.903	0.0021	7.5717	0.1321	7.1507
35	72.069	0.0139	546.681	0.0018	7.5856	0.1318	7.1998
40	132.782	0.0075	1013.704	0.0010	7.6344	0.1310	7.3888
45	244.641	0.0041	1874.165	0.0005	7.6609	0.1305	7.5076
50	450.736	0.0022	3459.507	0.0003	7.6752	0.1303	7.5811

TABLE D.9 14% Interest Factors for Discrete Annual Compounding

	Single Payment		Equal Payment Series				Uniform Gradient-Series Factor
	Compound-Amount Factor	Present-Worth Factor	Compound-Amount Factor	Sinking-Fund Factor	Present-Worth Factor	Capital-Recovery Factor	
	To find F Given P	To find P Given F	To find F Given A	To find A Given F	To find P Given A	To find A Given P	To find A Given G
n	$F/P, i, n$	$P/F, i, n$	$F/A, i, n$	$A/F, i, n$	$P/A, i, n$	$A/P, i, n$	$A/G, i, n$
1	1.140	0.8772	1.000	1.0000	0.8772	1.1400	0.0000
2	1.300	0.7695	2.140	0.4673	1.6467	0.6073	0.4673
3	1.482	0.6750	3.440	0.2907	2.3216	0.4307	0.9129
4	1.689	0.5921	4.921	0.2032	2.9137	0.3432	1.3370
5	1.925	0.5194	6.610	0.1513	3.4331	0.2913	1.7399
6	2.195	0.4556	8.536	0.1172	3.8887	0.2572	2.1218
7	2.502	0.3996	10.730	0.0932	4.2883	0.2332	2.4832
8	2.853	0.3506	13.233	0.0756	4.6389	0.2156	2.8246
9	3.252	0.3075	16.085	0.0622	4.9464	0.2022	3.1463
10	3.707	0.2697	19.337	0.0517	5.2161	0.1917	3.4490
11	4.226	0.2366	23.045	0.0434	5.4527	0.1834	3.7333
12	4.818	0.2076	27.271	0.0367	5.6603	0.1767	3.9998
13	5.492	0.1821	32.089	0.0312	5.8424	0.1712	4.2491
14	6.261	0.1597	37.581	0.0266	6.0021	0.1666	4.4819
15	7.138	0.1401	43.842	0.0228	6.1422	0.1628	4.6990
16	8.137	0.1229	50.980	0.0196	6.2651	0.1596	4.9011
17	9.276	0.1078	59.118	0.0169	6.3729	0.1569	5.0888
18	10.575	0.0946	68.394	0.0146	6.4674	0.1546	5.2630
19	12.056	0.0829	78.969	0.0127	6.5504	0.1527	5.4243
20	13.743	0.0728	91.025	0.0110	6.6231	0.1510	5.5734
21	15.668	0.0638	104.768	0.0095	6.6870	0.1495	5.7111
22	17.861	0.0560	120.436	0.0083	6.7429	0.1483	5.8381
23	20.362	0.0491	138.297	0.0072	6.7921	0.1472	5.9549
24	23.212	0.0431	158.659	0.0063	6.8351	0.1463	6.0624
25	26.462	0.0378	181.871	0.0055	6.8729	0.1455	6.1610
26	30.167	0.0331	208.333	0.0048	6.9061	0.1448	6.2514
27	34.390	0.0291	238.499	0.0042	6.9352	0.1442	6.3342
28	39.204	0.0255	272.889	0.0037	6.9607	0.1437	6.4100
29	44.693	0.0224	312.094	0.0032	6.9830	0.1432	6.4791
30	50.950	0.0196	356.787	0.0028	7.0027	0.1428	6.5423
31	58.083	0.0172	407.737	0.0025	7.0199	0.1425	6.5998
32	66.215	0.0151	465.820	0.0021	7.0350	0.1421	6.6522
33	75.485	0.0132	532.035	0.0019	7.0482	0.1419	6.6998
34	86.053	0.0116	607.520	0.0016	7.0599	0.1416	6.7431
35	98.100	0.0102	693.573	0.0014	7.0700	0.1414	6.7824
40	188.884	0.0053	1342.025	0.0007	7.1050	0.1407	6.9300
45	363.679	0.0027	2590.565	0.0004	7.1232	0.1404	7.0188
50	700.233	0.0014	4994.521	0.0002	7.1327	0.1402	7.0714

TABLE D.10   15% Interest Factors for Discrete Annual Compounding

	Single Payment		Equal Payment Series				Uniform Gradient-Series Factor
	Compound-Amount Factor	Present-Worth Factor	Compound-Amount Factor	Sinking-Fund Factor	Present-Worth Factor	Capital-Recovery Factor	
	To find F Given P	To find P Given F	To find F Given A	To find A Given F	To find P Given A	To find A Given P	To find A Given G
n	$F/P, i, n$	$P/F, i, n$	$F/A, i, n$	$A/F, i, n$	$P/A, i, n$	$A/P, i, n$	$A/G, i, n$
1	1.150	0.8696	1.000	1.0000	0.8696	1.1500	0.0000
2	1.323	0.7562	2.150	0.4651	1.6257	0.6151	0.4651
3	1.521	0.6575	3.473	0.2880	2.2832	0.4380	0.9071
4	1.749	0.5718	4.993	0.2003	2.8550	0.3503	1.3263
5	2.011	0.4972	6.742	0.1483	3.3522	0.2983	1.7228
6	2.313	0.4323	8.754	0.1142	3.7845	0.2642	2.0972
7	2.660	0.3759	11.067	0.0904	4.1604	0.2404	2.4499
8	3.059	0.3269	13.727	0.0729	4.4873	0.2229	2.7813
9	3.518	0.2843	16.786	0.0596	4.7716	0.2096	3.0922
10	4.046	0.2472	20.304	0.0493	5.0188	0.1993	3.3832
11	4.652	0.2150	24.349	0.0411	5.2337	0.1911	3.6550
12	5.350	0.1869	29.002	0.0345	5.4206	0.1845	3.9082
13	6.153	0.1625	34.352	0.0291	5.5832	0.1791	4.1438
14	7.076	0.1413	40.505	0.0247	5.7245	0.1747	4.3624
15	8.137	0.1229	47.580	0.0210	5.8474	0.1710	4.5650
16	9.358	0.1069	55.717	0.0180	5.9542	0.1680	4.7523
17	10.761	0.0929	65.075	0.0154	6.0472	0.1654	4.9251
18	12.375	0.0808	75.836	0.0132	6.1280	0.1632	5.0843
19	14.232	0.0703	88.212	0.0113	6.1982	0.1613	5.2307
20	16.367	0.0611	102.444	0.0098	6.2593	0.1598	5.3651
21	18.822	0.0531	118.810	0.0084	6.3125	0.1584	5.4883
22	21.645	0.0462	137.632	0.0073	6.3587	0.1573	5.6010
23	24.891	0.0402	159.276	0.0063	6.3988	0.1563	5.7040
24	28.625	0.0349	184.168	0.0054	6.4338	0.1554	5.7979
25	32.919	0.0304	212.793	0.0047	6.4642	0.1547	5.8834
26	37.857	0.0264	245.712	0.0041	6.4906	0.1541	5.9612
27	43.535	0.0230	283.569	0.0035	6.5135	0.1535	6.0319
28	50.066	0.0200	327.104	0.0031	6.5335	0.1531	6.0960
29	57.575	0.0174	377.170	0.0027	6.5509	0.1527	6.1541
30	66.212	0.0151	434.745	0.0023	6.5660	0.1523	6.2066
31	76.144	0.0131	500.957	0.0020	6.5791	0.1520	6.2541
32	87.565	0.0114	577.100	0.0017	6.5905	0.1517	6.2970
33	100.700	0.0099	664.666	0.0015	6.6005	0.1515	6.3357
34	115.805	0.0086	765.365	0.0013	6.6091	0.1513	6.3705
35	133.176	0.0075	881.170	0.0011	6.6166	0.1511	6.4019
40	267.864	0.0037	1779.090	0.0006	6.6418	0.1506	6.5168
45	538.769	0.0019	3585.128	0.0003	6.6543	0.1503	6.5830
50	1083.657	0.0009	7217.716	0.0002	6.6605	0.1501	6.6205

TABLE D.11   20% Interest Factors for Discrete Annual Compounding

	Single Payment		Equal Payment Series				Uniform Gradient-Series Factor
	Compound-Amount Factor	Present-Worth Factor	Compound-Amount Factor	Sinking-Fund Factor	Present-Worth Factor	Capital-Recovery Factor	
	To find F Given P	To find P Given F	To find F Given A	To find A Given F	To find P Given A	To find A Given P	To find A Given G
n	$F/P, i, n$	$P/F, i, n$	$F/A, i, n$	$A/F, i, n$	$P/A, i, n$	$A/P, i, n$	$A/G, i, n$
1	1.200	0.8333	1.000	1.0000	0.8333	1.2000	0.0000
2	1.440	0.6945	2.200	0.4546	1.5278	0.6546	0.4546
3	1.728	0.5787	3.640	0.2747	2.1065	0.4747	0.8791
4	2.074	0.4823	5.368	0.1863	2.5887	0.3863	1.2742
5	2.488	0.4019	7.442	0.1344	2.9906	0.3344	1.6405
6	2.986	0.3349	9.930	0.1007	3.3255	0.3007	1.9788
7	3.583	0.2791	12.916	0.0774	3.6046	0.2774	2.2902
8	4.300	0.2326	16.499	0.0606	3.8372	0.2606	2.5756
9	5.160	0.1938	20.799	0.0481	4.0310	0.2481	2.8364
10	6.192	0.1615	25.959	0.0385	4.1925	0.2385	3.0739
11	7.430	0.1346	32.150	0.0311	4.3271	0.2311	3.2893
12	8.916	0.1122	39.581	0.0253	4.4392	0.2253	3.4841
13	10.699	0.0935	48.497	0.0206	4.5327	0.2206	3.6597
14	12.839	0.0779	59.196	0.0169	4.6106	0.2169	3.8175
15	15.407	0.0649	72.035	0.0139	4.6755	0.2139	3.9589
16	18.488	0.0541	87.442	0.0114	4.7296	0.2114	4.0851
17	22.186	0.0451	105.931	0.0095	4.7746	0.2095	4.1976
18	26.623	0.0376	128.117	0.0078	4.8122	0.2078	4.2975
19	31.948	0.0313	154.740	0.0065	4.8435	0.2065	4.3861
20	38.338	0.0261	186.688	0.0054	4.8696	0.2054	4.4644
21	46.005	0.0217	225.026	0.0045	4.8913	0.2045	4.5334
22	55.206	0.0181	271.031	0.0037	4.9094	0.2037	4.5942
23	66.247	0.0151	326.237	0.0031	4.9245	0.2031	4.6475
24	79.497	0.0126	392.484	0.0026	4.9371	0.2026	4.6943
25	95.396	0.0105	471.981	0.0021	4.9476	0.2021	4.7352
26	114.475	0.0087	567.377	0.0018	4.9563	0.2018	4.7709
27	137.371	0.0073	681.853	0.0015	4.9636	0.2015	4.8020
28	164.845	0.0061	819.223	0.0012	4.9697	0.2012	4.8291
29	197.814	0.0051	984.068	0.0010	4.9747	0.2010	4.8527
30	237.376	0.0042	1181.882	0.0009	4.9789	0.2009	4.8731
31	284.852	0.0035	1419.258	0.0007	4.9825	0.2007	4.8908
32	341.822	0.0029	1704.109	0.0006	4.9854	0.2006	4.9061
33	410.186	0.0024	2045.931	0.0005	4.9878	0.2005	4.9194
34	492.224	0.0020	2456.118	0.0004	4.9899	0.2004	4.9308
35	590.668	0.0017	2948.341	0.0003	4.9915	0.2003	4.9407
40	1469.772	0.0007	7343.858	0.0002	4.9966	0.2001	4.9728
45	3657.262	0.0003	18281.310	0.0001	4.9986	0.2001	4.9877
50	9100.438	0.0001	45497.191	0.0000	4.9995	0.2000	4.9945

TABLE D.12 25% Interest Factors for Discrete Annual Compounding

	Single Payment		Equal Payment Series				Uniform
	Compound-Amount Factor	Present-Worth Factor	Compound-Amount Factor	Sinking-Fund Factor	Present-Worth Factor	Capital-Recovery Factor	Gradient-Series Factor
	To find F Given P	To find P Given F	To find F Given A	To find A Given F	To find P Given A	To find A Given P	To find A Given G
n	$F/P, i, n$	$P/F, i, n$	$F/A, i, n$	$A/F, i, n$	$P/A, i, n$	$A/P, i, n$	$A/G, i, n$
1	1.250	0.8000	1.000	1.0000	0.8000	1.2500	0.0000
2	1.563	0.6400	2.250	0.4445	1.4400	0.6945	0.4445
3	1.953	0.5120	3.813	0.2623	1.9520	0.5123	0.8525
4	2.441	0.4096	5.766	0.1735	2.3616	0.4235	1.2249
5	3.052	0.3277	8.207	0.1219	2.6893	0.3719	1.5631
6	3.815	0.2622	11.259	0.0888	2.9514	0.3388	1.8683
7	4.768	0.2097	15.073	0.0664	3.1611	0.3164	2.1424
8	5.960	0.1678	19.842	0.0504	3.3289	0.3004	2.3873
9	7.451	0.1342	25.802	0.0388	3.4631	0.2888	2.6048
10	9.313	0.1074	33.253	0.0301	3.5705	0.2801	2.7971
11	11.642	0.0859	42.566	0.0235	3.6564	0.2735	2.9663
12	14.552	0.0687	54.208	0.0185	3.7251	0.2685	3.1145
13	18.190	0.0550	68.760	0.0146	3.7801	0.2646	3.2438
14	22.737	0.0440	86.949	0.0115	3.8241	0.2615	3.3560
15	28.422	0.0352	109.687	0.0091	3.8593	0.2591	3.4530
16	35.527	0.0282	138.109	0.0073	3.8874	0.2573	3.5366
17	44.409	0.0225	173.636	0.0058	3.9099	0.2558	3.6084
18	55.511	0.0180	218.045	0.0046	3.9280	0.2546	3.6698
19	69.389	0.0144	273.556	0.0037	3.9424	0.2537	3.7222
20	86.736	0.0115	342.945	0.0029	3.9539	0.2529	3.7667
21	108.420	0.0092	429.681	0.0023	3.9631	0.2523	3.8045
22	135.525	0.0074	538.101	0.0019	3.9705	0.2519	3.8365
23	169.407	0.0059	673.626	0.0015	3.9764	0.2515	3.8634
24	211.758	0.0047	843.033	0.0012	3.9811	0.2512	3.8861
25	264.698	0.0038	1054.791	0.0010	3.9849	0.2510	3.9052
26	330.872	0.0030	1319.489	0.0008	3.9879	0.2508	3.9212
27	413.590	0.0024	1650.361	0.0006	3.9903	0.2506	3.9346
28	516.988	0.0019	2063.952	0.0005	3.9923	0.2505	3.9457
29	646.235	0.0016	2580.939	0.0004	3.9938	0.2504	3.9551
30	807.794	0.0012	3227.174	0.0003	3.9951	0.2503	3.9628
31	1009.742	0.0010	4034.968	0.0003	3.9960	0.2503	3.9693
32	1262.177	0.0008	5044.710	0.0002	3.9968	0.2502	3.9746
33	1577.722	0.0006	6306.887	0.0002	3.9975	0.2502	3.9791
34	1972.152	0.0005	7884.609	0.0001	3.9980	0.2501	3.9828
35	2465.190	0.0004	9856.761	0.0001	3.9984	0.2501	3.9858

APPENDIX E
RECOMMENDED BIBLIOGRAPHY

When addressing the subject of maintainability engineering and maintenance, one should become familiar not only with the available literature in this field itself, but also with some of the other related and closely aligned subject areas, particularly as related to the system/product engineering process. With this in mind, a number of key references have been included pertaining to reliability, human factors and safety, logistics, systems engineering, concurrent/simultaneous engineering, software systems (computer-aided systems), production operations, operations research, quality engineering, quality assurance, quality management, engineering economics, life-cycle costing and cost estimating, management, and expert systems. These subject areas, and how they relate to maintainability and maintenance engineering and management, have been discussed extensively throughout this text, and hopefully the references listed will prove to be beneficial. This list is certainly not to be considered as being all-inclusive.

A. MAINTAINABILITY AND MAINTENANCE

1. Anderson, R. T. and L. Neri, *Reliability-Centered Maintenance,* Elsevier Science Publishing, Ltd., London, England, 1990.
2. Blanchard, B. S. and E. E. Lowery, *Maintainability Principles and Practices,* McGraw-Hill Book Co., New York, 1969.
3. Bray, D. E. and D. McBride, Eds., *Nondestructive Testing Techniques,* John Wiley & Sons, Inc., New York, 1992.
4. Cunningham, C. E. and W. Cox, *Applied Maintainability Engineering,* John Wiley & Sons, Inc., New York, 1972.
5. Faulkenberry, L. M., Ed., *Systems Troubleshooting Handbook,* John Wiley & Sons, Inc., New York, 1986.
6. Goldman, A. and T. Slattery, *Maintainability—A Major Element of System Effectiveness,* John Wiley & Sons, Inc., New York, 1967.

7. Higgins, L. R., *Maintenance Engineering Handbook,* 4th ed., McGraw-Hill Book Co., New York, 1988.

8. Jardine, A.K.S., *Maintenance, Replacement, and Reliability,* A Halsted Press Book, John Wiley & Sons, Inc., New York, 1973.

9. Kelly, A. and M. J. Harris, *Management of Industrial Maintenance,* Newness-Butterworths, London, 1978.

10. Mann, L., *Maintenance Management,* Lexington Books, D.C. Heath and Co., Lexington, Massachusetts, 1976.

11. MIL-HDBK-472, Military Handbook, *Maintainability Prediction,* Department of Defense, Washington, DC.

12. MIL-STD-470B, Military Standard, *Maintainability Program for Systems and Equipment,* Department of Defense, Washington, DC.

13. MIL-STD-471A, Military Standard, *Maintainability Verification, Demonstration, Evaluation,* Department of Defense, Washington, DC.

14. MIL-STD-721C, Military Standard, *Definitions of Effectiveness Terms for Reliability, Maintainability, Human Factors, and Safety,* Department of Defense, Washington, DC.

15. MIL-STD-2084, Military Standard, *General Requirements for Maintainability,* Department of Defense, Washington, DC.

16. MIL-STD-2165, Military Standard, *Testability Program for Electronic Systems and Equipments,* Department of Defense, Washington, DC.

17. Mobley, R. K., *An Introduction to Predictive Maintenance,* Van Nostrand Reinhold, New York, 1990.

18. Moss, M. A., *Designing for Minimal Maintenance Expense: The Practical Application of Reliability and Maintainability,* Marcel Dekker, Inc., 1985.

19. Moubray, J., *Reliability-Centered Maintenance,* Butterworth-Heinemann Ltd., Boston, MA, 1991.

20. Nakajima, S., *Total Production Maintenance (TPM),* Productivity Press, Inc., Cambridge, MA, 1988.

21. Nakajima, S., Ed., *TPM Development Program: Implementing Total Productive Maintenance,* Productivity Press, Inc., Cambridge, MA, 1989.

22. Niebel, B. W., *Engineering Maintenance Management,* Marcel Dekker, Inc., New York, 1985.

23. Nowlan, F. S. and H. F. Heap, *Reliability-Centered Maintenance,* United Airlines (MDA 903-75-C-0349), San Francisco, CA, 1978.

24. Patton, J. D., *Maintainability and Maintenance Management,* 2nd ed., Instrument Society of America, Research Triangle Park, NC, 1988.

25. Patton, J. D., *Preventive Maintenance,* Instrument Society of America, Research Triangle Park, NC, 1983.

26. Wireman, T., *World Class Maintenance Management,* Industrial Press, New York, 1990.

B. RELIABILITY

1. Amstadter, B. L., *Reliability Mathematics: Fundamentals, Practices, Procedures,* McGraw-Hill Book Co., New York, 1971.

2. *Annual Reliability and Maintainability Symposium,* Proceedings, Evans Associates, Durham, NC.

3. Arsenault, J. E. and J. A. Roberts, *Reliability and Maintainability of Electronic Systems,* Computer Science Press, Inc., Potomac, MD, 1980.

4. Dhillon, B. S. and C. Singh, *Engineering Reliability: New Techniques and Applications,* John Wiley and Sons, Inc., New York, 1981.

5. Dhillon, B. S. and H. Reich, *Reliability and Maintainability Management,* Van Nostrand Reinhold Co., New York, 1985.

6. Doty, L. A., *Reliability for the Technologies,* 2nd ed., Industrial Press, Inc., New York, 1989.

7. Fugua, N. B., *Reliability Engineering for Electronic Design,* Marcel Dekker, Inc., New York, 1987.

8. Grosh, D. L., *A Primer of Reliability Theory,* John Wiley and Sons, Inc., New York, 1989.

9. Harr, M. E., *Reliability-Based Design in Civil Engineering,* McGraw-Hill Book Co., New York, 1987.

10. Henley, E. J. and H. Kumamoto, *Reliability Engineering and Risk Assessment,* Prentice-Hall, Inc., Englewood Cliffs, NJ, 1981.

11. Ireson, W. G., and C. F. Coombs, Eds., *Handbook of Reliability Engineering and Management,* McGraw-Hill Book Co., New York, 1988.

12. Kapur, K. C. and L. R. Lamberson, *Reliability in Engineering Design,* John Wiley & Sons, Inc., New York, 1977.

13. Kececioglu, D., *Reliability Engineering Handbook,* Vols. I and II, Prentice Hall, Inc., Englewood Cliffs, NJ, 1991.

14. Knezevic, J., *Reliability, Maintainability, and Supportability: A Probabilistic Approach,* McGraw-Hill Book Company, New York, 1993.

15. Lewis, E. E., *Introduction to Reliability Engineering,* John Wiley & Sons, Inc., New York, 1987.

16. Lloyd, D. K. and M. Lipow, *Reliability: Management, Methods, and Mathematics,* 2nd ed., published by the authors, Defense and Space Systems Group, TRW Systems and Energy, Redondo Beach, CA, 1984.

17. Locks, M. O., *Reliability, Maintainability, and Availability Assessment,* Hayden Book Company, Inc., Rochelle Park, NJ, 1973.

18. Mann, N. R., R. E. Schafer, and N. D. Singpurwalla, *Methods for Statistical Analysis of Reliability and Test Data,* John Wiley & Sons, Inc., New York, 1974.

19. MIL-HDBK-189, Military Handbook, *Reliability Growth Management,* Department of Defense, Washington, DC.

20. MIL-HDBK-217F, Military Handbook, *Reliability Predictions of Electronic Equipment,* Department of Defense, Washington, DC.

21. MIL-HDBK-338, Military Handbook, *Electronic Reliability Design Handbook,* Department of Defense, Washington, DC.

22. MIL-STD-721C, Military Standard, *Definitions of Terms for Reliability and Maintainability,* Department of Defense, Washington, DC.

23. MIL-STD-756B, Military Standard, *Reliability Modeling and Prediction,* Department of Defense, Washington, DC.

24. MIL-STD-781D, Military Standard, *Reliability Testing for Engineering Development, Qualification, and Production*, Department of Defense, Washington, DC.

25. MIL-STD-785B, Military Standard, *Reliability Program for Systems and Equipment Development and Production*, Department of Defense, Washington, DC.

26. MIL-STD-1629A, Military Standard, *Procedures for Performing a Failure Mode, Effects and Criticality Analysis*, Department of Defense, Washington, DC.

27. MIL-STD-2155, Military Standard, *Failure Reporting, Analysis, and Corrective Action System (FRACAS)*, Department of Defense, Washington, DC.

28. MIL-STD-2164, Military Standard *Environmental Stress Screening Process for Electronic Equipment*, Department of Defense, Washington, DC.

29. Modarres, M., *What Every Engineer Should Know About Reliability and Risk Analysis*, Marcel Dekker, Inc., New York, 1993.

30. Musa, J. D., A. Iannino, and K. Okumoto, *Software Reliability: Measurement, Prediction, Application*, McGraw-Hill Book Co., New York, 1987.

31. O'Connor, P.D.T., *Practical Reliability Engineering*, 3rd ed., John Wiley & Sons, Inc., New York, 1991.

32. *RADC Reliability Engineer's Toolkit*, Rome Air Development Center, RADC/RBE, Griffiss AFB, New York, 1988.

33. Raheja, D. G., *Assurance Technologies: Principles and Practices*, McGraw-Hill, Inc., New York, 1991.

34. Rao, S. S., *Reliability-Based Design*, McGraw-Hill, Inc., New York, 1992.

35. RDH-376, *Reliability Design Handbook*, Reliability Analysis Center, Griffiss AFB, New York, 1976.

36. Shooman, M. L., *Probabilistic Reliability: An Engineering Approach*, McGraw-Hill Book Co., New York, 1968.

37. Siewiorek, D. P. and R. B. Swarz, *The Theory and Practice of Reliable System Design*, Digital Press, Educational Services, Digital Equipment Corp., Bedford, MA, 1982.

38. Von Alven, W. H. Ed., *Reliability Engineering*, Prentice-Hall, Inc., Englewood Cliffs, NJ, 1964.

C. HUMAN FACTORS AND SAFETY

1. Cushman, W. H. and D. J. Rosenberg, *Human Factors in Product Design*, Elsevier Science Publishing Company, Inc., New York, 1991.

2. DeGreene, K. B., Ed., *Systems Psychology*, McGraw-Hill Book Co., New York, 1970.

3. Eastman Kodak Company, *Ergonomic Design for People at Work*, Van Nostrand Reinhold, Co., New York, 1983.

4. Hammer, W., *Occupational Safety Management and Engineering*, 4th ed., Prentice-Hall Inc., Englewood Cliffs, NJ, 1989.

5. Meister, D., *Behavioral Analysis and Measurement Methods*, John Wiley & Sons, Inc., New York, 1985.

6. MIL-HDBK-759A, Military Handbook, *Human Factors Engineering Design for Army Material*, Department of Defense, Washington, DC.

7. MIL-H-46855B, Military Specification, *Human Engineering Requirements for Military*

Systems, Equipment, and Facilities, US Army Missile R&D Command (DRDMI-ESD), Redstone Arsenal, Alabama.

8. MIL-STD-882B, Military Standard, *System Safety Program Requirements,* Department of Defense, Washington, DC.

9. MIL-STD-1472D, Military Standard, *Human Engineering Design Criteria for Military Systems, Equipment, and Facilities,* Department of Defense, Washington, DC.

10. MIL-STD, *Human Factors Engineering Performance Requirements for Systems,* Department of Defense, Washington, DC.

11. Roland, H. E. and B. Moriarty, *System Safety Engineering and Management,* 2nd ed., John Wiley & Sons, Inc., New York, 1990.

12. Sanders, M. S., and E. J. McCormick, *Human Factors in Engineering and Design,* 6th ed., McGraw-Hill Book Co., New York, 1987.

13. Salvendy, G., ed., *Handbook of Human Factors,* John Wiley & Sons, Inc., New York, 1987.

14. Van Cott, H. P. and R. G. Kinkade, Eds., *Human Engineering Guide to Equipment Design,* US Government Printing Office, Washington, DC, 1972.

15. Woodson, W. E., B. Tillman, and P. Tillman, *Human Factors Design,* 2nd ed., McGraw-Hill Book Co., New York, 1991.

D. LOGISTICS

1. Allen, M. K. and O. K. Helferich, *Putting Expert Systems to Work in Logistics,* Council of Logistics Management, Oak Brook, IL, 1990.

2. Ballou, R. H., *Business Logistics Management,* 3rd ed., Prentice-Hall, Inc., Englewood Cliffs, NJ, 1992.

3. Banks, J. and W. J. Fabrycky, *Procurement and Inventory Systems Analysis,* Prentice-Hall, Inc., Englewood Cliffs, NJ, 1987.

4. Barnes, T. A., *Logistic Support Training: Design and Development,* McGraw-Hill, Inc., New York, 1992.

5. Blanchard, B. S., *Logistics Engineering and Management,* 4th ed., Prentice-Hall, Inc., Englewood Cliffs, NJ, 1992.

6. Bowersox, D., D. Closs, And O. Helferich, *Logistical Management,* Macmillan Publishing Co., New York, 1986.

7. Council of Logistics Management, *Journal of Business Logistics,* Oak Brook, IL.

8. Council of Logistics Management, *Logistics Software,* Annual Edition, Anderson Consulting, New York, NY, 1993.

9. Coyle, J. J., E. J. Bardi, and C. J. Langley, *The Management of Business Logistics,* 4th ed., West Publishing Co., St. Paul, MN, 1988.

10. Coyle, J. J., E. J. Bardi, and J. L. Cavinato, *Transportation,* West Publishing Co., St. Paul, MN, 1982.

11. Defense Systems Management College (DSMC), *Integrated Logistics Support Guide,* DSMC, Fort Belvoir, VA.

12. DOD Instruction 5000.2, *Defense Acquisition Management Policies and Procedures,* Part 6N (CALS) and Part 7 (ILS), Department of Defense, February 23, 1991.

13. Glaskowsky, N. A., D. R. Hudson, and R. M. Ivie, *Business Logistics,* 3rd ed., The Dryden Press, Harcourt Brace Jovanovich Co., Orlando, FL, 1991.

14. Green, L. L., *Logistics Engineering,* John Wiley & Sons, Inc., New York, 1990.

15. Hutchinson, N. E., *An Integrated Approach to Logistics Management,* Prentice-Hall, Inc., Englewood Cliffs, NJ, 1987.

16. Johnson, J. C. and D. F. Wood, *Contemporary Logistics,* 4th ed., Macmillan Publishing Co., New York, 1990.

17. Jones, J. V., *Logistic Support Analysis Handbook,* Tab Books, Inc., Blue Ridge Summit, PA, 1989.

18. Jones, J. V., *Integrated Logistics Support Handbook,* Tab Books, Inc., Blue Ridge Summit, PA, 1987.

19. Magee, J. F., W. C. Copacino, and D. B. Rosenfield, *Modern Logistics Management,* John Wiley & Sons, Inc., New York, 1985.

20. MIL-HDBK-59A, Military Handbook, *Computer-Aided Acquisition Logistic Support (CALS) Implementation Guide,* Department of Defense, Washington, DC.

21. MIL-HDBK-226, Military Handbook, *Application of Reliability-Centered Maintenance to Naval Aircraft, Weapon Systems, and Support Equipment,* Department of Defense, Washington, DC.

22. MIL-STD-974, *Contractor Integrated Technical Information Service (CITIS),* Department of Defense, Washington DC, August 1993.

23. MIL-STD-1388-1A, Military Standard, *Logistic Support Analysis,* Department of Defense, Washington, DC.

24. MIL-STD-1388-2B, Military Standard, *Department of Defense Requirements for a Logistic Support Analysis Record,* Department of Defense, Washington, DC.

25. MIL-STD-1840A, Military Standard, *Automated Interchange of Technical Information,* Department of Defense, Washington, DC.

26. Military Standard, *Digital Representation for Communication of Product Data: IGES Subsets,* Department of Defense, Washington, DC.

27. O'Neil and Associates, Inc., *Integrated Logistics Support,* Dayton, OH.

28. Orsburn, Douglas K., *Introduction to Spares Management,* Academy Printing & Publishing Co., Paramount, CA, 1985.

29. Patton, Jr., Joseph D., *Logistics Technology and Management—The New Approach,* The Solomon Press, New York, 1986.

30. Peppers, J. G., *History of United States Military Logistics—1935 to 1985,* 438 Coronado Drive, Fairborn, OH, 1987.

31. Shapiro, Roy D. and James L. Heskett, *Logistics Strategy: Cases and Concepts,* West Publishing Co., St. Paul, MN, 1985.

32. Society of Logistics Engineers (SOLE), *Annals,* New Carrollton, MD.

33. Society of Logistics Engineers (SOLE), *Logistics Spectrum,* New Carrollton, MD.

34. Society of Logistics Engineers (SOLE), Proceedings, *Annual Symposium,* New Carrollton, MD.

35. Stock, James R. and Douglas M. Lambert, *Strategic Logistics Management,* 2nd ed., Richard D. Irwin, Inc., Homewood, IL. 1987.

36. Air Force, *Air Force Journal of Logistics,* US Government Printing Office, Washington, DC.

37. Air Force, *Compendium of Authenticated System and Logistics Terms, Definitions, and Acronyms,* AU-AFIT-LS-3-81, US Air Force Institute of Technology, Wright-Patterson AFB, Dayton, Ohio, April 1981.
38. Army, Annual Department of Defense Bibliography of Logistics Studies and Related Documents, *Defense Logistics Studies Information Exchange (DLSIE),* US Army Logistics Management Center, Fort Lee, VA.

E. SYSTEMS ENGINEERING

1. Beam, W. R., *Systems Engineering: Architecture and Design,* McGraw-Hill, Inc., New York, 1990.
2. Belcher, R. and E. Aslaksen, *Systems Engineering,* Prentice-Hall of Australia, Sydney, Australia, 1992.
3. Blanchard, B. S., *System Engineering Management,* John Wiley & Sons, Inc., New York, 1991.
4. Blanchard, B. S. and W. J. Fabrycky, *Systems Engineering and Analysis,* 2nd ed., Prentice-Hall, Inc., Englewood Cliffs, NJ, 1990.
5. Boardman, J., *Systems Engineering: An Introduction,* Prentice-Hall International, London, England, 1990.
6. Chase, W. P., *Management of System Engineering,* John Wiley & Sons, Inc., New York, 1974.
7. Chestnut, H., *Systems Engineering Methods,* John Wiley & Sons, Inc., New York, 1967.
8. Chestnut, H., *Systems Engineering Tools,* John Wiley & Sons, Inc., New York, 1965.
9. Defense Systems Management College (DSMC), *Systems Engineering Management Guide,* DSMC, Fort Belvoir, VA.
10. DODD 5000.1, *Defense Acquisition,* Department of Defense, Washington, DC, February 1991.
11. DODD 5000.1, *Defense Acquisition Management Policies and Procedures,* Department of Defense, Washington, DC, February 1991.
12. Drew, D. R. and C. H. Hsieh, *A Systems View of Development: Methodology of Systems Engineering and Management,* Cheng Yang Publishing Co., Taipei, ROC 1984.
13. Forrester, J. W., *Principles of Systems,* The MIT Press, Cambridge, MA, 1968.
14. Gheorghe, A., *Applied Systems Engineering,* John Wiley & Sons, Inc., New York, 1982.
15. Grady, J. O., *System Requirements Analysis,* McGraw-Hill, Inc., New York, 1993.
16. Hall, A. D., *A Methodology for Systems Engineering,* D. Van Nostrand Co., Ltd., Princeton, NJ, 1962.
17. Machol, R. E., Ed., *System Engineering Handbook,* McGraw-Hill Book Co., New York, 1965.
18. MIL-STD-499B, Draft Military Standard, *System Engineering,* Headquarters, US Air Force/Air Force Systems Command, Andrews AFB, Maryland.
19. Ostrofsky, B., *Design, Planning and Development Methodology,* Prentice-Hall, Inc., Englewood Cliffs, NJ, 1977.

20. Pugh, S., *Total Design: Integrated Methods for Successful Product Engineering*, Addison-Wesley Publishing Company, Inc., Reading, MA, 1991.

21. Rechtin, E. *Systems Architecting*, Prentice-Hall, Inc., Englewood Cliffs, NJ, 1991.

22. Rouse, W. B., *Systems Engineering Models of Human-Machine Interaction*, Elsevier/North Holland, Inc., New York, 1980.

23. Sage, A. P., *Decision Support Systems Engineering*, John Wiley & Sons, Inc., New York, 1991.

24. Sage, A. P., *Economic System Analysis: Microeconomics for Systems Engineering, Engineering Management, and Project Selection*, Elsevier Science Publishing Co., New York, 1983.

25. Sage, A. P., *Methodology for Large Scale Systems*, McGraw-Hill Book Co., New York, 1977.

26. Sage, A. P., *Systems Engineering*, John Wiley & Sons, Inc., New York, 1992.

27. Shinners, S. M., *A Guide to Systems Engineering and Management*, Lexington Books, Lexington, MA, 1976.

28. Singh, M. G., Ed., *Systems and Control Encyclopedia: Theory, Technology, Applications*, Pergamon Press, Fairview Park, Elmsford, NY, 1989.

29. Thomé, Bernhard, Ed., *Systems Engineering: Principles and Practice of Computer-Based Systems Engineering*, John Wiley & Sons, New York, 1993.

30. Truxal, J. G., *Introductory System Engineering*, McGraw-Hill Book Co., New York, 1972.

31. Von Bertalanffy, L., *General Systems Theory*, George Braziller Publisher, New York, 1968.

32. Weinberg, G. M., *An Introduction to General Systems Thinking*, John Wiley & Sons, NY, 1975.

33. Weinberg, G. M., *Rethinking Systems Analysis and Design*, Dorset House Publishing, NY, 1988.

34. Wymore, A. W., *Systems Engineering Methodology for Interdisciplinary Teams*, John Wiley & Sons, Inc., New York, 1976.

35. Wymore, A. W., *Model-Based Systems Engineering*, CRC Press, Inc., Boca Raton, FL 1993.

F. CONCURRENT/SIMULTANEOUS ENGINEERING

1. Hartley, J. R., *Concurrent Engineering: Shortening Lead Times, Raising Quality, and Lowering Costs*, Productivity Press, Cambridge, MA, 1992.

2. Kusiak, A., *Concurrent Engineering: Automation, Tools, and Techniques*, John Wiley & Sons, Inc., New York, 1992.

3. Miller, L.C.G., *Concurrent Engineering Design: Integrating the Best Practices for Process Improvement*, Society of Manufacturing Engineers, Dearborn, MI, 1993.

4. Nevins, J. A. and D. E. Whitney, Ed., *Concurrent Design of Products and Processes*, McGraw Hill Book Co., New York, 1989.

5. Darsaei, H. R. and W. G. Sullivan (Ed.,), *Concurrent Engineering: Contemporary Issues and Modern Design Tools*, Chapman and Hall, NY, 1993.

6. Shina, S. G., *Concurrent Engineering and Design for Manufacture of Electronics Products,* Van Nostrand Reinhold, New York, 1991.

7. Turino, J., *Managing Concurrent Engineering: Buying Time to Market,* Van Nostrand Reinhold, New York, 1992.

8. Winner, R. I., J. P. Pennell, H. E. Bertrand, and M.M.G. Slusarczuk, *The Role of Concurrent Engineering in Weapons Systems Acquisition,* Report R-338, Institute for Defense Analysis, 1988.

G. SOFTWARE SYSTEMS/COMPUTER-AIDED SYSTEMS

1. Boehm, B. W., *Software Engineering Economics,* Prentice-Hall, Inc., Englewood Cliffs, NJ, 1981.

2. Defense Systems Management College (DSMC), *Mission Critical Computer Resources Management Guide,* DSMC, Fort Belvoir, VA.

3. DOD-STS-2167, Defense Standard, *Defense System Software Development,* Department of Defense, Washington, DC.

4. Eisner, H., *Computer-Aided Systems Engineering,* Prentice-Hall, Inc., Englewood Cliffs, NJ, 1988.

5. Fairley, R., *Software Engineering Concepts,* McGraw-Hill Book Co., New York, 1985.

6. General Electric Co., *Software Engineering Handbook,* McGraw-Hill Book Co., New York, 1986.

7. Hordeski, M., *Computer Integrated Manufacturing,* TAB Books, Inc., Blue Ridge Summit, PA, 1988.

8. Krouse, J. K., *What Every Engineer Should Know About Computer-Aided Design and Computer-Aided Manufacturing,* Marcel Dekker, Inc., New York, 1982.

9. MIL-HDBK-59A, Military Handbook, *Computer-Aided Acquisition and Logistic Support (CALS) Program Implementation Guide,* Department of Defense, Washington, DC.

10. MIL-STD-1679A, Military Standard, *Software Development,* Department of Defense, Washington, DC.

11. Musa, J. D., A. Iannino, and K. Okumoto, *Software Reliability: Measurement, Prediction, Application,* McGraw-Hill Book Co., New York, 1987.

12. Pressman, R. S., *Software Engineering: A Practitioner's Approach,* McGraw-Hill Book Co., New York 1982.

13. Sage, A. P. and J. D. Palmer, *Software Systems Engineering,* John Wiley & Sons, Inc., New York, 1990.

14. Shere, K. D., *Software Engineering and Management,* Prentice-Hall, Inc., Englewood Cliffs, NJ, 1988.

15. Shooman, M., *Software Engineering Design, Reliability and Management,* McGraw-Hill Book Co., New York, 1983.

16. Teicholz, E., Ed., *CAD/CAM Handbook,* McGraw-Hill Book Co., New York, 1985.

17. Vick, C. R. and C. V. Ramamcorthy, *Handbook of Software Engineering,* Van Nostrand Reinhold Co., New York, 1984.

H. PRODUCTION OPERATIONS

1. Buffa, E. S., *Modern Production and Operations Management*, 5th ed., John Wiley & Sons, Inc., New York, 1987.
2. Defense Systems Management College (DSMC), *Manufacturing Management: Guide for Program Managers*, DSMC, Fort Belvoir, VA.
3. Nakamura, S., *Standardization*, Productivity Press, Portland, OR, 1993.
4. Northley, P. and N. Southway, *Cycle Time Management*, Productivity Press, Portland, OR, 1993.
5. Robinson, A., *Modern Approaches to Manufacturing Improvement*, Productivity Press, Cambridge, MA, 1990.
6. Starr, M. K., *Operations Management*, Prentice-Hall, Inc., Englewood Cliffs, NJ, 1978.

I. OPERATIONS RESEARCH AND FUZZY SET THEORY

1. Churchman, C. W., R. L. Ackoff, and E. L. Arnoff, *Introduction to Operations Research*, John Wiley & Sons, Inc., New York, 1957.
2. Dubois, D. and H. Prade, *Fuzzy Sets and Systems: Theory and Applications*, Academic Press, New York, 1980.
3. Hillier, F. S. and G. J. Lieberman, *Introduction to Operations Research*, 5th ed., Holden-Day, Inc., San Francisco, CA, 1990.
4. Kaufmann, A. and M. M. Gupta, *Introduction to Fuzzy Arithmetic: Theory and Applications*, Van Nostrand Reinhold, New York, 1985.
5. Klir, G. J. and T. A. Folger, *Fuzzy Sets: Uncertainty and Information*, Prentice Hall, Englewood Cliffs, NJ, 1988.
6. Taha, H. A., *Operations Research: An Introduction*, 3rd ed., Macmillan Publishing Co., Inc., New York, 1982.
7. Zimmermann, H. J., *Fuzzy Set Theory—And Its Applications*, Kluwer, Nijhoff Publishing, Dordrecht, 1985.

J. QUALITY ENGINEERING, QUALITY ASSURANCE, QUALITY MANAGEMENT

1. American Productivity and Quality Center, *The Benchmarking Management Guide*, Productivity Press, Portland, OR, 1993.
2. Asaka, T. and K. Ozeki, *Handbook of Quality Tools*, Productivity Press, Cambridge, MA, 1988.
3. Besterfield, D. H., *Quality Control*, 3rd ed., Prentice-Hall, Inc., Englewood Cliffs, NJ, 1990.
4. Crosby, P. B., *Quality Is Free*, The New American Library, Inc., New York, 1979.
5. Deming, W. E., *Out of the Crisis*, Massachusetts Institute of Technology Press, Cambridge, MA, 1986.

6. Duncan, A. J., *Quality Control and Industrial Statistics,* 5th ed., Richard D. Irwin, Inc., Homewood, IL, 1986.

7. Feigenbaum, A. V., *Total Quality Control,* 3rd ed., McGraw-Hill Book Co., New York, 1983.

8. Garvin, D. A., *Managing Quality: The Strategic and Competitive Edge,* the Free Press, Macmillan Publishing Co., Inc., New York, 1988.

9. Grant, E. L. and R. S. Leavenworth, *Statistical Quality Control,* 6th Edition, McGraw-Hill Book Co., New York, 1988.

10. Ishikawa, K., *Introduction to Quality Control, Chapman and Hall, London, England, 1991.*

11. Juran, J. M., Ed., *Quality Control Handbook,* 4th ed., McGraw-Hill Book Co., New York, 1988.

12. Juran, J. M. and F. M. Gryna, *Quality Planning and Analysis,* 3rd ed., McGraw-Hill, Inc., New York, 1993.

13. MIL-Q-9858A, Military Standard, *Quality Program Requirements,* Department of Defense, Washington, DC.

14. Pall, G. A., *Quality Process Management,* Prentice-Hall, Inc., Englewood Cliffs, NJ, 1987.

15. RAC SOAR-7, *A Guide for Implementing Total Quality Management,* Rome Air Development Center, Griffis AFB, New York, 1990.

16. Ross, P. J., *Taguchi Techniques for Quality Engineering,* McGraw-Hill Book Co., New York, 1988.

17. Saylor, J. H., *TQM Field Manual,* McGraw-Hill, Inc., New York, 1992.

18. Scherkenbach, W. W., *The Deming Route to Quality and Productivity: Road Maps and Roadblocks,* CEE Press Books, George Washington University, Washington, DC, 1986.

19. Taguchi, G., E. A. Elsayed, and T. C. Hsiang, *Quality Engineering in Production Systems,* McGraw-Hill Book Co., New York, 1989.

K. ENGINEERING ECONOMICS, LIFE-CYCLE COSTING, COST ESTIMATING

1. Berliner, C. and J. Brimson, *Cost Management for Today's Advanced Manufacturing— The CAM-I Conceptual Design,* Harvard Business School Press, Inc., Boston, MA, 1988.

2. Brown, R. J. and R. R. Yanuck, *Introduction to Life Cycle Costing,* AEE Energy Books, Atlanta, GA, 1985.

3. Canada, J. R. and W. G. Sullivan, *Economic and Multiattribute Evaluation of Advanced Manufacturing Systems,* Prentice-Hall, Inc., Englewood Cliffs, NJ, 1989.

4. DARCOM P700-6 (Army), NAVMAT P5242 (Navy), AFLCP/AFSCP 800-19 (Air Force), *Joint-Design-to-Cost Guide, Life Cycle Cost as a Design Parameter,* Departments of the Army/Navy/Air Force, Washington, DC.

5. Dhillon, B. S., *Life Cycle Costing: Techniques, Models and Applications,* Gordon and Breach Science Publishers, New York, 1989.

6. DOD Guide LCC-1, *Life Cycle Costing Procurement Guide,* Department of Defense, Washington, DC.

7. DOD Guide LCC-2, *Casebook, Life Cycle Costing in Equipment Procurement*, Department of Defense, Washington, DC.

8. DOD Guide LCC-3, *Life Cycle Costing Guide for System Acquisitions*, Department of Defense, Washington, DC.

9. DOD-HDBK-766, Military Handbook, *Design To Cost*, Department of Defense, Washington, DC.

10. Fabrycky, W. J. and B. S. Blanchard, *Life-Cycle Cost and Economic Analysis*, Prentice-Hall, Inc., Englewood Cliffs, NJ, 1991.

11. Fabrycky, W. J. and G. J. Thuesen, *Economic Decision Analysis*, 2nd ed., Prentice-Hall, Inc., Englewood Cliffs, NJ, 1980.

12. Fisher, G. H., *Cost Considerations in System Analysis*, Prentice-Hall, Inc., Englewood Cliffs, NJ, 1971.

13. Fowler, T. C., *Value Analysis in Design*, Van Nostrand Reinhold, New York, 1990.

14. Grant, E. L., W. G. Ireson, and R. S. Leavenworth, *Principles of Engineering Economy*, 7th ed., The Ronald Press Co., New York, 1982.

15. Jelen, F. C. and J. H. Black, *Cost and Optimization Engineering*, McGraw-Hill Book Co., New York, 1983.

16. Michaels, J. V. and W. P. Wood, *Design to Cost*, John Wiley & Sons, Inc., New York, 1989.

17. MIL-HDBK-259, Military Handbook, *Life Cycle Cost in Navy Acquisitions*, Department of Defense, Washington, DC.

18. MIL-STD-337, Military Standard, *Design to Cost*, Department of Defense, Washington, DC.

19. MIL-STD-1390C, Military Standard, *Level of Repair*, Department of Defense, Washington, DC.

20. OMB Circular A-76, *Cost Comparison Handbook*, Office of the Management of the Budget, Washington, DC.

21. Ostwald, P. F., *Engineering Cost Estimating*, 3rd ed., Prentice-Hall, Inc., Englewood Cliffs, NJ, 1992.

22. Shillito, M. L. and D. J. De Marle, *Value: Its Measurement, Design and Management*, John Wiley & Sons, New York, 1992.

23. Stewart, R. D. and A. L. Stewart, *Cost Estimating with Microcomputers*, McGraw-Hill Book Co., New York, 1980.

24. Stewart, R. D., *Cost Estimating*, 2nd ed., John Wiley & Sons, Inc., New York, 1990.

25. Stewart, R. D. and R. M. Wyskida, *Cost Estimator's Reference Manual*, John Wiley & Sons, Inc., New York, 1987.

26. Thuesen, G. J. and W. J. Fabrycky, *Engineering Economy*, 8th ed., Prentice-Hall, Inc., Englewood Cliffs, NJ, 1994.

27. Witt, P. R., *Cost Competitive Products: Managing Product Concept to Marketplace Reality*, Reston Publishing Co., Reston, VA, 1986.

L. MANAGEMENT AND SUPPORTING AREAS

1. Blanchard, B. S., *Engineering Organization and Management*, Prentice-Hall, Inc., Englewood Cliffs, NJ, 1976.

2. Cleland, D. I. and W. R. King, *Project Management Handbook,* 2nd ed., Van Nostrand Reinhold Co., Inc., New York, 1989.

3. Defense Systems Management College (DSMC), *Test and Evaluation Management Guide,* DSMC, Fort Belvoir, VA.

4. Dieter, G. E., *Engineering Design: A Materials and Processing Approach,* McGraw-Hill Book Co., New York, 1983.

5. Griffin, R. W. and G. Moorhead, *Organizational Behavior,* Houghton Mifflin Co., Boston, MA, 1986.

6. Johnson, R. A., F. E. Kast, and J. E. Rosenzweig, *The Theory and Management of Systems,* 3rd ed., McGraw-Hill Book Co., New York, 1973.

7. Kerzner, H., *Project Management: A Systems Approach to Planning, Scheduling, and Controlling,* 3rd ed., Van Nostrand Reinhold, New York, 1989.

8. Koontz, H., C. O'Donnell, and H. Weihrich, *Essentials of Management,* 4th ed., McGraw-Hill Book Co., New York, 1986.

9. Moder, J. J., C. R. Phillips, and E. W. Davis, *Project Management with CPM, PERT and Precedence Diagramming,* 3rd ed., Van Nostrand Reinhold, New York, 1983.

10. Rosenau, M. D., *Successful Project Management,* Lifetime Learning Publications, Belmont, CA, 1981.

11. Stewart, R. D. and A. L. Stewart, *Proposal Preparation,* John Wiley & Sons, Inc., New York, 1984.

12. Ullmann, J. E., D. A. Christman, and B. Holtje eds., *Handbook of Engineering Management,* John Wiley & Sons, Inc., New York, 1986.

13. Wall, W. C., *Proposed Preparation Guide,* John Wiley & Sons, Inc., New York, 1990.

M. EXPERT SYSTEMS

1. Dym, C. L. and R. E. Levitt, *Knowledge-Based Systems in Engineering,* McGraw-Hill, Inc., New York, 1991.

2. Giarratano, J. and G. Riley, *Expert Systems: Principles and Programming,* PWS-KENT Publishing Company, Boston, 1989.

3. Harmon, P. and B. Sawyer, *Creating Expert Systems for Business and Industry,* John Wiley & Sons, Inc., New York, 1990.

INDEX